HANDBOOK OF CHEMICAL SPECIALTIES

HANDBOOK OF CHEMICAL SPECIALTIES

Textile Fiber Processing, Preparation, and Bleaching

JOHN E. NETTLES

Prescott, Inc.
Charlotte, North Carolina

A Wiley-Interscience Publication

JOHN WILEY & SONS

New York Chichester Brisbane Toronto Singapore

Library of Congress Cataloging in Publication Data:

Nettles, John E.
Handbook of chemical specialties.

"A Wiley-Interscience publication."
Bibliography: p.
Includes index.
1. Textile chemicals. 2. Textile finishing.
I. Title.

TS1449.N47 1983 677'.02835 83-5862
ISBN 0-471-89318-8

Printed in the United States of America

10 9 8 7 6 5 4 3 2 1

This book is gratefully dedicated to my wife, Joy, for without her help, guidance, understanding, and most of all, patience, it would never have been finished. I shall forever be in her debt for being both friend and companion during the days of the preparation of this manuscript and the one who managed to keep her sanity when typing into the wee hours of the night so that it would become a reality and finished on time.

I also dedicate this book to my children—listed by age so as to prevent dissension among their ranks—Johnny, Laury, and Catherine, for their efforts at being patient and for giving up both father and mother for the lengthy period of time that it took to complete this book.

I hereby dedicate this book to each of the above and hope that they will be as proud of it as I am of them.

ACKNOWLEDGMENTS

It would have been impossible for me to have written this book had it not been for the spoken and unspoken inspiration of the following people; I would truly be remiss if I did not acknowledge them with sincere and grateful appreciation:

Dr. C. S. Black of Wake Forest College, Chemistry Department, for teaching me to enjoy analytical chemistry and for showing me that it could be an interesting and fascinating career.

Dr. Harry Miller of Wake Forest College, for introducing me to the world of organic synthesis and for making me be thorough and exact in my synthesis experiments.

Professor K. T. Raynor of Wake Forest College, for teaching me that without mathematics, science would be an impossibility, and for proving to me beyond any shadow of a doubt that in mathematics, although there may be more than one way to arrive at an answer, there is only one correct answer.

L. W. Stevens of High Point Chemical Corporation, for seeing in me the potential to be an industrial chemist and for giving me the opportunity to share this experience in an exciting and rewarding manner.

John D. Honeycutt of High Point Chemical Corporation, who, having forgotten more about textile specialties than I will ever know, had the kindness and patience to teach me the "prior arts" as well as the chemical technology for the production and use of textile specialty chemicals.

Daryl Von Cannon of High Point Chemical Corporation, for patiently indoctrinating an "alchemist" into the fine arts of manufacturing with all its eccentricities.

Dr. Steve Koenig of Henkel, Inc., for permission to use some of his work on fiber lubricants and spin finishing.

Carl Blumenstein of Texchem Consulting Labs, for invaluable assistance in sizing formulations.

R. C. Forrester of Eastman Chemical Products, for assistance in chemistry of fiber technology.

Duane Alewine of Intex, for assistance in warp size formulations and general size chemistry.

George Heath of Vectra Fibers, for chemistry of polypropylene fibers.

To the many people of FMC, E. I. DuPont, Emery Industries, and Olin Corporation, for their invaluable assistance and guidance in the preparation of this manuscript.

Special thanks are also due to the many persons and companies whom, if listed, would fill a book all by themselves.

J.E.N.

PREFACE

The textile industry utilizes many manufacturing processes in the production of fibers, yarns, and fabrics. Since any of these may be finished as white goods or dyed to many different shades, a multi-million-dollar business has evolved as a large number of natural and synthetic fibers currently available to stylists and fabric designers necessitated a highly sophisticated technology.

Textile specialty chemicals comprise chemical entities that range from simple blends to exotically synthesized molecules and encompass the entire ionic range of anionic, nonionic, cationic, amphoteric, and blends. The development of a multitude of textile chemical specialties was necessary to augment the technology that has developed over the last 30 to 40 years.

At the present time it is estimated that there are over 8000 specialty chemicals, exclusive of dyes, being offered textile mills in the United States alone. When neophytes look at the specialty chemical field, they are at first baffled, frustrated, and then usually severely confused when they find that a particular entity is not specific for a given end use but can, in fact, be used for more than one purpose. If one takes into account the fact that many of the products offered to the marketplace are, in reality, duplicates of each other, it will be noted that "prior arts" is probably of more economic importance to this field of endeavor than is actual scientific acumen.

This is a result of the fact that products of basically similar composition can have enough subtle differences that significantly different results can be obtained from their use. Bearing this peculiarity in mind, it is quite easy to see how mystique and mystery have surrounded and consumed the specialty chemical field. As "prior arts" have been handed down from parent to offspring, employer to employee, and chemist to technician, there has been a constant necessity to change formulations to stay competitive in the marketplace.

According to Ed Hochberg, of Hochberg Associates and Company, Inc., in the July 12, 1978 issue of *Purchasing* magazine, independent formulators control 85% of the specialty chemical market, and 80–85% of the annual dollar volume is controlled by suppliers who have less than $5 million in annual sales. Mr. Hochberg credits the success of the independent formulator to the fact that he can custom-tailor compounds to meet specific technical needs, can make fast deliveries of small quantities, can maintain an inventory close to the processor–buyer, and can assist in the financing of purchases. In the November 1979 issue of *American Dyestuff Reporter,* K. E. Tanzer of the G. A. F. Corporate Market Research Department estimates the specialty chemical market for dyebath auxiliaries alone at $110 million.

As a practicing textile chemist at the mill, manufacturing, and sales levels for approximately 25 years, I have often been confused, baffled, and disappointed that there was no one reference book to which I could turn for guidance and/or understanding of specialty chemicals. It was necessary, therefore, for me to compile my own notes as I searched the literature in many different places in my quest for answers.

At the request of numerous co-workers and peer groups whom I have encountered while lecturing both domestically and internationally, I have been persuaded to publish a collection of processes, procedures, and specialty formulations so that anyone might have access to this information without the laborious and time-consuming efforts that were required of me.

As a reference source, I have used the patent literature where available, personal contacts, and published source information where possible. As many actual use formulations are in reality proprietary information and rarely, if ever, published, I have made an effort to use only generic names or common chemical names. Formulations are given generically and in ranges that are consistent with existing theoretical technology. A genuine effort has been made to avoid the use of registered trademarks, but where it was necessary, they have been documented if at all possible.

If an understanding is to be developed of the textile chemical specialty field and we are to be able to make an orderly classification of the proliferation of products that serve the industry, it will be necessary to devise a systematic scheme to catalog these entities. Inasmuch as specialty chemicals are used in practically all stages of the processing of natural and synthetic fibers, by beginning with fiber synthesis and continuing in the normal flow pattern of processing, we can develop a very logical, systematic way to catalog these auxiliaries. As we proceed through our investigation of these auxiliaries, as textile specialty chemicals are called in the trade, I trust that the reader will develop an appreciation of why these compounds are known to the trade as "soapy water" or "snake oil" chemicals and why the field itself has often been called the "hoe and barrel" game.

It is my sincere hope that you will enjoy my book, and I welcome your comments and suggestions for future revisions.

NORMAL FLOW PATTERN OF TEXTILE PROCESSING

Fiber processing
↓
Yarn Processing
↓
Preparation
↓
Bleaching
↓
Dyeing
↓
Finishing
↓
Rework

JOHN E. NETTLES

Charlotte, North Carolina
June 1983

CONTENTS

HANDBOOK OF CHEMICAL SPECIALTIES

CHAPTER ONE

FIBER CHEMISTRY

To understand how and why various chemical entities or auxiliaries are chosen for use, it is necessary to have at least a general understanding of the substrate to which they are to be applied. Therefore, it will be necessary to allow ourselves a cursory look at generic fiber chemistry and particularly at those fibers that have found a measure of commercial acceptance. It is not the intention of this text to provide a complete treatise on fiber technology, as there are already many good books on this subject on the market.[1] We merely present fiber chemistry in reference form and then proceed to discuss how "proprietary formulations" are chosen for use as fiber finishes. We ac-

TABLE 1.1. Common Textile Fibers, by Class

Natural Fibers	Regenerated Natural Fibers	Synthetic Fibers
Cotton	Rayon	Acrylic
Hemp	Viscose	Modacrylic
Linen	Secondary acetate	Nylon
Ramie	Triacetate	Olefin
Hair	Protein	Polyester
Wool		Glass
Silk		Polyvinyl chloride
Alginate		Elastomeric
Casein		
Lignin		

complish this by first reviewing some of the current patent literature on the subject and then proceeding on to review formulations for use on fibers as they are converted to yarn.

Fibers are generally classified as either one of three types: (1) natural; (2) regenerated natural; or (3) synthetic. Table 1.1 represents a list of the more common fibers typical of each class.

CELLULOSE FIBERS

Cotton

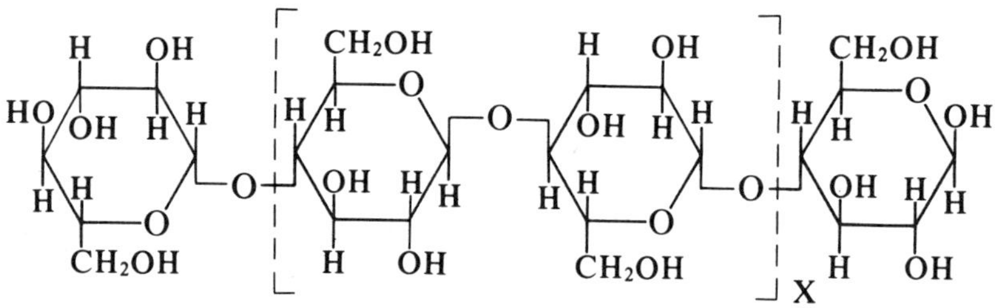

Linen (Flax)

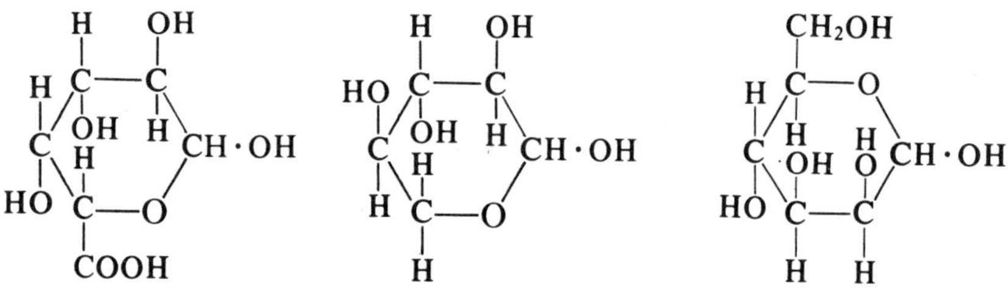

64% Cellulose + 16% Hemicellulose

Alginate

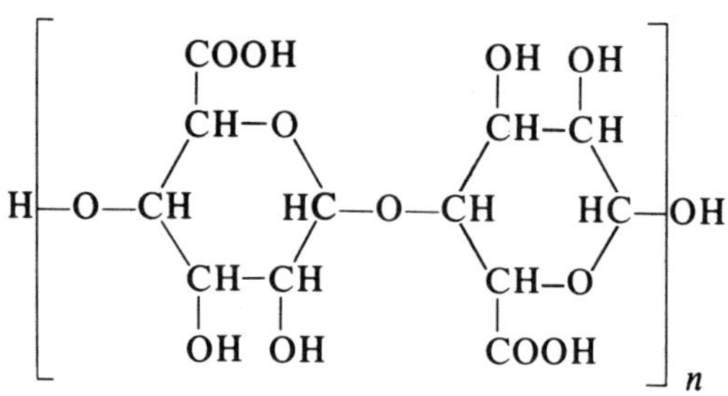

PROTEIN FIBERS

Protein fibers (wool and silk) contain 16–18 different α-amino acids.

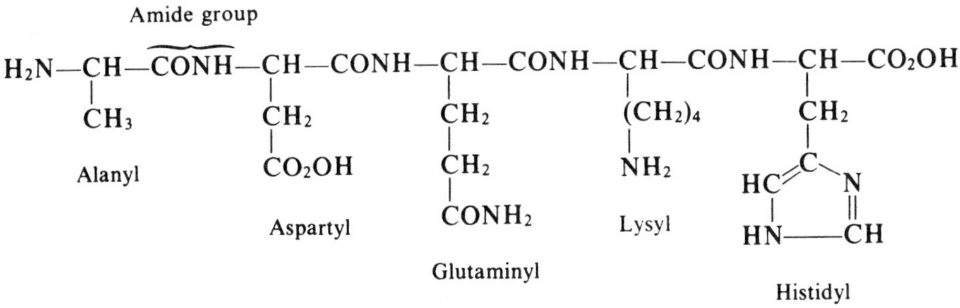

	Alanyl	Aspartyl	Glutaminyl	Lysyl	Histidyl
Basic dye site		X			X
Acid dye site			X	X	X

Wool

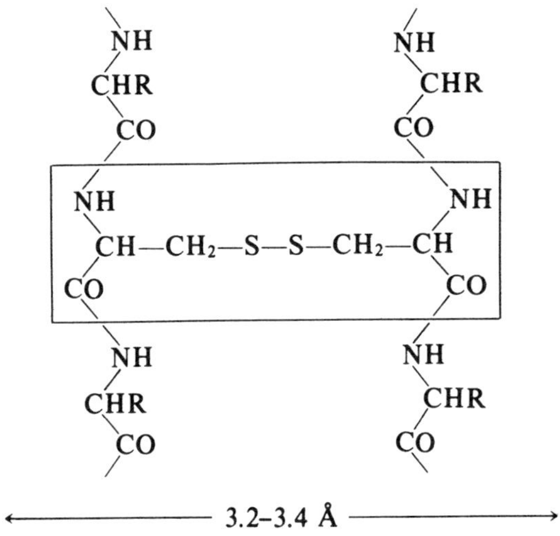

Silk

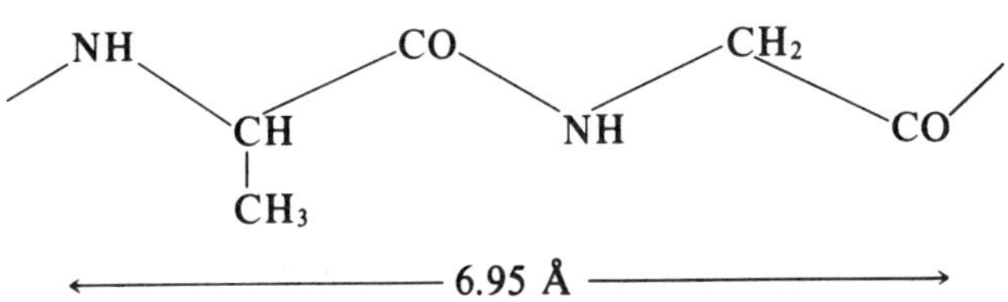

REGENERATED NATURAL FIBERS

Viscose

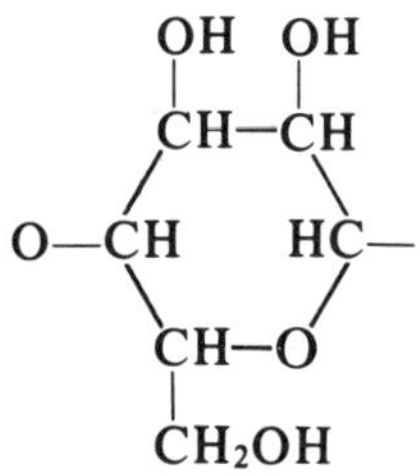

Rayon (Cuprammonium Process)

$$C_6H_7O_2(OH)_3 + Cu(NH_4OH) \rightarrow$$
$$C_6H_7O_2(OH){:}Cu(NH_4OH) + H_2SO_4 \rightarrow$$
$$C_6H_7O_2(OH)_3$$

Protein (Vicara®)

$$[—NH + CH_2O + NH—] \rightarrow [—HN^+—CH_2O—{}^+NH—]$$

Casein

$$RNH_2 + HCHO + H_2NR \rightarrow RNHCH_2NHR$$

SYNTHETIC FIBERS

Triacetate (Arnel®)

Acetate is produced by the reaction of natural cellulose and glacial acetic acid, acetic anhydride, and sulfuric acid. It is reacted for 8 hours at 20–30°C.

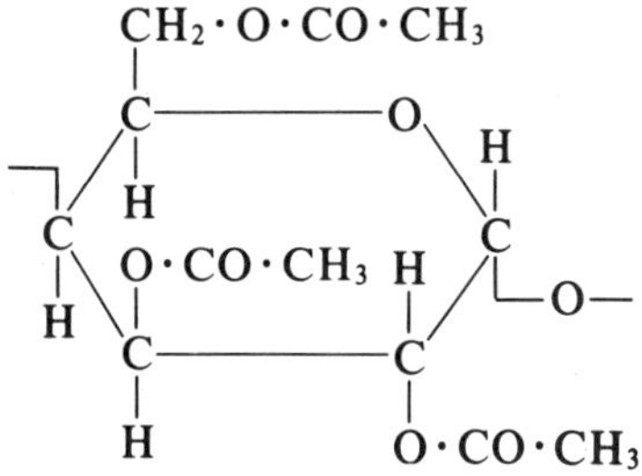

Courtesy of Eastman Chemical Products

Secondary Acetate

Cellulose triacetate as formed above is not soluble in acetone. By acid hydrolysis, acetyl content is reduced from 3.0 to 2.0–2.5. Secondary acetate then can be precipitated as a flake into water. Flakes can be dissolved in acetone for spinning.

Formula of Secondary Acetate

$$C_6H_{7.7}O_{2.7}(OCOCH_3)_{2.3}$$

Courtesy of Eastman Chemical Products

Nylon 6

Nylon 6 is a polymer of ε-caprolactam.[3]

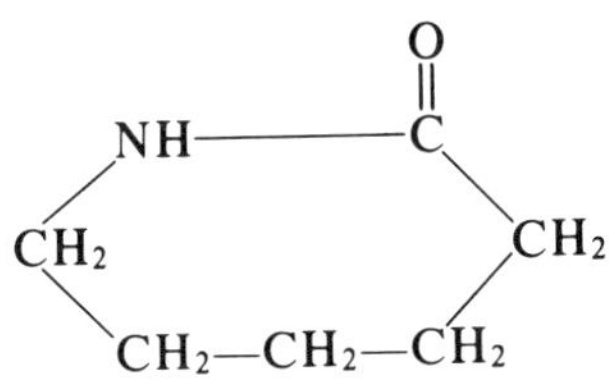

ε-Caprolactam

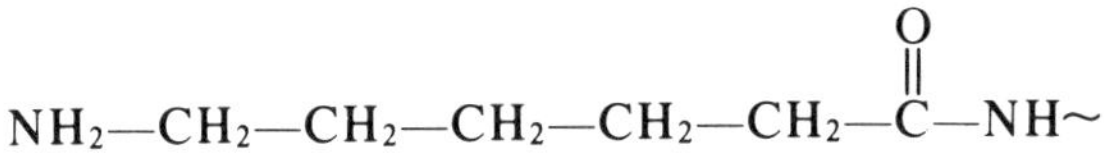

Nylon 6

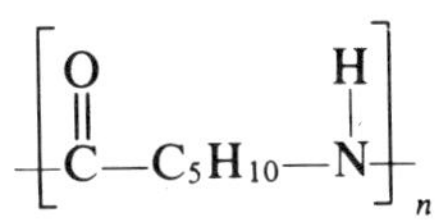

Alternate structure nylon 6

Nylon 6.6

Nylon 6.6 is a condensate of adipic acid and hexamethylene diamine.

$$HO-\overset{O}{\overset{\|}{C}}-CH_2-CH_2-CH_2-CH_2-\overset{O}{\overset{\|}{C}}-OH +$$

Adipic acid

$$H_2N-CH_2-CH_2-CH_2-CH_2-CH_2-CH_2-NH_2 \longrightarrow$$

Hexamethylenediamine

$$H_2N-(CH_2)_6-NH-\overset{O}{\overset{\|}{C}}-CH_2-CH_2-CH_2-CH_2-\overset{O}{\overset{\|}{C}}-NH\sim$$

Nylon 6.6

"Basic" dyeable at carboxylic acid end groups; "acid" dyeable at amine and at low pH (<2), amide groups.

$$HO-\overset{O}{\overset{\|}{C}}-C_4H_8-\overset{O}{\overset{\|}{C}}\left[-\overset{H}{\overset{|}{N}}-C_6H_{12}-\overset{H}{\overset{|}{N}}-\overset{O}{\overset{\|}{C}}-C_4H_8-\overset{O}{\overset{\|}{C}}\right]_n-\overset{H}{\overset{|}{N}}-C_6H_{12}-NH_2$$

Alternate structure nylon 6.6

Polycarbonamide (Qiana®)

Qiana is a registered trademark and is patented by DuPont company. It is a condensate of dodecanoic acid and 4.4′-diaminodicyclohexylmethane.

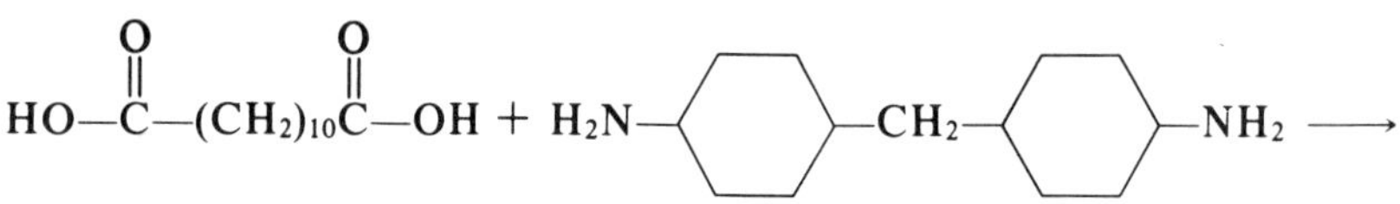

Dodecanoic acid 4,4′-Diaminodicyclohexylmethane

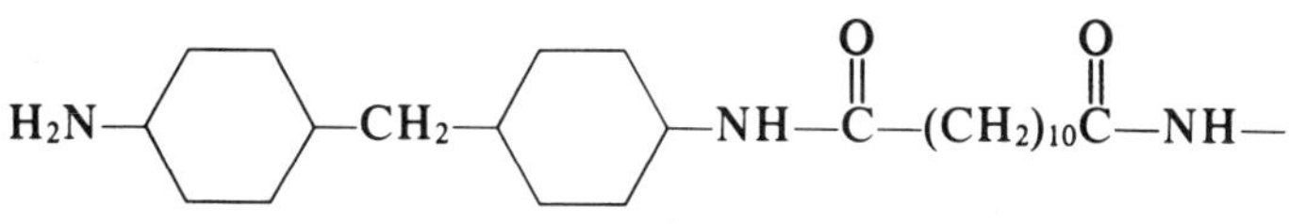

Qiana®

Nylon 4

$$+(CH_2)_3CONH(CH_2)_3CONH+_n 2H_2O$$

Nylon 11

$$H+NH(CH_2)_{10}CO+_n OH$$

Orlon®

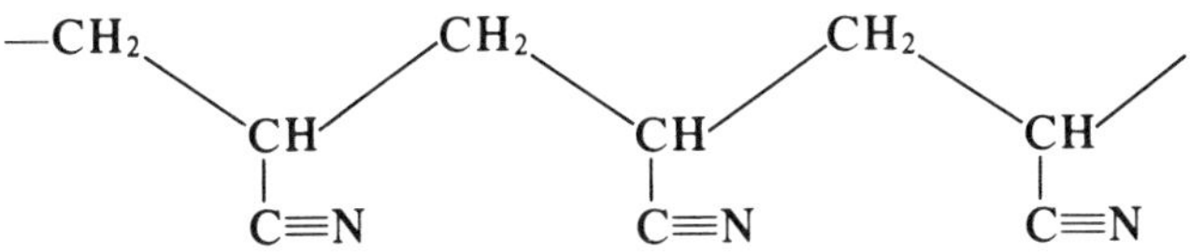

Orlon® is a registered trademark of E. I. DuPont and Company.

Acrylic

Acrylics are condensates of acrylonitrile containing 15% or less copolymer.[4]

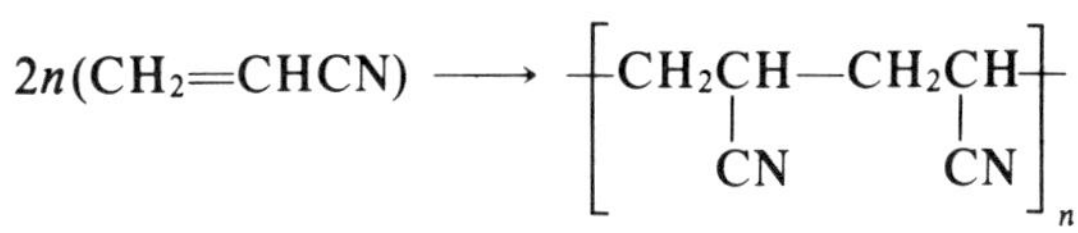

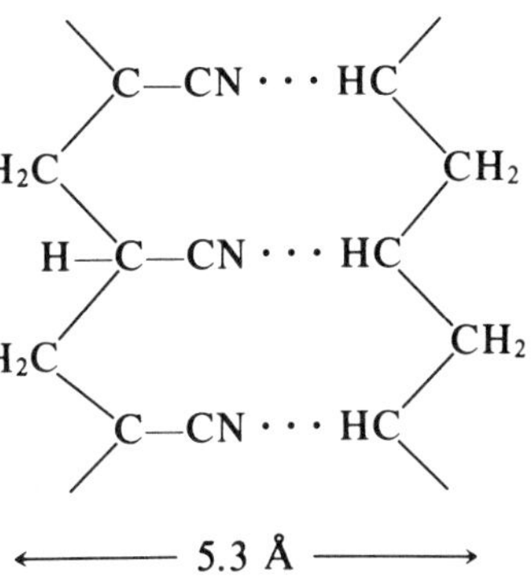

Acrylic fiber

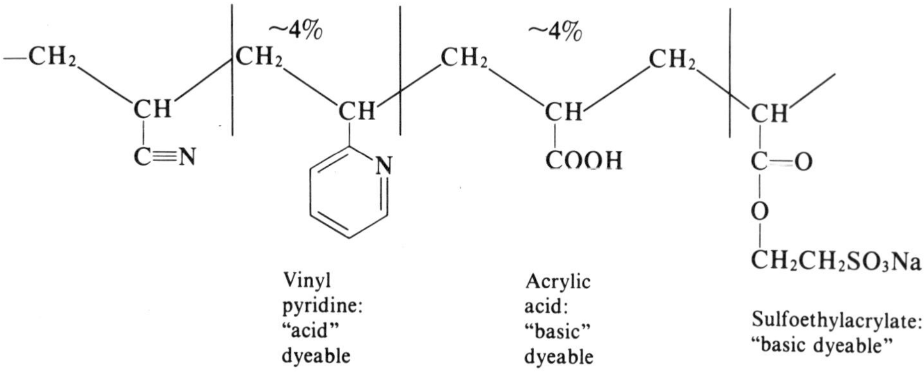

Creslan®

Creslan® is a registered trademark of American Cyanamid and is a condensate of vinyl cyanide and acrylic acid amide.

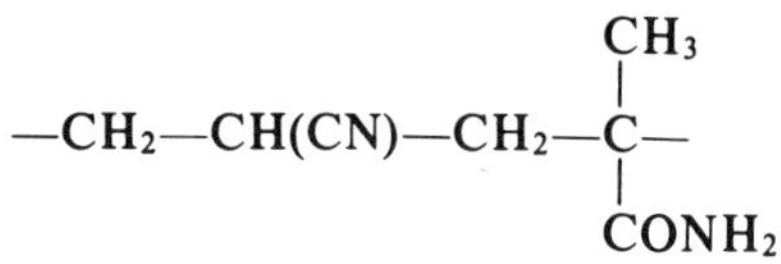

Acrilan®

Acrilan® is a registered trademark of Chemstrand Corporation and is a condensate of a copolymer of acrylonitrile–vinyl acetate and a constituent of a mildly basic character, usually vinylpypyridine.[5]

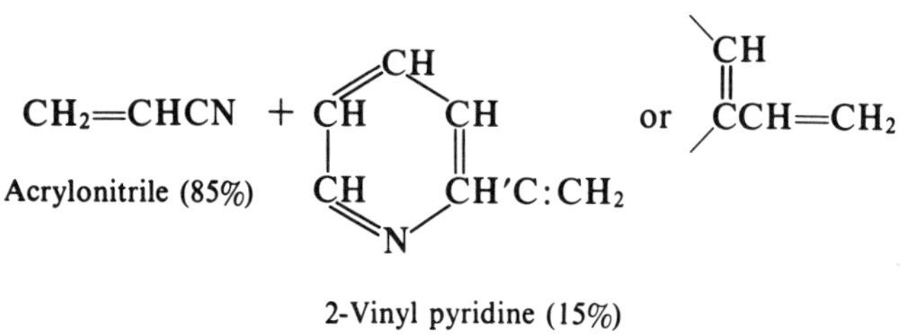

Modacrylic

Modacrylic contains 35–85% acrylonitrile.

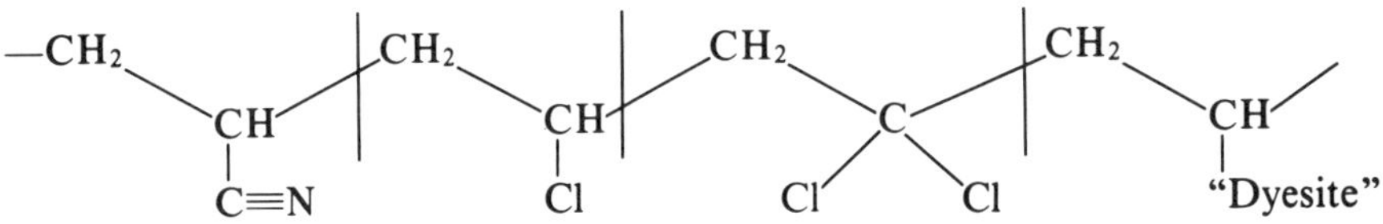

Dynel®

Dynel® is a registered trademark of Union Carbide and is a copolymer of 60% vinyl chloride and 40% vinyl cyanide.

$$CH_2{=}CHCl + CH_2{=}CHCN + CH_2{=}CHCl \rightarrow$$

Vinyl chloride — Vinyl cyanide — Vinyl chloride

$$—CH_2CHClCH_2CHCNCH_2CHCl—$$ or

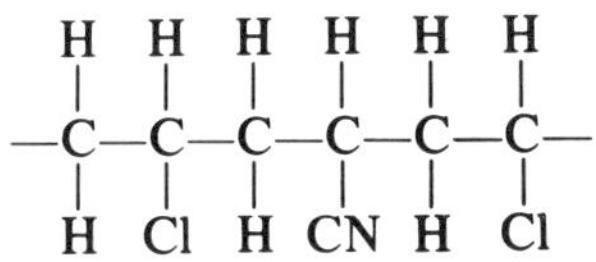

Teflon®

Teflon® is a registered trademark of the DuPont Company for tetrafluoroethylene polymers.

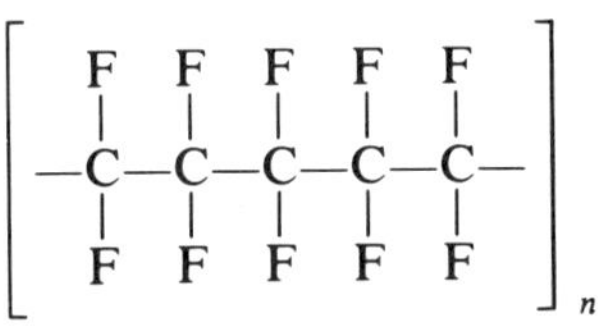

Saran®

Saran® is a registered trademark of Dow Chemical Corporation and a copolymer of vinylidene chloride and vinyl chloride.[7]

$$CH_2{=}CCl_2 + CH_2{=}CHCl + CH_2{=}CCl_2 \rightarrow$$

Vinylidene chloride — Vinyl chloride — Vinylidene chloride

$$-CH_2CCl_2CH_2CHClCH_2CCl_2-$$

Copolymer

Elastomeric

$$\left[NH{-}CO{-}O{-}(CH_2)_6{-}NHCOO(CH_2)_4 \right]_n$$

or

$$\left[NHCOO(CH_2)_6NHCOO(CH_2)_4 \right]_n$$

Polyester (Disperse Dyeable)

A typical polyester is a condensation polymer of ethylene glycol and a terephthalate derivative:[8]

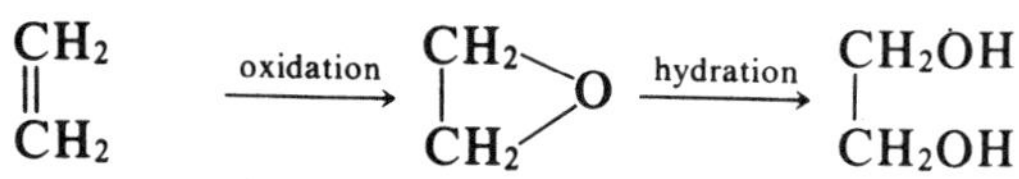

Ethylene (from a "cracking" plant) — Ethylene oxide — Ethylene glycol

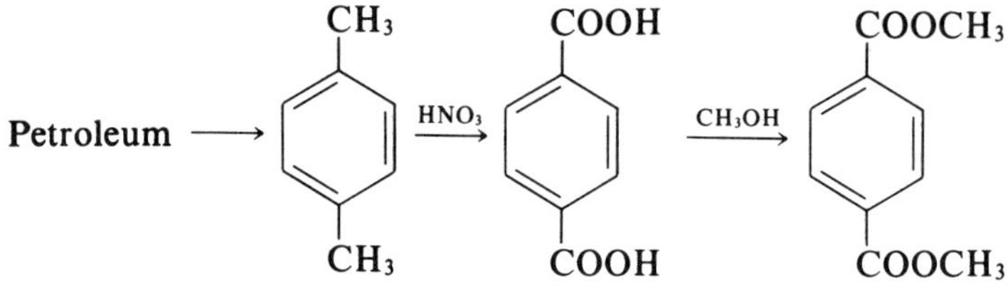

p-Xylene — Terephthalic acid — Dimethyl terephthalate

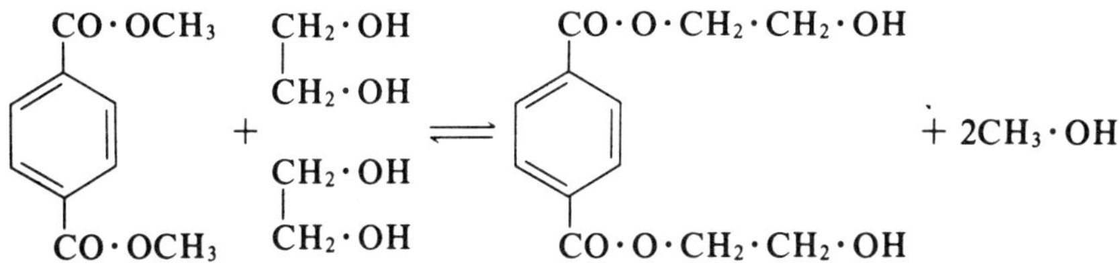

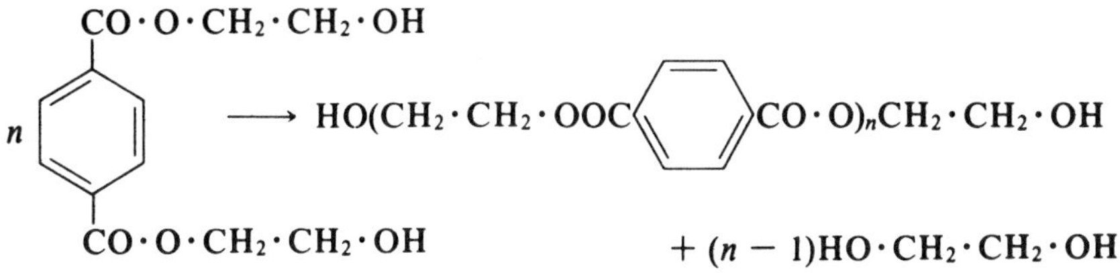

Courtesy of Eastman Chemical Products

Kodel® is a registered trademark of Eastman Chemical Products and is covered under U.S. Patent 2,901,466. Structures typical of the various Kodel® series are shown in the following subsections.

Type 200 Series

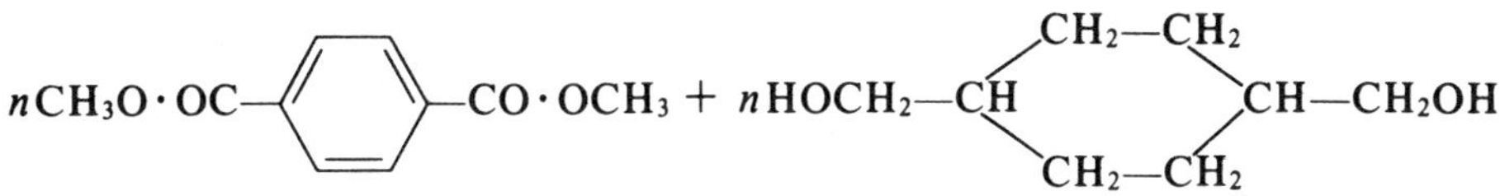

Dimethyl terephthalate + 1,4-Cyclohexanedimethanol

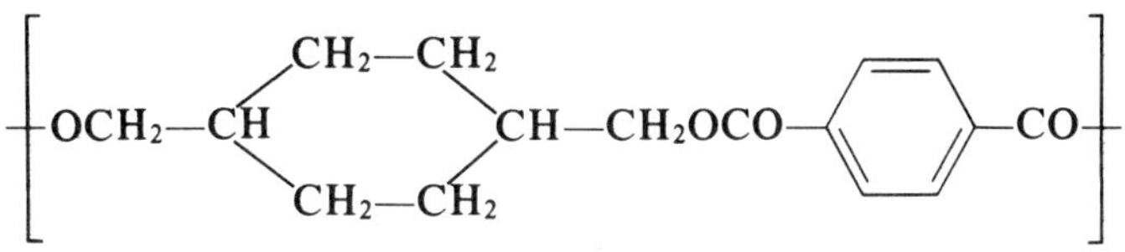

Cyclohexanedimethanol terephthalate

Type 400 Series

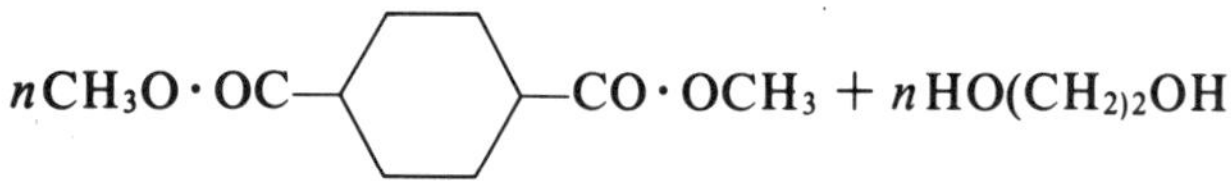

Dimethyl terephthalate + ethylene glycol

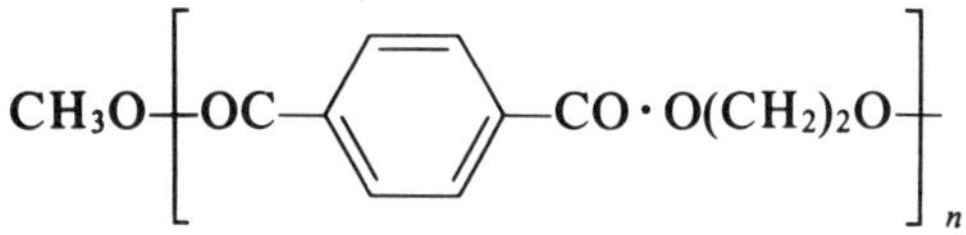

Polyethylene terephthalate
Courtesy of Eastman Chemical Products

Polyester (Basic Dyeable)

Linear high-molecular-weight polyesters such as polyethylene terephthalate and poly(1.4-cyclohexylene dimethylene terephthalate) are made dyeable with basic dyes by incorporating in the polymer chain alkali metal salts of aromatic sulfonic acids or esters.[9] The structure of a basic dyeable polyester fiber that was synthesized by the addition of 3–5% 5-sulfoisophthalic co-monomer is as follows:

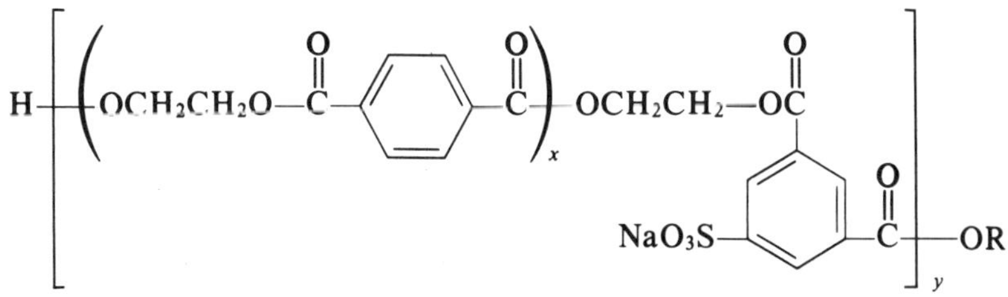

Polypropylene*

These are olefins consisting primarily of polymerized polypropylene with molecular-weight ranges of 250,000–500,000. They may contain either stabi-

* Structures and information regarding polypropylene fibers courtesy of George A. Heath, Vectra Corporation.

lizers to inhibit degradation to ultraviolet (UV) light and oxidation or pigments to provide color or delustering.

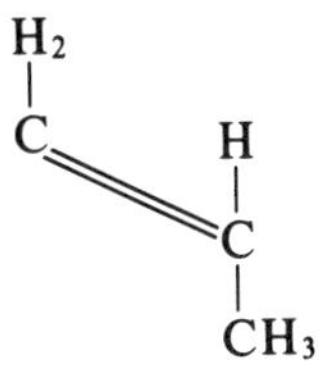

Polypropylene monomer

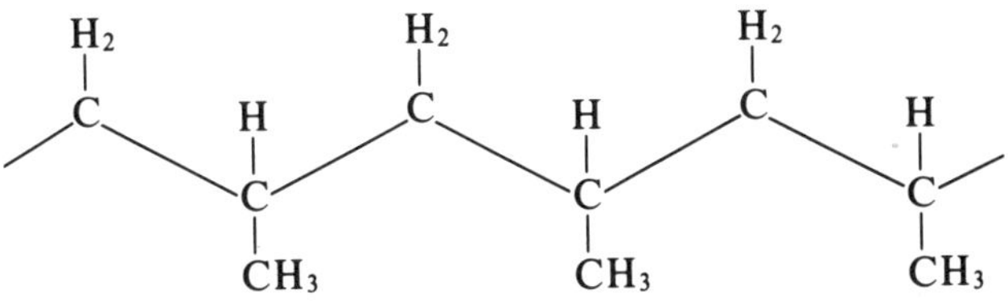

Polymerized isotactic polypropylene

CHAPTER TWO

FIBER FINISHING

Once the fiber—whether natural or synthetic—has been produced, it must be treated so that it can be transformed into either a yarn for use as is or for weaving or knitting. The fiber may be manufactured in the staple form (short lengths of fiber) or in a filament form (continuous length). Auxiliaries are necessary to make this transition proceed smoothly and to increase the efficiency of the operation.

The types of product that are used in auxiliaries for fiber finishing are as follows:[10]

AUXILIARY COMPOUNDS USED IN FIBER FINISHING

1. Oils, waxes, and fats.
2. Starches and polymers.
3. Tints.
4. Esters: butyl stearate; tridecyl stearate; oleates; palmitates; laurates; trimethylolpropane trispelargonate; pentaerythritol tetrapelargonate; di(2-ethylhexyl) adipate; azelates; and sebacates.
5. Polyoxyalkylene glycols and poly(oxyethylene-oxypropylene) ethers.
6. Silicone and modified silicone fluids.
7. Emulsifiers: anionic; nonionic; cationic; and amphoteric.
8. Antistatic agents: anionic; nonionic; and cationic.

As can be seen from this list, there is a wide diversity of products that can be chosen, and one factor to consider is that anything that is put on at this stage of fiber processing will, of necessity, at this stage of our technology, have to be removed prior to bleaching, printing, dyeing, or finishing.

FACTORS TO CONSIDER IN CHOOSING FIBER FINISH

In choosing which products to use, there are several criteria that make selection easier. Factors that must be considered, and the reasons for consideration, are shown in Table 2.1.

Once the fiber has been treated and shipped to the yarn processor, whether it is natural or synthetic, it must undergo additional processing. In the case of polyamide fibers, a coning oil is usually applied, whereas with polyester, texturizing oils and lubricants are applied. The finishes listed in

TABLE 2.1. Criteria for Selection of Finish Components[11]

Property	Performance Characteristic
Cohesion	To ensure uniform cones of yarn in winding operations; to minimize fiber slippage on cone
Lubrication	To reduce coefficient of friction between fiber and metal rolls; to prevent fiber abrasion; to provide uniform tension during processing
Oxidation resistance	To prevent discoloration of yarn; to prevent microbial attack; to prevent formation of nonsoluble residues due to oxidation
Antistatic properties	To reduce and dissipate any buildup of static charge during processing
Controlled viscosity	To provide uniform processing on prolonged running
Corrosion resistance	Must not cause damage to rollers, guides, or needles
Nonvolatility	Must be high-temperature stable to prevent buildup on machinery during processing; must not gum or break down on prolonged running or smoke or create health hazards
Toxicity	Must take into consideration environmental concerns; must not create health hazards
Heat buildup	Should not create heat buildup in processing machinery; should not sling off as heat increases
Ease of removal	Whatever is applied should be capable of being removed easily, even after prolonged storage

TABLE 2.2. Applications of Auxiliaries

	Synthetic Fiber	
Natural Fiber	Staple	Filament
Opening	Chopping	Drawing
Picking	Blending	Texturizing
Roving	Carding	Coning
Carding	Twisting	Winding
Twisting	Winding	Beaming
Winding	Beaming	Weaving
Beaming	Slashing	Knitting
Slashing	Weaving	
Weaving	Knitting	
Knitting	After waxing	
After waxing		

Table 2.1 are usually applied to filament or continuous yarn. Spun yarn, during and after yarn formation, is treated with a mixture of winding emulsions, spinning lubricants, and warp sizes.

AREAS OF AUXILIARY USE

Table 2.2 lists places where auxiliaries may be used.

In each of the areas listed in Table 2.2, problems can occur that will create difficulty at later stages of processing. Some of the areas that can create problems are (1) nonuniform areas of application, (2) nonfugitive tint, (3) nonuniform ease of removal, (4) wrong chemical for process, and (5) unclean machinery.

SPIN FINISH COMPONENTS

Spin finishes must be developed for each specific fiber and may even have to be modified for a particular mill condition. Spin finishes consist, as a general rule, of a lubricant, a fiber-to-fiber friction aid, an antistatic agent, and an emulsifier. Chemicals that are used in each category are listed in Table 2.3.

TABLE 2.3. Spin Finish Components[12]

Component	Products Commonly Used
Lubricant	Mineral oils and waxes; fatty acid esters; dicarboxylic acid esters; vegetable oils and waxes; neoalcohol esters; polyoxyalkylene derivatives; silicone and modified silicone fluids
Fiber-to-fiber friction aids	Mineral oils; linear esters; branched esters; polyalkylene glycols; polyoxyalkylene monoethers
Antistatic agents	
Anionics	Alkyl alcohol sulfates; alkyl alcohol phosphates; alkyl alcohol ethoxylates (sulfate and phosphate)
Nonionics	Alcohol ethoxylates; alkyl phenol ethoxylates; alkyl amine ethoxylates; polyoxyalkylene glycol derivatives
Cationics	Imidazoline quaternaries; alkyl quaternaries
Emulsifiers	
Anionics	Alkylbenzene sulfonates; alkyl alcohol phosphates; alkyl alcohol sulfates; alkyl alcohol ethoxylate (sulfate and phosphate); sulfosuccinates; fatty acid soaps
Nonionics	Alkyl alcohol ethoxylates; alkyl phenol ethoxylates; fatty acid ethoxylate; sorbitol–sorbitan;[13] alkyl amine ethoxylates; alkanolamides; alkanolamide ethoxylates
Cationics	Quaternary fatty amines; quaternary fatty amine ethoxylates; quaternary imidazolines

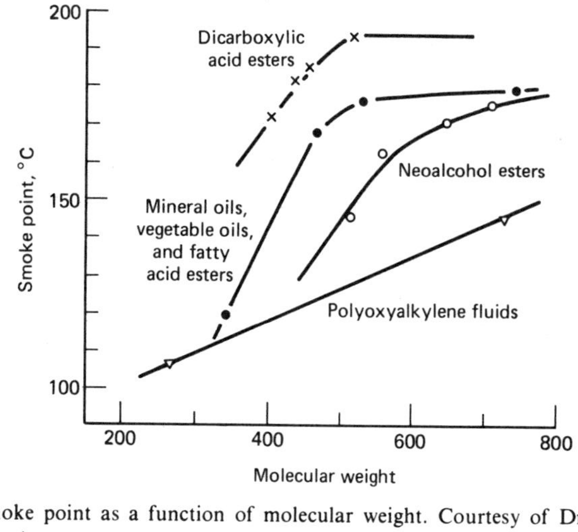

FIG. 2.1. Smoke point as a function of molecular weight. Courtesy of Dr. Steve Koenig, Henkel Corporation.

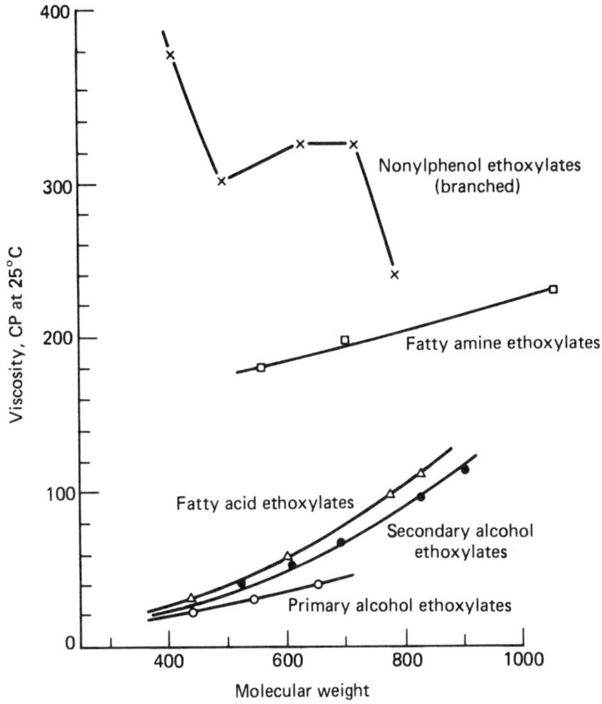

FIG. 2.2. Viscosity–molecular weight relationship of typical nonionic surfactants. Courtesy of Dr. Steve Koenig, Henkel Corporation.

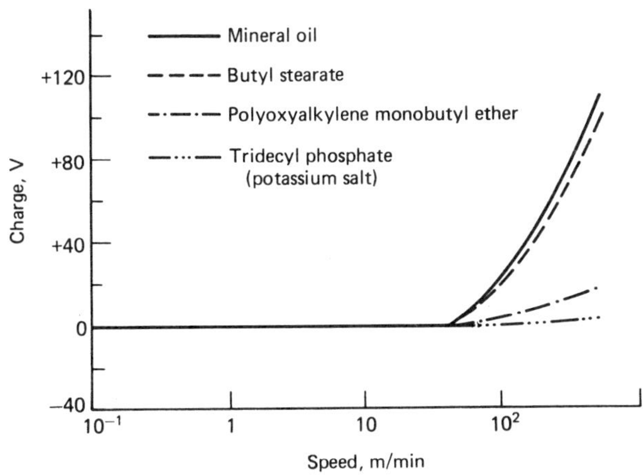

FIG. 2.3. Yarn-to-metal friction charge versus speed. Courtesy of Dr. Steve Koenig, Henkel Corporation.

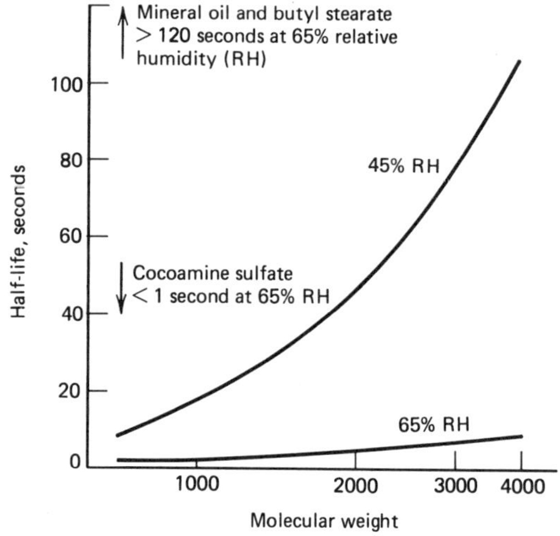

FIG. 2.4. Half-life of charge dissipation versus molecular weight. Courtesy of Dr. Steve Koenig, Henkel Corporation.

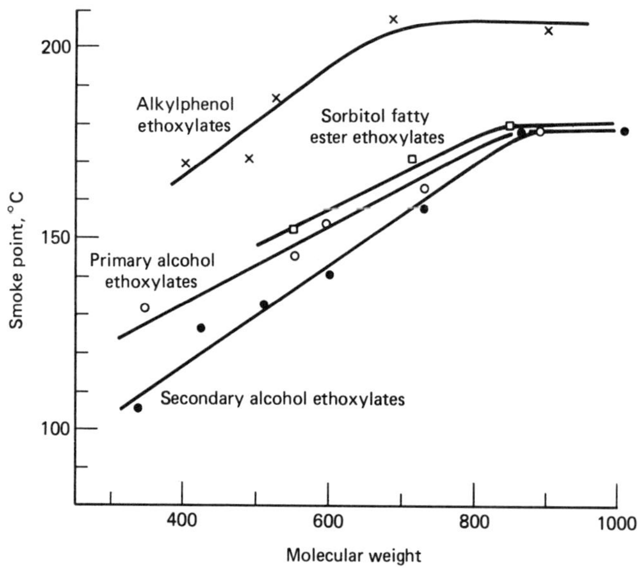

FIG. 2.5. Smoke point versus molecular weight. Courtesy of Dr. Steve Koenig, Henkel Corporation.

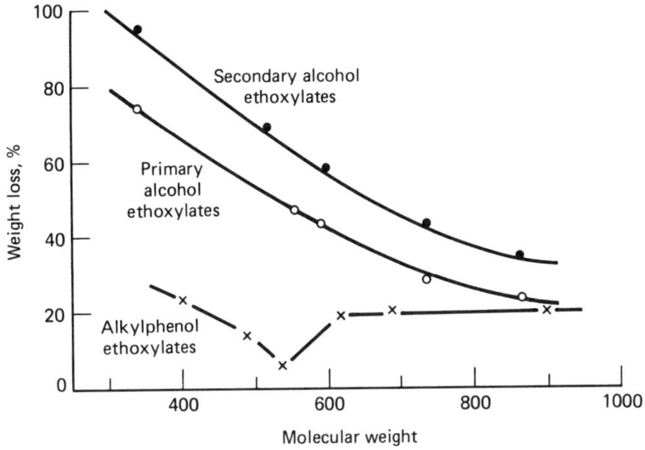

FIG. 2.6. Percentage of weight loss (in isothermal degrees Celsius) versus molecular weight—alcohol and alcohol ethoxylates. Courtesy of Dr. Steve Koenig, Henkel Corporation.

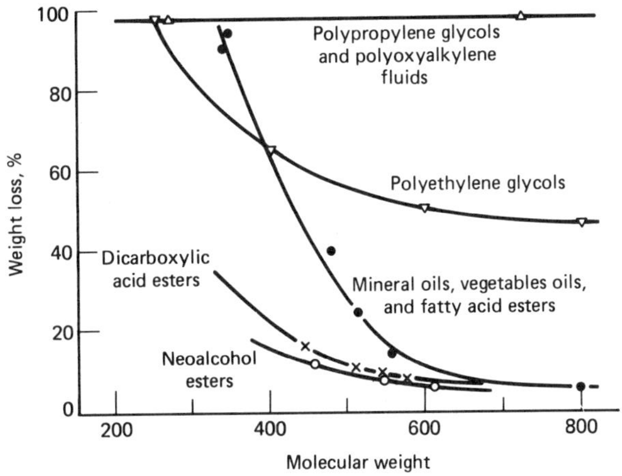

FIG. 2.7. Percentage of weight loss (in isothermal degrees Celsius) versus molecular weight—polypropylene glycols and polyoxyalkylene fluids. Courtesy of Dr. Steve Koenig, Henkel Corporation.

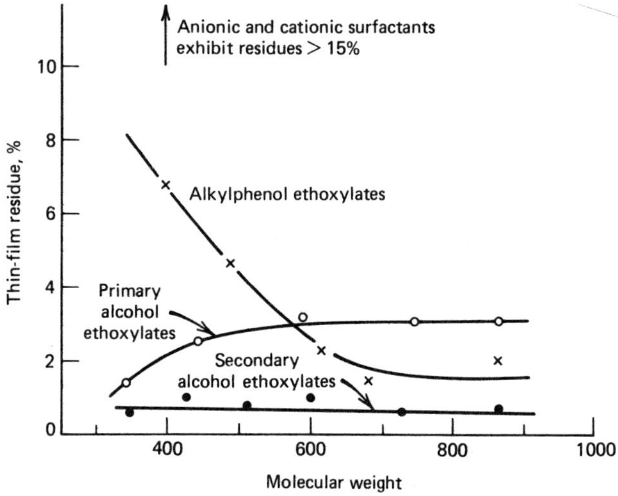

FIG. 2.8. Percentage of thin film residue of ionic surfactants as a function of molecular weight. Courtesy of Dr. Steve Koenig, Henkel Corporation.

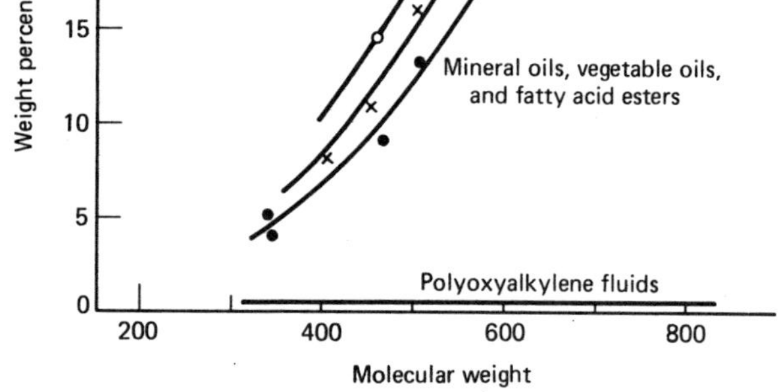

FIG. 2.9. Weight percent residue as a function of molecular weight. Courtesy of Dr. Steve Koenig, Henkel Corporation.

PROPERTIES OF FIBER COMPONENTS

Figures 2.1–2.9 show some of the characteristic properties of various components of fiber finishes. These drawings are the courtesy of Steve Koenig of Henkel, Inc.

SPIN FINISH COMPOSITIONS

An examination of the patent literature gives a good foundation for spin finish technology; for example:

1. *U.S. Patent 4,051,299(77)*—Fiber Industries: polyamide false twist

Mineral oil	20.00%
Butyl stearate	30.00%
Sorbitol (ethylene oxide)$_{40}$ hexaoleate	30.00%
Potassium salt 01-decyl alcohol (ethylene oxide)$_{6}$ phosphate	20.00%

2. *U.S. Patent 3,338,830*—DuPont: textured nylon

Polyoxyethylene–oxypropylene glycol	50–80%
Nonyl phenol (ethylene oxide)$_{10}$	10–40%
Pentaerythritol tetrastearate	10–40%

3. *U.S. Patent 3,954,631*—Allied Chemical: carpet yarn

Tridecyl stearate	40–60%
Oleic (ethylene oxide)$_{10}$	20–30%
Sodium petroleum sulfate	20–30%

4. *U.S. Patent 3,850,819(74)*—ICI: nylon filament yarn

Trimethylol propane tris pelargonate	50.00%
Nonyl phenol (ethylene oxide)$_8$	25.00%
Imidazoline quaternary	25.00%

5. Japan 54,547(73)—Teijin: nylon filament yarn

Octyl stearate	70.00%
Lauryl alcohol (ethylene oxide)$_x$	20.00%
Nonyl phenol (ethylene oxide)$_x$	9.70%
Potassium salt of oleic acid	0.30%

These patents are fairly representative of current technology in nylon processing. Listed below is technical information obtained from the patents indicated on polyester spin finishes:

1. *U.S. Patent 3,704,225(72)*—ICI: texturizing lubricant—polyester

Butyl stearate (lubricant)	58.0%
Potassium salt of hexyl phosphate (antistat)	13.0%
Isopropyl amine salt of dodecyl benzene sulfonic acid (DDBSA) (emulsifier)	5.0%
Sorbitol (ethylene oxide)$_4$ tetraoleate (lubricant)	22.0%
Antioxidant	2.0%

2. *German Patent 528,734(75)*—Hoechst Fibers: texturized polyester

Mineral oil (lubricant)	30.0%
Oleyl alcohol (ethylene oxide)$_7$ (emulsifier)	20.0%
Potassium salt of lauryl alcohol (ethylene oxide)$_4$ phosphate (antistat)	20.0%
Trimethylol propane tris pelargonate	30.0%

3. *U.S. Patent 3,859,122(75)*—DuPont: draw-textured polyester

Bistridecyl stearate	53.0%
Sorbitol® (ethylene oxide)$_{30}$ tetraoleate monolaurate	40.0%
Nonyl phenol (ethylene oxide)$_5$	5.0%
Polyoxyethylene–oxypropylene glycol	2.0%

4. *U.S. Patent 3,563,892(71)*—DuPont: high-temperature filament processing

Coconut oil	63–67%
Sodium glycerol trioleate sulfate	13–17%
Nonyl phenol (ethylene oxide)$_5$	8–12%
Mono- and diglycerol oleate	8–12%

5. *U.S. Patent 3,549,530(70)*—DuPont: staple processing

$C_{11}C_{15}$–secondary alcohol	48–52%
Triethanolamine (TEA) salt decyl alcohol phosphate	33–37%
Sorbitol® (ethylene oxide)$_4$ monolaurate	13–17%

Typical of coning oil formulations and the requirements for formulating the same are well represented by the information contained in *U.S. Patent 4,105,569(78)*—Crossfield:

80 Saybolt Universal Seconds (SUS) white mineral oil (lubricant)	83.45%
Polyisobutylene (sling control)	0.45%
Silicone polymer (wetting agent)	0.10%
(Ethylene oxide)$_3$ alcohol (emulsifier)	12.00%
(Ethylene oxide)$_{6.5}$ alcohol (emulsifier)	3.20%
Water (clarifier)	0.80%

Winding emulsions and lubricants generally consist of emulsified mineral oils and paraffins. A typical formulation might be:[14]

Mineral oil (lubricant)	89.00%
Industrol LAS 4.2 (emulsifier)	10.00%
Water (clarifier)	1.00%

A formulation that uses 135° melting-point paraffin might be represented by:[14]

Paraffin	4.00%
Mineral oil	4.00%
Stearic acid	8.00%
TEA	5.00%
Water	78.50%
Silicone defoamer	0.50%

Robert M. Marshall of Allied Chemical Corporation was issued *U.S. Patent 4,134,839* on January 16, 1979, which discloses a soil-resistant spin finish for nylon feeder yarn to be processed at high temperatures, which is composed of an oil–water emulsion containing sorbitan monooleate, polyoxyethylene tallow amine, and an alkyl stearate. In addition to the above, Marshall's composition contained an alkanolamide, a fluorochemical, and hydroxyl-containing organic radicals. Yarn prepared with the above formulation is claimed to show excellent resistance to soiling by oily materials.

U.S. Patent 4,137,181, issued to Roland L. Hawkins of Hoechst Fibers, discloses a composition for the treatment of nonwoven yarns of polymeric fibers in a short-cut form. The composition of this patent is based on ethoxylated castor oil, sorbitol hexoleate, and various esters of adipic acid. Further claims for this composition are low-foaming properties and no adverse effects on the bond ability of the fiber.

Owens Corning Fiberglas, through Richard A. Karpik and David H. Griffiths, were assigned *U.S. Patent 4,166,872* for a composition of an aqueous size that contained a starch solution of swollen but unburst partially cooked starch granules that were made from an amylaceous material. Additional compositions in this patent disclose a formulation consisting of cornstarch, hydrogenated vegetable oil, polyoxyethylene sorbitan monooleate, tetraethylene pentamine distearate, and polyethylene glycol.

Eastman Kodak Company and Bobby C. Carver and Raymond E. Donaldson were issued *U.S. Patent 4,169,061* for a composition to treat partially oriented polyester fiber that was to be texturized by the friction twist method. This composition contained as the lubricant a blend of random copoly (oxyethyleneoxypropylene) butanols. In addition to the lubricant, the composition contained phosphated lauryl alcohol and phosphated polyoxyethylene (POE-8) dinonyl phenol as an antistat.

Peter I. Lee and Brian Holt of CIBA Geigy Corporation were issued *U.S. Patent 4,178,253* for a formulation that contained lubricating oils of a cycloalkylene structure and a quaternary ammonium complex.

U.S. Patent 4,169,062 discloses additional fatty lubricants that are random copolymers of polyoxyethylene and polyoxypropylene glycol. These fatty esters are water-soluble and biodegradable and are claimed to exhibit superior lubricating properties when applied to synthetic fibers. This patent was issued to Eugene A. Weipert of Southern Sizing Company.

U.S. Patent 4,192,649, to Kato, et al., discloses a warp size from an aqueous solution containing a partially hydrolyzed polyvinyl acetate and an anionic surface-active agent (surfactant) capable of solubilizing the hydrolyzed polyvinyl acetate. It is claimed that the material used as a sizing agent provides high weaving efficiency, ease of desizing, and ease of waste water treatment.

U.S. Patent 4,199,319, to Lifentsev et al., discloses a method for the treatment of yarn produced by pneumatic spinning with liquid ammonia.

U.S. Patent 4,328,108, to William O. Deeken of the Goodyear Tire and Rubber Company, discloses a method for the elimination of circumferential stress cracks that occur during the production of spun polyesters.

U.S. Patent 4,278,552, to Hisamoto of Daikin Kogyo Co. Ltd., discloses the use of fluorine containing betaine compounds and the production and use thereof.

U.S. Patent 4,278,555, to Zaweski of the Ethyl Corporation, discloses the use of lubricating oils with improved anticorrosion properties imparted by a block oxyalkylated trimethylol alkane and an oxyethylated alkylphenol.

U.S. Patent 4,316,929, to McIntire of Eastman Kodak Company, discloses a size composition comprised of a terpolymer that is useful when applied as a hot melt to a variety of filaments, yarns, and fabrics.

U.S. Patent 4,278,553, to Sung of Texaco, Inc., discloses a lubricating oil base that contains a corrosion inhibitor comprised of benzotriazole, *n*-alkyl-1,3,propanediamine, and formaldehyde condensation products.

U.S. Patent 4,259,078, to Kleber of Hoechst, discloses an antistatic heat-stable fiber dressing produced by the reaction of quaternary ammonium compounds with phosphate groups.

U.S. Patent 3,925,589, to Struwold of Emery Industries, Inc., discloses an ester lubricant derived from polyoxyalkylene glycols with a dibasic acid mixture consisting of dimer acids and short-chain dibasic acids.

U.S. Patent 3,853,607, to Iyengar of E. I. DuPont, discloses a lubricating finish for filament yarns used as reinforcement in rubber articles and is composed of a polyorganosiloxane oil and a microcrystalline wax dispersed in a dialkylphthalate.

CHAPTER THREE

WARP SIZING

PRINCIPLES OF WARP SIZING

The preceding formulations related primarily to filament fiber finishing, coning oils for filaments, or lubricants for use on yarn to be knitted. A second process to which fibers can be subjected is weaving.[15] In weaving, a "filling" or weft yarn is interlaced at right angles to a warp in a set pattern. Warp yarns run lengthwise of the fabric in long continuous rows, and for efficiency purposes, are pretreated with a "size." This size prevents the shuttle, which is used to interlace the filling yarns with the warp yarns, from breaking or seriously abrading them during processing. As a general rule, filling yarns are not sized, as the stresses and strains of weaving are applied to the warps and very little is applied to the filling.

The warp size, once applied to the yarn and dried, increases the tensile strength and the abrasion resistance of the yarn. This minimizes down time, or "breakouts," as they are called.

The slashing operation, which is the name for the application of size to warp yarns, consists of coating the yarns in parallel rows with a formulation consisting of natural and synthetic polymers, gums, starches, metal-to-fiber lubricants, preservatives, and defoamers. The solution, after impregnation into the yarn at a level of 10–15% on the weight of the goods[16] for cotton and cotton blends, and at a level of 3–5% on filament synthetics,[17] is dried in continuous form over drying cylinders called "cans" and then wound onto large spools called "beams." The beams are then threaded into weaving machines, called "looms," where the filling yarn is interlaced to form the fabric. In simplistic terms, this is weaving.

The major cost in weaving, which can be controlled, is that of the size formulation. As the size is needed only for processing, and in reality must be removed prior to dyeing, printing, or finishing, it is necessary that the cheapest starch, gum, or polymer that is obtainable be used.

In the early days of weaving, where only cellulosic fibers were important, starch—modified with gums such as locust bean, guar, trag, and lubricants composed of sulfonated tallow blended with raw tallow—was the primary size. It was customary at this time to load warps heavily with starch, because fabrics were sold by weight, and not by performance specifications. The invention of shrinkage control necessitated better constructions in weaving and less manipulation in the size room. Consumer education further required a better piece of fabric than was being processed at that time. A good example of heavy sizing is the backfilling of sheets, which when washed, bore no resemblance to the original material. This, of course, was prior to the advent of polyester–cotton sheeting.

E. F. Patterson of Seydel Woolley and Company is quoted in the November 1973 issue of *America's Textiles Reporter/Bulletin* as stating that the magic ingredient in sizing is the one that reduces shedding and thereby improves weaving efficiency. He further states that a synthetic sizing material is probably one whose elongation properties are similar to those of the fiber being sized. Additional factors to be considered are adhesion, strength, and flexibility, but elongation is the major requirement. Shown below are compounds commonly used for filament sizing and spun yarn sizing:[18]

FILAMENT SIZING MATERIALS

Fiber	Basic Size
Nylon	Polyacrylic acid
	Polyvinyl alcohol (PVA)
Glass	PVA
	Dextrins
	Amylose derivatives
	Blends

Fiber	Basic Size
Acetate	Stymer (styrene–maleic anhydride copolymer)
	Gelatin
Viscose rayon	PVA
	Amylose derivatives
	Carboxymethylcellulose (CMC)
	Blends
Polyester	Acrylic copolymers
	Alkali-soluble polyvinyl acetate
	Linear polyester

SPUN YARN SIZING MATERIALS

Spun yarn sizing materials are as follows:

Fiber (100% or blends)	Basic Size
Cotton	Starches: corn, potato, tapioca
Rayon	Unmodified—pearl
Nylon	Acid modified, 20–60 fluidity
Polyester	Oxidized, several fluidities
Acetate	Dextrinized—British gum
Acrylic	Derivatized
Wool	Acetate
	Hydroxyethyl ether
	Acrylate
	Styrene
	Crosslinked
	Cationic
	High amylose
	Low amylose (waxy)
	Polyvinyl alcohol (PVA)
	Fully hydrolyzed
	Partially hydrolyzed
	Carboxymethylcellulose (CMC)
	Blends
	Other Polymers

Sizing materials for both filament and spun yarn usually contain additional film modifiers to produce a product with improved sizing qualities. One such compound is lubricants. Lubricants can be divided into two basic groupings, saponifiable and unsaponifiable. The saponifiables are easily removed and can, in fact, act as emulsifiers for the unsaponifiables. The unsaponifiables, such as paraffin waxes, are considered to be the best friction reducers. Typical of fats and fatty components are fancy tallow, bleached tallow, hydrogenated tallow glycerides, and fatty esters. Typical of unsaponifiable materials are crude-scale paraffin wax, slack paraffin wax, refined paraffin wax, and some polyethylenes. The best products are mixtures of the above formulated to a specific saponification value and melting point determined by the yarn construction and previous history of the mill.

Additional additives that may or may not be needed are antistats, defoamers, softeners, humectants, penetrants, preservatives, emulsifiers, tints, binders, and release agents.

In commercial practice, there are three general categories of lubricants, or kettle waxes:

1. *Water soluble,* which are completely soluble and form homogeneous solutions in hot or cold water.
2. *Water dispersible,* which form uniform dispersions in hot, agitated solutions but generally separate on cooling or when agitation is removed.
3. *Size dispersible,* which disperse uniformly in hot water, with agitation, in the presence of a size film former and remain dispersed when cooled or when agitation is stopped. This type is the most prevalent in slashing or weaving operations.

Wax concentrations on cotton yarn increase weave efficiency when used at a concentration of up to 5%. Above 10–15%, no increase in efficiency is achieved and desize problems can occur.

SIZE COMPONENTS—NATURAL GUMS AND STARCHES

Structures and comments concerning the natural gums and polymers used for the sizing of cotton and cotton blends are presented in the following paragraphs.

Starch

The structure for starch is generally as follows:

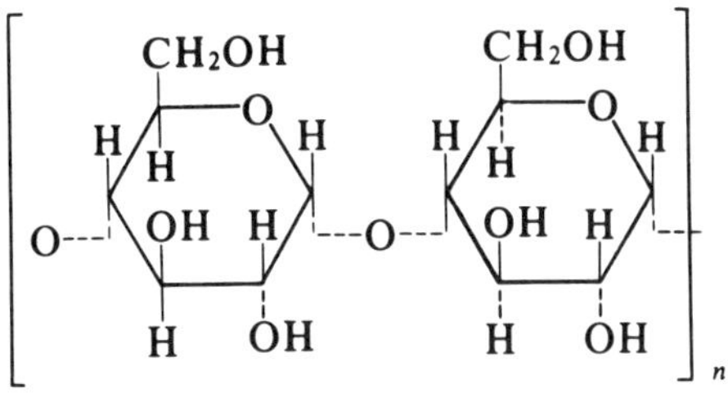

When starches are hydrolized by acid, they yield mixtures of saccharides or glucoses; when they are hydrolized by enzymes such as β-amylase, which yields maltose, or amyloglucosidase, which yields D-glucose, they give different-colored residues on treatment with iodine. Starch apparently contains two fractions that are soluble to varying degrees when treated with hot water. A soluble component known as *amylose* (10–20%) and an insoluble residue, *amylopectin* (80–90%), account for the violet color and the blue color that are yielded on treatment with iodine. Amylose gives the blue color, and amylopectin gives the violet to red violet color. In enzyme hydrolysis of starch, amylose is completely hydrolized by amylase,[19] whereas only 60% of amylopectin can be hydrolyzed.

Because of the hydroxyl content of the cellulose, reaction can occur with many different compounds. An excellent discussion of starch reactions can be found in the *Handbook of Water-Soluble Resins,* Chapter 22, and numerous references on the chemistry and use of starch are included at the end of that chapter.

Some typical reactions that starch may undergo are as follows[15]:

Starch with phosphorus oxychloride:

$$\mathrm{StOH + POCl_3 \xrightarrow{NaOH} StO\overset{\overset{\displaystyle O}{\|}}{\underset{\underset{\displaystyle ONa}{|}}{P}}OSt + NaCl}$$

Starch with sodium timetaphosphate
yields sodium dihydrogen pyrophosphate:

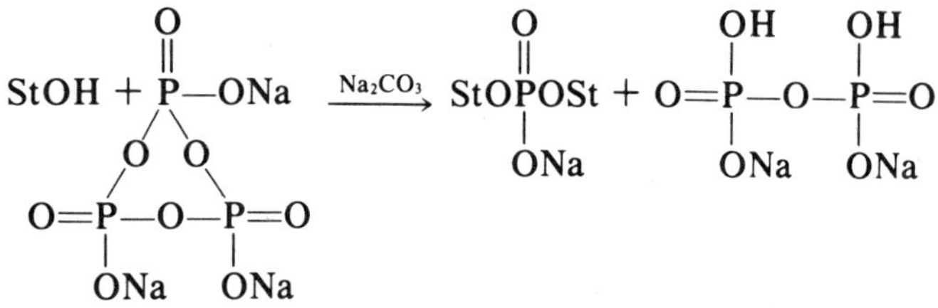

Starch with epichlorohydrin:

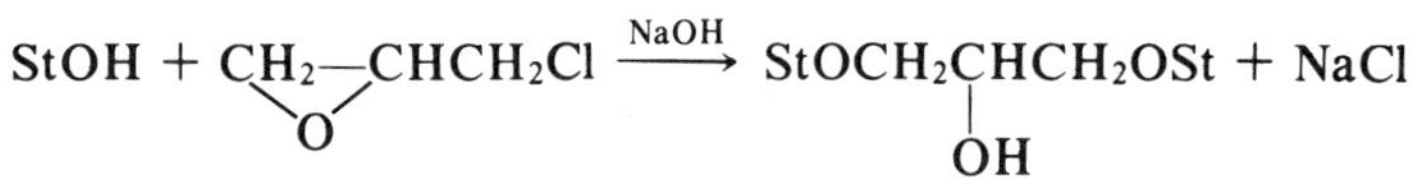

Starch with adipic–acetic anhydride:

$$\mathrm{StOH + CH_3\overset{O}{\overset{\|}{C}}O\overset{O}{\overset{\|}{C}}(CH_2)_4\overset{O}{\overset{\|}{C}}O\overset{O}{\overset{\|}{C}}CH_3 \xrightarrow[\text{pH 8}]{\text{NaOH}} StO\overset{O}{\overset{\|}{C}}(CH_2)_4\overset{O}{\overset{\|}{C}}OSt + CH_3\overset{O}{\overset{\|}{C}}ONa}$$

Starch with cyanuric chloride:

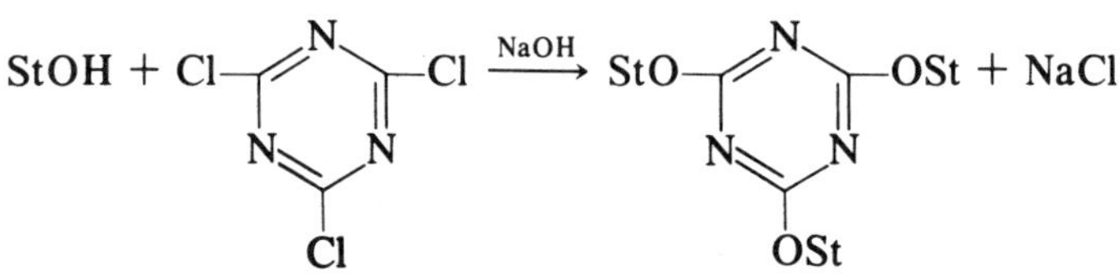

Starch with acetic anhydride:

$$\mathrm{StOH + (CH_3\overset{O}{\overset{\|}{C}})_2O \xrightarrow[\text{pH 8}]{\text{NaOH}} StO\overset{O}{\overset{\|}{C}}CH_3 + CH_3\overset{O}{\overset{\|}{C}}ONa}$$

Starch with vinyl acetate:

$$\mathrm{StOH + CH_2{=}CHO\overset{O}{\overset{\|}{C}}CH_3 \xrightarrow{Na_2CO_3} StO\overset{O}{\overset{\|}{C}}CH_3 + CH_3\overset{O}{\overset{\|}{C}}H}$$

Starch with alkylene oxides:

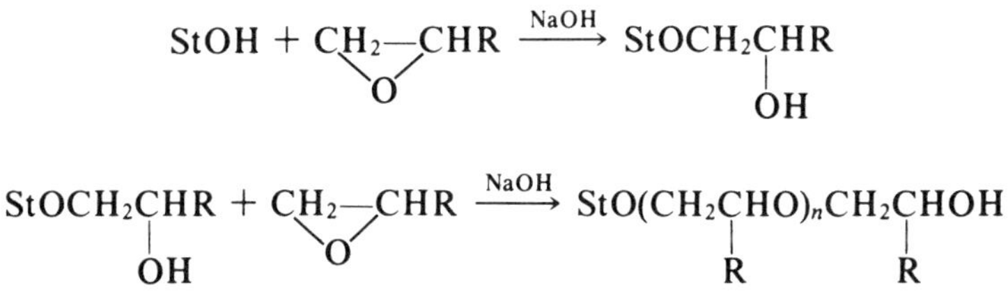

$(R = H \text{ or } CH_3)$

Starch with cyclic anhydrides—maleic anhydride:

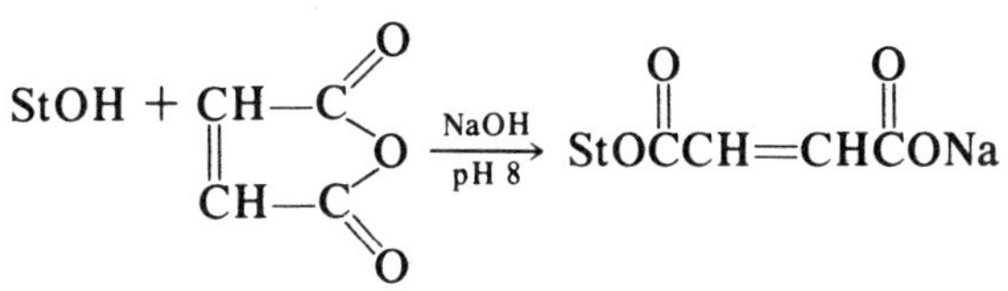

Starch with cyclic anhydrides—succinic anhydrides:

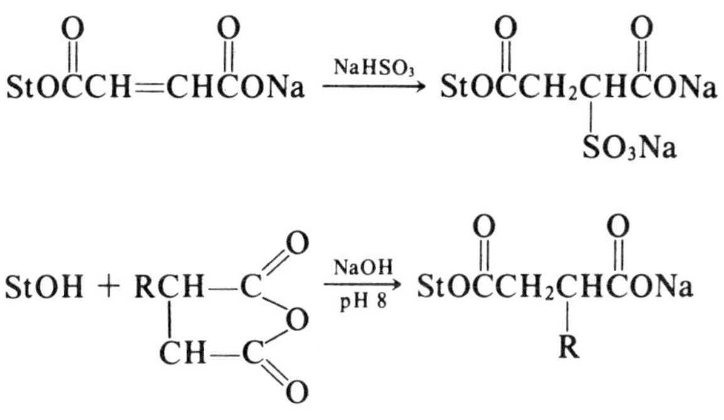

R is H or $CH_3(CH_2)_nCH{=}CHCH_2{-}$

Starch with trimethylamine–sulfatrioxide complex:

$$StOH + (CH_3)_3N \cdot SO_3 \xrightarrow{NaOH} StOSO_3Na + (CH_3)_3N$$

Starch with sodium phosphates—orthophosphates:

$$\mathrm{StOH + NaH_2PO_4 \xrightarrow[\text{pH 5–6.5}]{\Delta} StOP(=O)(OH)ONa + H_2O}$$

Starch with sodium phosphates—tripolyphosphates:

$$\mathrm{StOH + Na_5P_3O_{10} \xrightarrow[\text{pH 5–6.5}]{\Delta} StOP(=O)(ONa)_2 + Na_3HP_2O_7}$$

Starch with chloroacetic acid:

$$\mathrm{StOH + ClCH_2COOH \xrightarrow[\text{pH} > 11.8]{NaOH} StOCH_2COONa + NaCl}$$

Starch with diethylaminoethyl chloride:

$$\mathrm{StOH + (C_2H_5)_2NCH_2CH_2Cl \xrightarrow{NaOH} StOCH_2CH_2N(C_2H_5)_2 + NaCl}$$

$$\mathrm{\xrightarrow{H^{\oplus}} StOCH_2CH_2\overset{\oplus}{N}H(C_2H_5)_2\ Cl^{\ominus}}$$

Starch with 2,3-epoxypropyltrimethylammonium chloride:

$$\mathrm{StOH + \underbrace{CH_2—CH}_{O}CH_2\overset{\oplus}{N}(CH_3)_3Cl^{\ominus} \xrightarrow{NaOH}}$$

$$\mathrm{StOCH_2CH(OH)CH_2\overset{\oplus}{N}(CH_3)_3Cl^{\ominus}}$$

Starch with 4-chloro-2-butenyl trimethylammonium chloride:

$$\mathrm{StOH + ClCH_2CH{=}CHCH_2\overset{\oplus}{N}(CH_3)_3Cl^{\ominus} \xrightarrow{NaOH}}$$

$$\mathrm{StOCH_2CH{=}CHCH_2\overset{\oplus}{N}(CH_3)_3\ Cl^{\ominus}}$$

Starch with cyanamides:

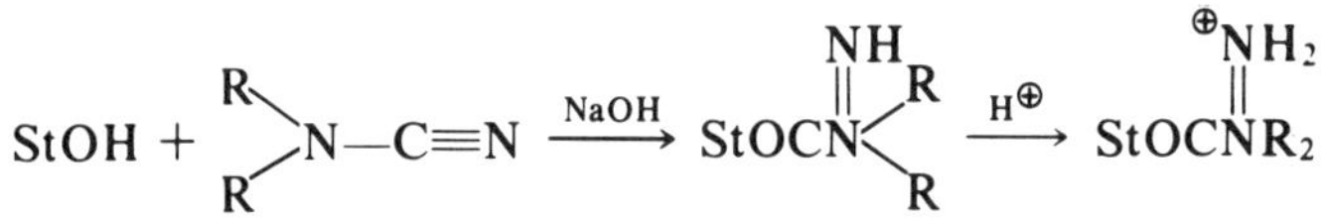

$R{=}H, CH_3{—}, C_2H_5{—}$

Starch polymerization:

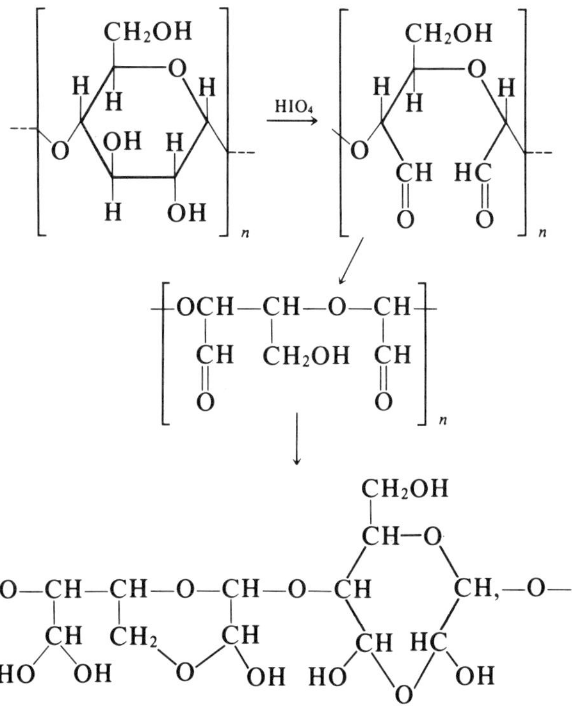

1 2 3

where 1 = hydrated carbonyl, 2 = hemiacetal form (intramolecular; can be intermolecular), and 3 = hemialdal form.

Starch with xanthates:

$$\mathrm{StOH + NaOH + CS_2 \longrightarrow StO\overset{\overset{\displaystyle S}{\|}}{C}SNa}$$

Xanthate

$$\mathrm{2\ StO\overset{\overset{\displaystyle S}{\|}}{C}SNa \xrightarrow{[O]} StO\overset{\overset{\displaystyle S}{\|}}{C}S{-}S\overset{\overset{\displaystyle S}{\|}}{C}OSt}$$

Xanthide

$$\mathrm{2\ StO\overset{\overset{\displaystyle S}{\|}}{C}SNa \xrightarrow{Zn^{2+}} StO\overset{\overset{\displaystyle S}{\|}}{C}S{-}Zn{-}S\overset{\overset{\displaystyle S}{\|}}{C}OSt}$$

Zinc starch xanthate

(Starch structures courtesy of McGraw-Hill Book Company, from Davidson, ed., *Handbook of Water-Soluble Resins,* Chapter 22, copyright 1980.)

Guar Gum

The structure of guar gum is as follows:

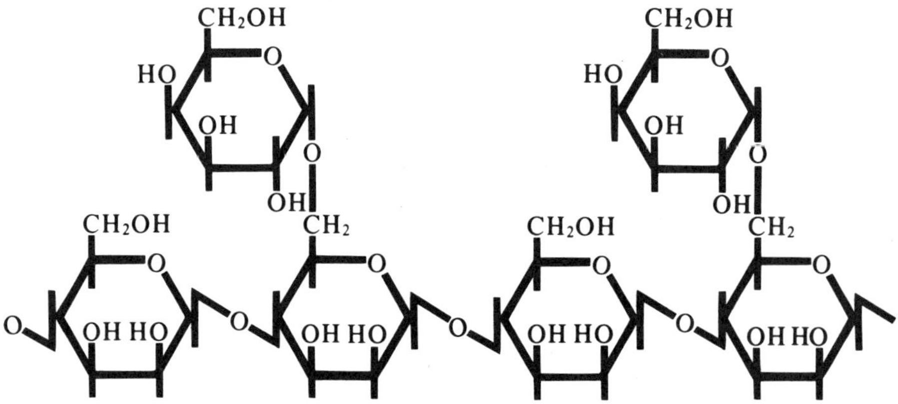

(Reproduced from the Kelco Xanthan Gum Book with permission of proprietor Merck & Co., Inc., Kelco Division, San Diego, California.)

Guar and guar derivatives are used as viscosity additives and to improve film formation and promote dry strength of sizes. Guar gums and locust bean gums are highly competitive, and the choice of which one to use is dictated by economics rather than chemistry. Guars find extensive use as printing agents, and their use is discussed further in this section.

Carrageenan[20]

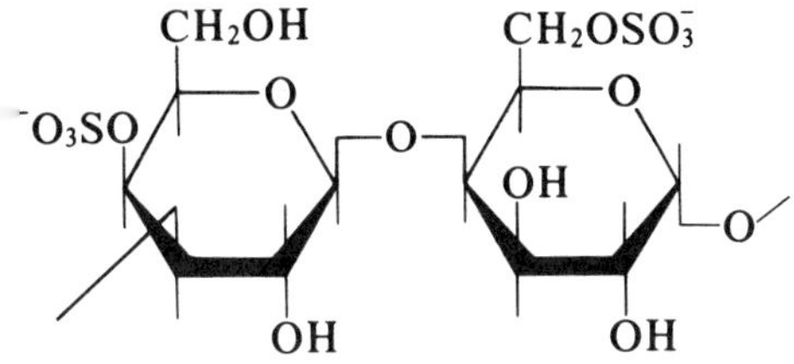

μ

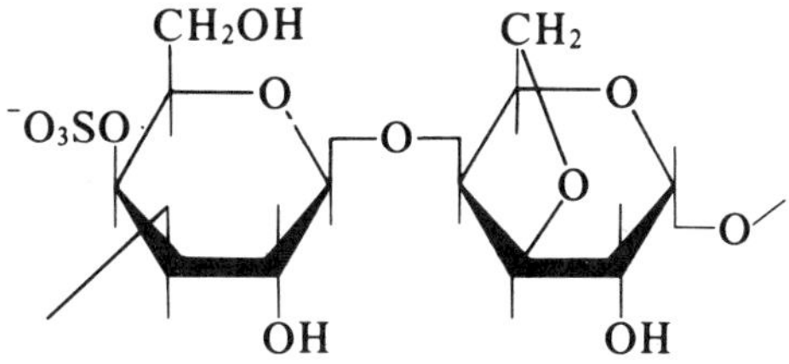

κ

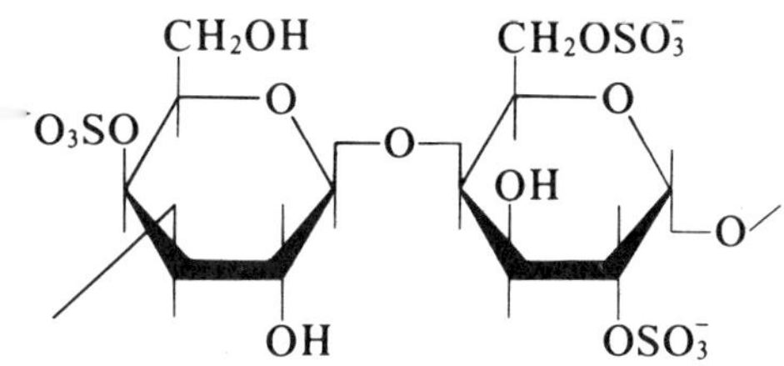

ν

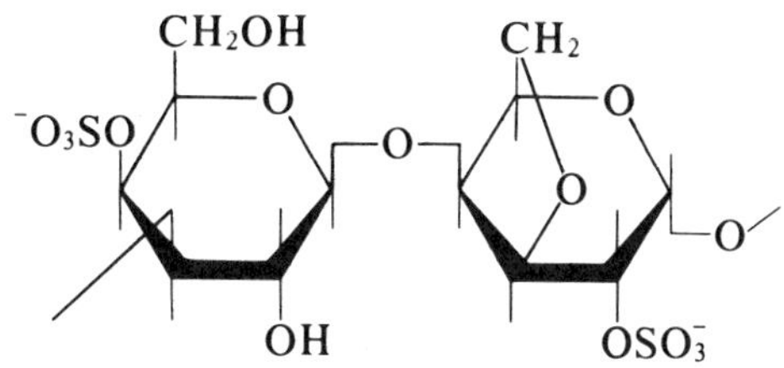

ι

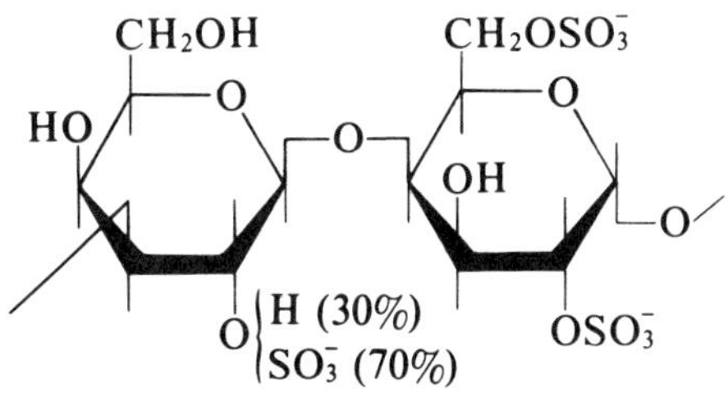

λ

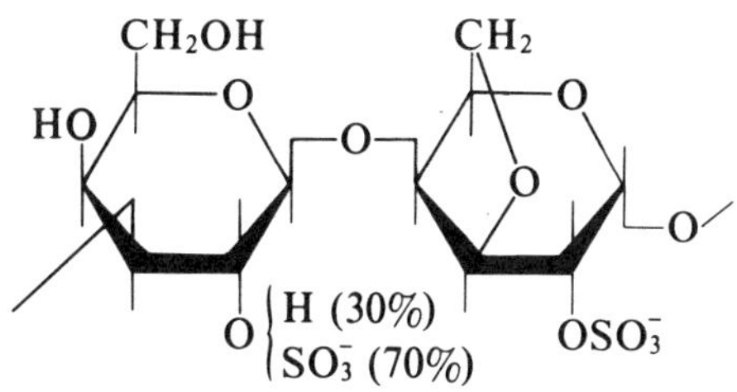

θ

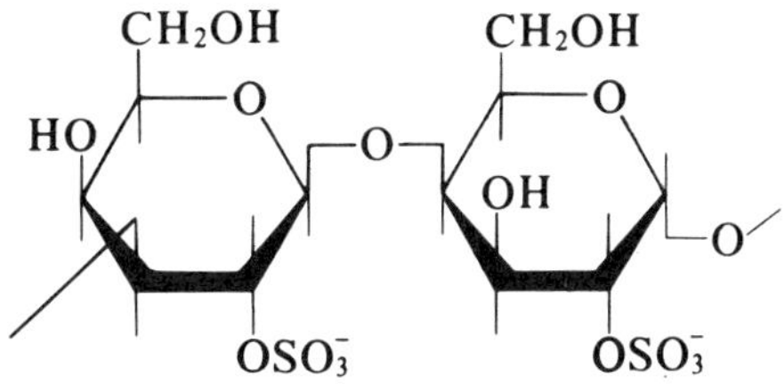

χ

Carrageenan is used as a tint dispersant and viscosity control agent and is usually used in conjunction with locust bean gum.

Gum Arabic

Gum arabic is the most widely used gum in industry. It is unique in that it forms solutions greater than 50% concentration. It has a comparatively low viscosity and is an excellent protective colloid and emulsifier. Gum arabic is produced from the sap of trees and is very expensive. It is used by the textile industry, both as a sizing and finishing agent, and in printing formulations. Its use in finishing, however, is as a nondurable component that is removed in washing. Low-cost sizes are generally manufactured out of rye starch[21] but can be improved by the addition of gum arabic.

Gum Karaya

Gum karaya is a partially acetylated complex polysaccharide with a high molecular weight.[22] Its use in the textile industry was considered unsatisfactory until a method of increasing its solubility was developed.[23] It is used more in printing than in sizing.

Gum Tragacanth[24]

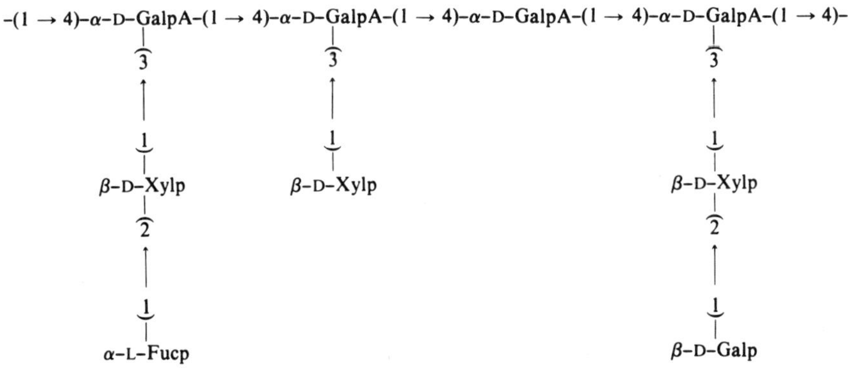

where D-GalpA is D-galactopyranuronic acid, D-Galp is D-galactopyranose, D-Xylp is D-xylopyranose, and L-Fucp is L-fucopyranose.

Gum trag, as this structure is called, is a natural vegetable gum that has excellent resistance to low pH. It has the ability to emulsify oil in water and can increase the aqueous-phase viscosity while lowering the interfacial tension between the oil–water emulsion.[25] Gum trag is very expensive and is used in textile operations primarily where acid stability and extended shelf life are required.

Locust Bean Gum

The structure of locust bean is as follows:

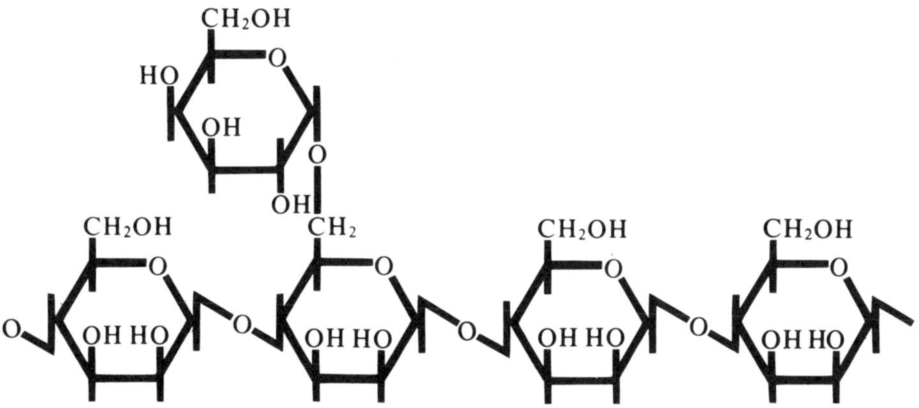

(Reproduced from the Kelco Xanthan Gum Book with permission of proprietor Merck & Co., Inc., Kelco Division, San Diego, California.)

Locust bean gum is the common name applied to the gum found in the seeds of the carob tree.[26] Derivatives of the locust bean gum are used more widely than the actual gum itself, with derivatives for the most part being either the carboxymethyl or hydroxyalkyl ethers of the gum.[27] Its use is to control viscosity and the degree of penetration into the fiber bundle. Locust bean gums compete directly with guar gum products, and the choice is usually made for economic reasons.[28] Locust bean gum is used extensively in printing and is discussed in more detail in the section on printing.

Tamarind Gum[29]

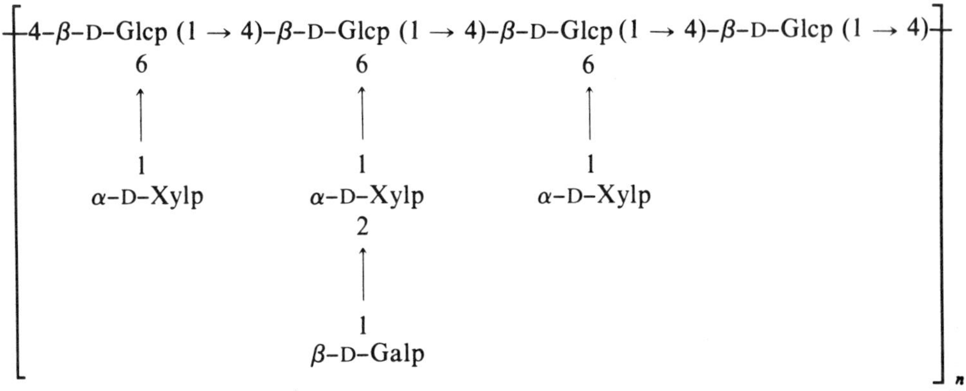

Tamarind gum is obtained from the seeds of the tamarind tree and was introduced into commercial production as a replacement for starch used in cotton sizing in Indian textile mills in 1943.[30] This gum disperses easily in cold water, but a uniform solution is obtained only on heating. The presence of a small amount of fat causes a tendency for the gum to lump and to have a characteristic fatty odor.[31] Tamarind gum is used either alone or in combination with gum trag and may be used for either sizing or print paste.[32] When used in sizing weaving for jute and cotton fabrics, it has excellent binding, film, and in modified form, thickening properties. In India it is reported to have replaced starch in many formulations because of its lower cost and its equivalent or better efficiency.[33]

Xanthan Gum[34]

Xanthan gum is a high-molecular-weight natural carbohydrate. It is a polysaccharide produced in culture fermentation by microorganisms originally isolated from vegetable plants.[35] These gums find more use in the food industry than the textile industry because viscosity contributed to a solution will remain constant regardless of whether the product is stored hot or cold or even frozen. Its flow characteristics will thus be the same in all climate zones.[36] Its use in the textile industry is primarily in the field of printing, where it aids in the suspension of dye pigments and in the control in the rheology, or tackiness, of the dye solution during printing.[37]

With the discovery and invention of synthetic fibers and their transformation into yarns for knitting and weaving, the sizing operation had to be modi-

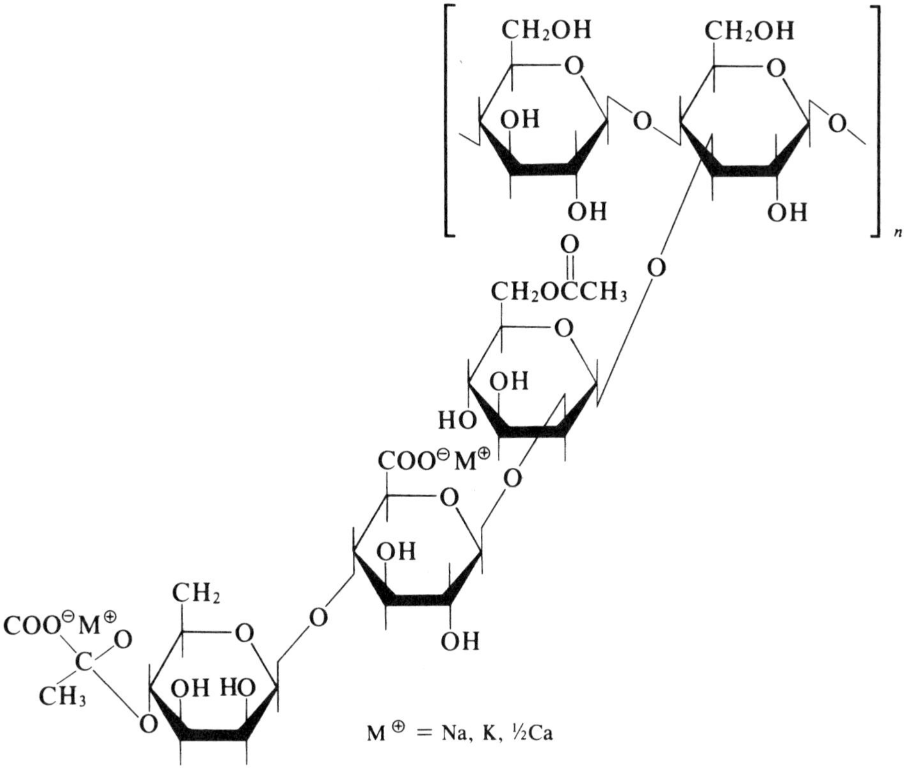

fied. For example, in the production of rayon fabrics and in rayon–cotton blends, gelatin replaced all or a portion of the starch size. As synthetic fabrics such as nylon, polyester, and acrylics became popular and achieved commercial significance, it was necessary to develop a new warp size technology. Prior to this time, "prior arts" was more important than chemistry. Now, however, chemical technology became an essential part of warp size processing.

SIZE COMPONENTS—SYNTHETIC POLYMERS

Typical synthetic polymers that are used as warp size additives are[38] (1) polyvinyl alcohol, (2) polyacrylates, (3) acrylic copolymers, (4) styrene–maleic acid copolymers, (5) polyester polymers, (6) butadiene-maleic acid copolymers, (7) vinyl derivatives, (8) carboxymethyl cellulose, (9) methyl cellulose,

(10) hydroxymethyl cellulose, (11) hydroxypropyl cellulose, (12) hydroxypropyl ethyl cellulose, (13) polyethylene glycol esters, and (14) polyvinyl pyrrolidone. A disscussion of, and structures for, the more common forms of synthetic sizing materials are presented in the following sections.

Polyvinyl Alcohol

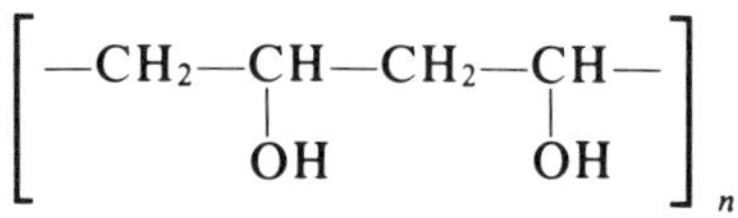

Polyvinyl alcohol is described as a polyhydric alcohol with secondary hydroxyl groups on alternate carbon atoms.[39] Polyvinyl alcohol cannot be synthesized directly but is made by first producing polyvinyl acetate from vinyl acetate and then hydrolyzing the polyvinyl acetate to polyvinyl alcohol.[40]

Production of Polyvinyl Alcohol[41]

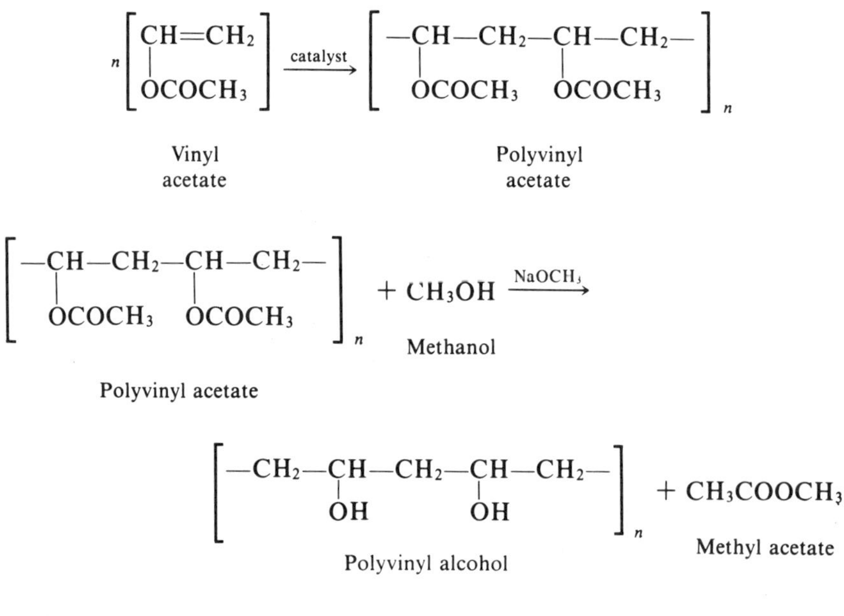

As it is supplied to the textile industry, polyvinyl alcohol is available in two forms. If the material is 88–89% hydrolyzed, it is considered to be a partial hydrolysis. If it is hydrolyzed to 98–99%, it is considered fully hydrolyzed.

Polyvinyl Alcohol—Fully and Partially Hydrolyzed

The structures for fully and partially hydrolyzed polyvinyl alcohol are:

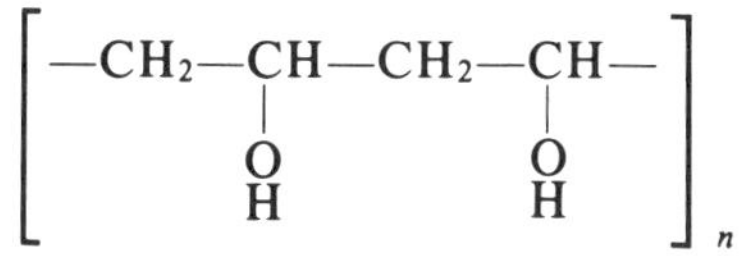

Fully hydrolyzed

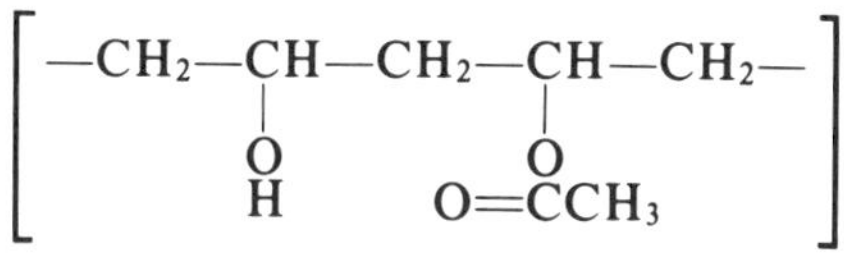

Partially hydrolyzed

Because polyvinyl alcohol may be gelled with certain inorganic acids, control of penetration of polyvinyl alcohol solutions into porous substrates and control of migration properties may be achieved.[42] Examples of gelling caused by boric acid and borax are shown below.

Boron Complexes with Polyvinyl Alcohol[43]

Polyvinyl alcohol warp sizes are characterized by high film strength, film flexibility, abrasion resistance, and good adhesion to the fiber.[44] Ease of removal of fully hydrolyzed forms and its good water solubility make polyvinyl alcohol a preferred size additive for textile warp sizing. Important performance characteristics of polyvinyl alcohol as a warp size are as follows:[45]

1. Low add-on—used at approximately one-half level required with starch.
2. High weaving efficiency.
3. Weave room comfort—weaving efficiency is high over a wide range of humidity but does not require high humidity for good weave performance.
4. Size bath stability—can be maintained at elevated temperatures for long periods of time without viscosity variations.
5. Ease of removal—dissolves easily in hot water and does not require costly enzymes.

Figure 3.1 shows the five-day biochemical oxygen demand for polyvinyl alcohol.

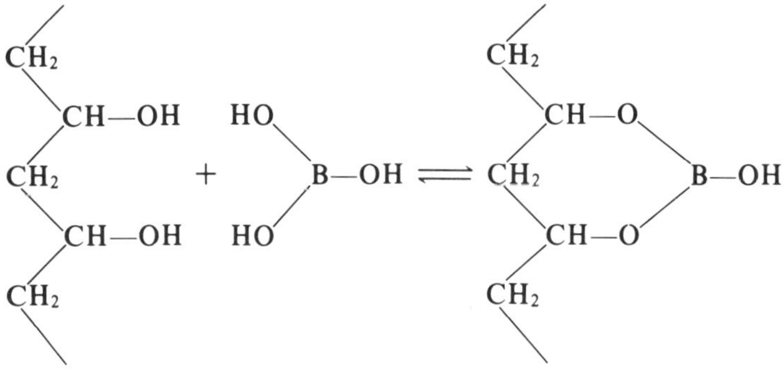

Polyvinyl alcohol Boric acid Monodiol complex (solution)

+

Base

⇌

Na^+

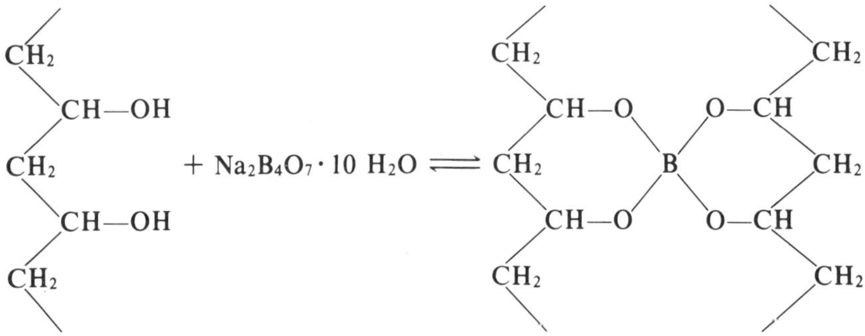

Polyvinyl alcohol Borax Didiol complex (gel)

Carboxymethyl Cellulose

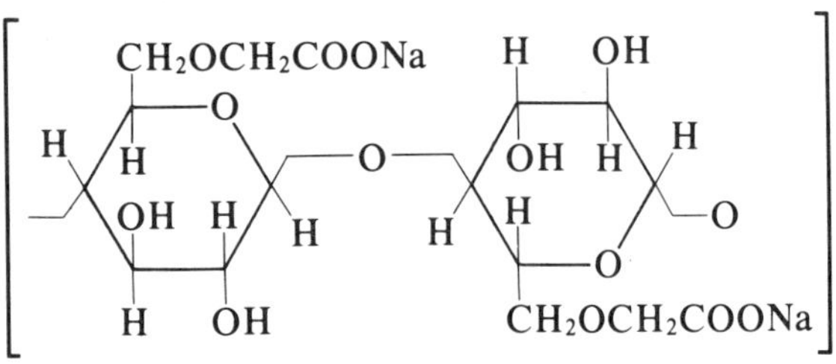

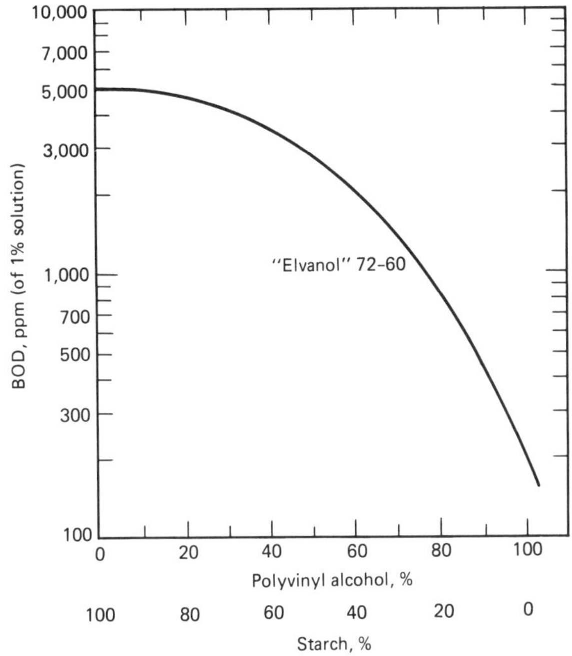

FIG. 3.1. Five-day biochemical oxygen demand (BOD) values for polyvinyl alcohol. Courtesy of E. I. DuPont.

Carboxymethyl cellulose (CMC) is a water-soluble anionic linear polymer.[46] It is prepared by the reaction of the cellulose hydroxyls with sodium monochloroacetate, as shown below:[47]

$$R_{cell}OH + NaOH + ClCH_2COONa \rightarrow R_{cell}OCH_2COONa + NaCl + H_2O$$

Carboxymethyl cellulose is usually used as a 6% solution with small amounts of textile wax, or kettle wax, as it is known in the trade. Sizing add-on is at a 5–10% level as compared to 15–20% for starches.[48] Carboxymethyl cellulose exhibits excellent film-forming properties but can be affected by moisture, which makes control of weave room humidity necessary. Moisture regain plasticizes the CMC film and aids in the minimizing of "shedding" on high-speed looms.[49] With government and environmental controls ever on mill management minds, the cleanliness factor is a part of overall mill efficiency and is one of the benefits of using CMC.[50] Additional advantages of CMC warp size are:[51]

1. Cold-water soluble—lower energy costs; does not need high water temperatures for dissolving or desizing; no cooking or homogenizing required; easy clean-up of equipment.
2. Forms tough, flexible film—warp stops reduced; shedding on the loom 30–40% lower than that of starch; cleaner weave room and lower clean-up costs; fuzzy yarn minimized.
3. Low add-on—1 lb of CMC does the work of nearly 3 lb of starch; slasher efficiency increased; yarn is more compact and splits easily at the lease rods (10–15% more yarn can be put on the warp beam); reduced shedding; lower drawing and tying-in costs; decrease in loom beam changes; lower pollution load; reduced storage and handling costs for size.
4. Stable solutions—can be stored over shifts, weekends, and shutdowns.
5. Available in pellet or powder form—dust-free operations with pellets or powders; dispersible in water without lumping.

Table 3.1 compares CMC, starch, and polyvinyl alcohol warp sizes. Figures 3.2 through 3.9 shows various relationships between CMC, PVA, and high-amylose starches when used in warp sizing.

Hydroxyethyl Cellulose[54]

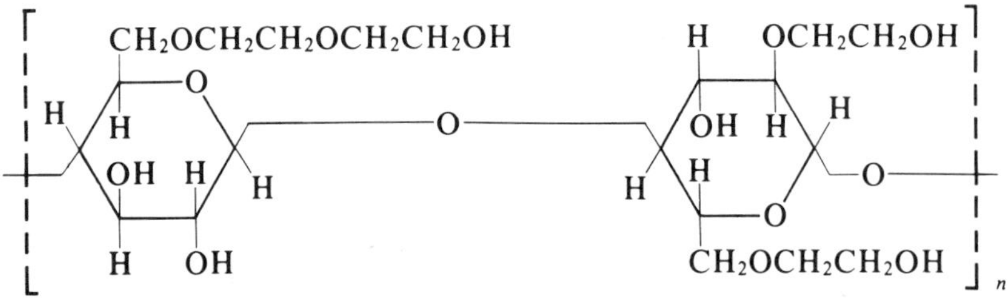

Hydroxypropyl Cellulose[55]

$$\left[C_6H_{10-n}O_5(CH_2CHOHCH_3)_n \right]_x$$

TABLE 3.1. Comparison of CMC Warp Size with Starch and Polyvinyl Alcohol[52] (Courtesy of Hercules Incorporated)

Size Preparation	Starch: Must Be Cooked to a Viscosity	CMC: Cold-Water Soluble Pellets; Dust Free Pellets or Powder	Fully Hydrolyzed PVA: Agitation and Heat Required; Dusty
Builds viscosity	Yes	Yes	No
Film strength	Weak	Strong	Strong
Film flexibility (pure state)	Brittle	Flexible	Stiff
Moisture regain	Low	Similar to cotton	Low
Ease of handling at slasher	Satisfactory	Easy restart of slasher operation after shutdown; easy clean-up	Difficult to restart slasher operation after shutdown
Desizing	Requires enzyme	Cold-water soluble	Requires high-temperature water

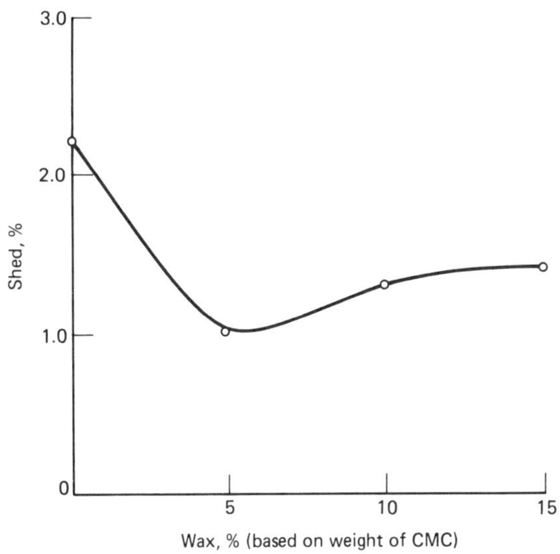

FIG. 3.2. Effect of wax concentration on CMC warp size efficiency. Courtesy of Hercules Incorporated.

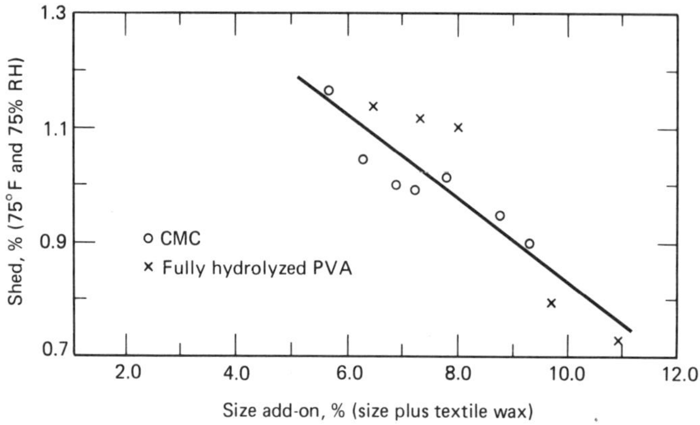

FIG. 3.3. Sizing efficiency of CMC versus PVA. Courtesy of Hercules Incorporated.

Polyethylene Glycol

$$HOH + nC_2H_4O \rightarrow HO\leftarrow C_2H_4O \rightarrow_n H$$

The oxyethylene distribution grouping for various polyethylene glycols is shown in Fig. 3.10.[56]

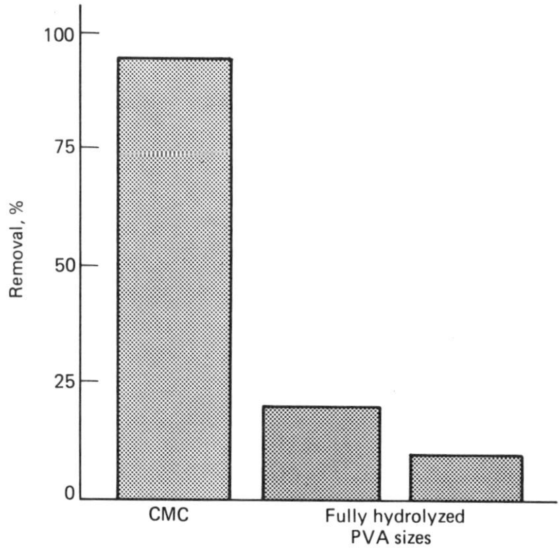

FIG. 3.4. Desizing efficiency of CMC versus PVA. Courtesy of Hercules Incorporated.

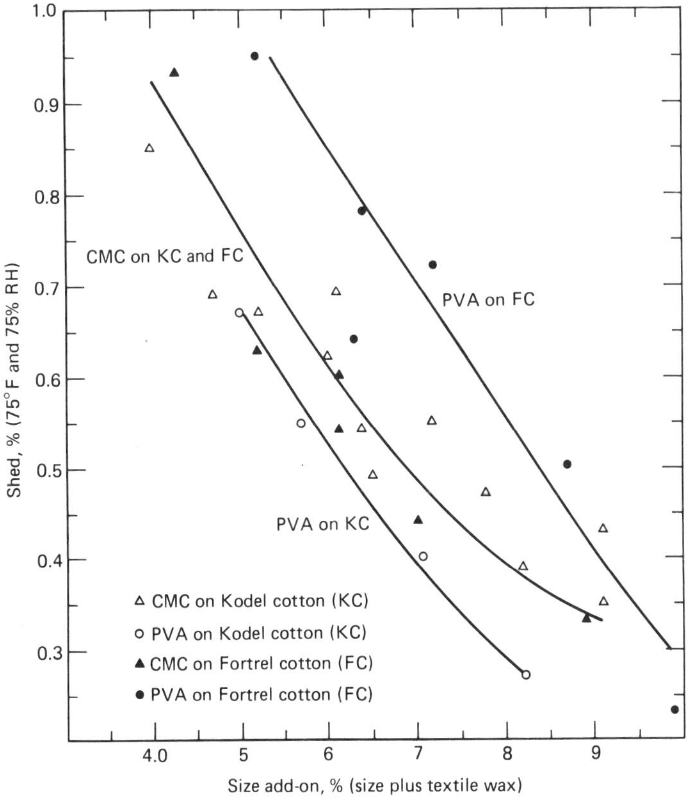

FIG. 3.5. Sizing efficiency of CMC and PVA on polyester–cotton. Courtesy of Hercules Incorporated.

Polyvinylpyrrolidone

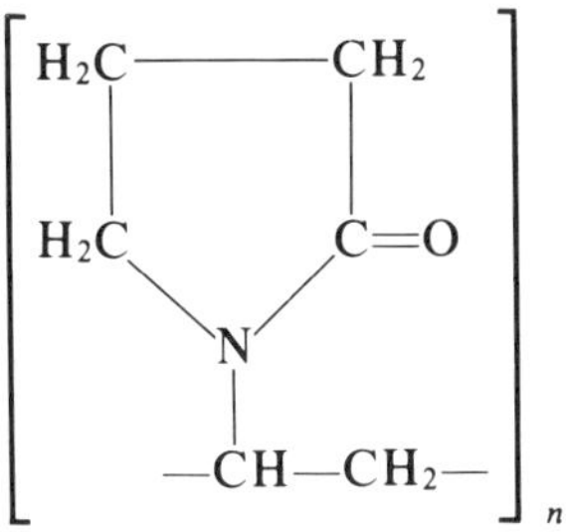

(Courtesy of GAF Corporation)

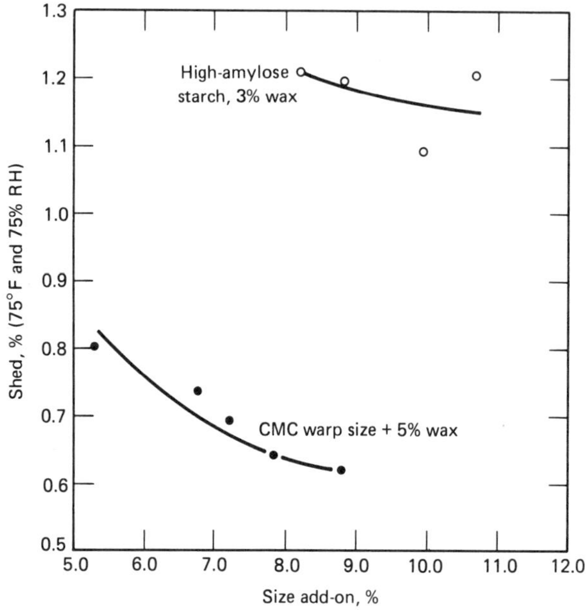

FIG. 3.6. Shed test comparison between CMC warp size and high-amylose starch. Courtesy of Hercules Incorporated.

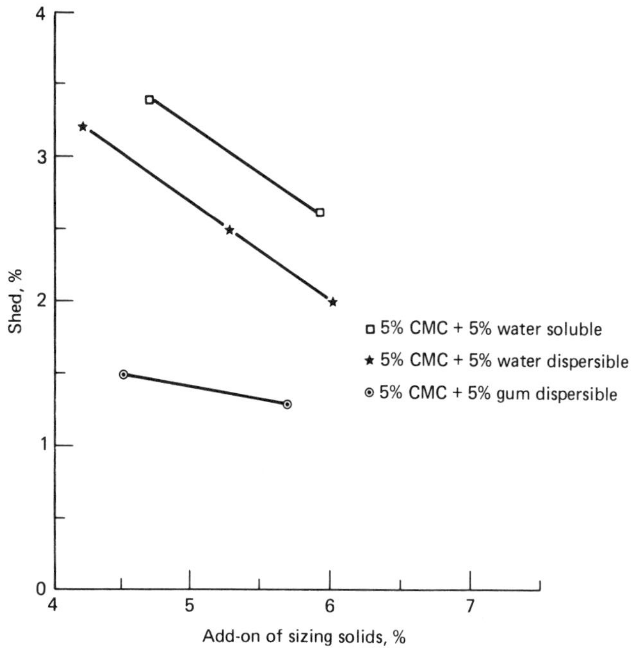

FIG. 3.7. Effect of waxes on cotton sizing efficiency of CMC warp size. Courtesy of Hercules Incorporated.

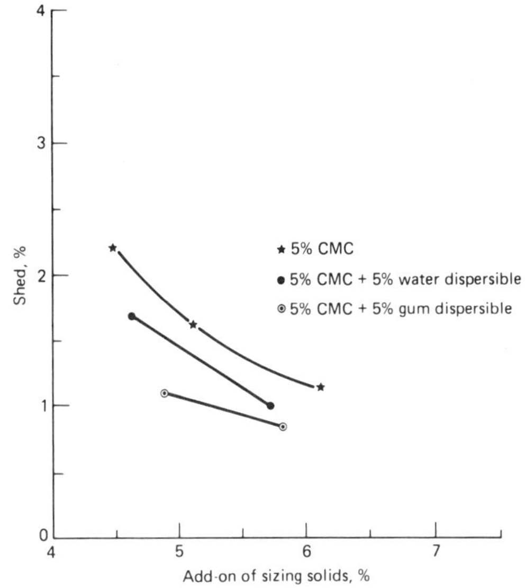

FIG. 3.8. Effect of waxes on cotton sizing efficiency of CMC warp size. Courtesy of Hercules Incorporated.

Polyacrylic Acid and Its Homologs

$$[CH_2{=}CHCOOH]$$

Polyacrylic acid and its water-soluble polymers possess a wide range of physical and chemical properties.[57] These polymers, produced as either homopolymers or copolymers, are derived from esters of acrylic acid. Film formation from hardest to softest proceeds in the following order: methyl methacrylate; methyl acrylate; ethyl acrylate; butyl acrylate; and 2-ethylhexyl acrylate. This order is shown schematically as follows:[58]

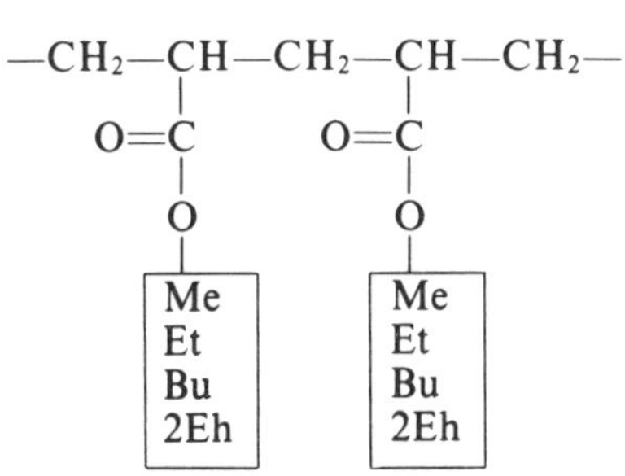

(Courtesy of Texchem Consulting Labs)

A typical schematic for a copolymer is:

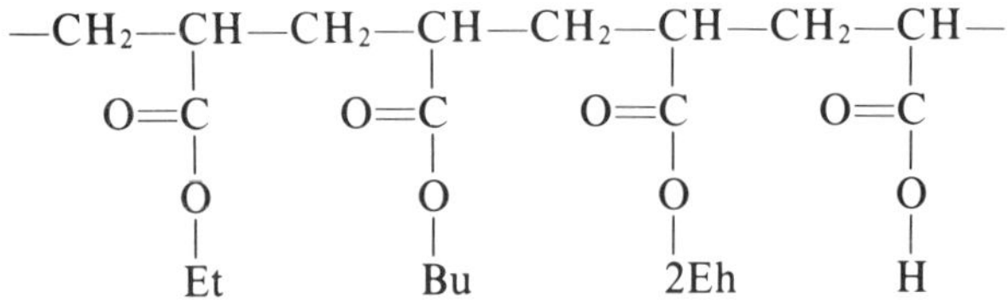

(Courtesy of Texchem Consulting Labs)

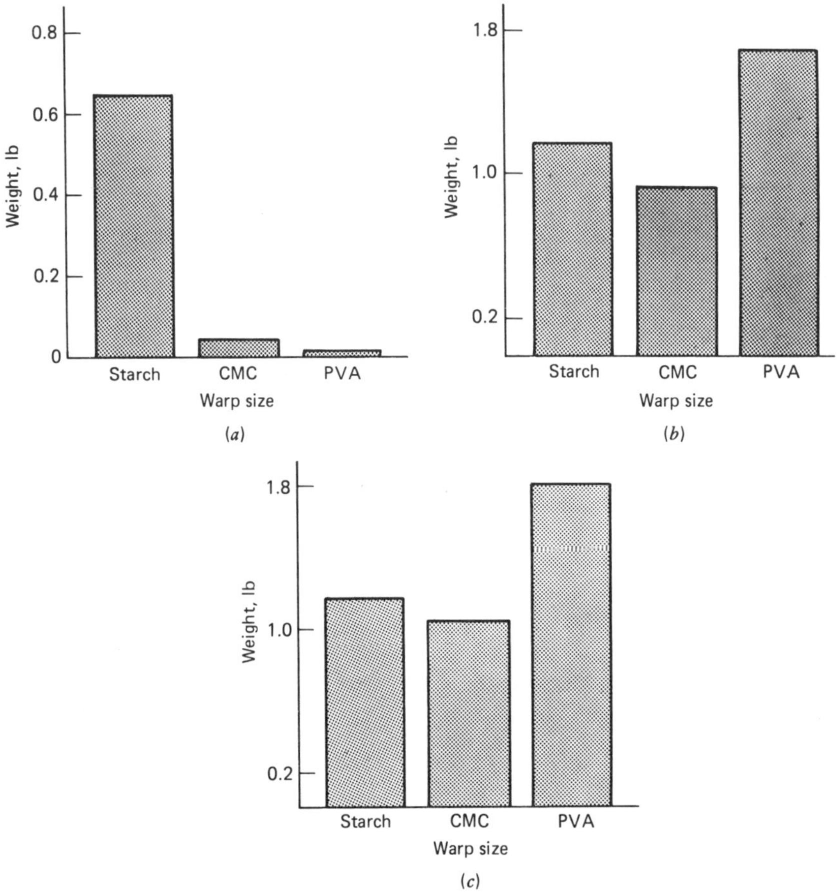

FIG. 3.9. Waste treatment requirements for starch, CMC, and polyvinyl alcohol warp sizes: (*a*) biological oxygen demand (after 5 days), pounds of oxygen required per pound of warp size; (*b*) chemical oxygen demand, pounds of oxygen required per pound of warp size; (*c*) total oxygen demand, pounds of oxygen required per pound of warp size. Courtesy of Hercules Incorporated.

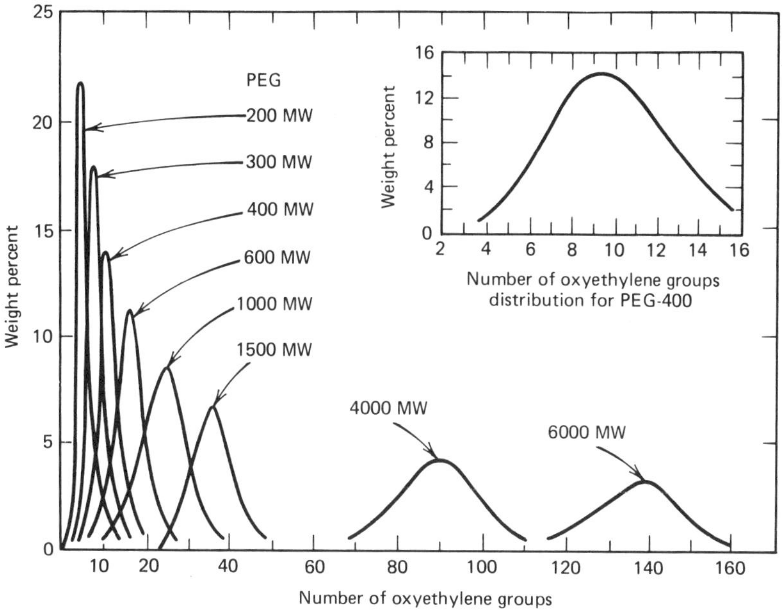

FIG. 3.10. Oxyethylene distribution grouping for polyethylene glycols (MW = molecular weight). Used by permission of McGraw Hill.

Other polymers that find acceptance for special applications are listed in the following sections.

Styrene-Maleic Acid Copolymer

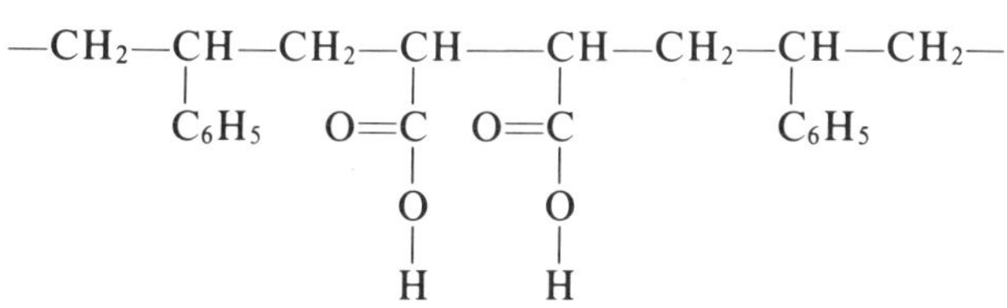

(Used as sodium salt)

(Courtesy of Texchem Consulting Labs)

Polyester Polymer

$$HO(C_2H_4O)_nOC\ C_6H_4CO(OC_2H_4)—O—$$

(PEG 4000); terephthalic acid)

(Courtesy of Texchem Consulting Labs)

Butadiene-Maleic Acid Copolymer

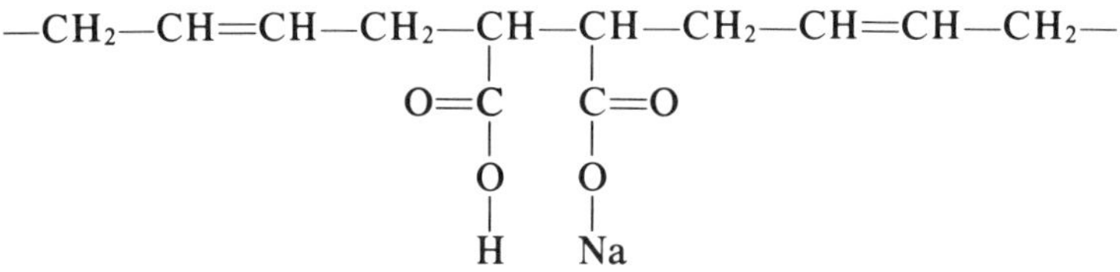

(Courtesy of Texchem Consulting Labs)

(Maleic acid = one-half sodium salt for solubility.)

Vinyl Warp Sizes

Although these polymer reactions seem complicated, they are really just a matter of terminology and can probably be best classified and understood by making a chart that shows that most warp size polymers are, in reality, formed from a single vinyl monomer, which in its simplest form may be represented as $CH_2{=}CHX$, where X =[59]

Cl	Vinyl chloride
OH	Vinyl alcohol
$OOCCH_3$	Vinyl acetate
COOH	Acrylic acid
COOR	Acrylic ester (methyl, cthyl, butyl, 2-ethylhexyl)
CN	Acrylonitrile
C_6H_5	Styrene
or	Vinyl ether

(If the hydrogen in CHX is substituted with a methyl group, a new series of esters is formed that are methyacrylic derivatives.)

Alginates

Algin, the polysaccharide extracted from the brown seaweed, *Phaeophyceae,* has only recently been identified chemically. Although purified alginic acid had its structure first reported in 1930 by Nelson and Cretcher, who named it as D-*mannuronic acid polymer,* Fischer and Dörfel reported in 1955 that by using paper chromatography L-*guluronic acid* was also present as a major component in alginic acid. The chemical composition of alginic acid varies with source and may, in fact, contain a third segment that consists of alternating D-mannuronic acid and L-guluronic acid residues. Shown below are structures representing the various forms of alginic acid components.

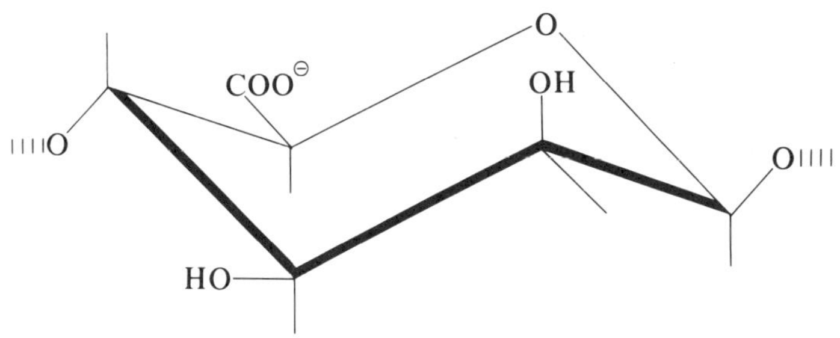

β-D-Mannopyranosyluronic
acid unit—C1 conformation

Conformation of mannuronic acid

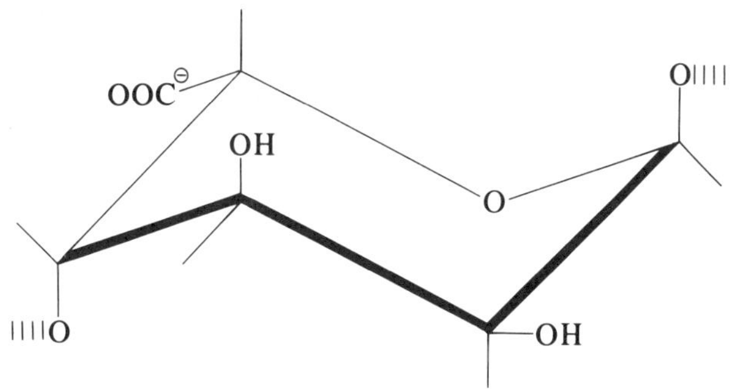

α-L-Gulopyranosyluronic
acid unit—1C conformation

Conformation of guluronic acid

5-A
-M-M-M-M-

5-B
-G-G-G-G-

5-C
-M-G-M-G-

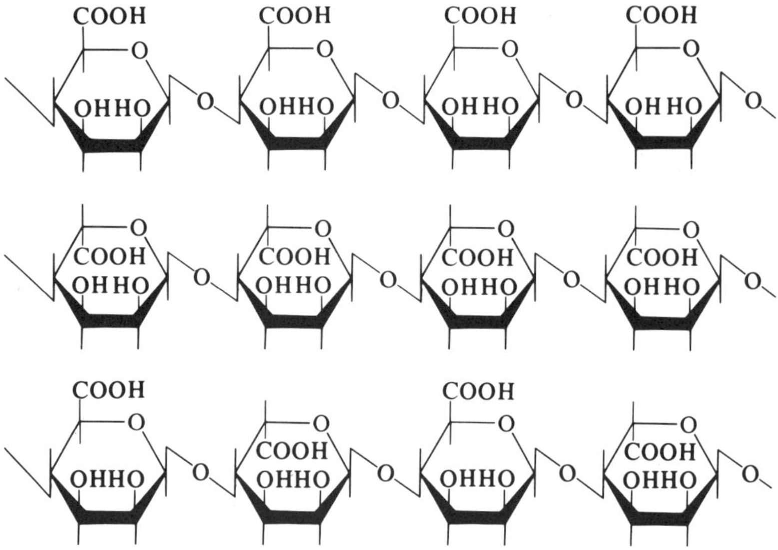

Structure of the polymer segments
contained in alginic acid

(Alginate structures reproduced from the Kelco Algin Book with permission of proprietor Merck & Co., Inc., Kelco Division, San Diego, California.)

Carbopol® Water-Soluble Resins

Carbopol® resins are 100% acrylic acid polymers. The carboxyl groups in these polymers, when neutralized to an alkaline pH, are anionic and are used for a wide variety of textile processes. A typical use is as thickening agent for production of specific viscosities and flow properties, to suspend insoluble ingredients, and to stabilize emulsions. Carbopol® resins in the unneutralized state have an average equivalent weight of 76 and can be represented by the following structure:

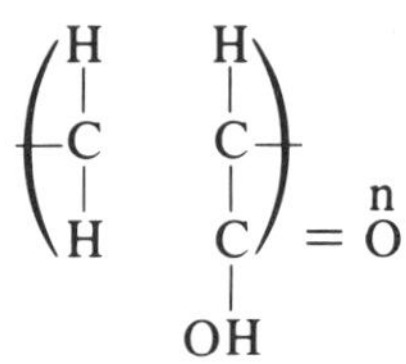

(Courtesy of B. F. Goodrich Chemical Division)

Carbopol® resins are depicted in the following diagrams:

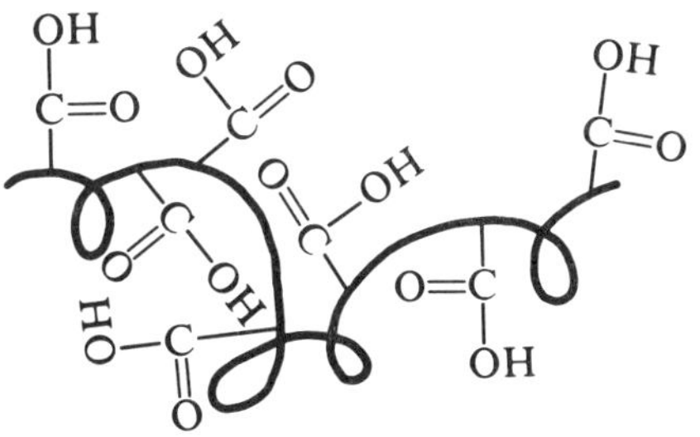

Relaxed state

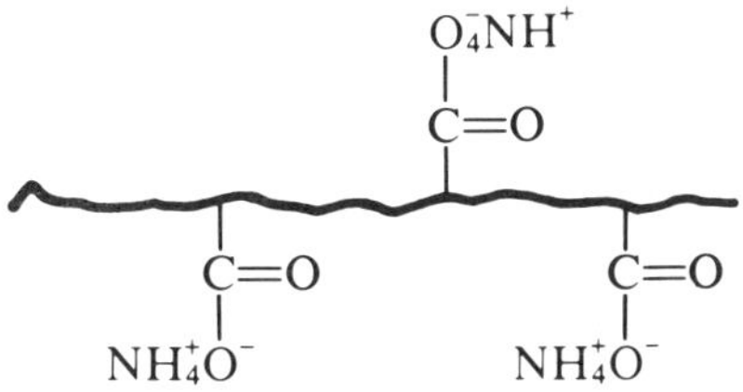

Uncoiled state

(Courtesy of B. F. Goodrich Chemical Division)

Carbopol® resins offer the following benefits:

Thickening efficiency—high viscosities at low concentrations.

Uniform performance—reproducibility unattainable with natural gums.

History—years of successful use in drug and cosmetic applications.

Temperature stability—only slight effect on mucilage properties after extended heating or cooling.

Resistance to aging—excellent shelf life.

Microbial resistance—resistance to bacterial attack and mold growth.

Versatility—although Carbopol® is used primarily in aqueous systems with neutralization, it can also be used in solvent systems, with or without neutralization.

Generating Viscosities

Carbopol® resins thicken these basic systems:

System	Properties of Carbopol® Resins
Aqueous systems	Neutralized with inorganic or water soluble organic bases (amines).
Aqueous systems	Modified with polyhydroxy and polyethoxy reagents.
Solvent systems	Neutralized with amines.

System	Properties of Carbopol® Resins
Solvent systems	Modified with polyhydroxy and polyethoxy reagents.
Emulsions	Neutralized and act as primary emulsifiers.
Emulsions	Modified with polyhydroxy and polyethoxy reagents.

Courtesy of B. F. Goodrich Chemical Division.

COMMERCIAL WARP SIZE FORMULATIONS

From a practical point of view, we should now consider conventional commercial size formulations:

100% Cotton—200-gal Mix[60]

A.	Starch	150 lb
	Kettle wax	12 lb
	Water	To make 200 gal
B.	Starch	125 lb
	Acrylic terpolymer	20 lb
	Kettle wax	5 lb
	Water	To make 200 gal

50/50 Polyester/Cotton Sheeting—330-gal Mix[61]

A.	Starch	330 lb
	Polyvinyl alcohol	150 lb
	Kettle wax	30 lb
	Water	To make 330 gal
B.	Starch	300 lb
	Polyvinyl alcohol	75 lb
	Acrylic terpolymer	75 lb
	Kettle wax	12 lb
	Water	To make 330 gal

Note: Use of formula B increased weave efficiency from 85–88% to 90–92%.

Cotton Fabrics[62]

Type of Fabric	Warp Sley Ends per Inch	Yarn Count	CMC, lb/100 fin. gal. (Finished Gallons)	Add-on, %
Bedspreads	40–60	18/1	30–40	4–5
Blankets	35–40	15/1	20–30	2–4
Broadcloth	136	40/1	40–45	6.5–7
Corduroy	84	30/1	35–40	6–7
Denim	60–100	8–14/1	20–30	3–5
Drapery	60–100	13–20/1	30–40	4–5
Gingham	80	30–40/1	30–35	5–6.5
Heavy gingham	100	30–40/1	35–40	6.5–7
Muslin sheeting	64	32/1	30–35	5–6
Percale sheeting	94	32/1	35–40	6.5–7.5
Print goods	80	30/1	30–35	5–6
Rainwear	100	20/1	40–45	6.5–7
Terry pile	30–60	10–20/1	10–15	1–3
Terry ground	40–60	30/1	20–30	3–5
Twills, cords, poplins	100–110	15/1	40–45	6.5–7
Upholstery	40–80	15/1	25–40	3.5

Note: Five percent of textile wax based on weight of CMC is recommended. (Courtesy of Hercules Incorporated).

Cotton Blends with Synthetics (35-50% Cotton)[63]

Type of Fabric	Warp Sley Ends per Inch	Yarn Count	CMC, lb/100 fin. gal. (Finished Gallons)	Add-on, %
Corduroy	30–40	15–20/1	40	6–7
Plain	40–60	15–30/1	35	5–6
Plain	60–80	30–50/1	40	6–7

Type of Fabric	Warp Sley Ends per Inch	Yarn Count	CMC, lb/100 fin. gal. (Finished Gallons)	Add-on, %
Corduroy	70–80	30–35/1	45	6.5–7
Plain	80–90	30–35/1	45	6.5–7
Plain	90–100	30–50/1	45	7–8
Denin	60–100	8–14/1	25–35	4–6
Sheeting	90–100	30–45/1	45–50	7.5–8.5
Terry ground	40–60	10–20/1	45–50	7–8
Twill, poplins	100–130	15–30/1	50–55	8.5–9.5
Plain	120–136	40–50/1	50–55	8–9

Note: Five percent of textile wax based on weight of CMC is recommended. (Courtesy of Hercules Incorporated).

Rayons and Rayon Blends[64]

Type of Fabric	Warp Sley Ends per Inch	Yarn Count	CMC, lb/100 fin. gal. (Finished Gallons)	Add-on, %
Spun rayon	40–60	4–15/1	10–15	2
	60–80	15–30/1	15–20	2–3
	80–100	30–40/1	25–40	3–6
Polyester–modified rayon	60–80	15–30/1	40	6–7
	80–90	30–40/1	45	7–7.5
	90–100	30–40/1	50	8–9
	Total Ends	Denier		
Rayon filament	9,000–9,600	150	10–12	0.5–1.0
	9,000–10,000	75	12–15	1–1.5

Note: Five percent of textile wax based on weight of CMC recommended. (Courtesy of Hercules Incorporated).

Double-Dip Slashers[65]

Type of Fabric Fiber Content	Warp Sley Ends per Inch	Yarn Count	CMC, lb/100 fin. gal. (Finished Gallons)	Add-on, %
Batiste (poly-ester–cotton)	96	50/1	30–35	6–7
Broadcloth (poly-ester–cotton	128	50/1	35–40	8–9
Poplin (poly-ester–cotton)	106	25/1	40	7–8
Sheeting (poly-ester–cotton)	90–95	35/1	35–40	8–10
Sheeting (poly-ester–cotton)	104	45/1	40–45	9–10
Triblend (poly-ester–modified rayon)	100–110	15–20/1	38–42	9–10

Note: Five percent of textile wax based on weight of CMC is recommended. (Courtesy of Hercules Incorporated).

In spun warp sizing, polyvinyl alcohol may be replaced on a pound-for-pound basis with CMC. On filament sizing, the use of polyester size produces excellent weaving efficiencies; however, like polyvinyl alcohol, if not properly handled, it can create problems in the preparation process. Use of the proper size formulation will result in a more even deposit and pick-up of size and will reduce fiber dusting and shedding, thereby producing a smoother surface and an increase in size efficiency.

HIGH-PRESSURE SIZING[68]

Water has been the traditional medium in which sizing products have been dissolved and applied to yarn. Evaporation of that water during processing requires an enormous amount of energy to be expended. Sizing compounds available today can be dissolved in greater concentrations than normally used; therefore, an effort has been made to develop machinery to apply these compounds from more concentrated solutions to minimize the amount of

Nylon-100-gal Mix[66]

A.	Polyacrylic acid solution	20 gal
	Water	To make 100 gal
B.	Polyacrylic acid solution	15 gal
	Acrylic terpolymer	5 gal
	Water	To make 100 gal

Acetate Filament—100-gal Mix[67]

A.	Stymer®	50 lb
	Nopco oil	10 lb
	Water	To make 100 gal
B.	Stymer®	40 lb
	Acrylic terpolymer	10 lb
	Water	To make 100 gal

Note: Use of terpolymer minimizes formation of water repellency during creep speed. Acrylic terpolymers appear to minimize shuttle marks, chafing, and crow's feet. Stymer® is a registered trademark of Monsanto Company.

energy needed for drying. The major limitation on sizing compounds has been viscosity, which must remain sufficiently low to allow proper application to the yarn. By utilizing high-pressure rolls and pneumatic configurations, West Point Foundry and Machine Company has been able to develop a high-pressure squeeze box, which is specifically designed for use with high-pressure squeeze roll technology. The potential economic benefits of this type of sizing are significant. In fact, it is reported that savings and performance improvements exist at several levels:[69]

1. The first level of savings is that of reduced energy cost, due to reduced drying requirements.

2. If drying capacity is a problem, high pressure can allow higher speeds without increasing the number of dry cans.

3. A third level of savings was realized only when actual production had begun, in that an improvement in size efficiency had occurred by applying a reduced amount of yarn while getting better performance from the smaller amount of size. In some cases, this increased efficiency allowed the use of less expensive components while maintaining optimum slashing efficiency. This technique will, no doubt, become more important as the cost of energy and chemicals increases.

CHAPTER FOUR

INFRARED SPECTRA OF FIBER FINISH AND SIZE COMPONENTS

Many problems encountered in later stages of processing may be attributed to the incomplete removal of finishing or sizing products. It is, therefore, often desirable to be able to identify components in fiber finishing and sizing compounds.

Typical components used in spin finishing are:

POE(16) Castor Oil
POE(10) Castor Oil
POE(39) Castor Oil
POE(4) Sorbitan Monooleate
POE(9) Linear Alcohol
POE(20) Coconut Oil
Polybutylene Glycol 500
Pentaerythritol Tetrapelargonate

POE(50) Sorbitol Hexoleate
n-Octyl–*n*-Decyl Adipate
POE(5) Castor Oil
POE(25) Castor Oil
POE(50) Hexaoleate
POE(10) Di-1,4-Butylene Glycol
POE(4) Sorbitan Monostearate
POE(10) Coconut Oil
Butyl Stearate
Tridecyl Stearate
POE(4) Lauryl Alcohol
POE(12) Coconut Fatty Acid
EO/PO Molecular Weight 2750

POE(40) Sorbitol Septoleate
POE(4) Sorbitan Monolaureate
Carboxymethyl Cellulose
POE(30) Sorbitol Laureate
Polyvinyl Alcohol
Polyvinyl Pyrrolidone

Infrared spectra typical of the majority of the components used in fiber and yarn processing are shown in Figs. 4.1 through 4.67.

Once a group of fibers has been transformed into yarn or has been made into a state where they can be further processed for dyeing, weaving, or knitting, fiber and yarn processing is considered to be ended. Fiber or yarn in this state is complete and can be left alone, stored, sold, or used for some later textile operation.

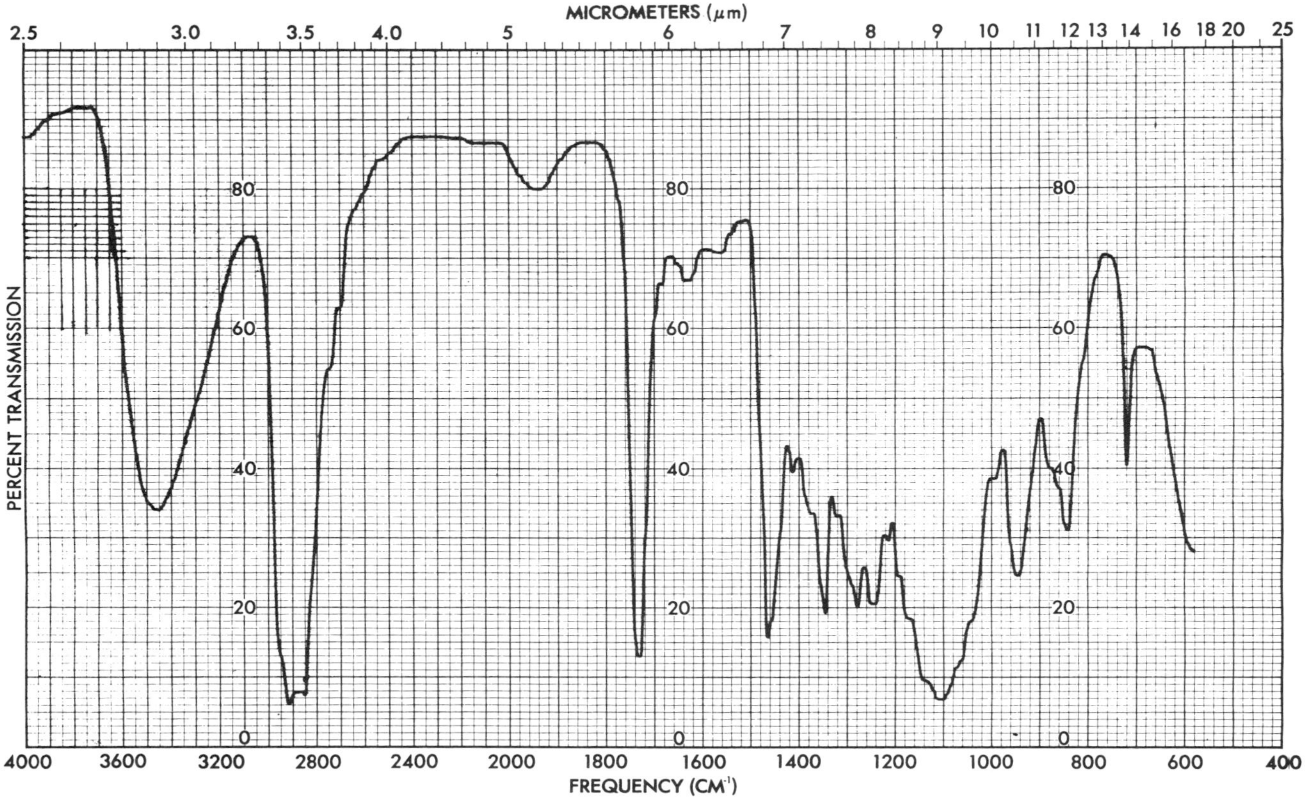

FIG. 4.1. PEG-400 monostearate.

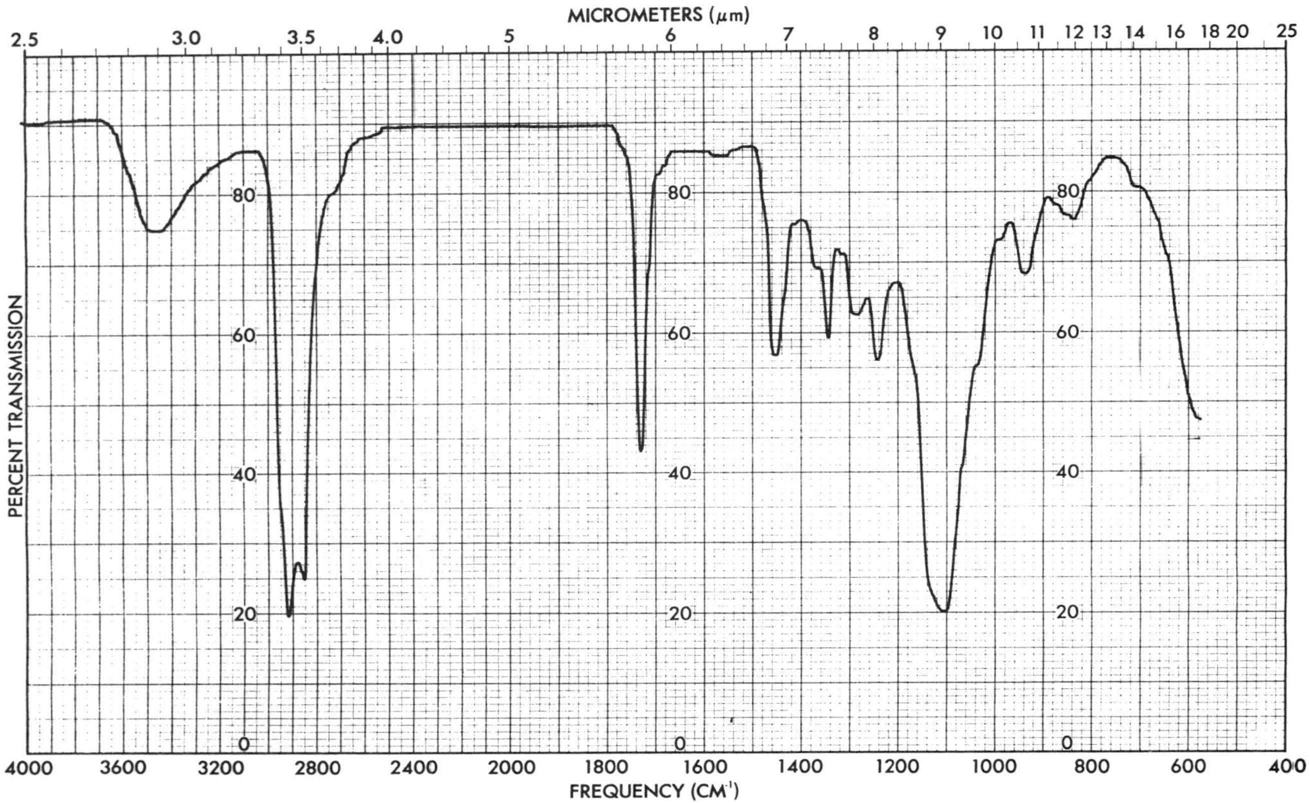

FIG. 4.2. PEG-400 isostearate.

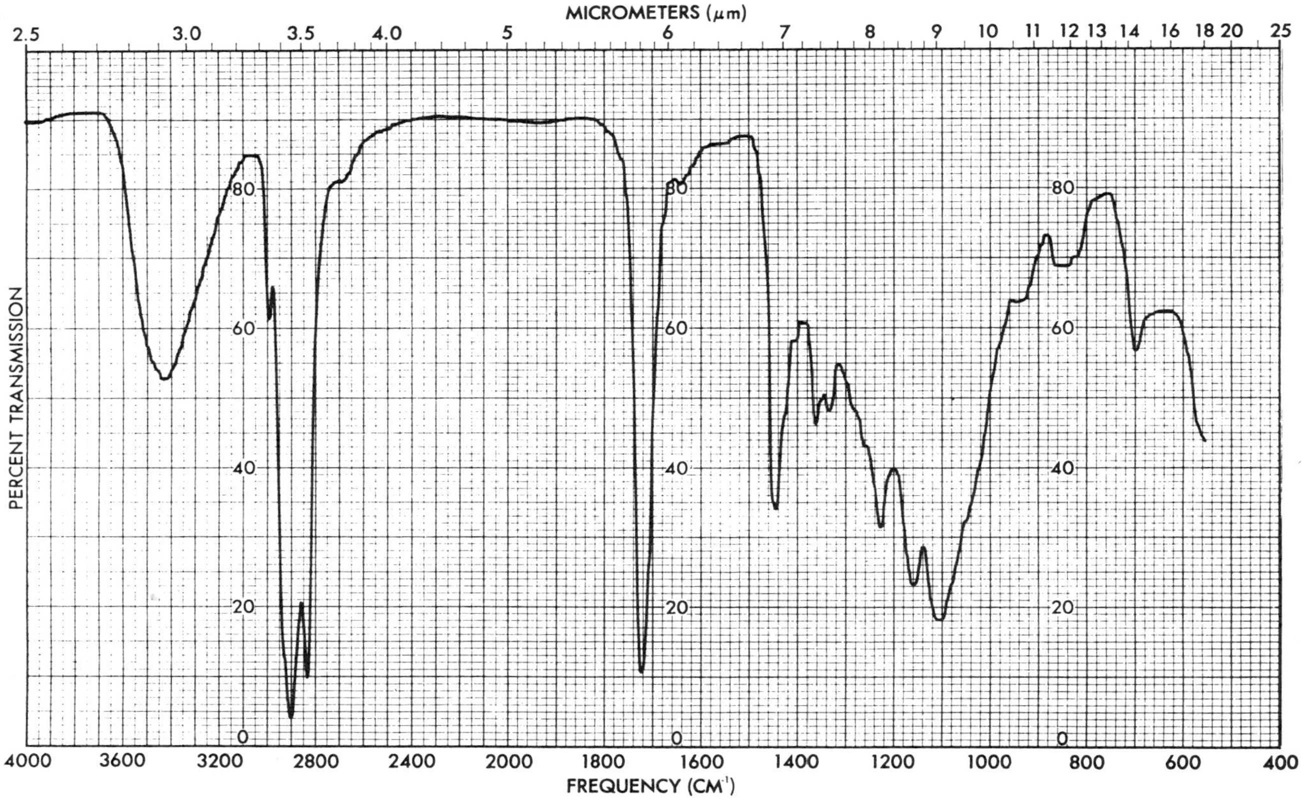

FIG. 4.3. $(EO)_5$ castor oil.

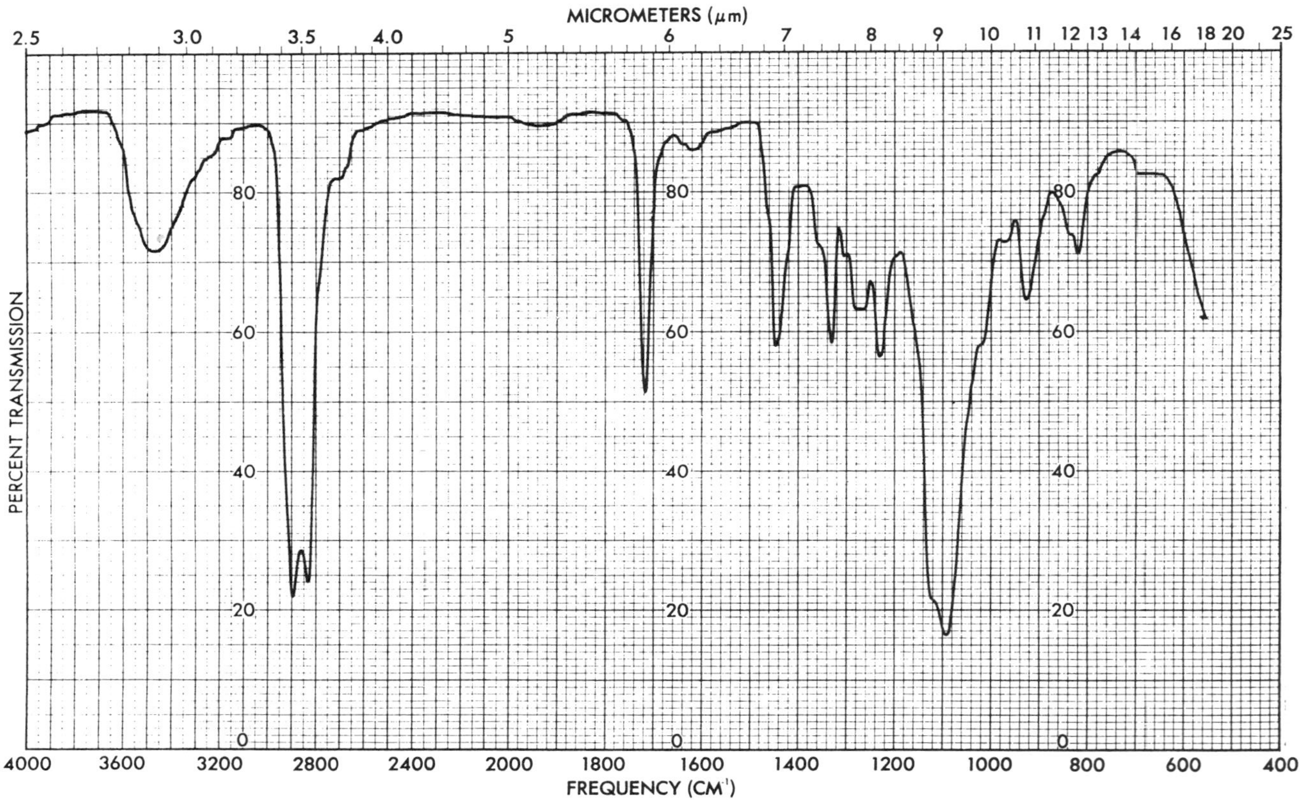

FIG. 4.4. $(EO)_{40}$ castor oil.

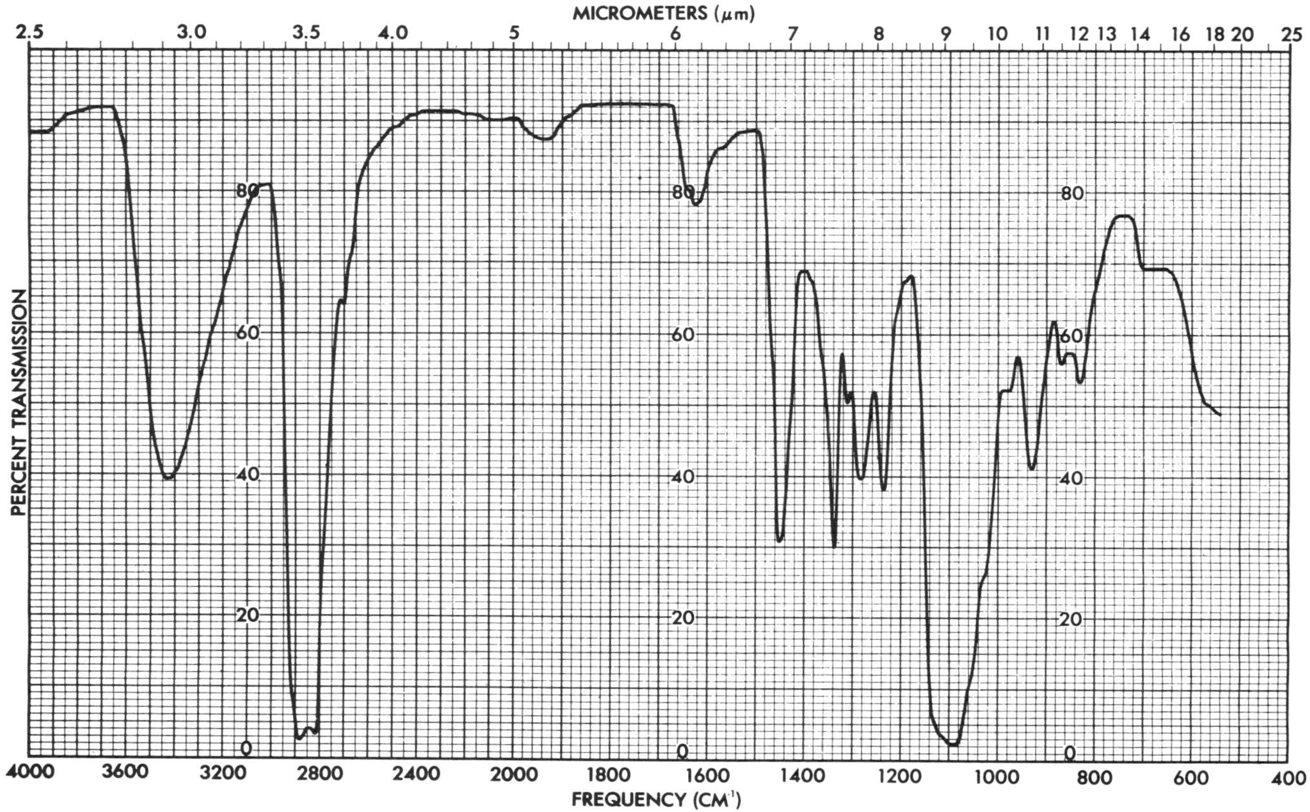

FIG. 4.5. $(EO)_{15}$ tallow diamine.

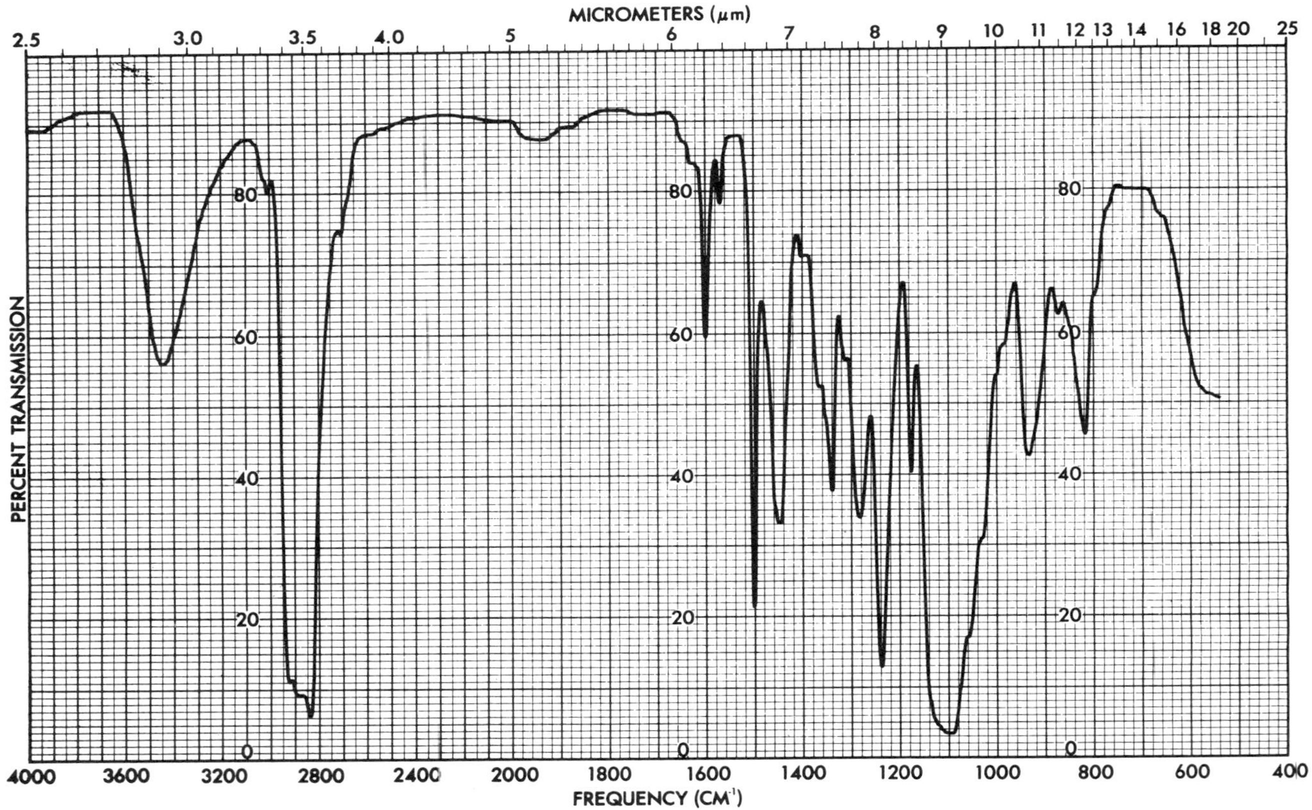

FIG. 4.6. $(EO)_{10}$ nonyl phenol.

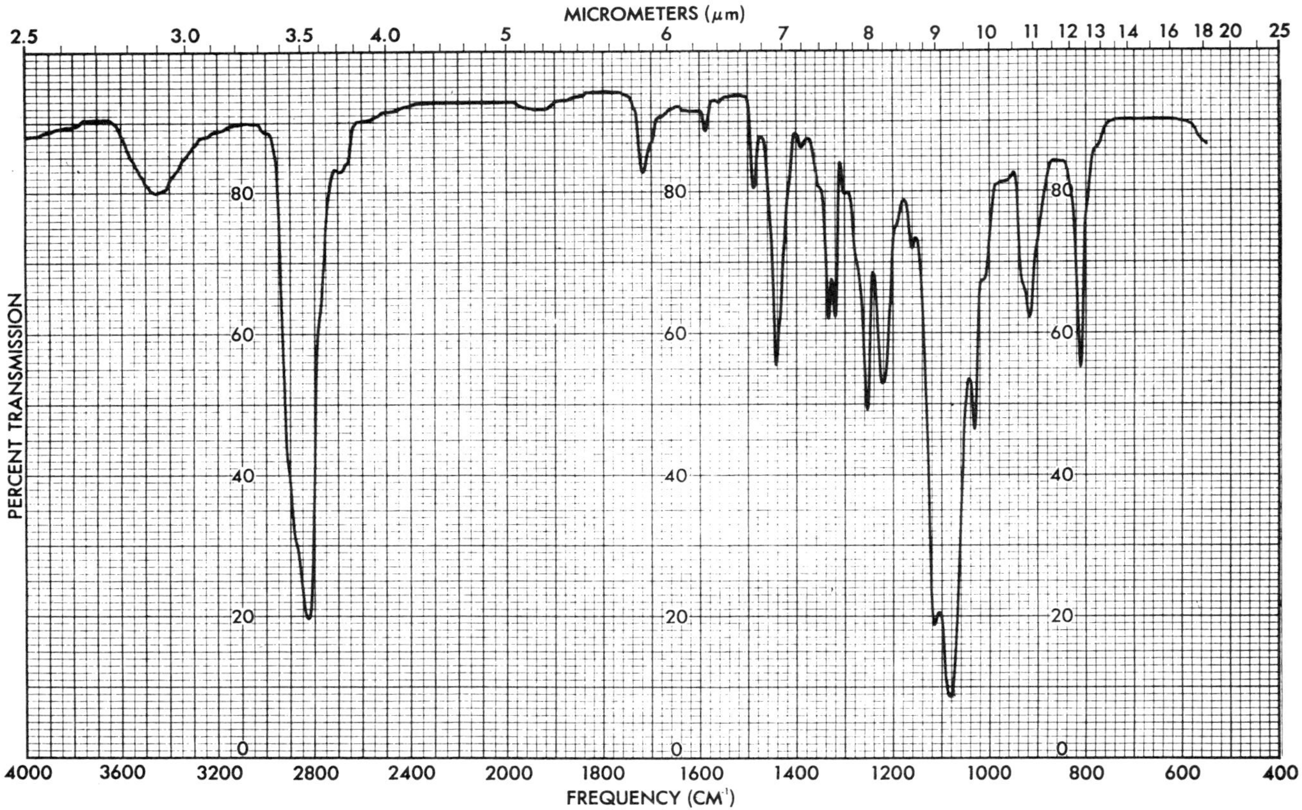

FIG. 4.7. $(EO)_{30}$ nonyl phenol.

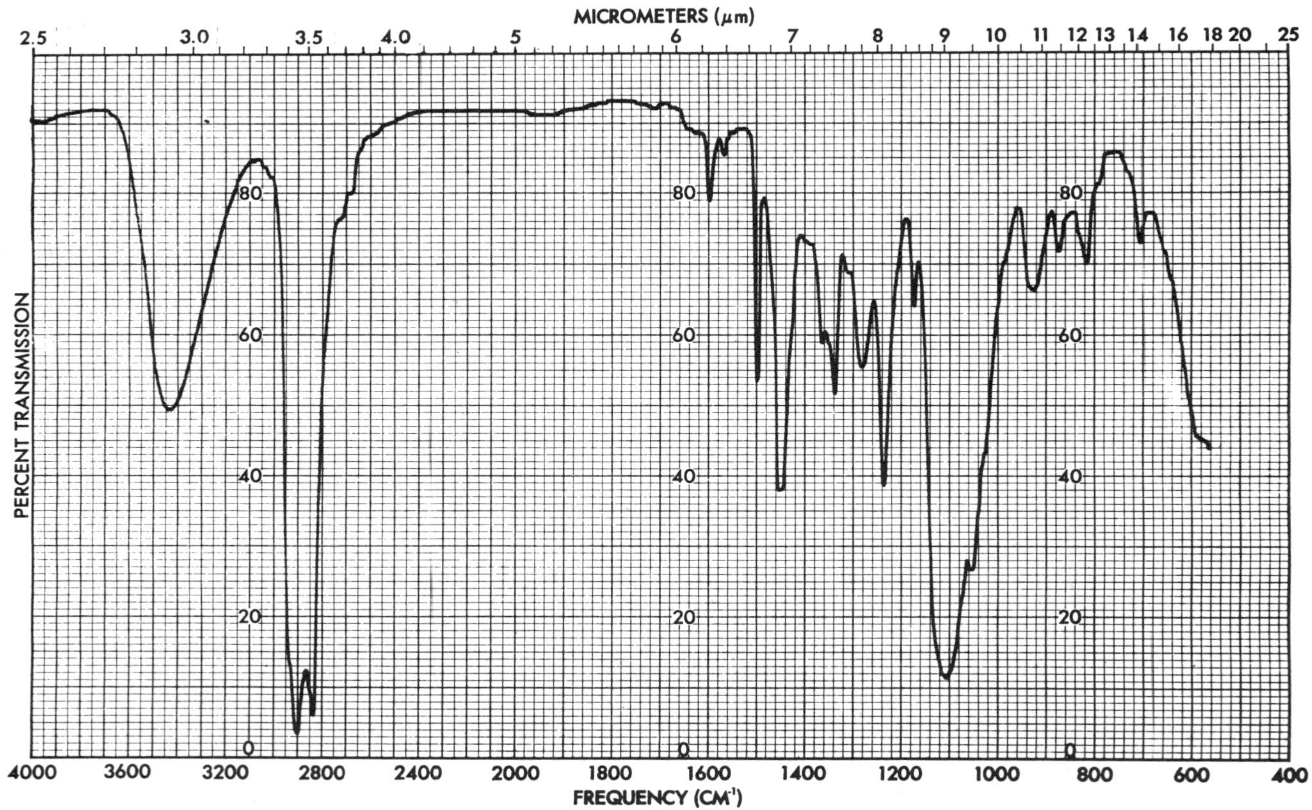

FIG. 4.8. $(EO)_{4.2}$ nonyl phenol.

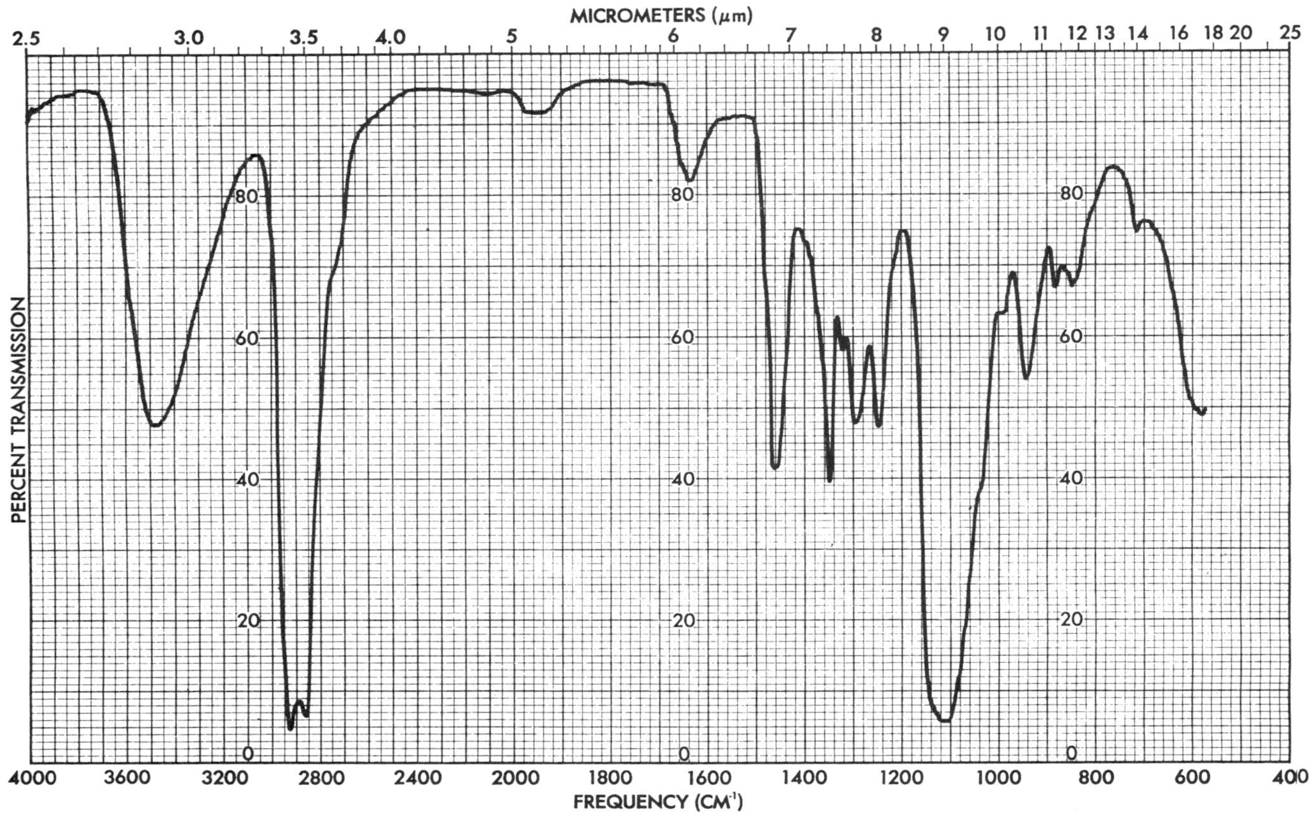

FIG. 4.9. $(EO)_{15}$ tallow amine.

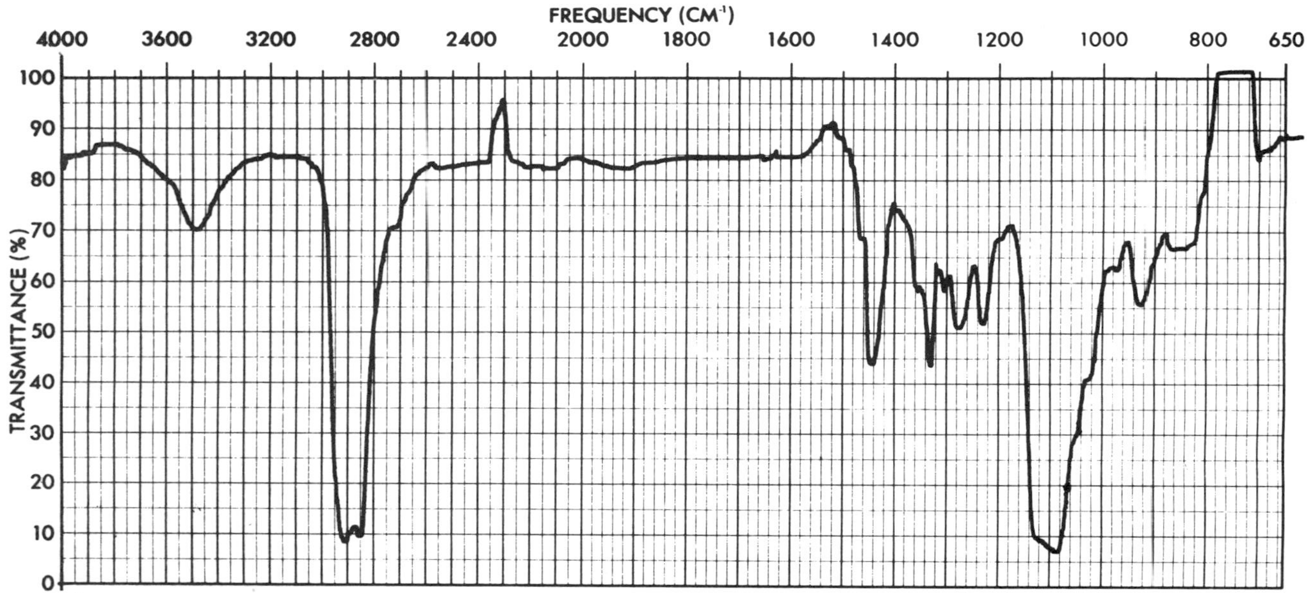

FIG. 4.10. $(EO)_9$ secondary alcohol.

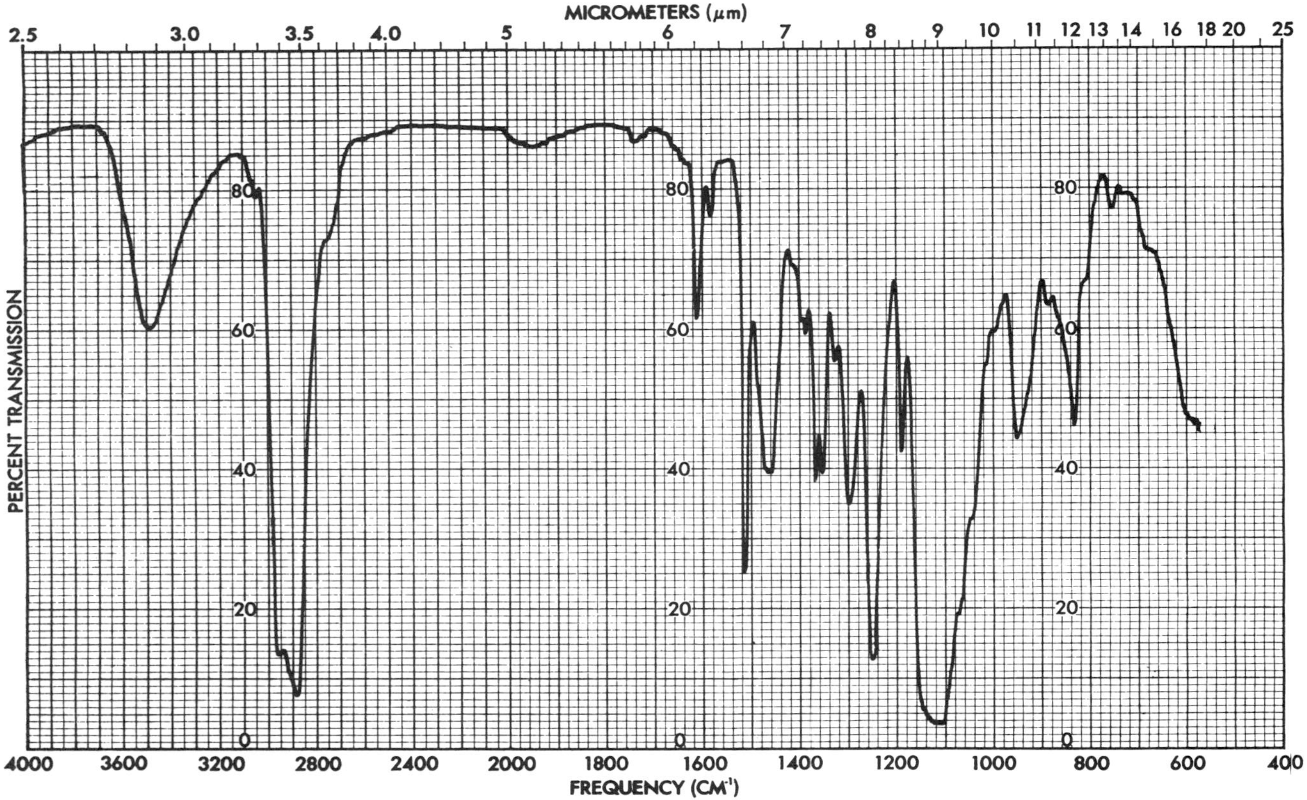

FIG. 4.11. $(EO)_{10}$ octyl phenol.

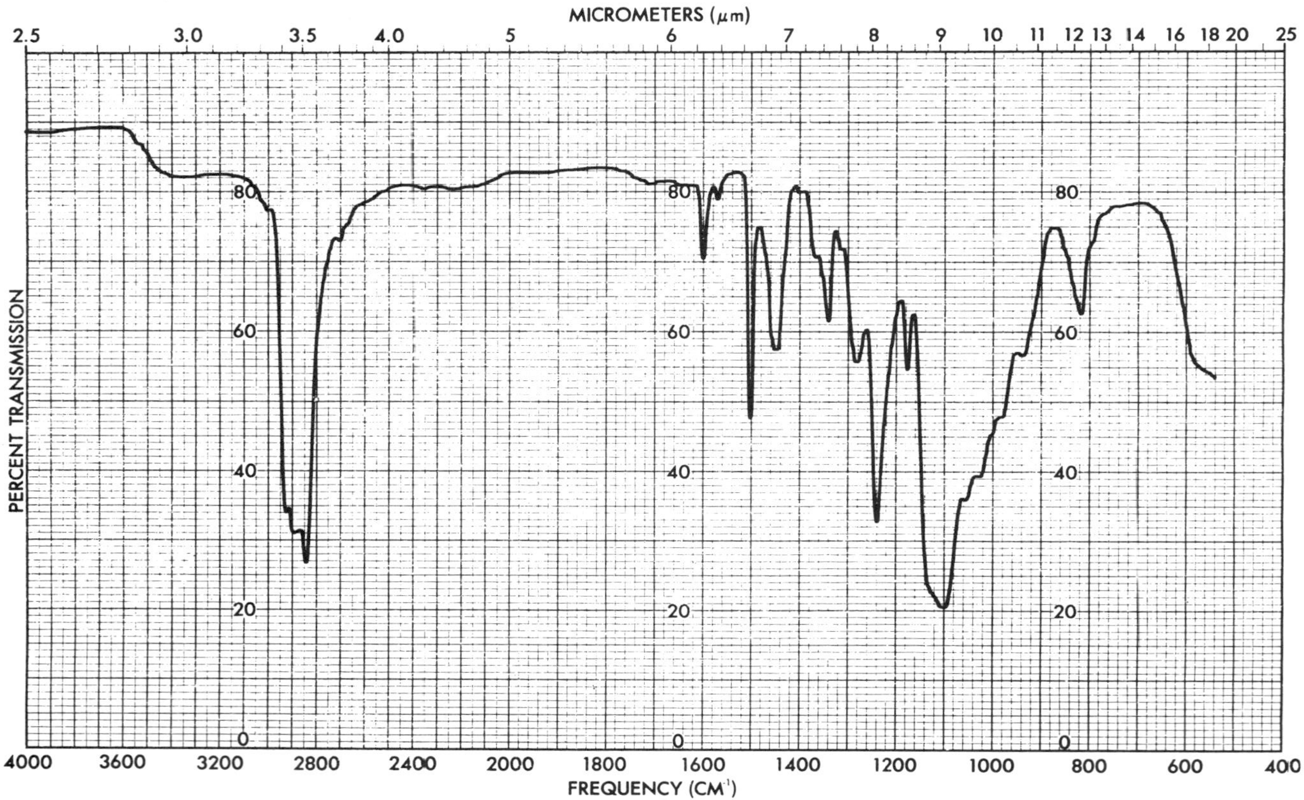

FIG. 4.12. (EO)$_9$ nonyl phenol.

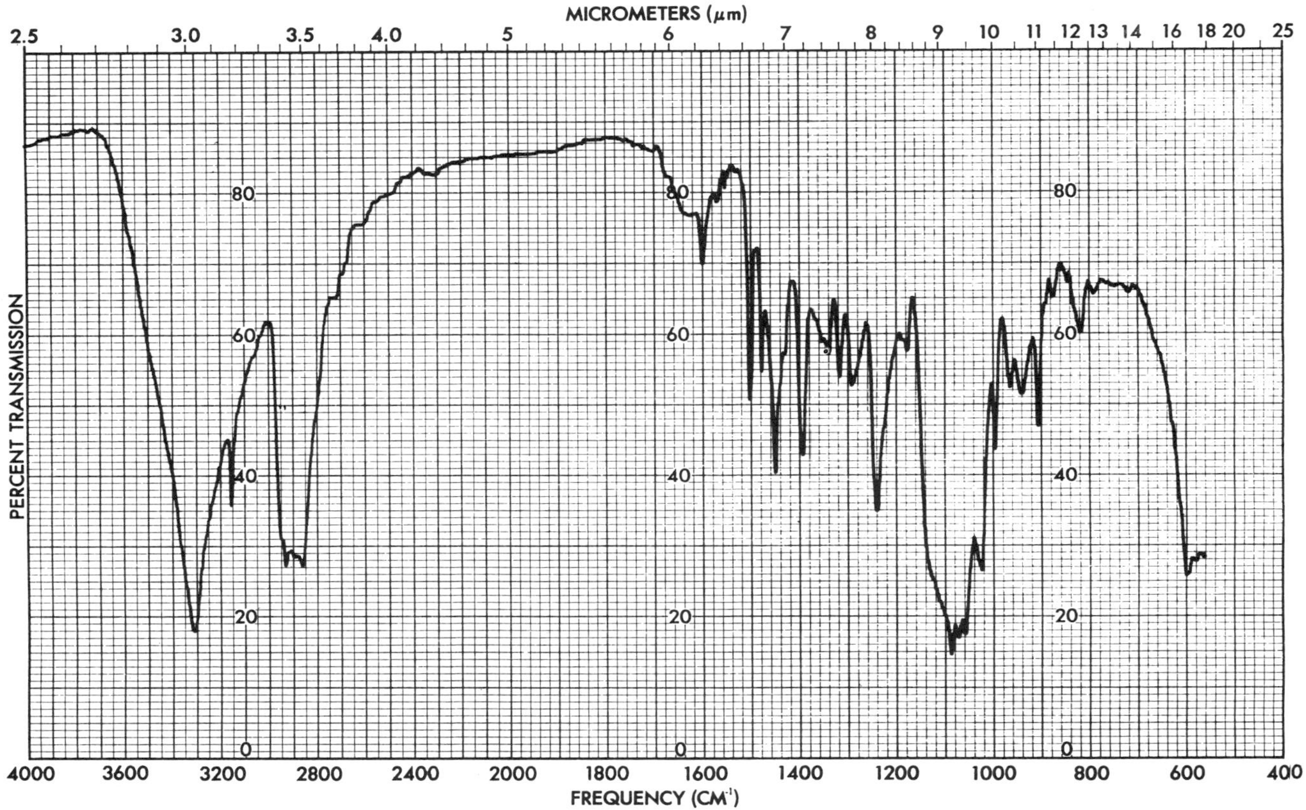

FIG. 4.13. TEA salt–$(EO)_9$ nonyl phenol P_2O_5.

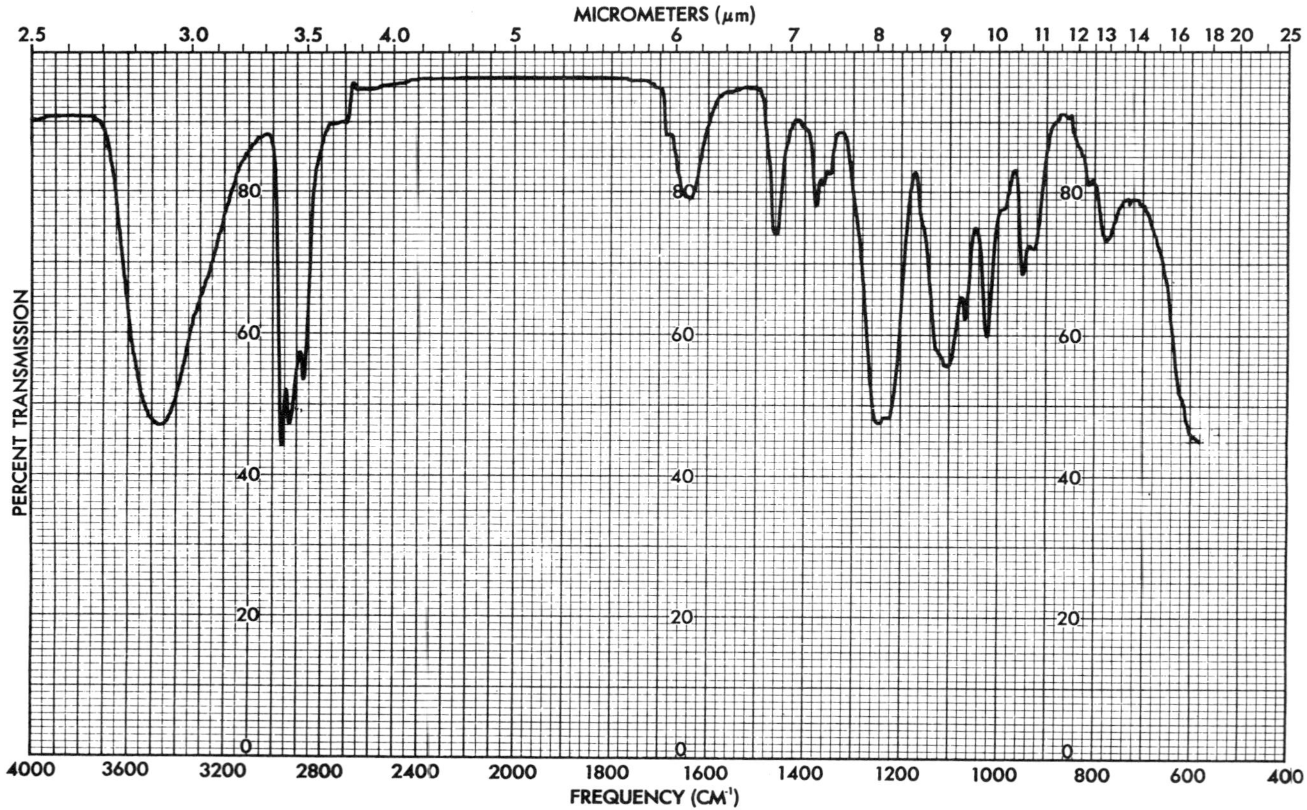

FIG. 4.14. Sodium salt–tridecyl ether sulfate.

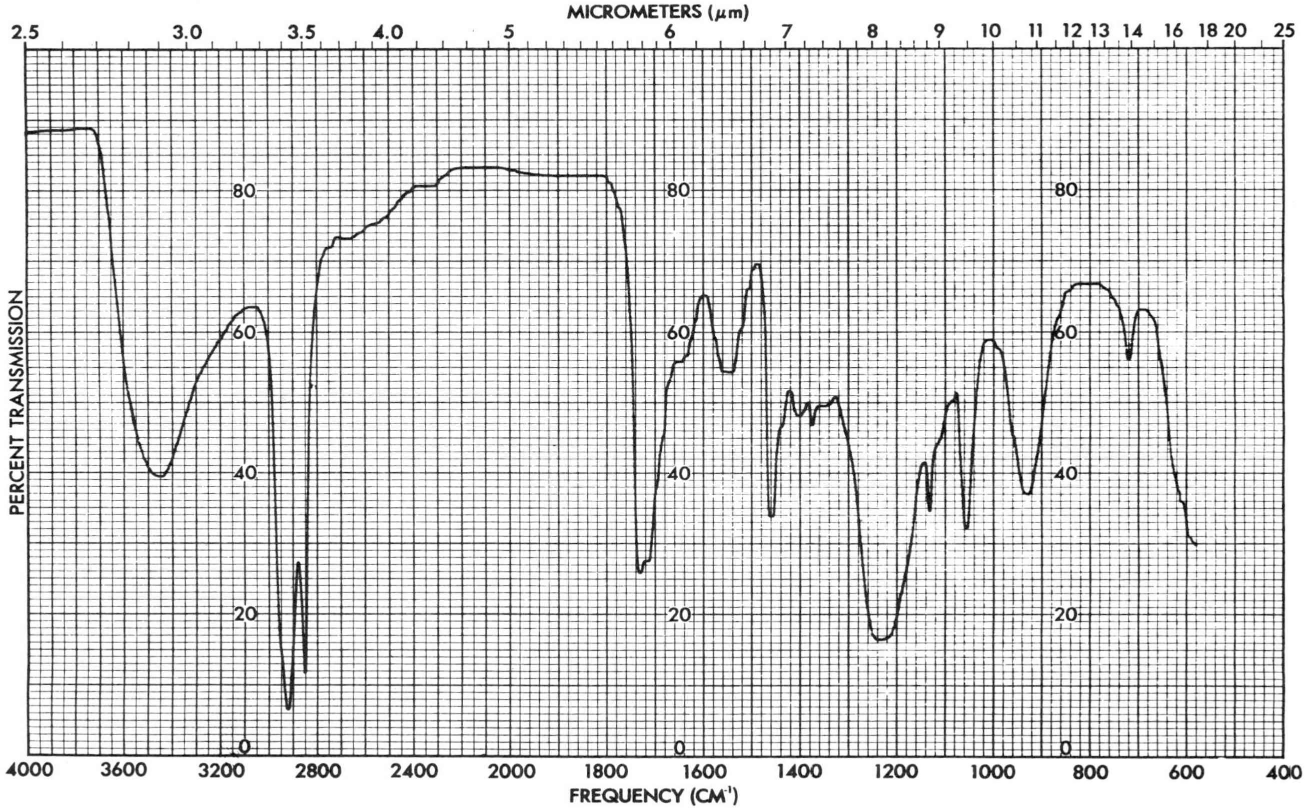

FIG. 4.15. Sodium salt–glycerol trioleate sulfate.

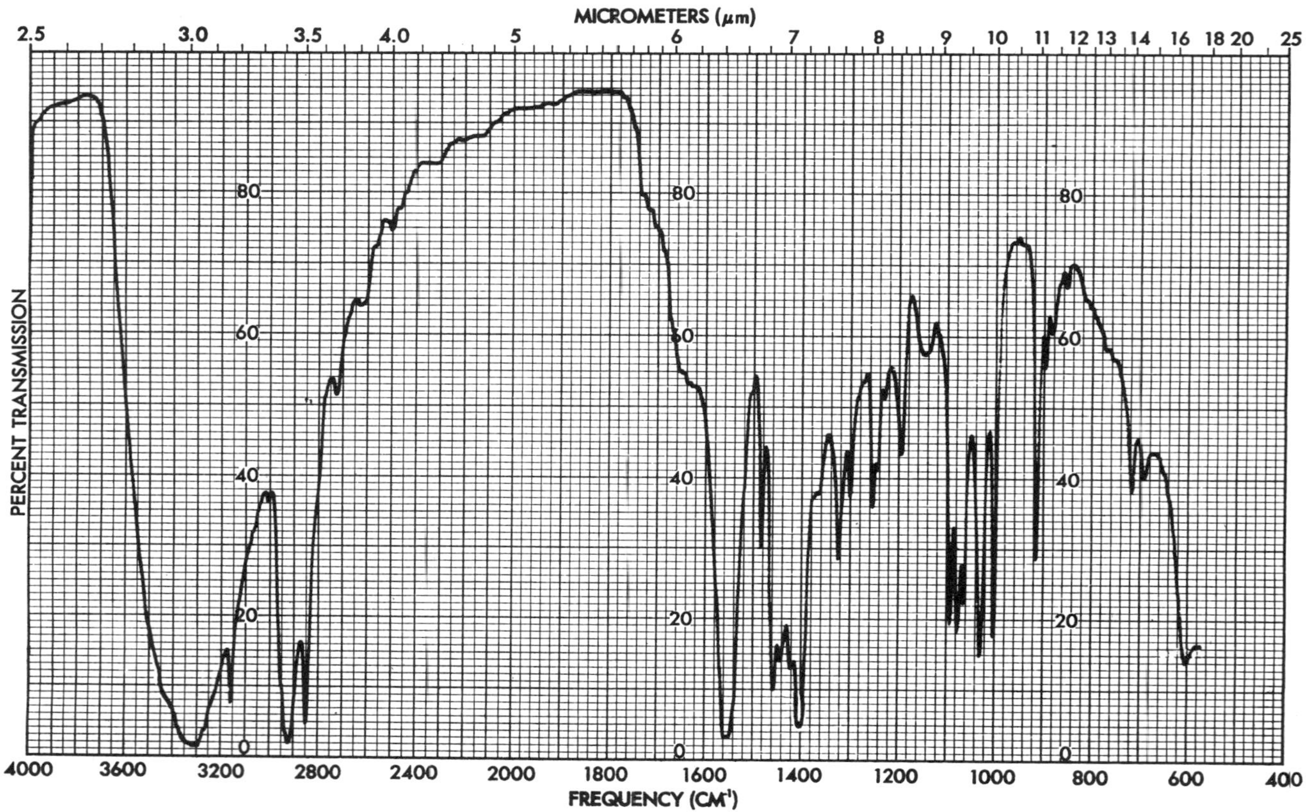

FIG. 4.16. TEA–fatty acid soap.

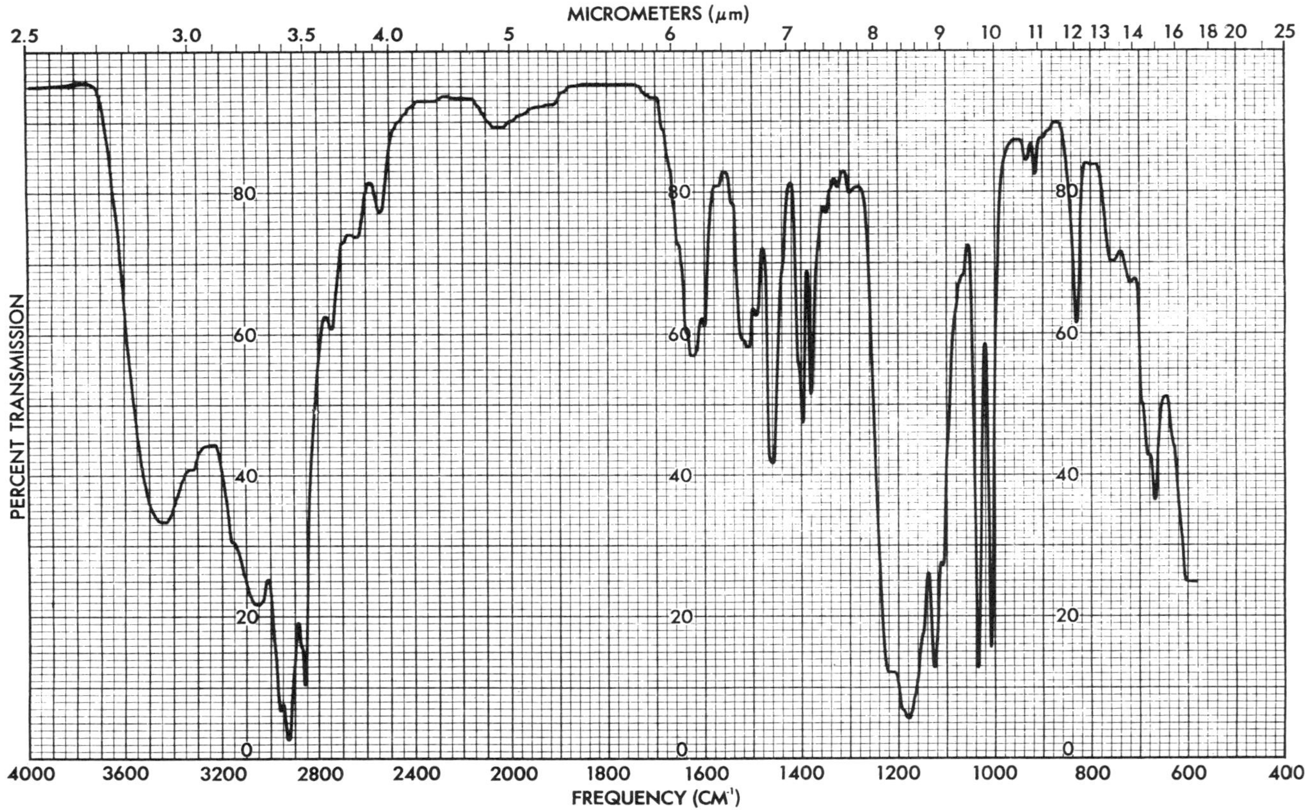

FIG. 4.17. Isopropyl amine salt–DDBSA.

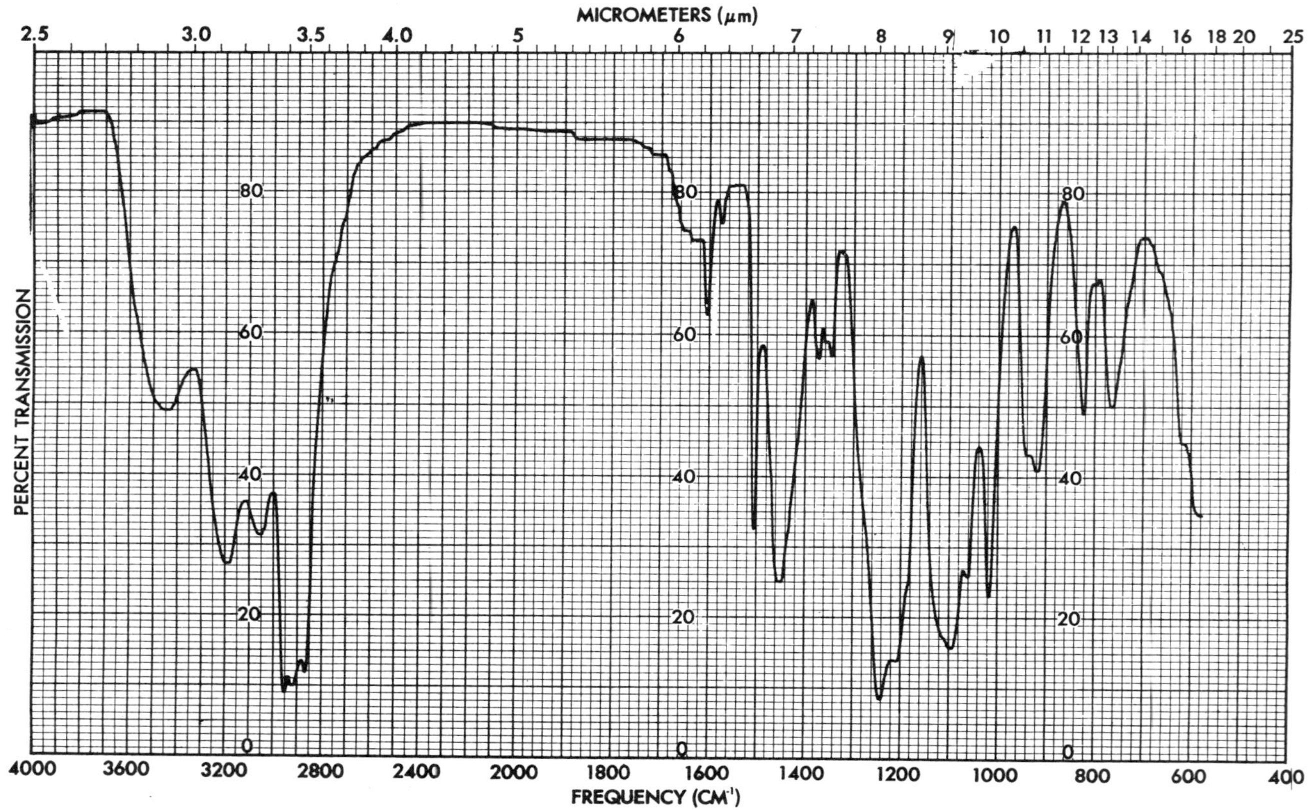

FIG. 4.18. Ammonium salt–$(EO)_4$ nonyl phenol sulfate.

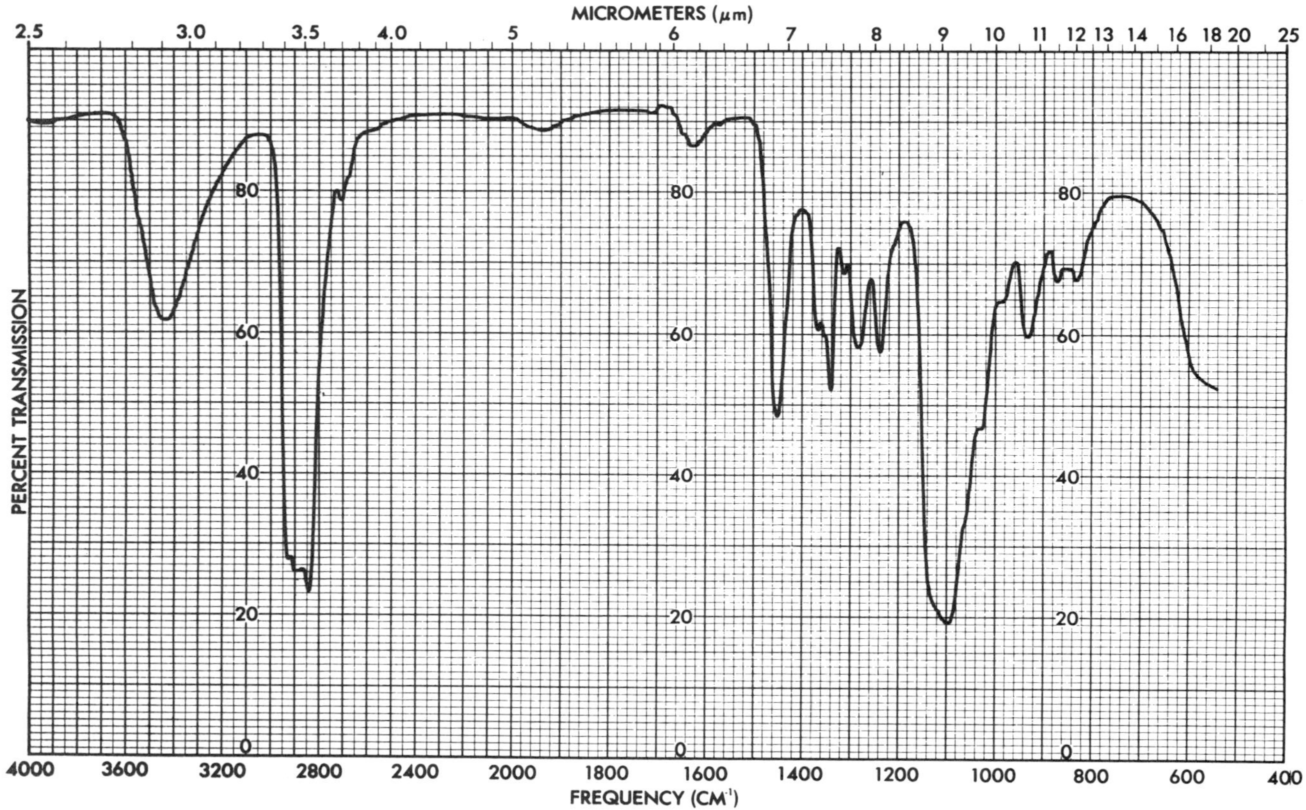

FIG. 4.19. $(EO)_6$ thioether (phosphate).

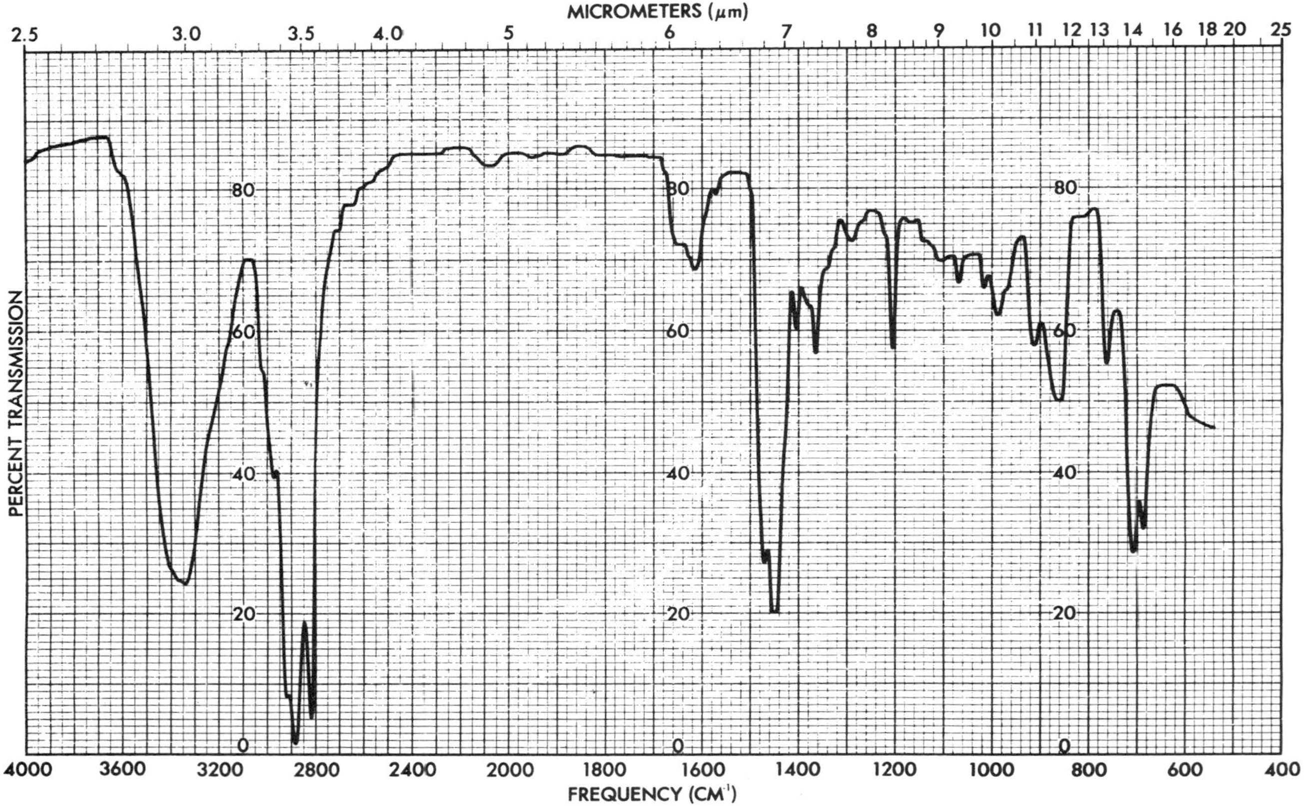

FIG. 4.20. Dodecyl benzyl trimethyl ammonium chloride.

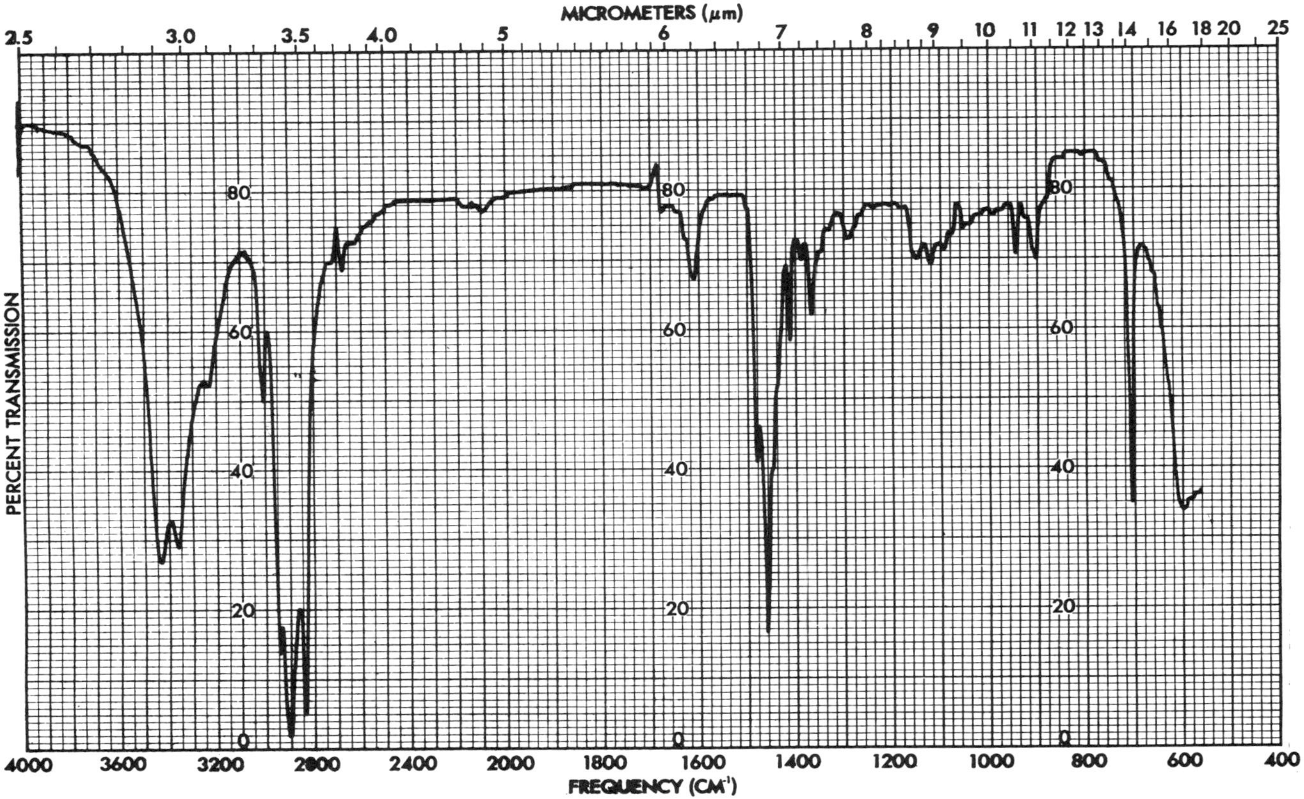

FIG. 4.21. Ditallow dimethyl ammonium chloride.

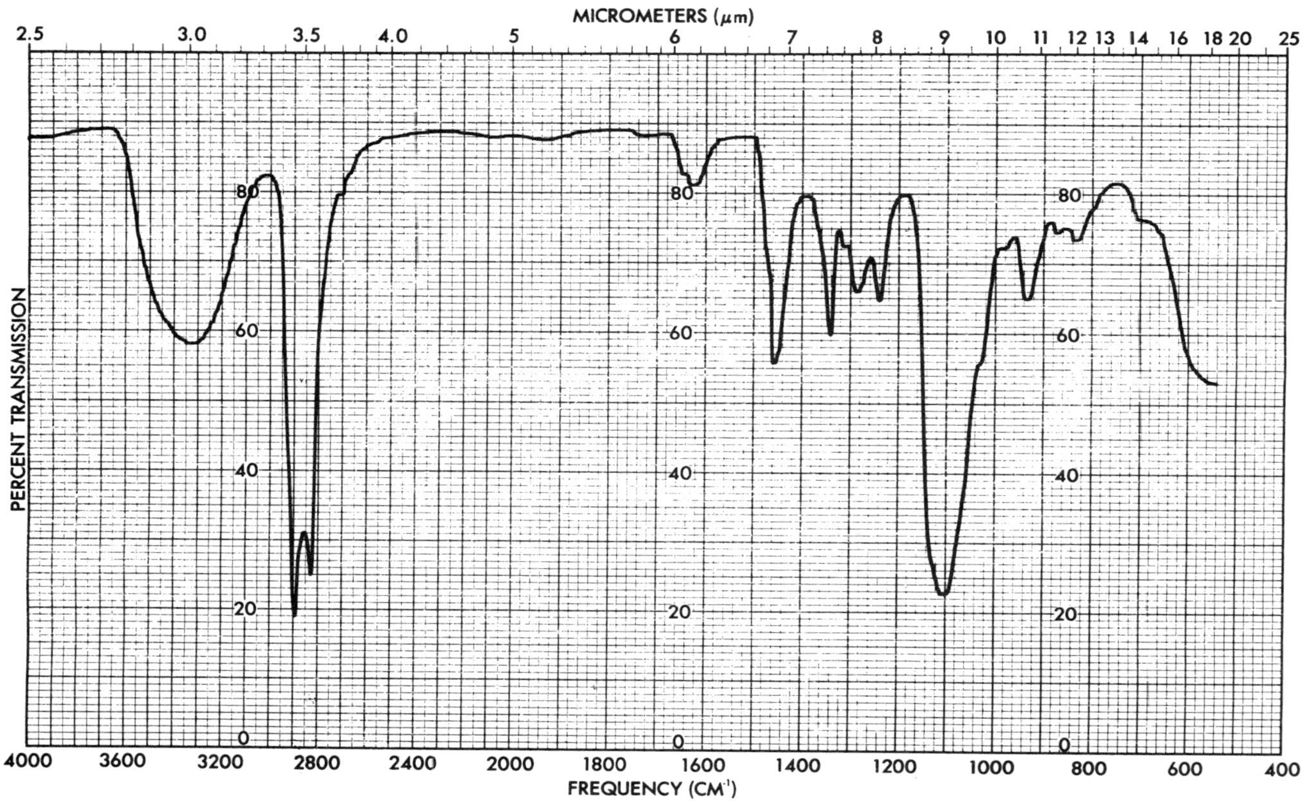

FIG. 4.22. Alkoxy alkyl quaternary.

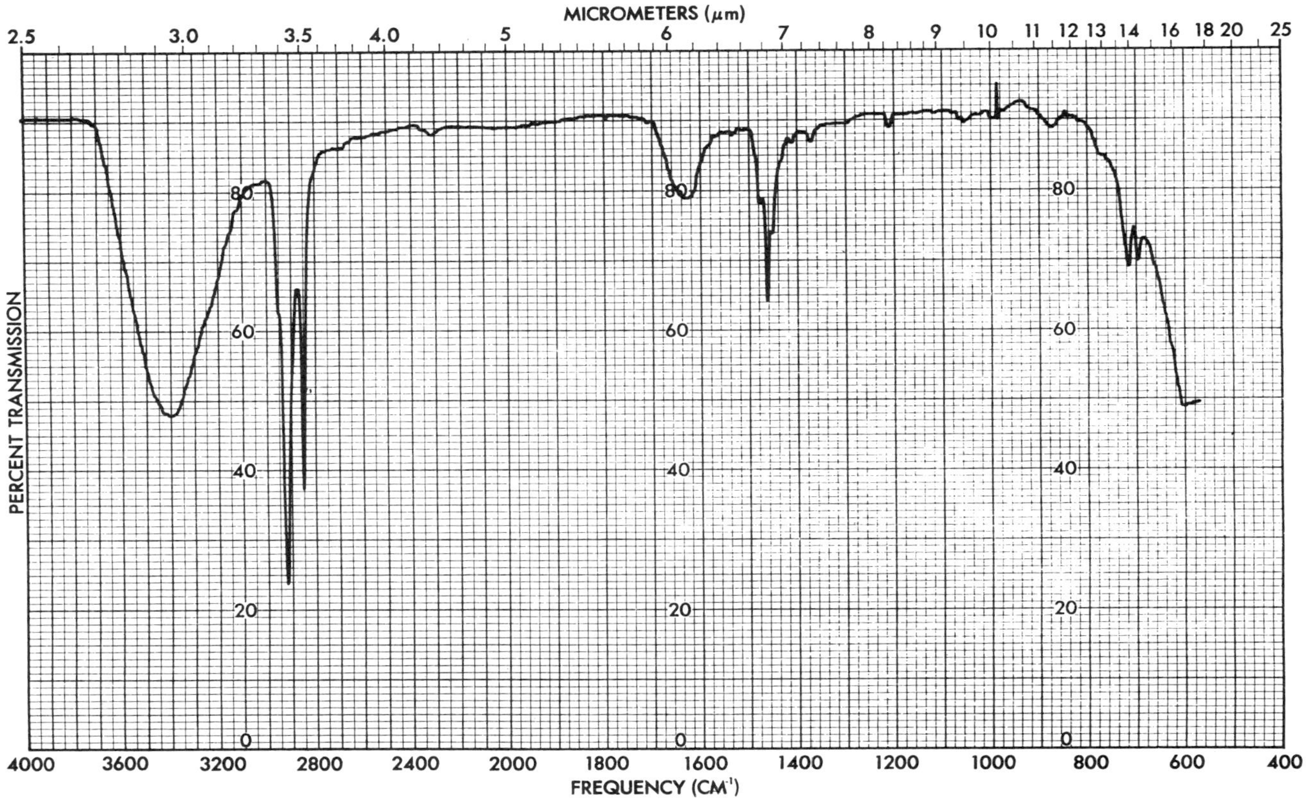

FIG. 4.23. Stearyl dimethyl benzyl ammonium chloride.

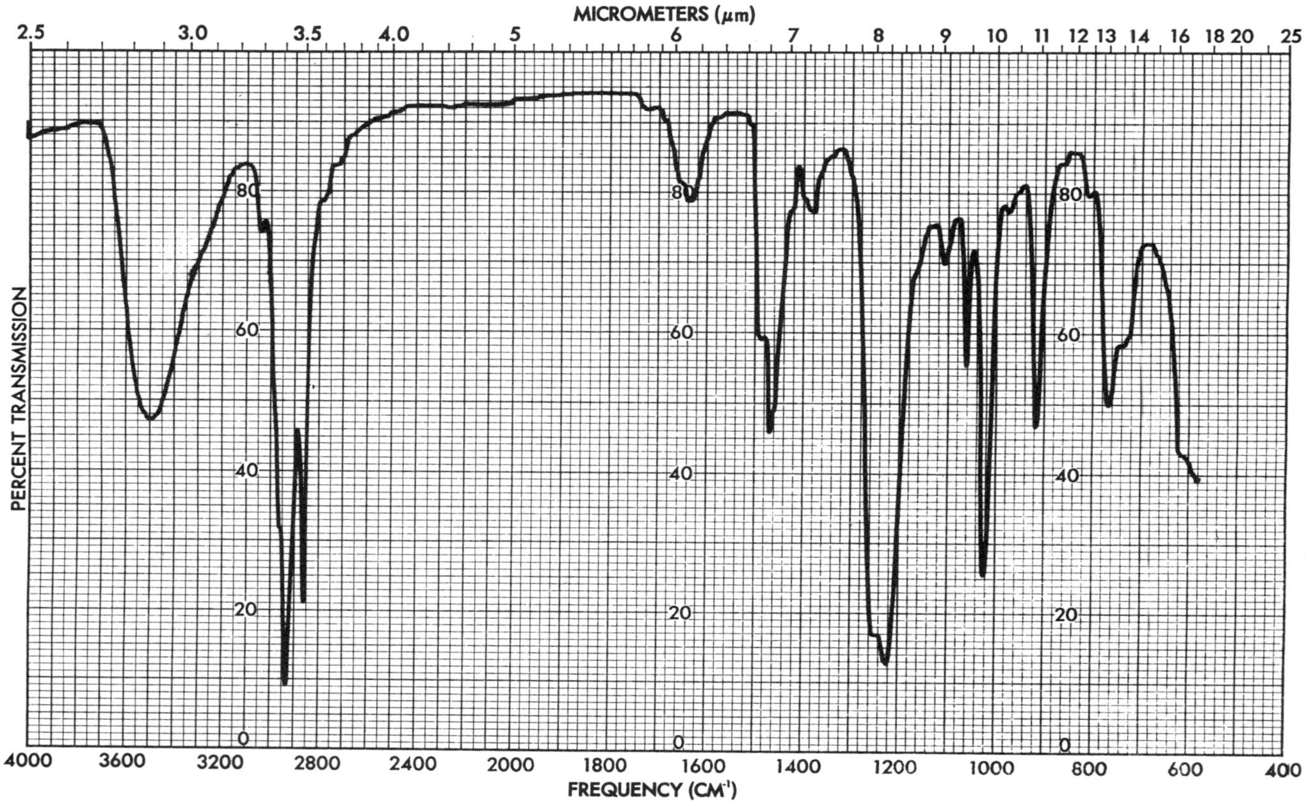

FIG. 4.24. Coconut dimethyl ethosulfate.

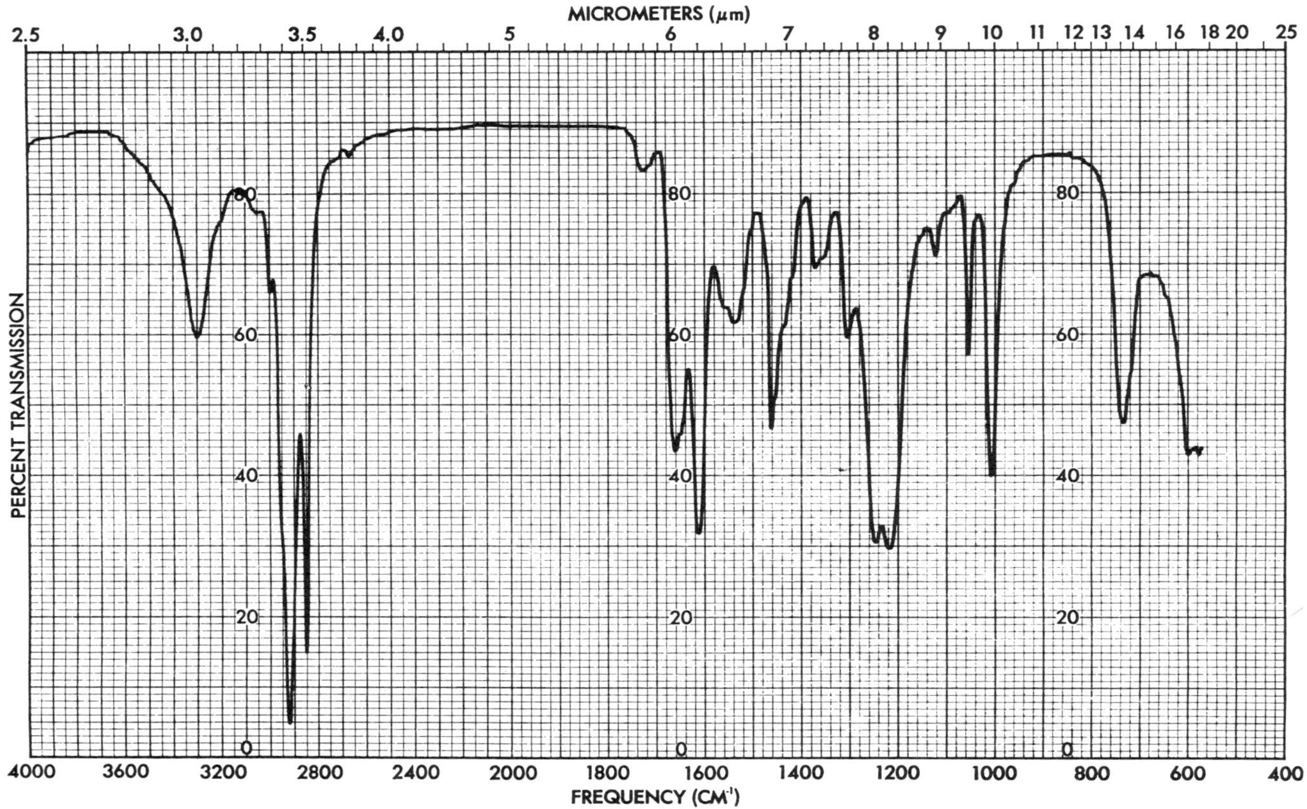

FIG. 4.25. Imidazolinium methyl sulfate quaternary.

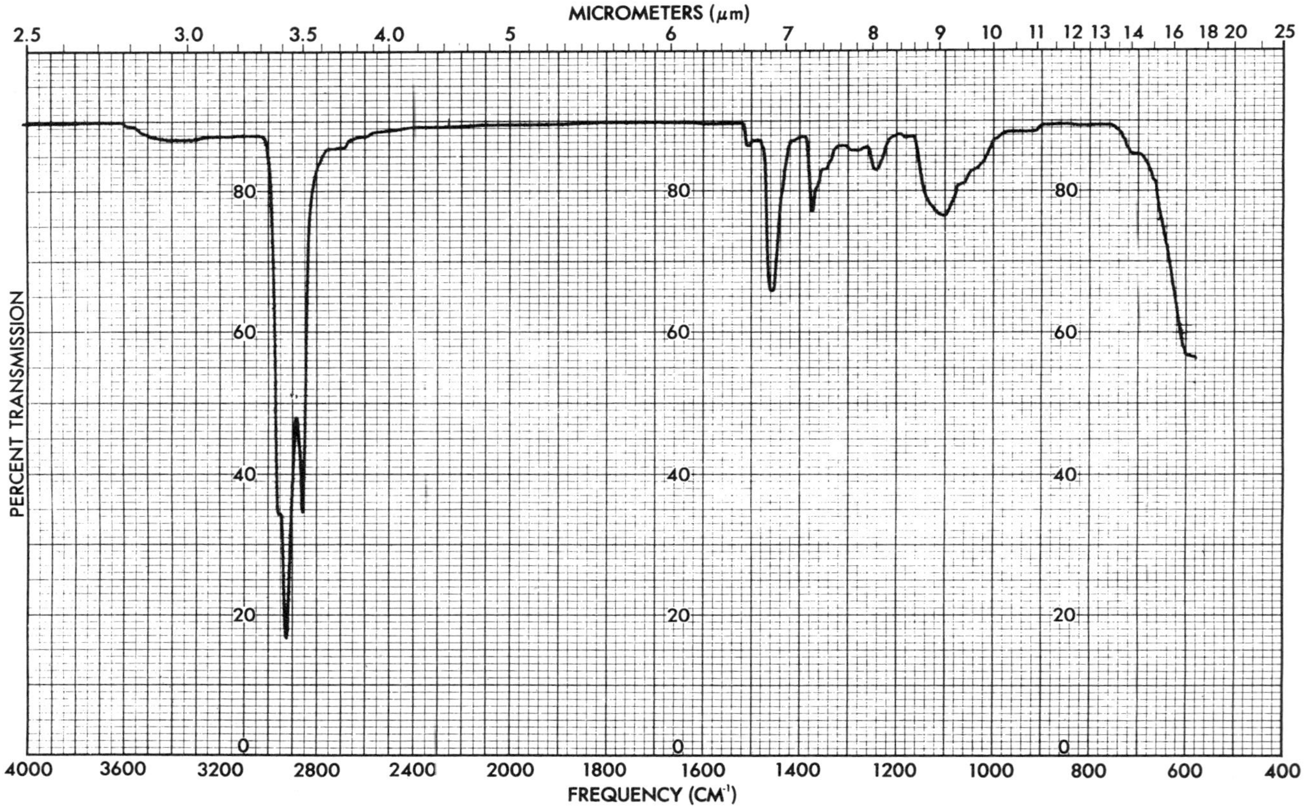

FIG. 4.26. Emulsified mineral oil.

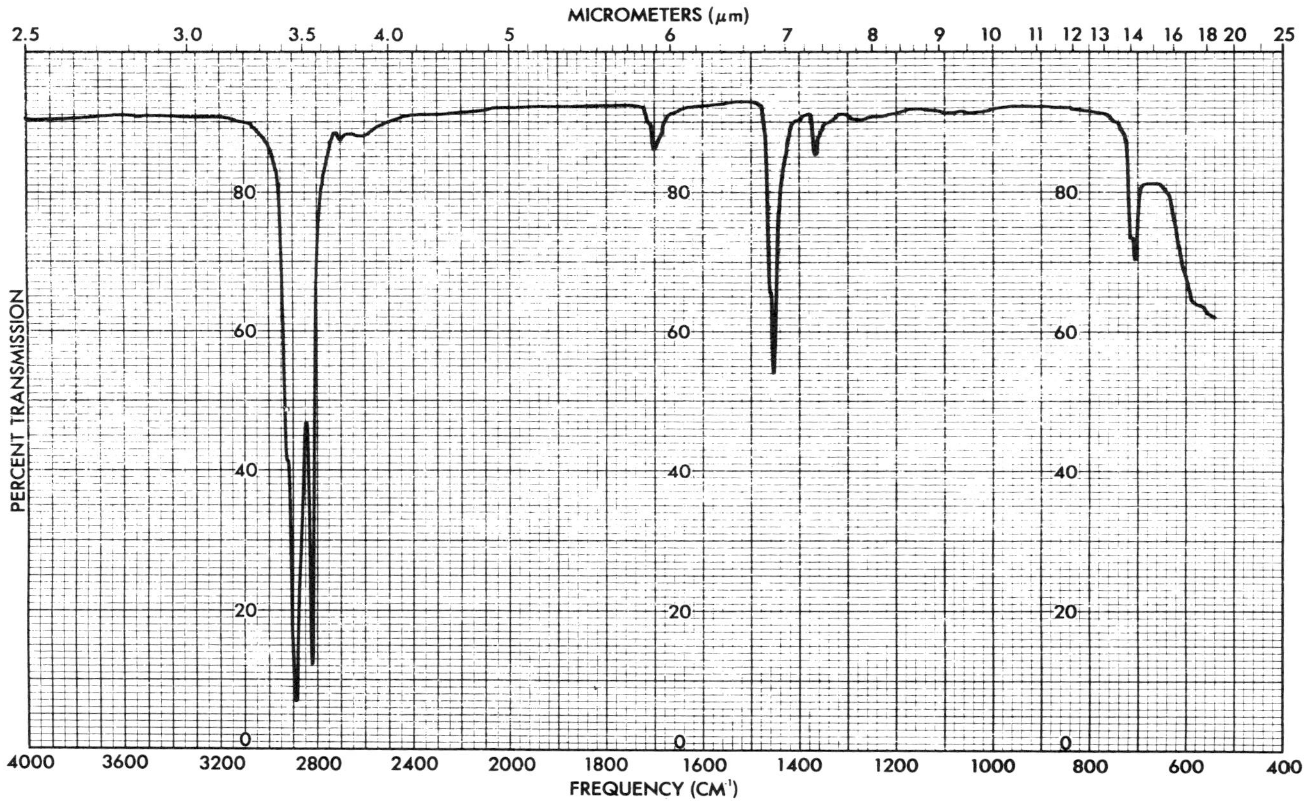

FIG. 4.27. Process wax.

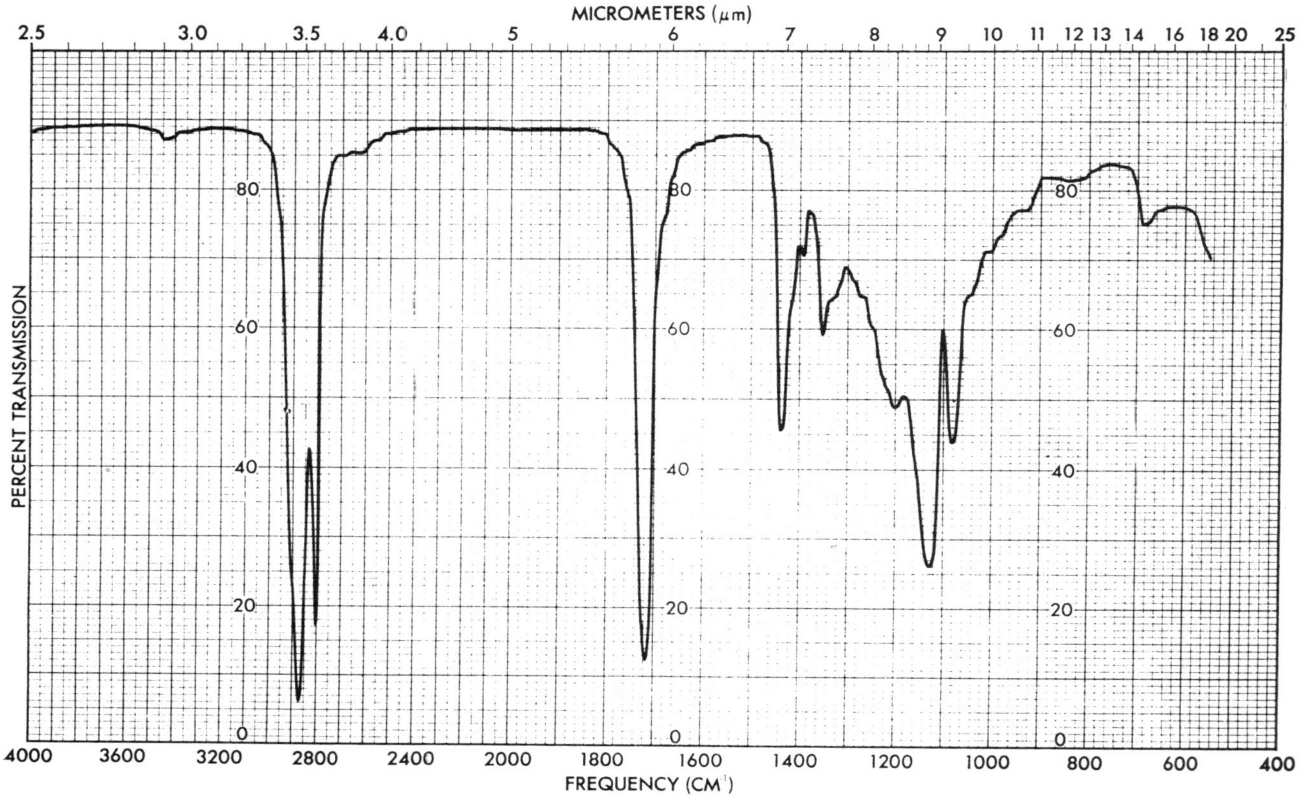

FIG. 4.28. Coconut oil.

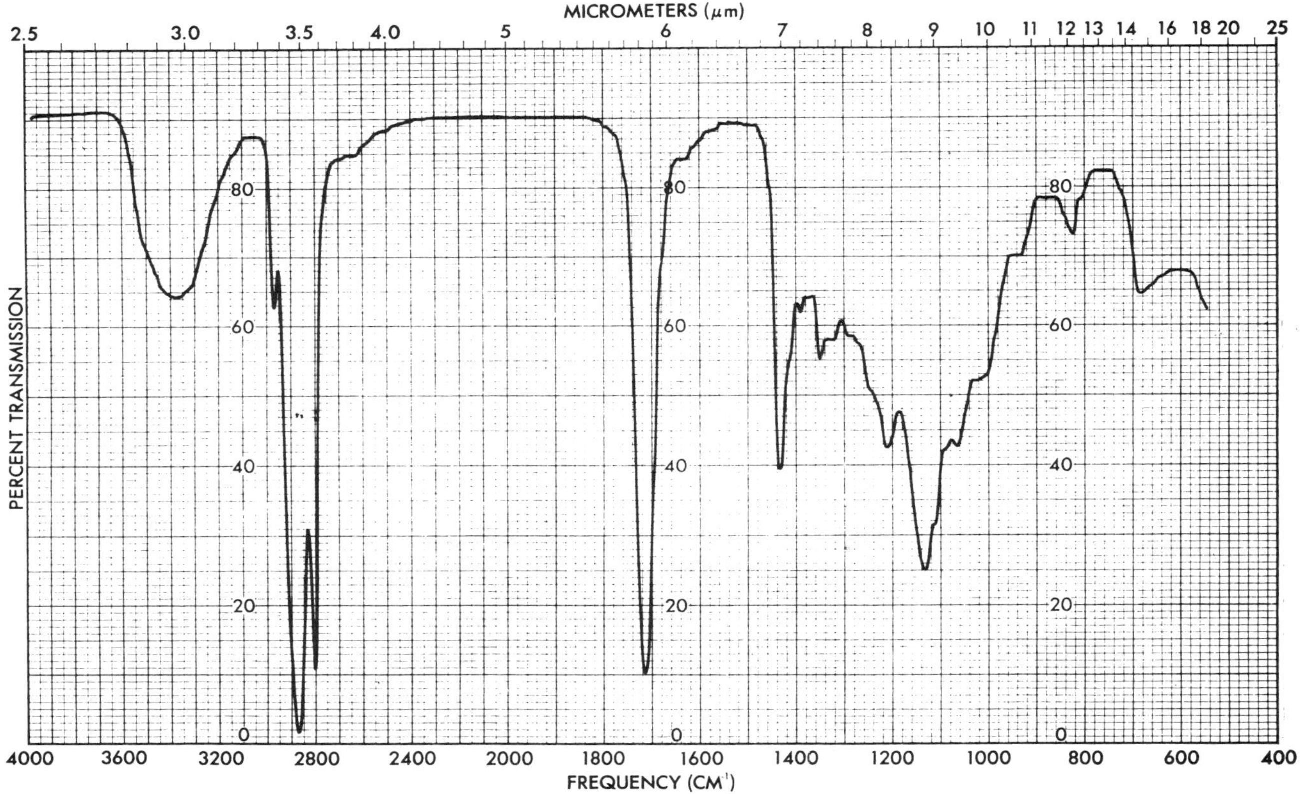

FIG. 4.29. Castor oil.

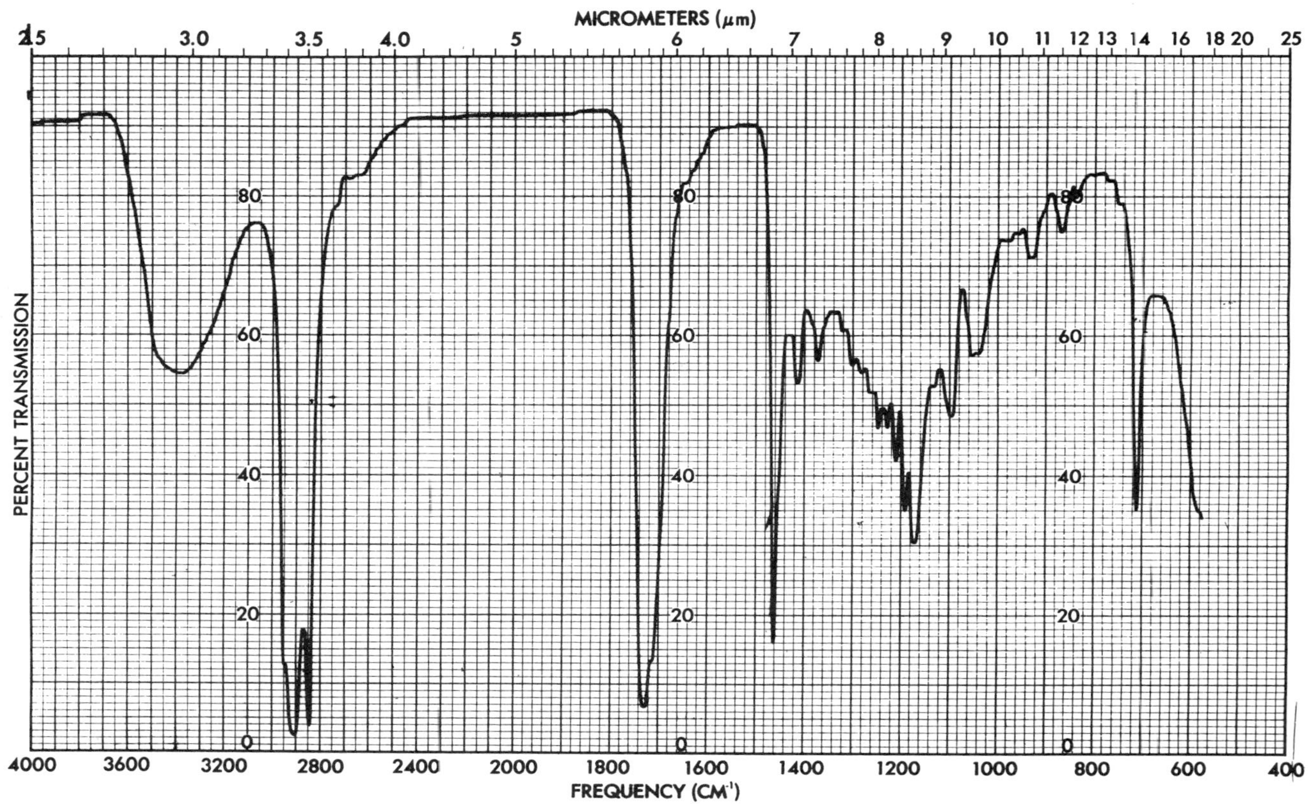

FIG. 4.30. Glycerol monostearate.

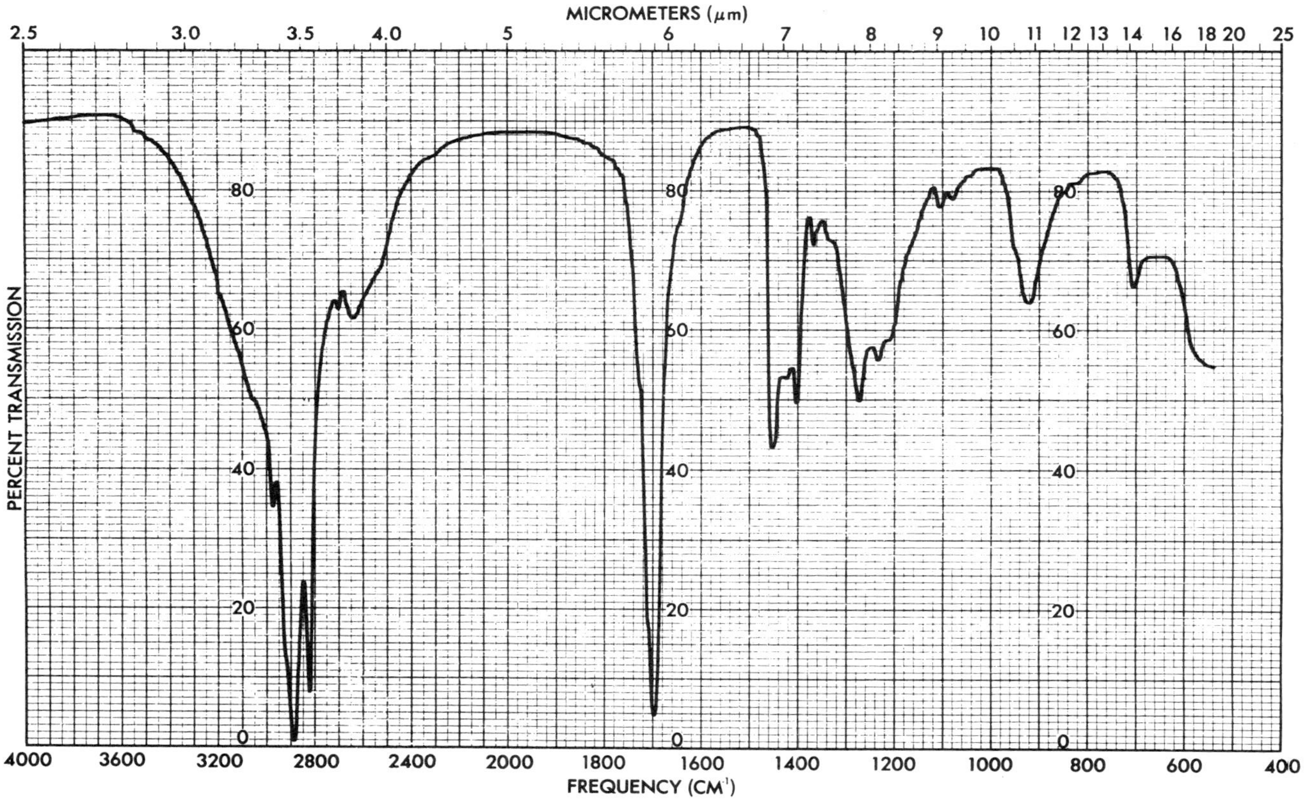

FIG. 4.31. Oleic acid.

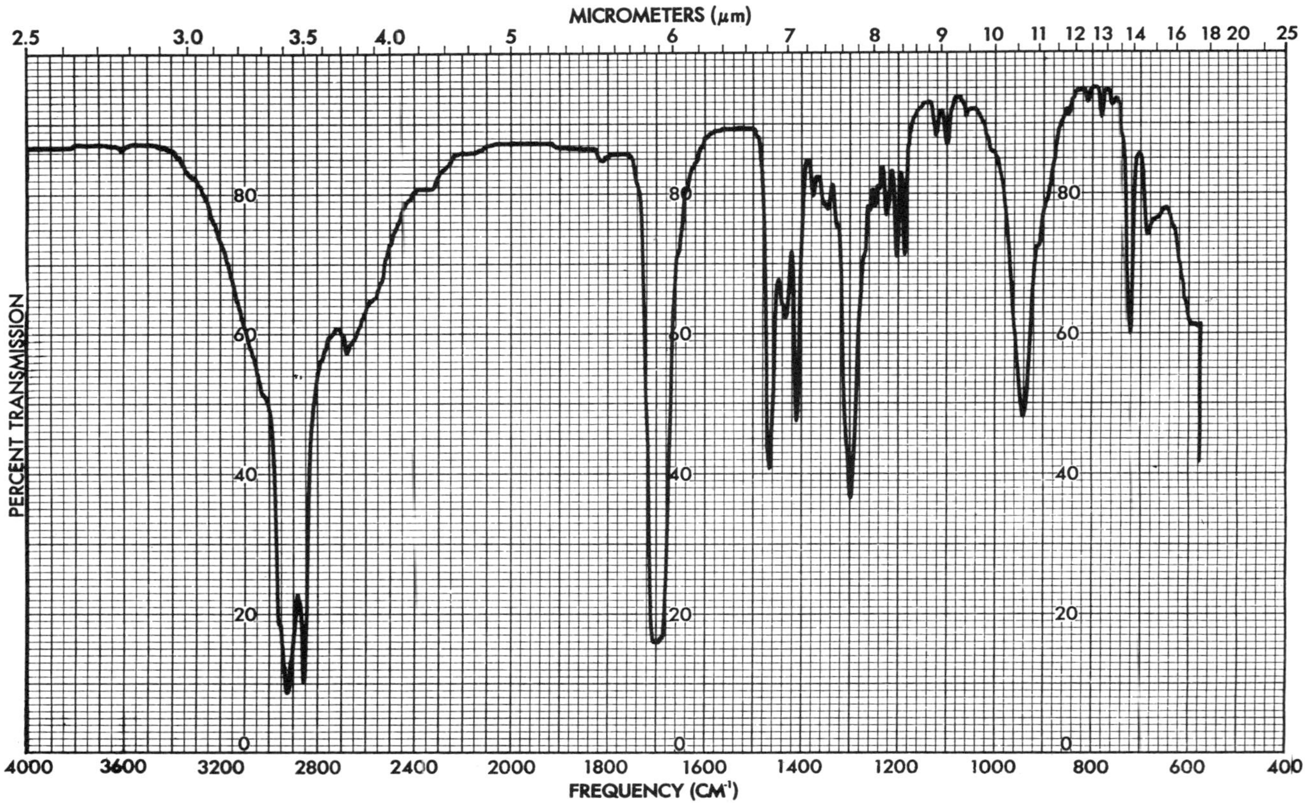

FIG. 4.32. Stearic acid.

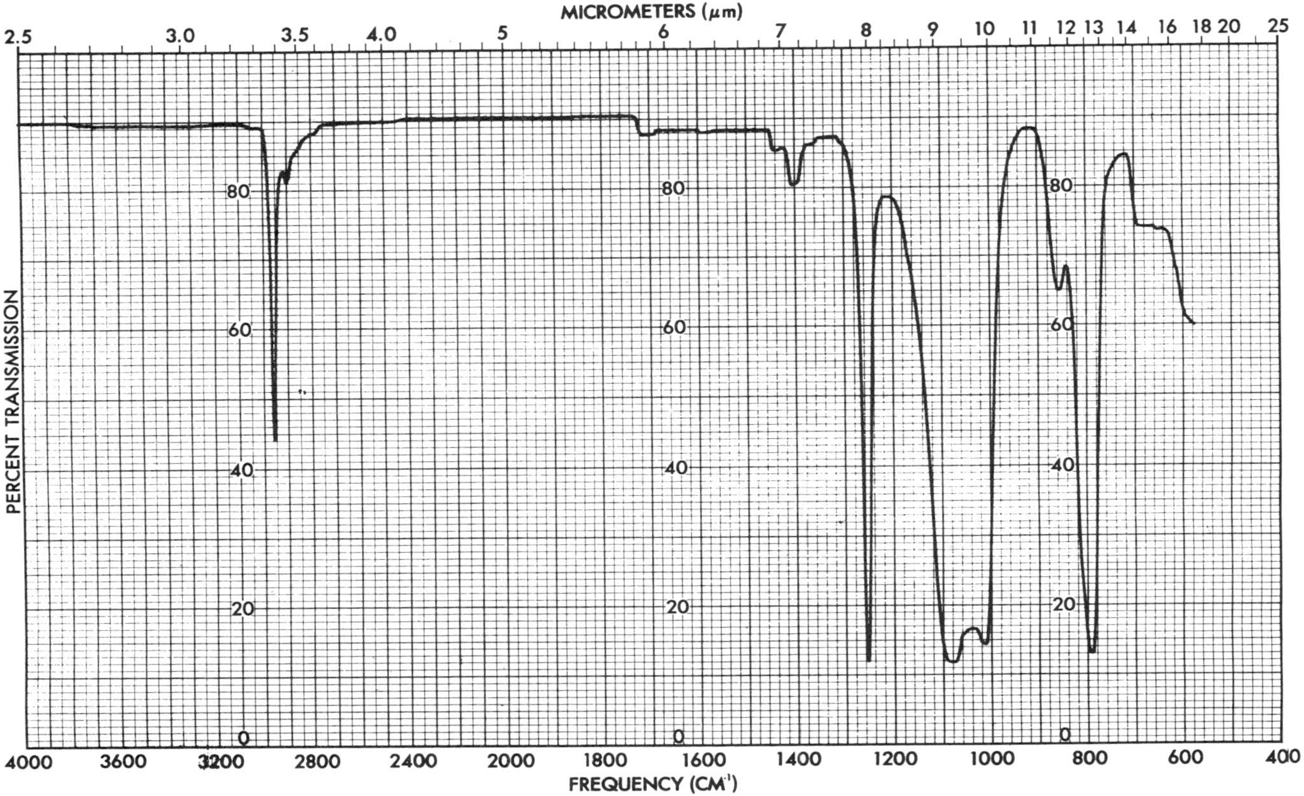

FIG. 4.33. Silicone compound.

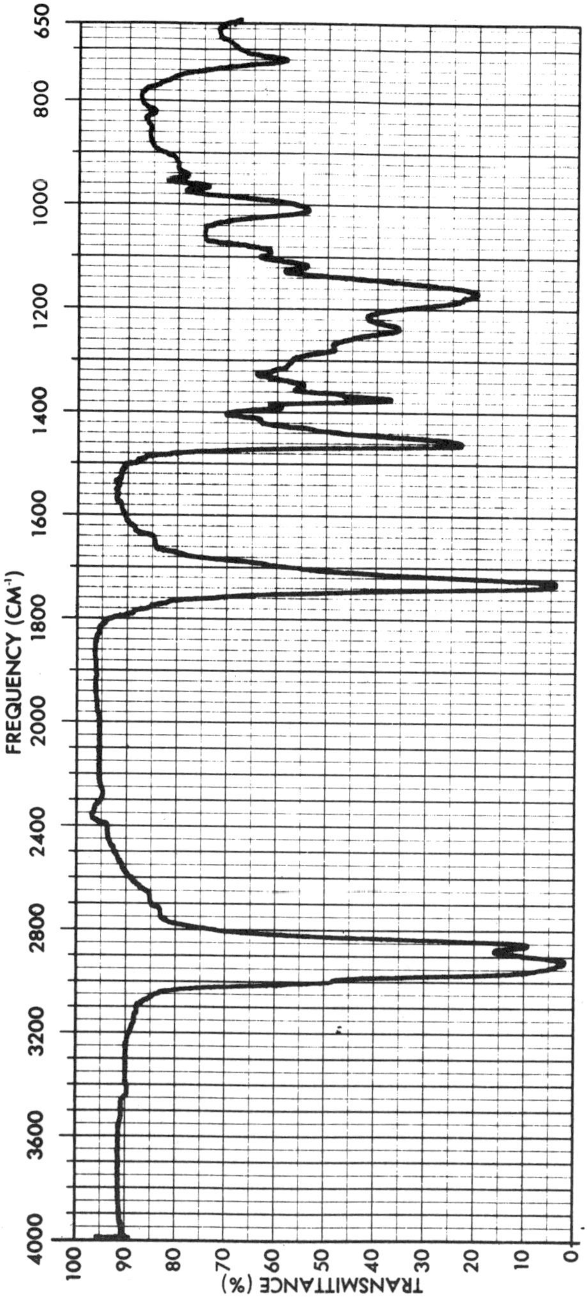

FIG. 4.34. Butyl oleate.

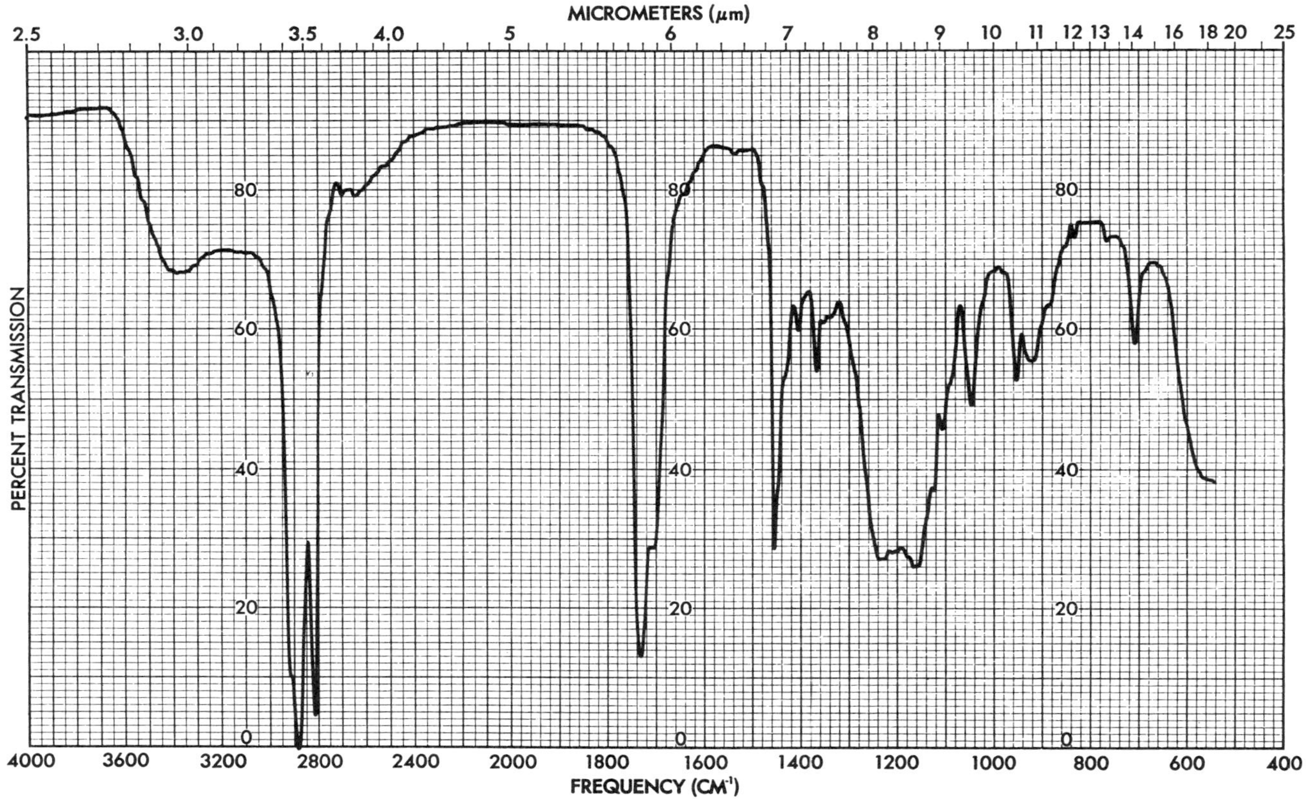

FIG. 4.35. Sulfonated tallow.

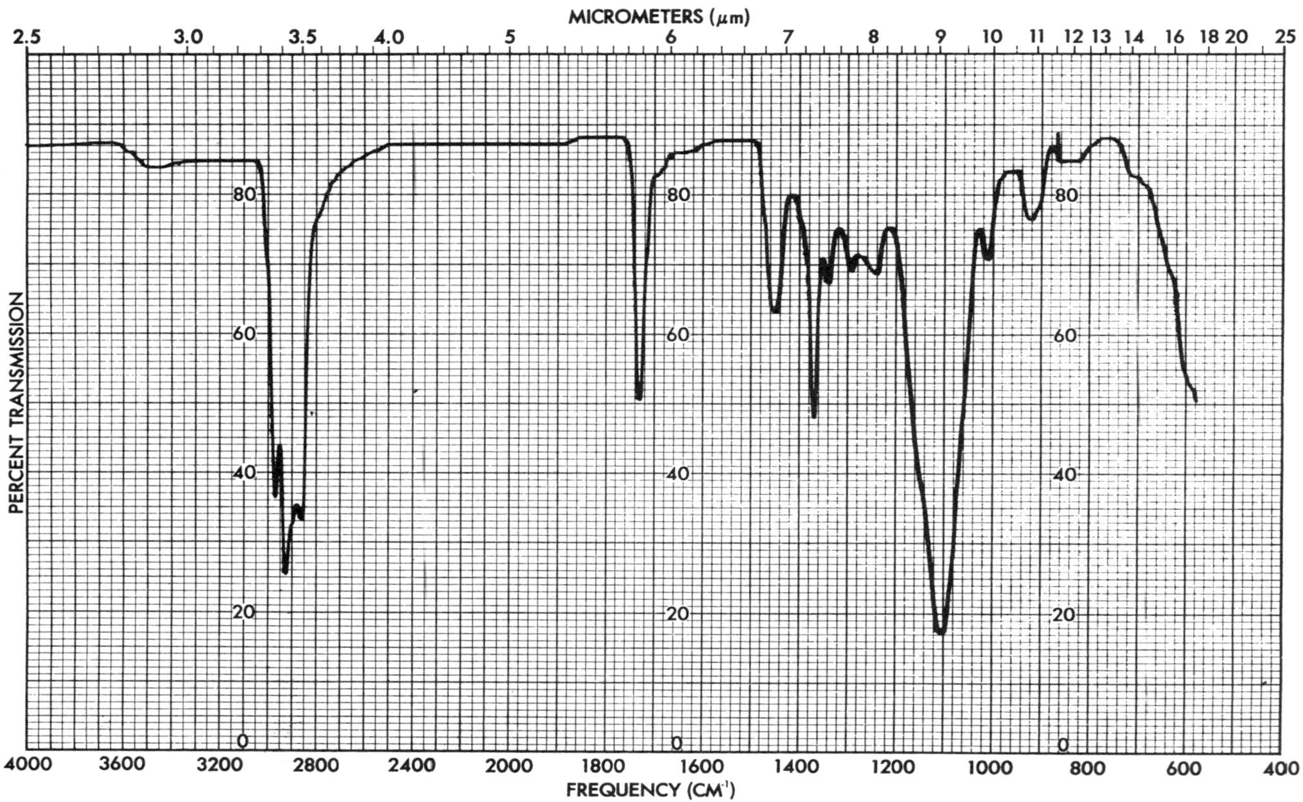

FIG. 4.36. PPG ditallate.

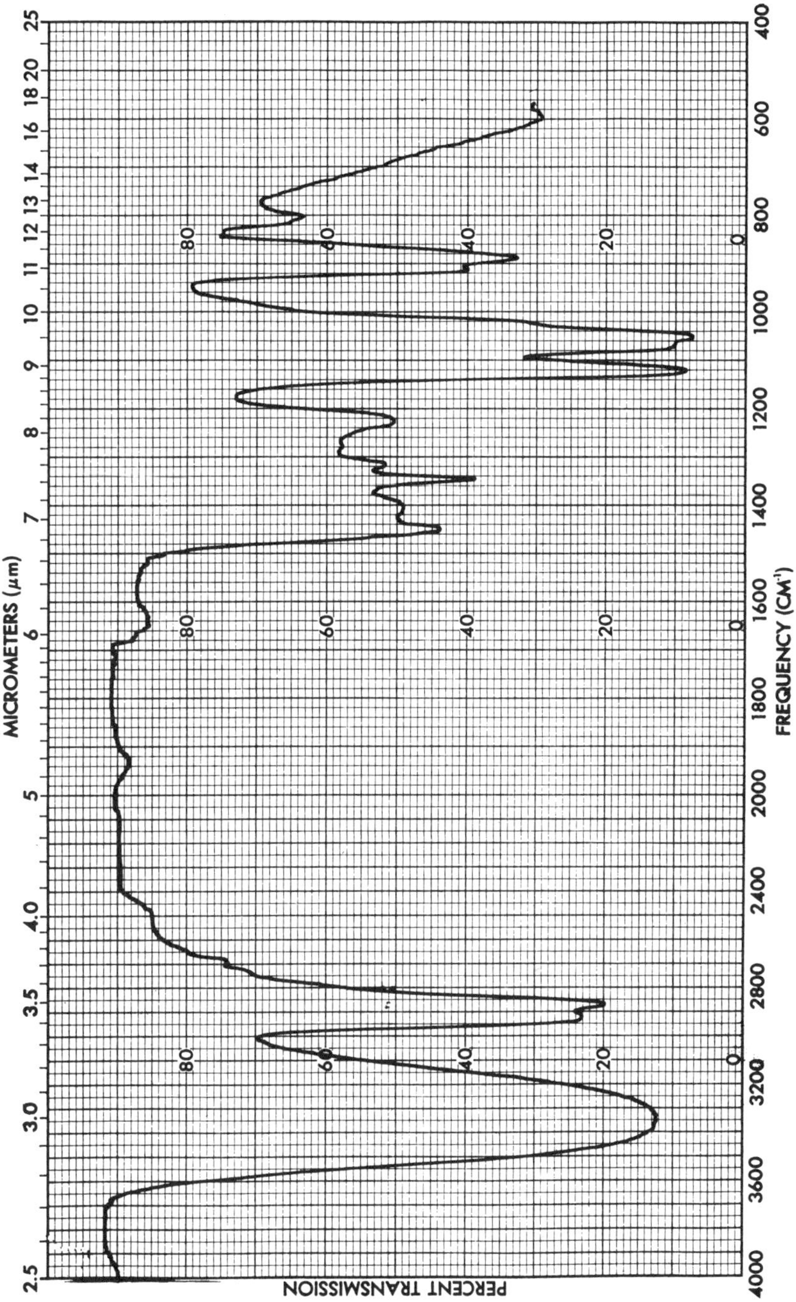

FIG. 4.37. Diethylene glycol.

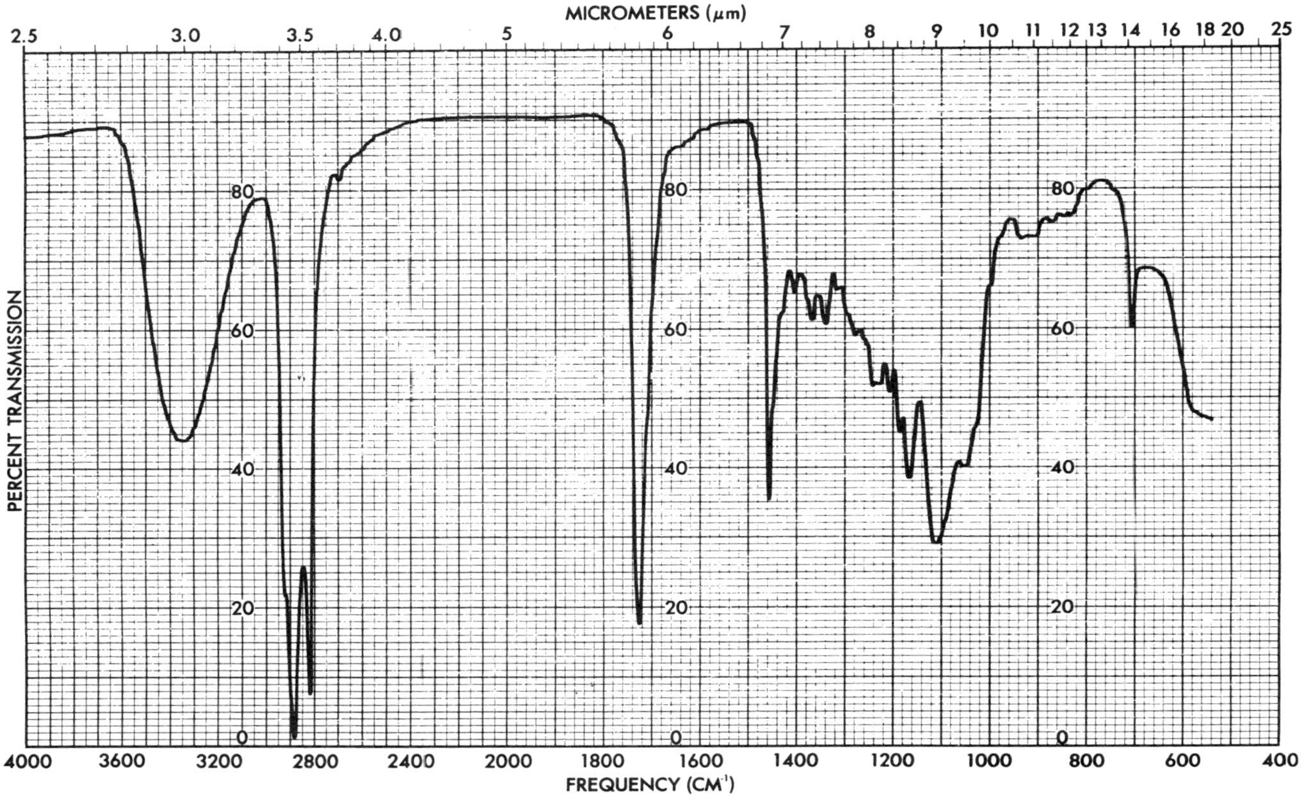

FIG. 4.38. Diglycol stearate.

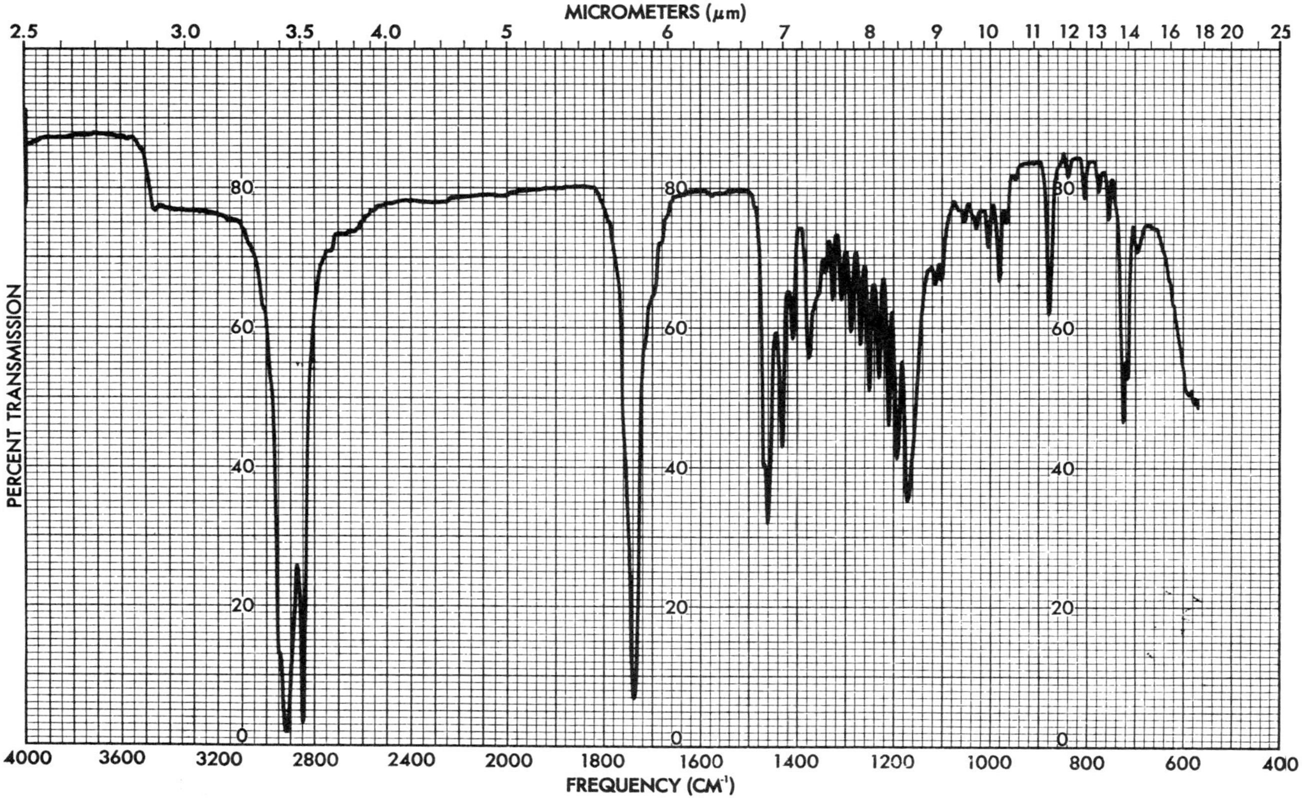

FIG. 4.39. Methyl stearate.

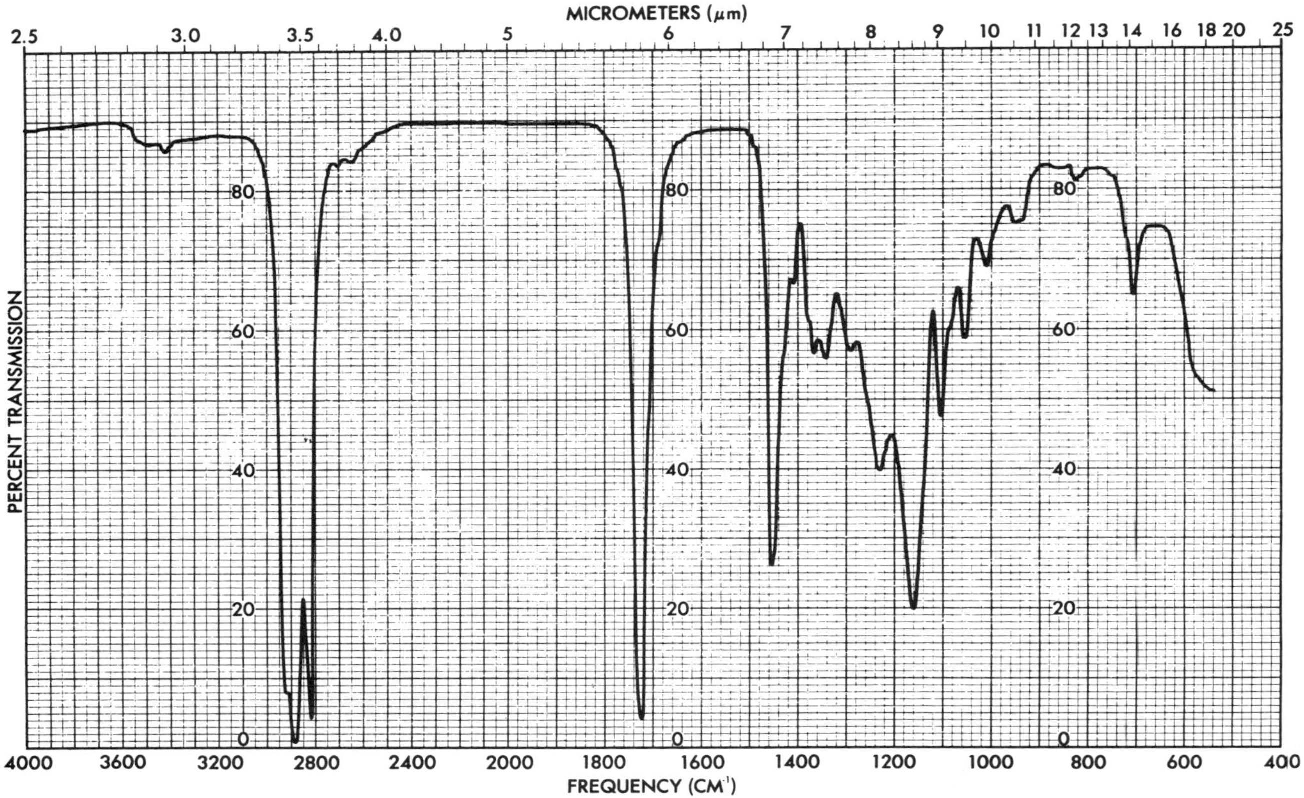

FIG. 4.40. Butyl stearate.

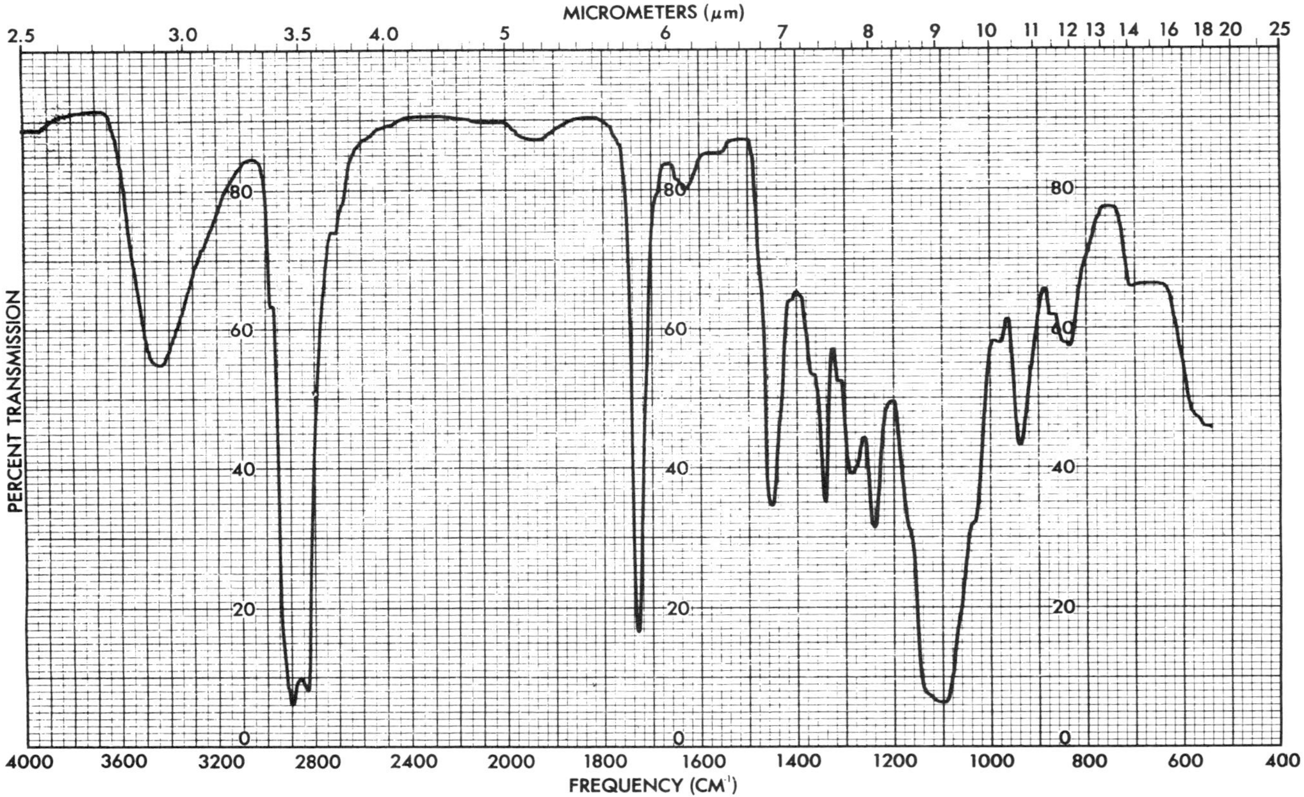

FIG. 4.41. PEG-400 monotallate.

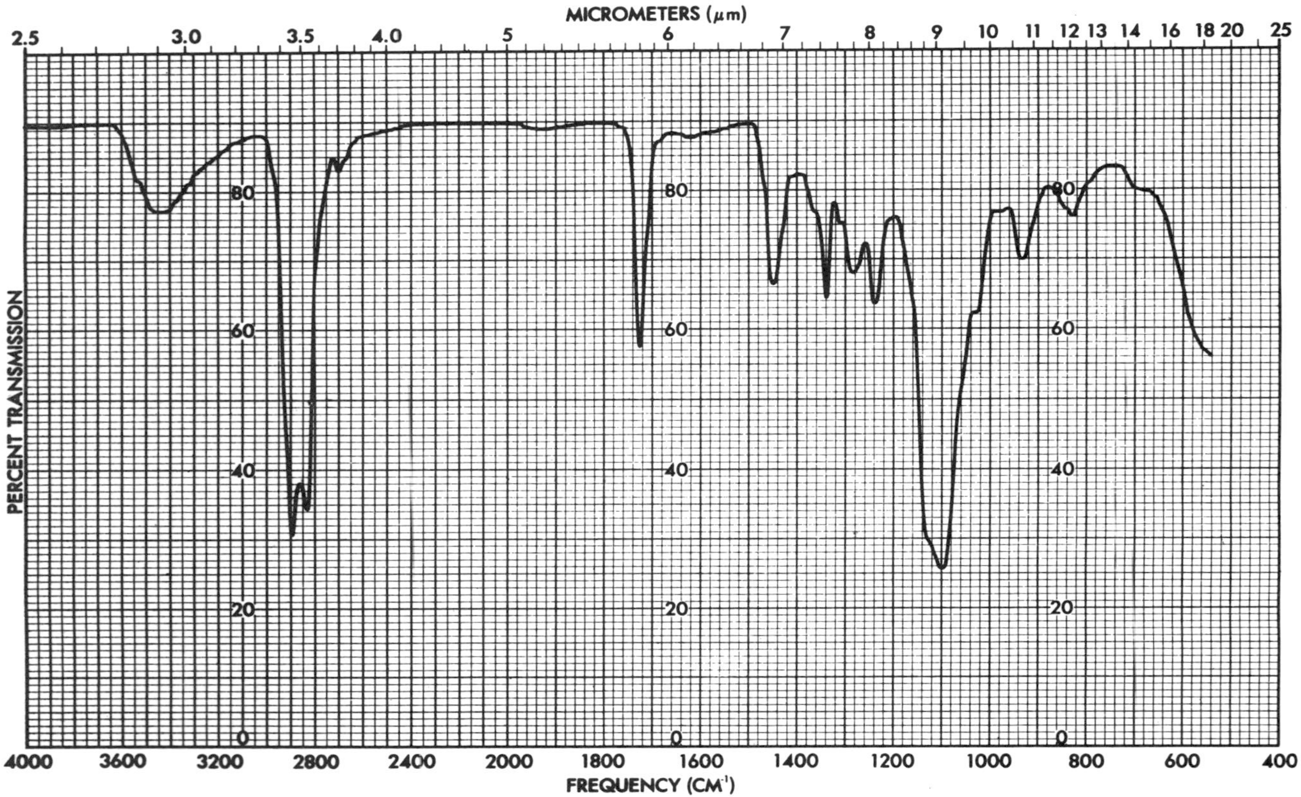

FIG. 4.42. PEG-600 monooleate.

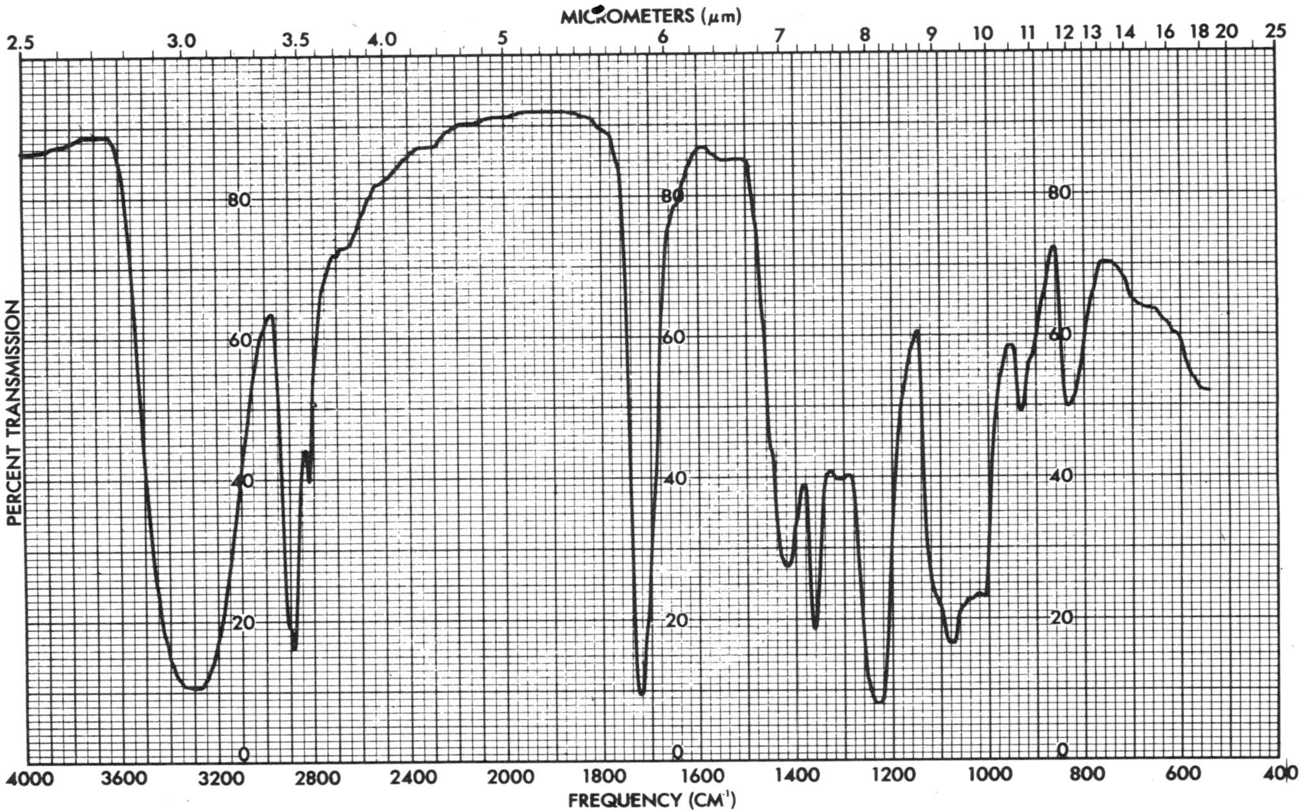

FIG. 4.43. Polyvinyl alcohol.

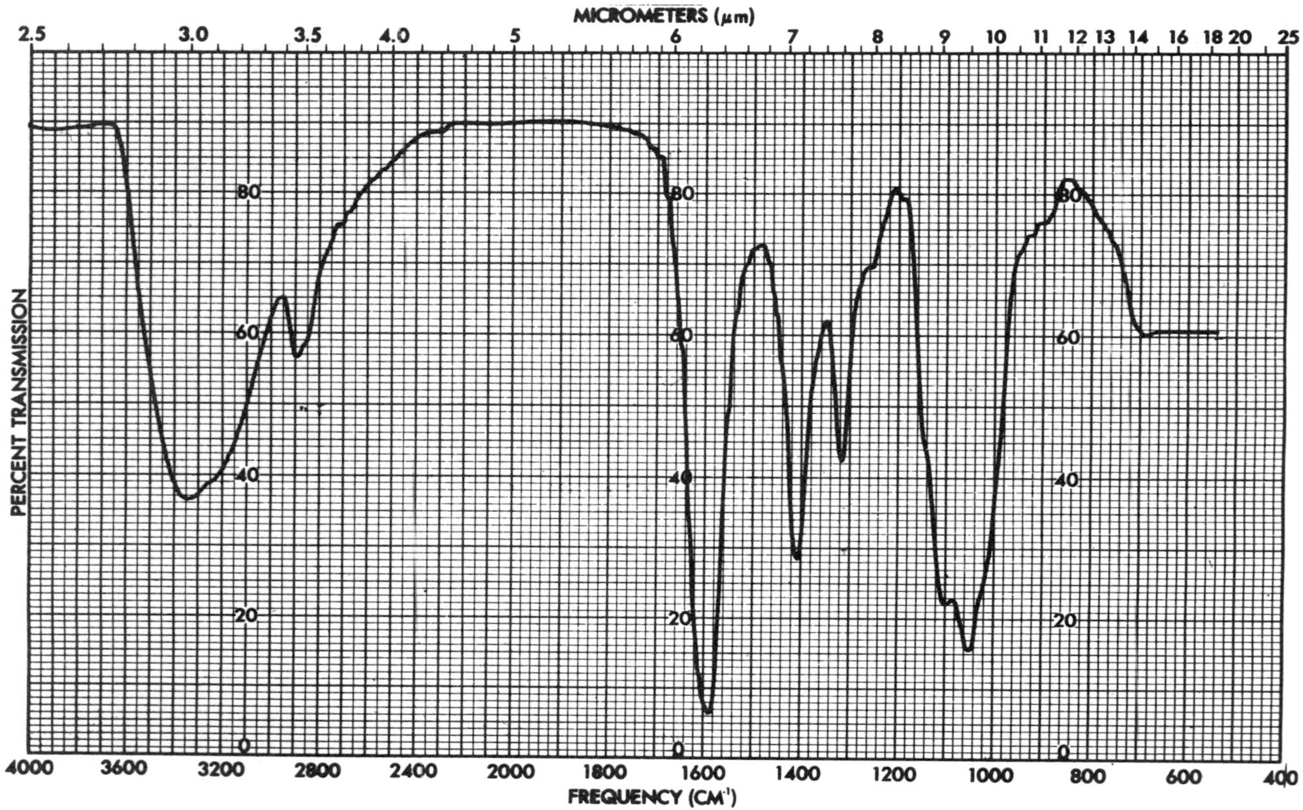

FIG. 4.44. Carboxymethyl cellulose.

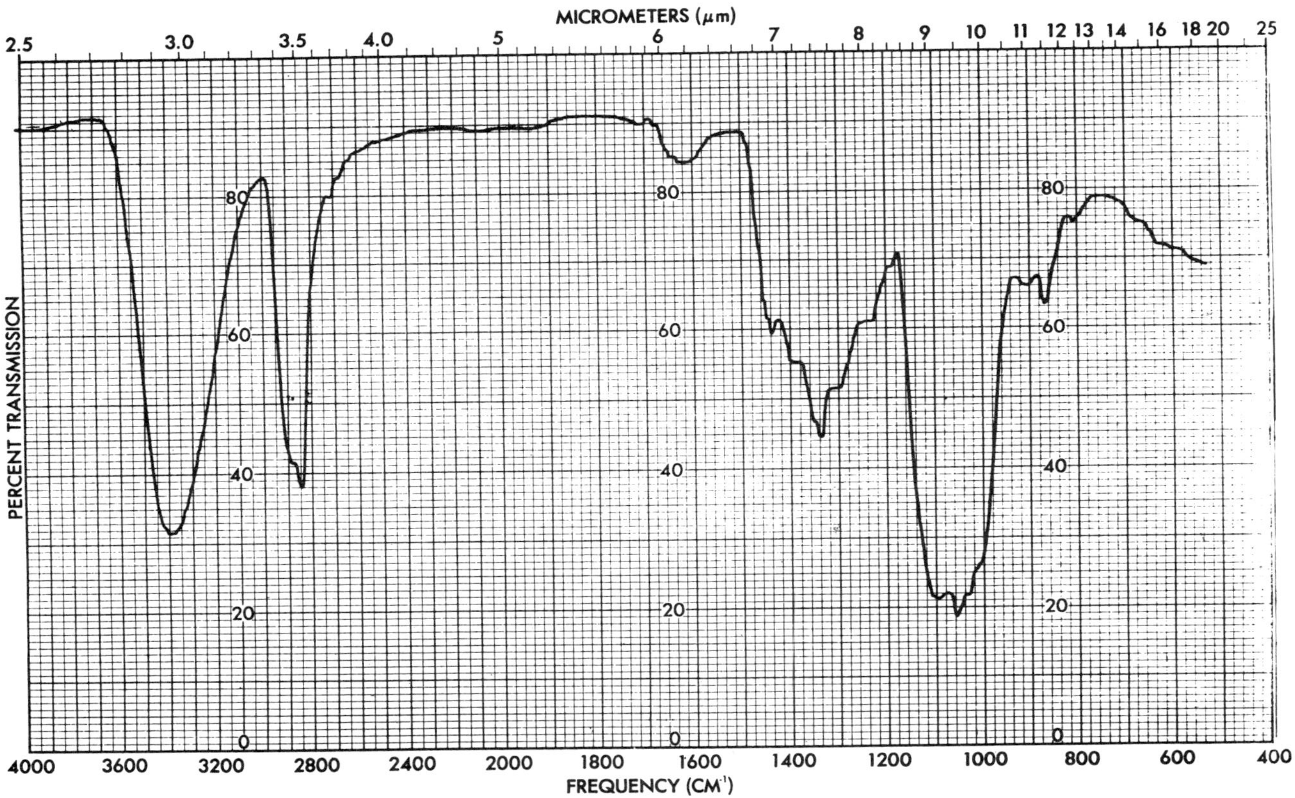

FIG. 4.45. Hydroxyethyl cellulose.

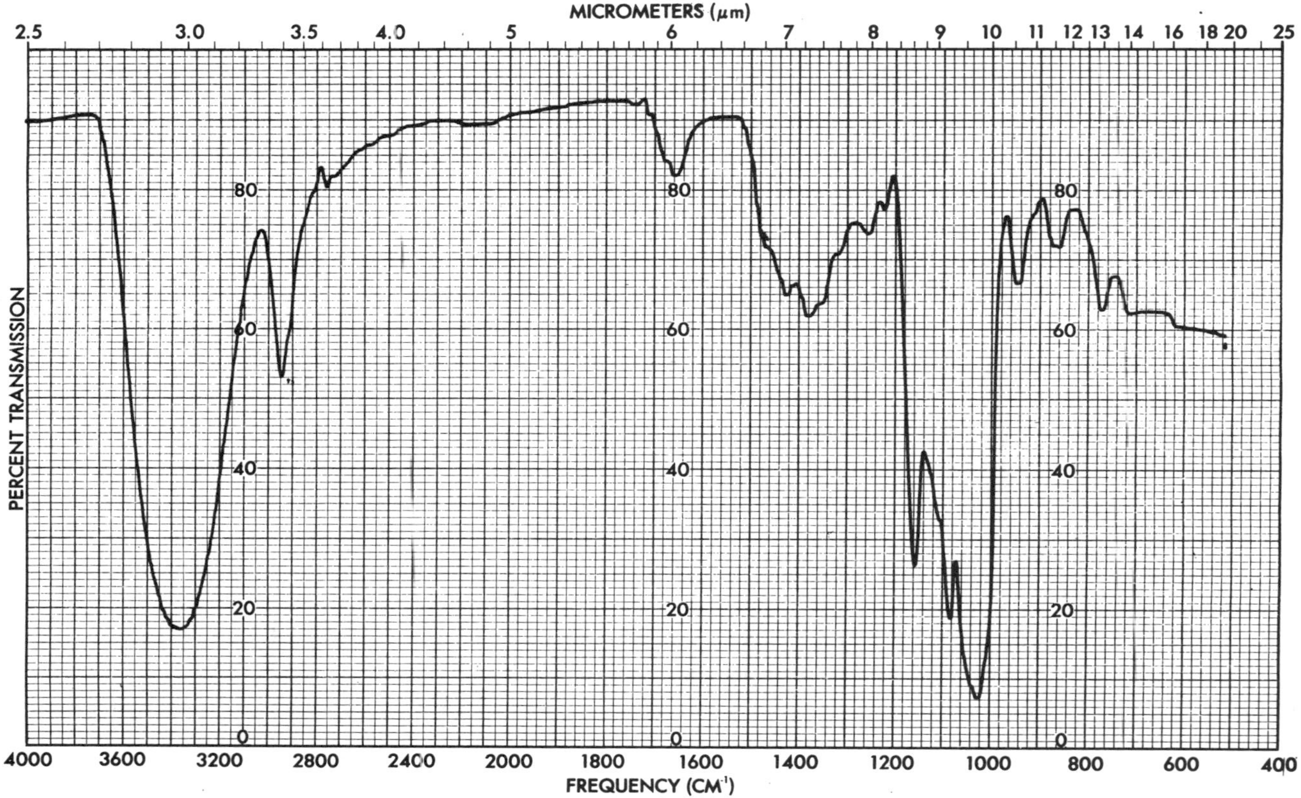

FIG. 4.46. Starch.

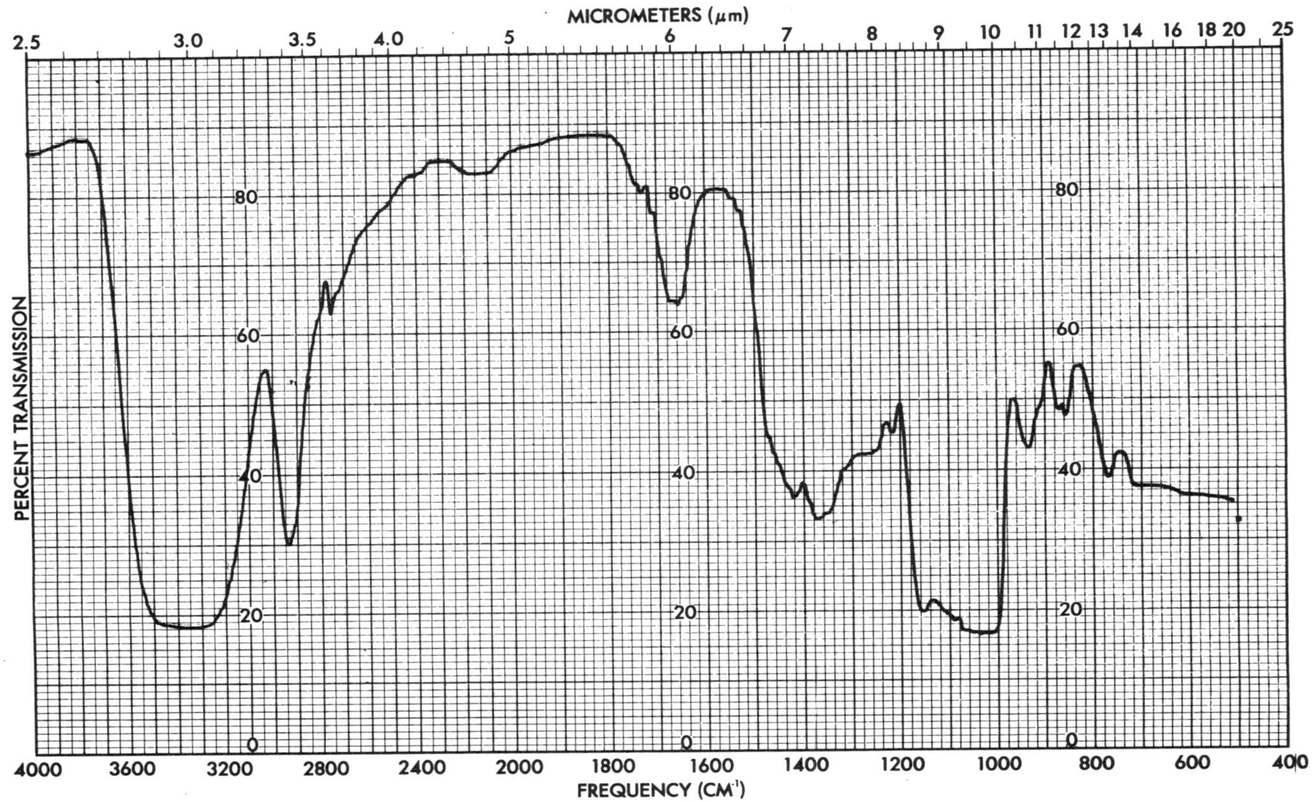

FIG. 4.47. Dextrin.

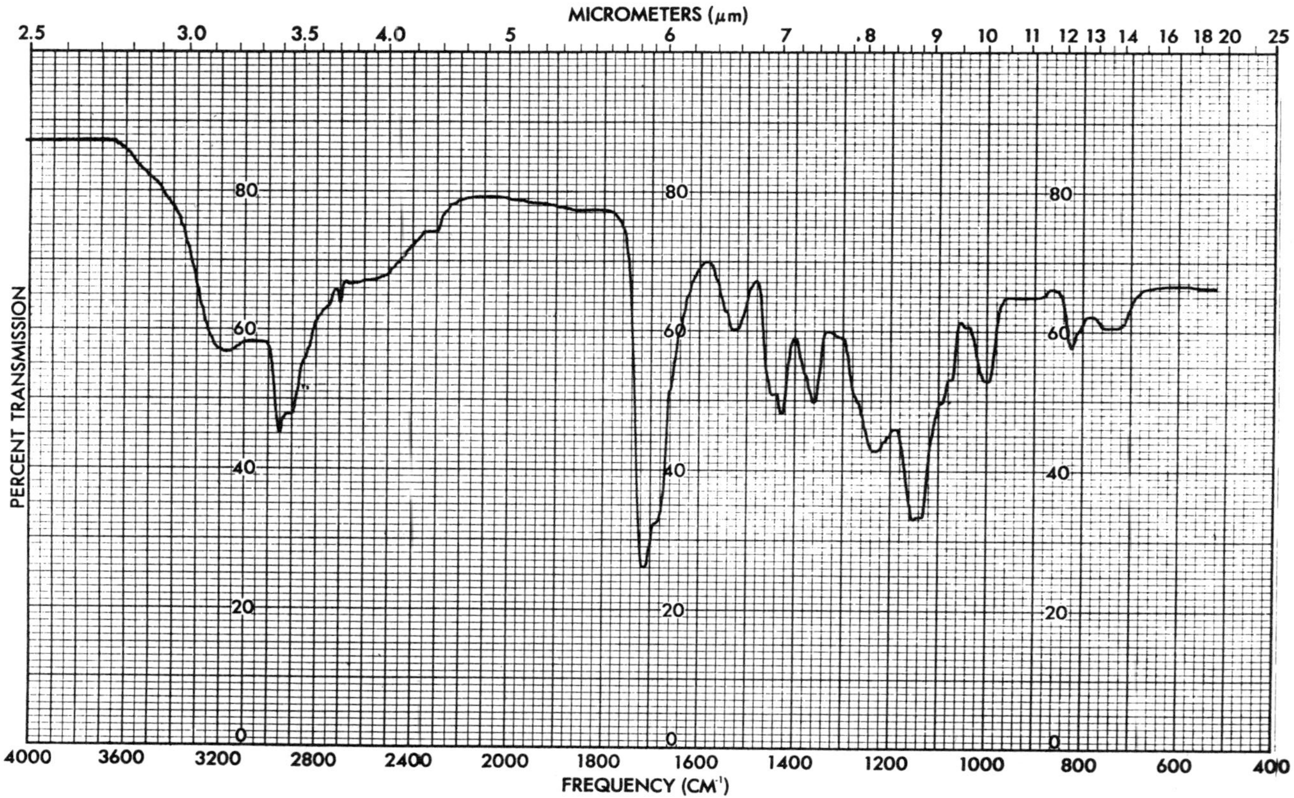

FIG. 4.48. Acrylic.

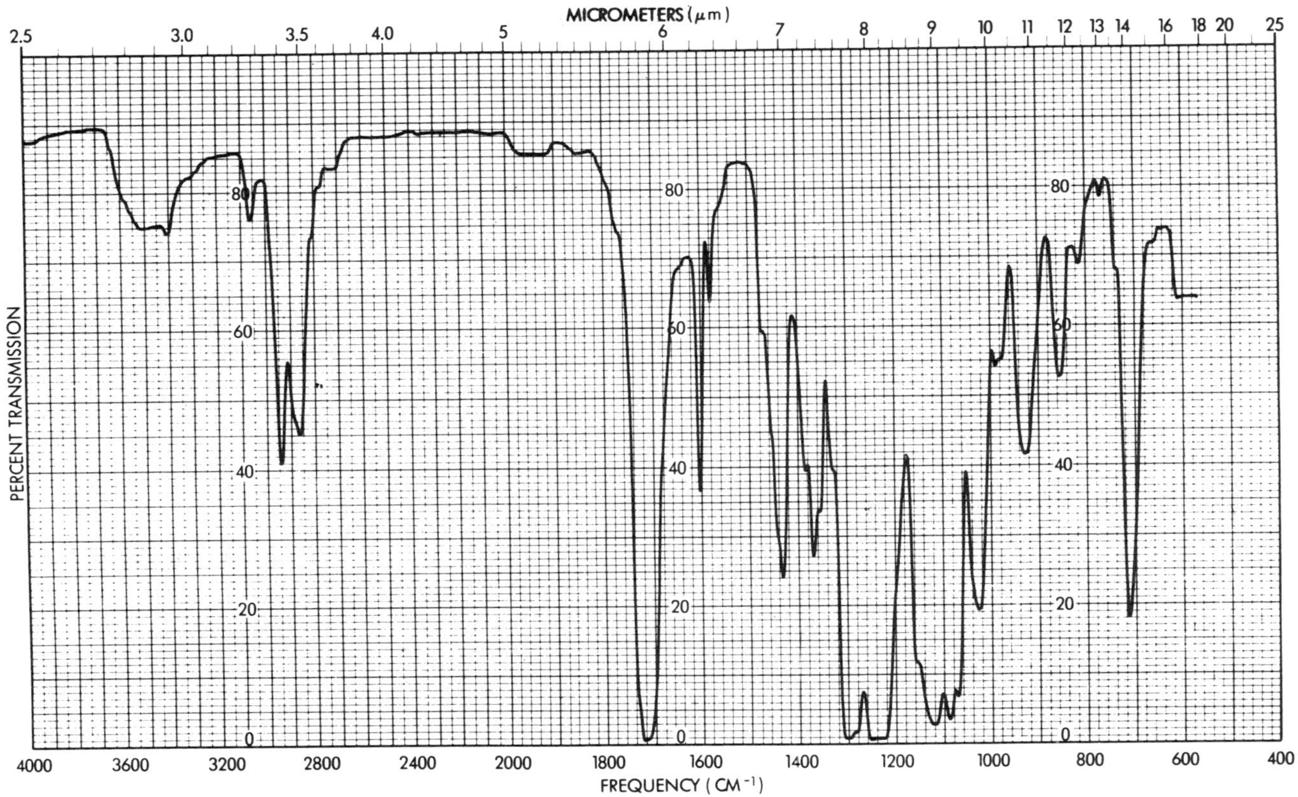

FIG. 4.49. Polyester.

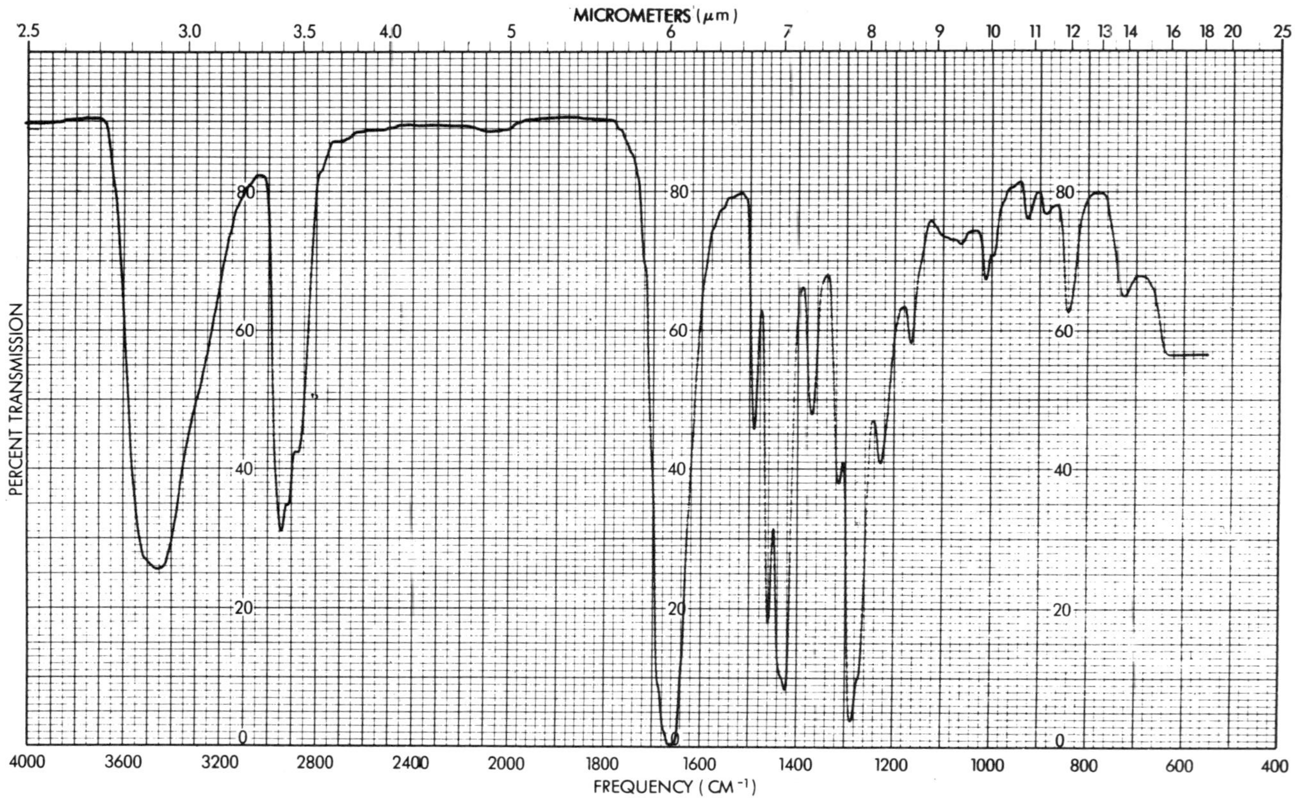

FIG. 4.50. Polyvinyl pyrrolidone.

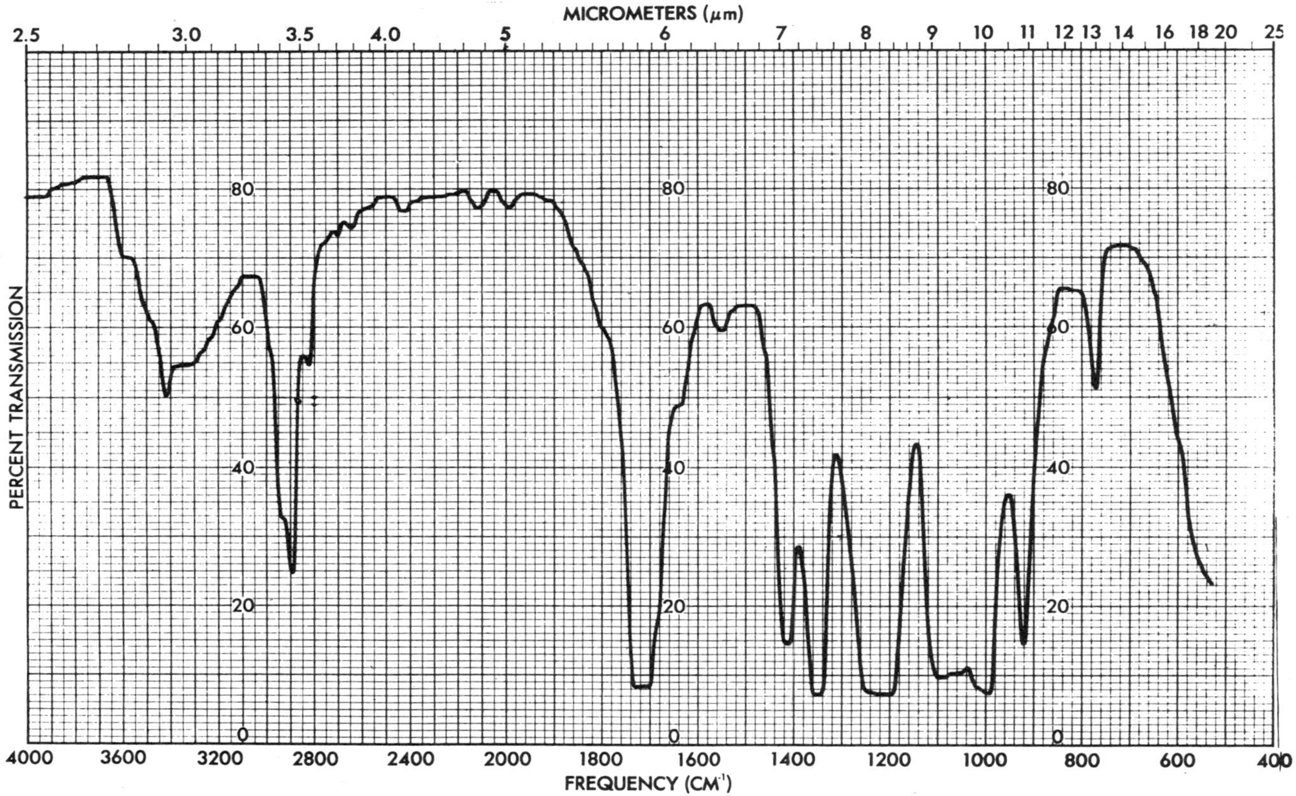

FIG. 4.51. Polyvinyl acetate.

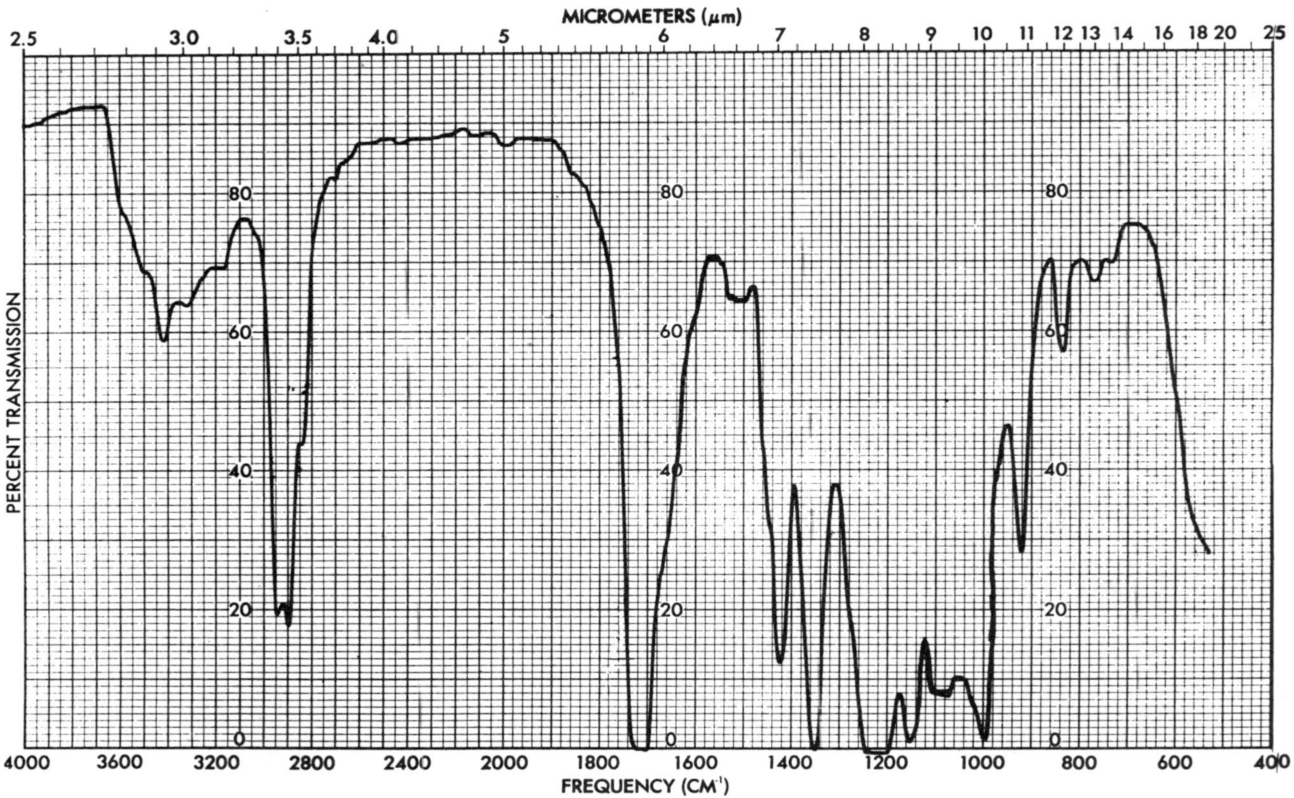

FIG. 4.52. Acrylic.

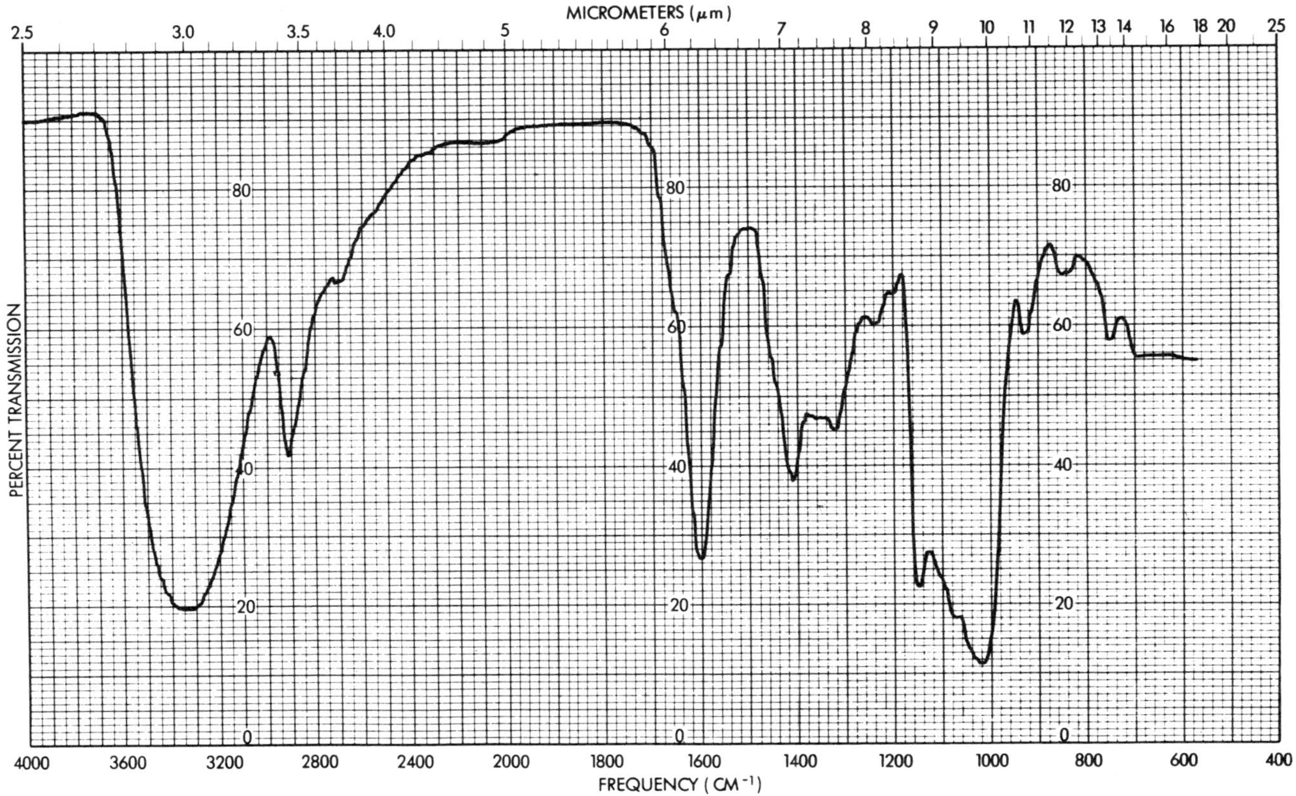

FIG. 4.53. Blend of starch and gum.

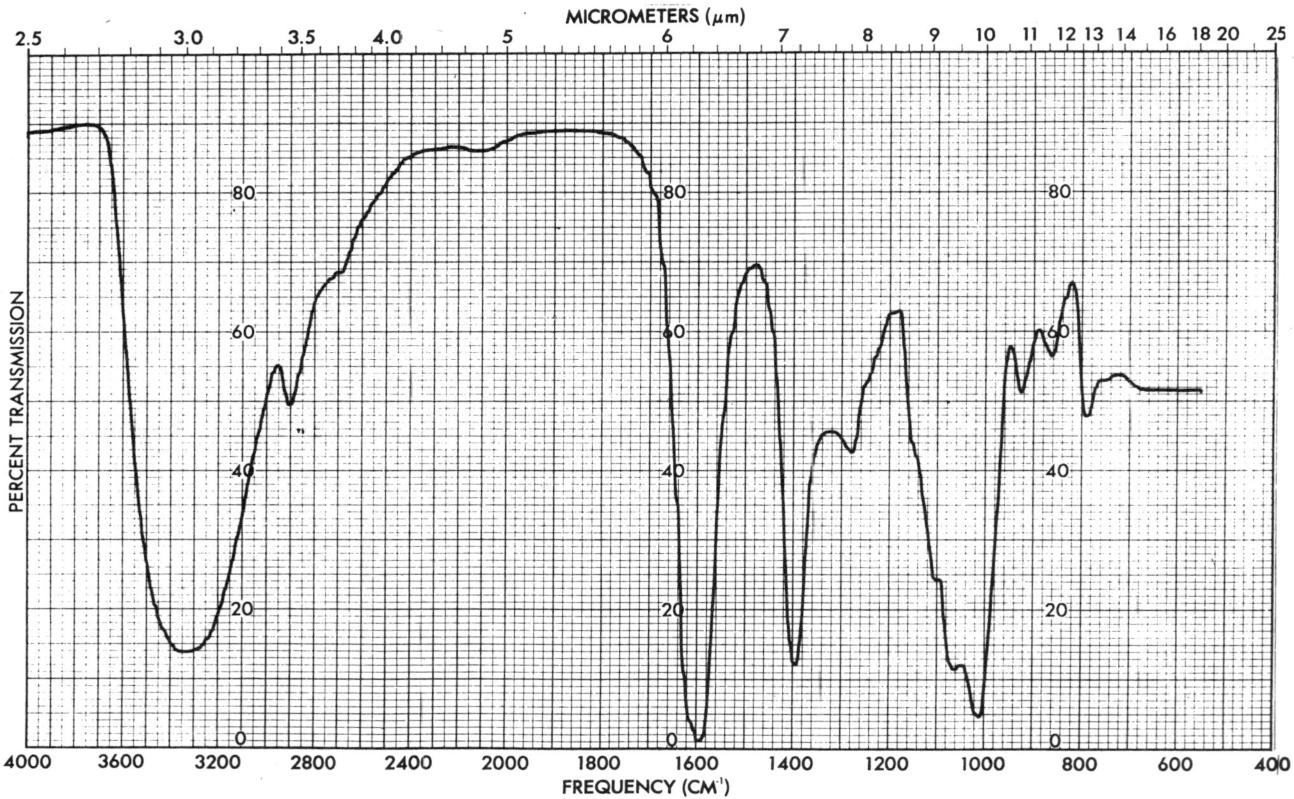

FIG. 4.54. Alginate.

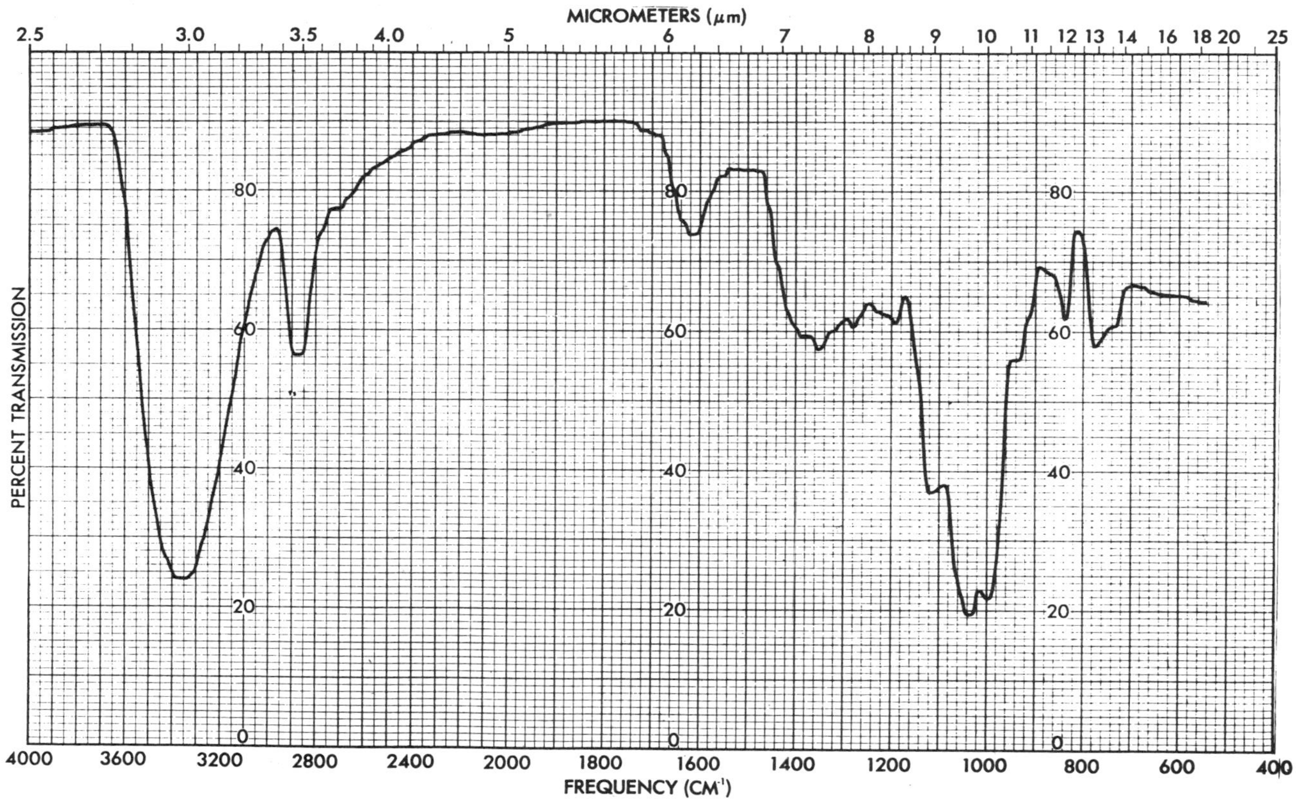

FIG. 4.55. Guar (high viscosity).

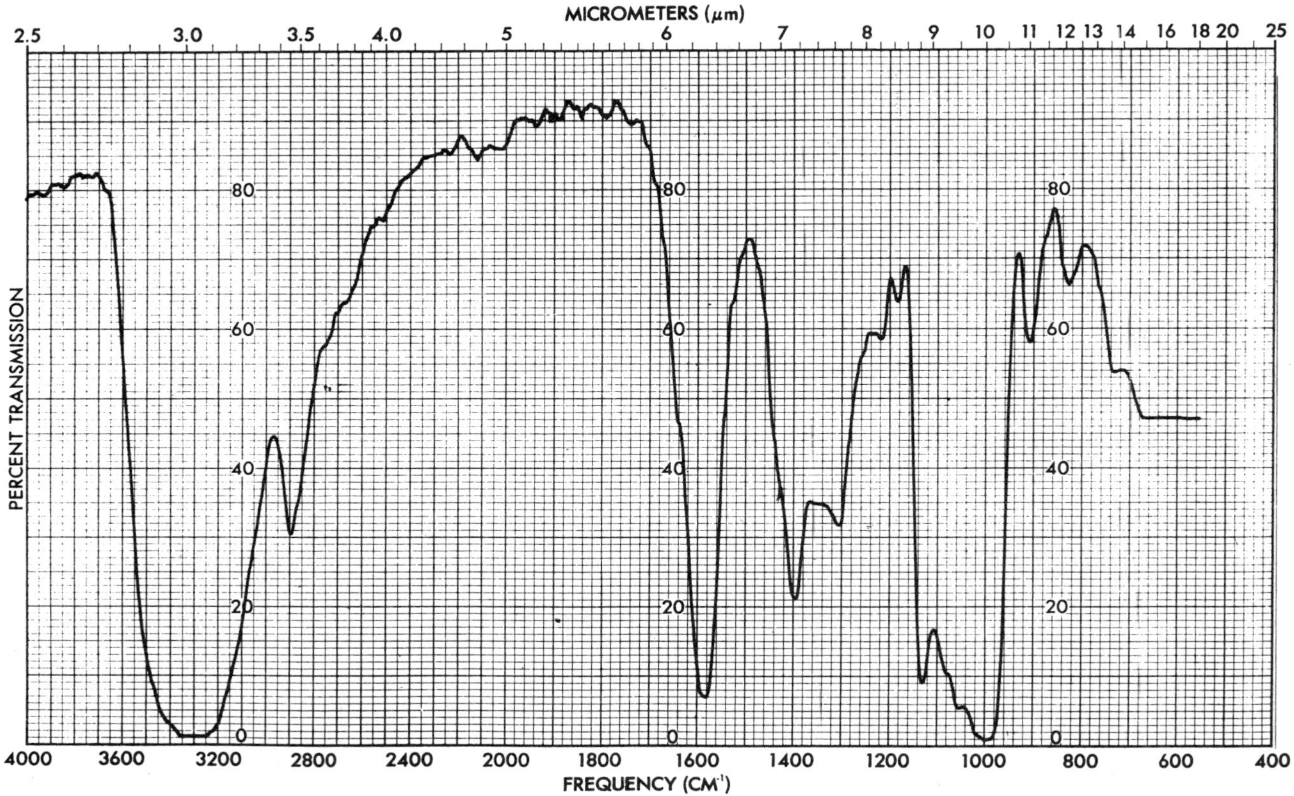

FIG. 4.56. Carboxymethylated starch.

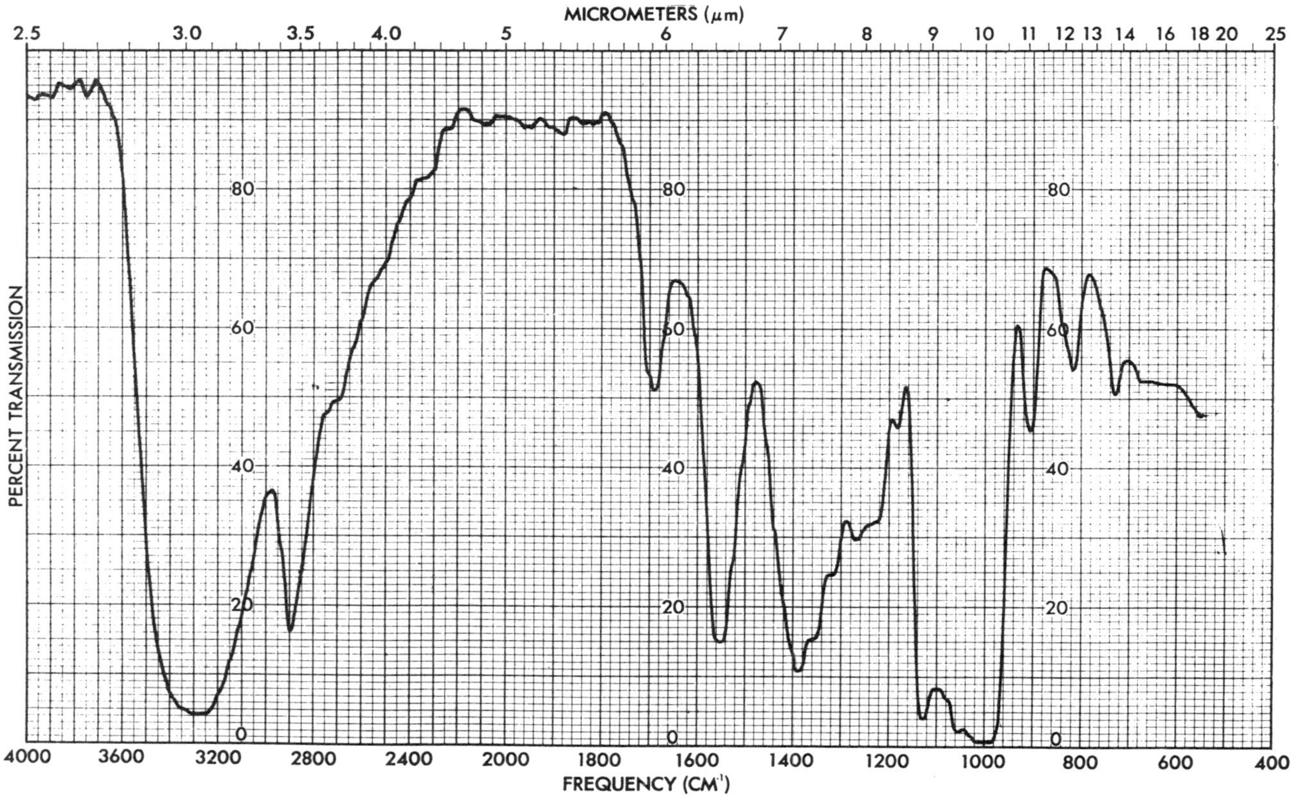

FIG. 4.57. Propyoxylated starch.

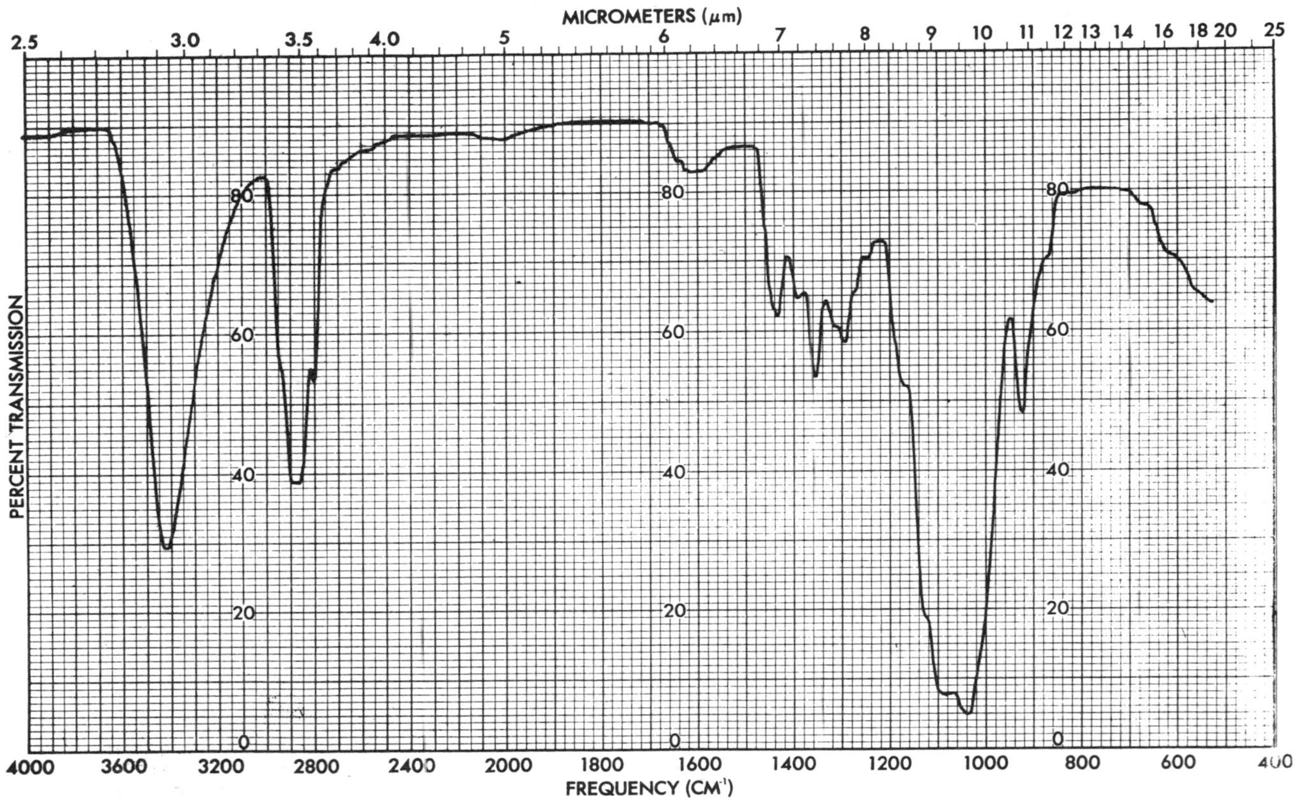

FIG. 4.58. Methocel®.

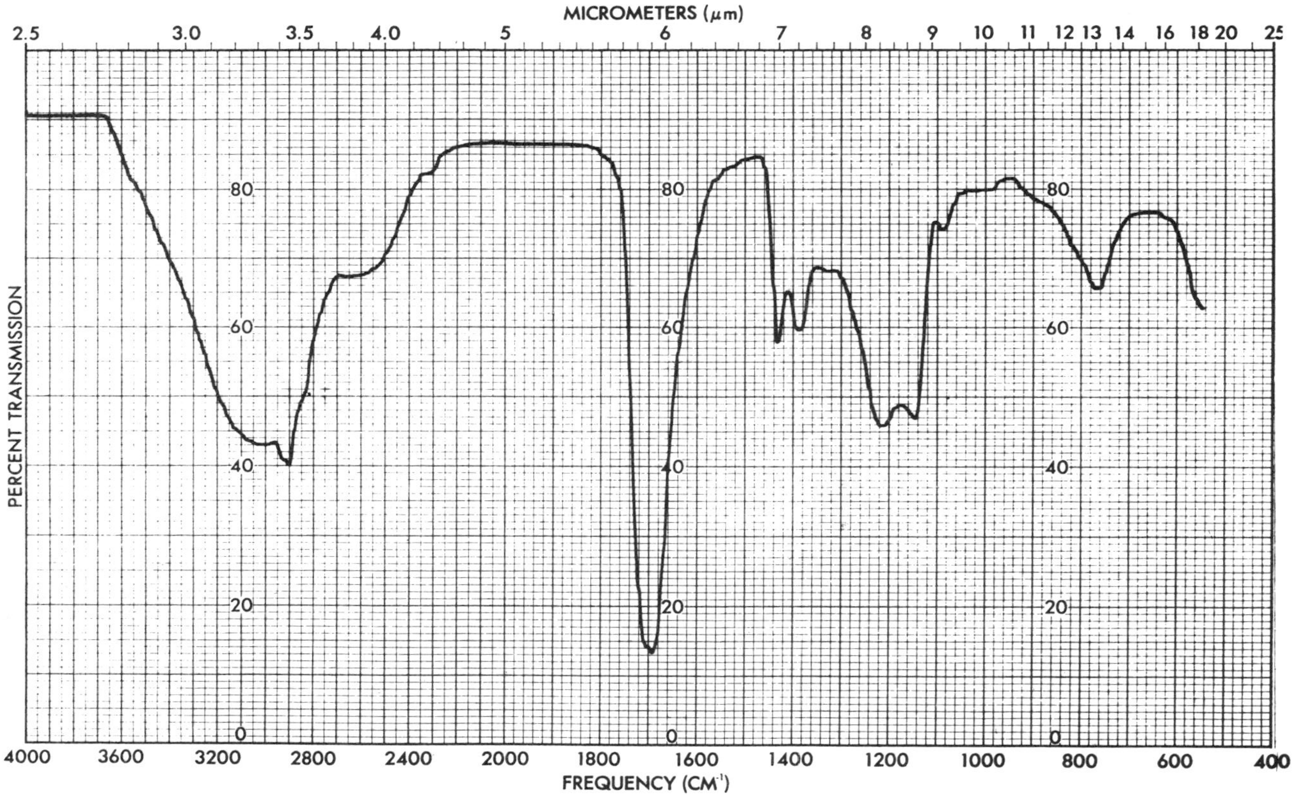

FIG. 4.59. Carbopol® 820.

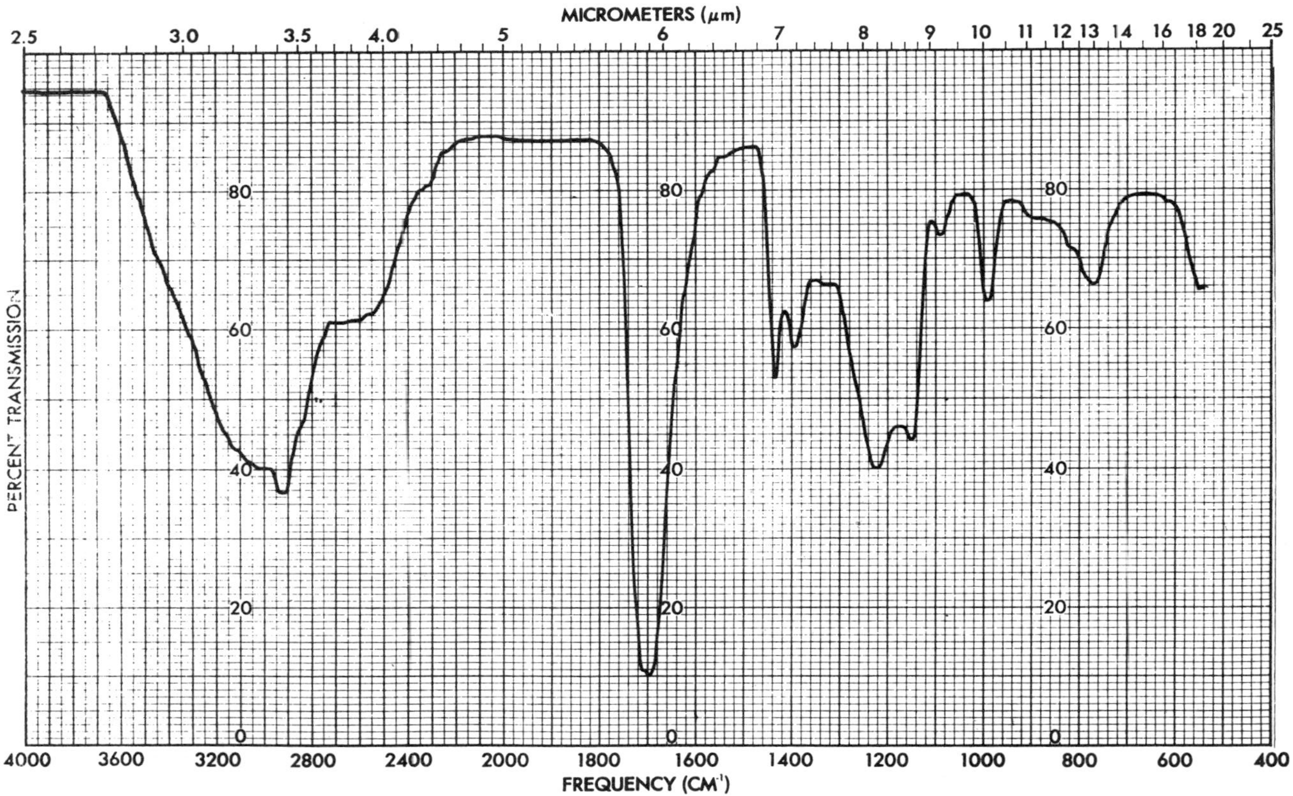

FIG. 4.60. Carbopol® 845.

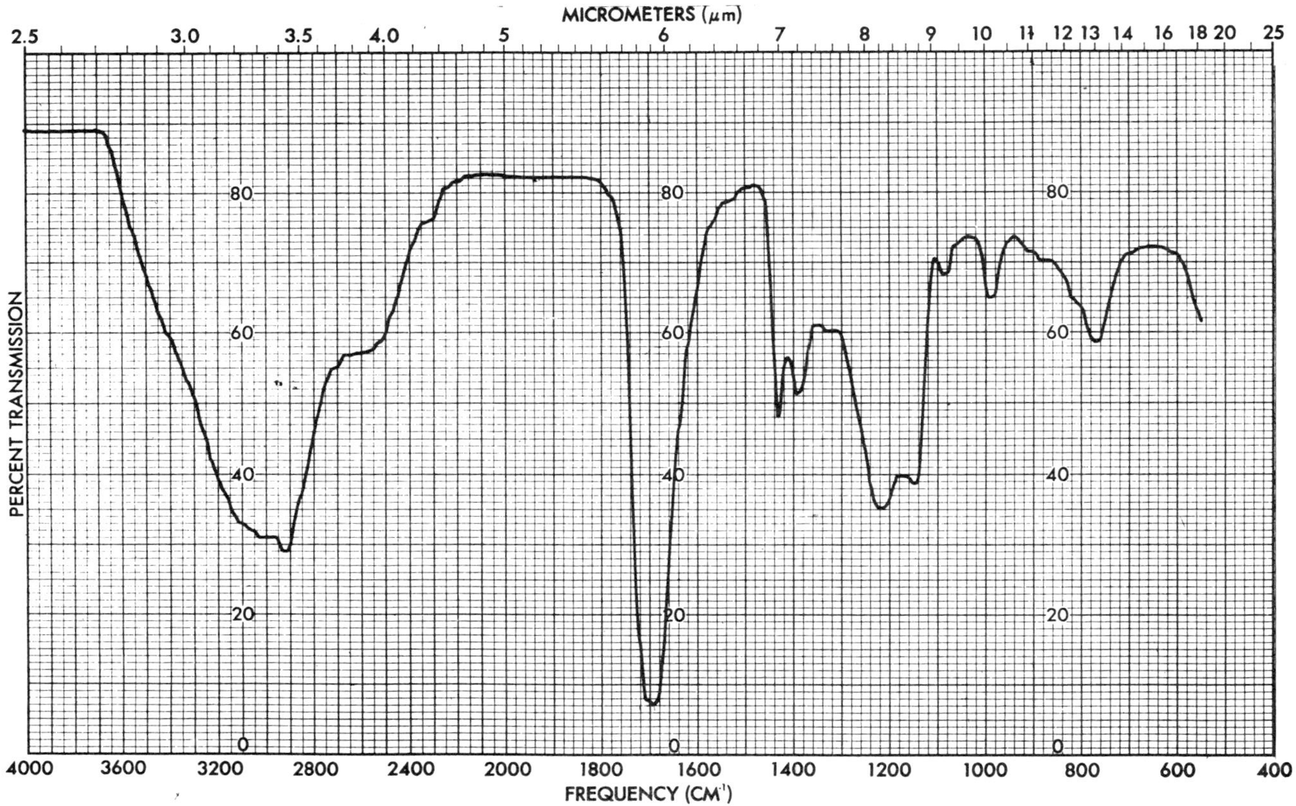

FIG. 4.61. Carbopol® 846.

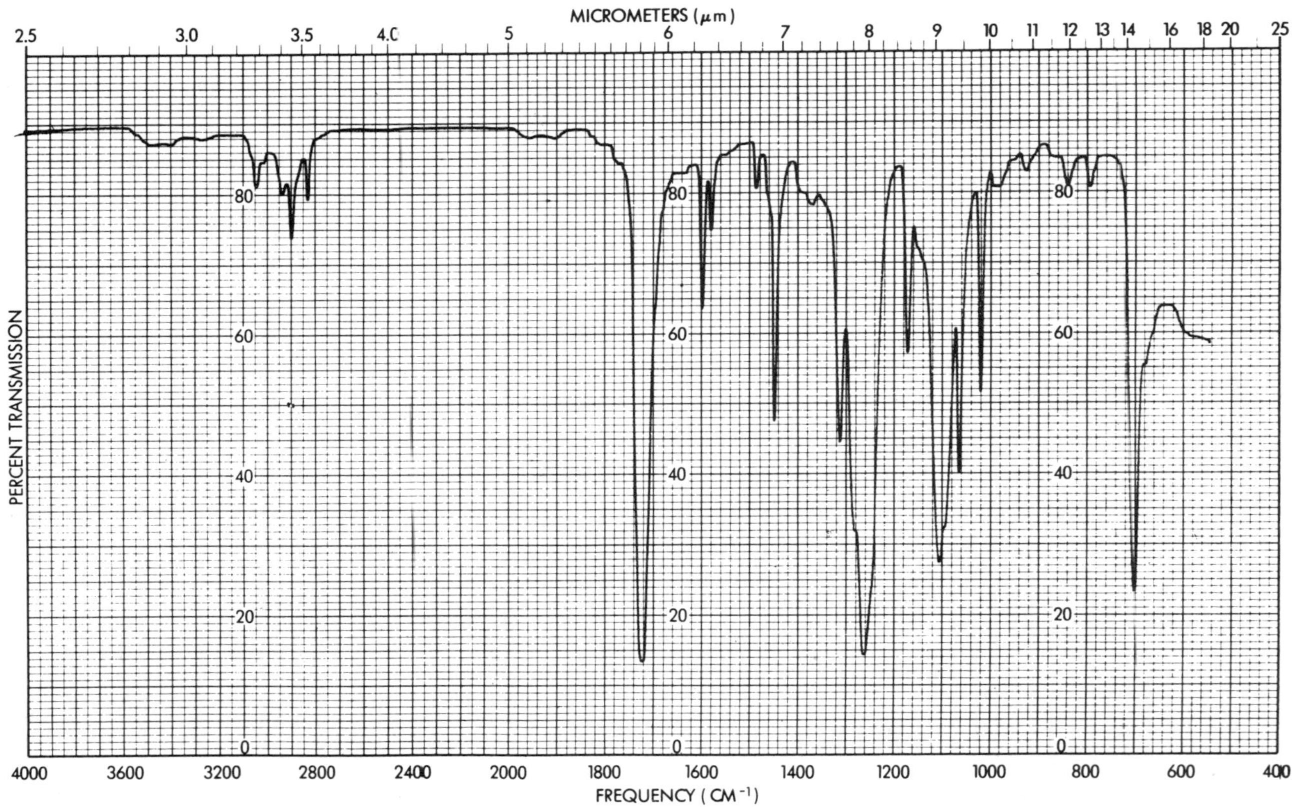

FIG. 4.62. Glycerol tribenzoate.

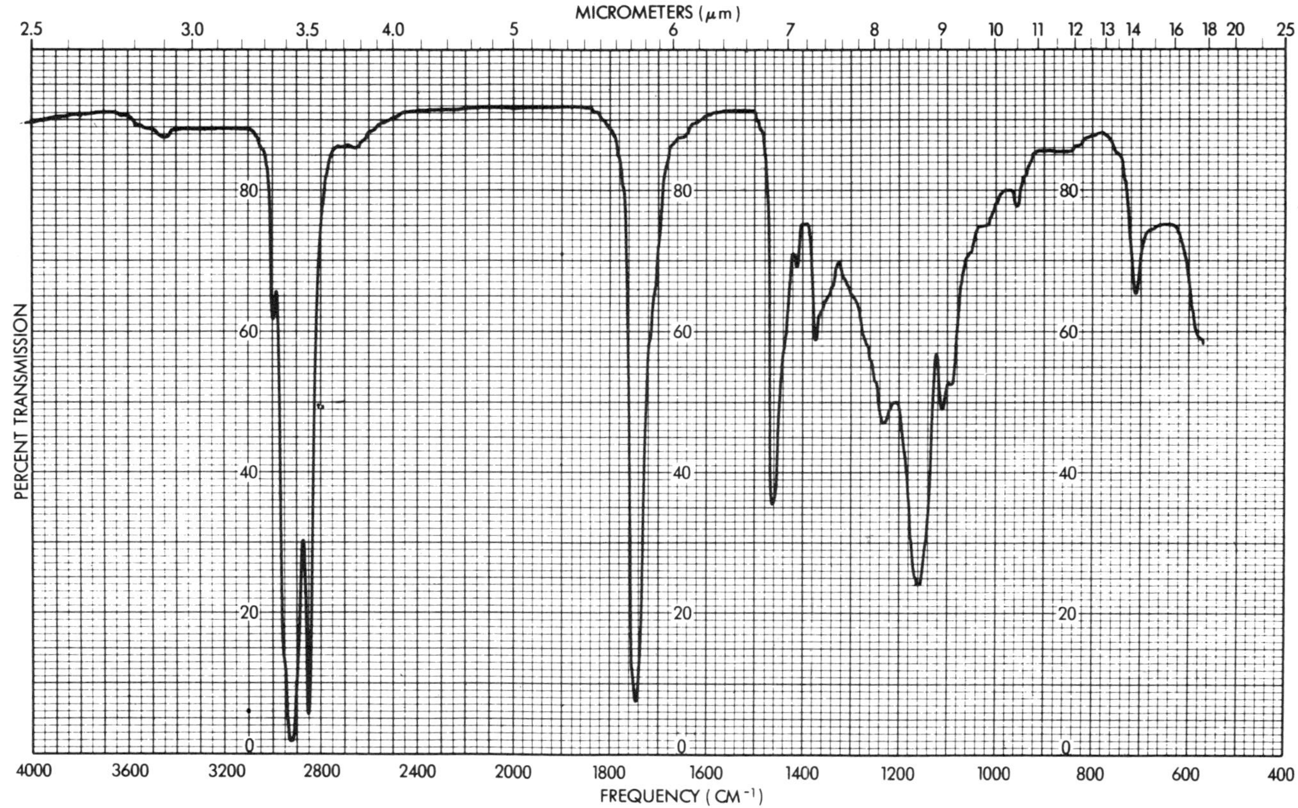

FIG. 4.63. Fancy bleached tallow.

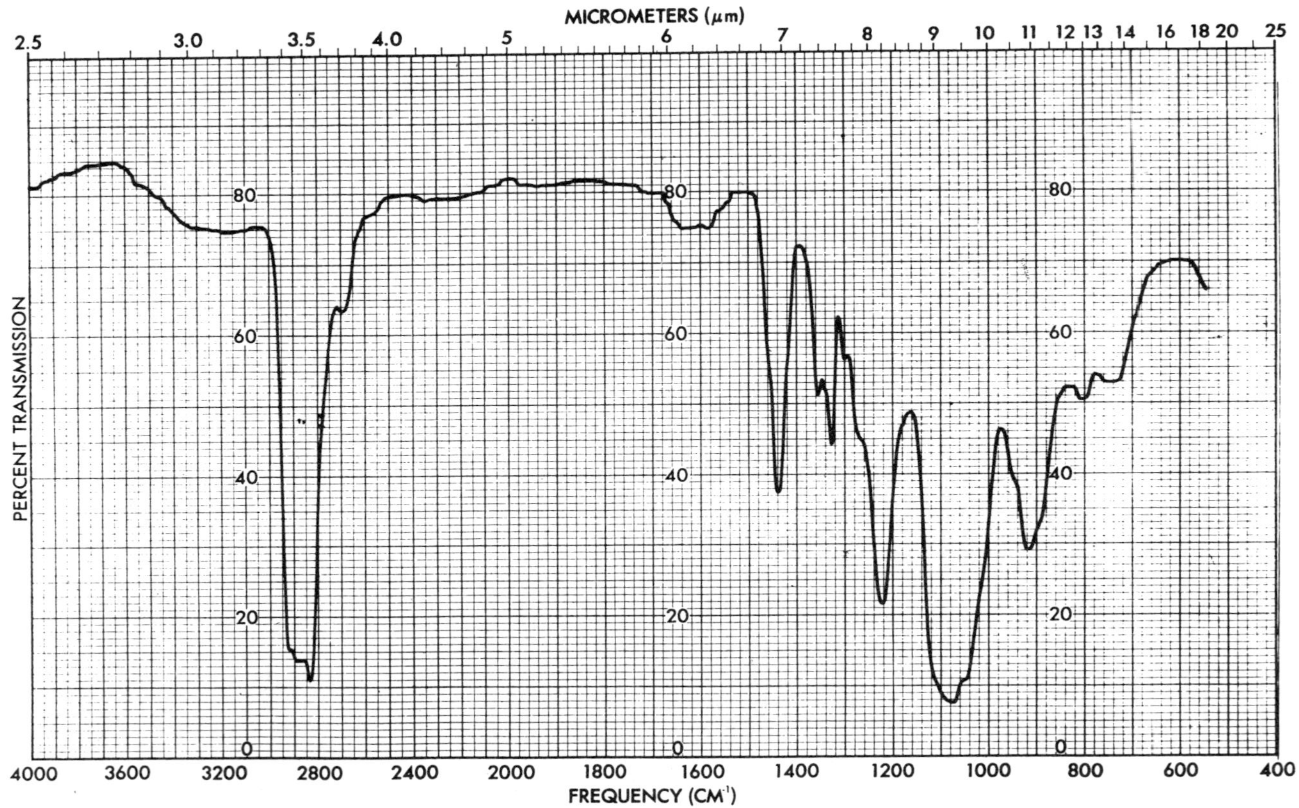

FIG. 4.64. Isodecyl alcohol $(EO)_3$ phosphate.

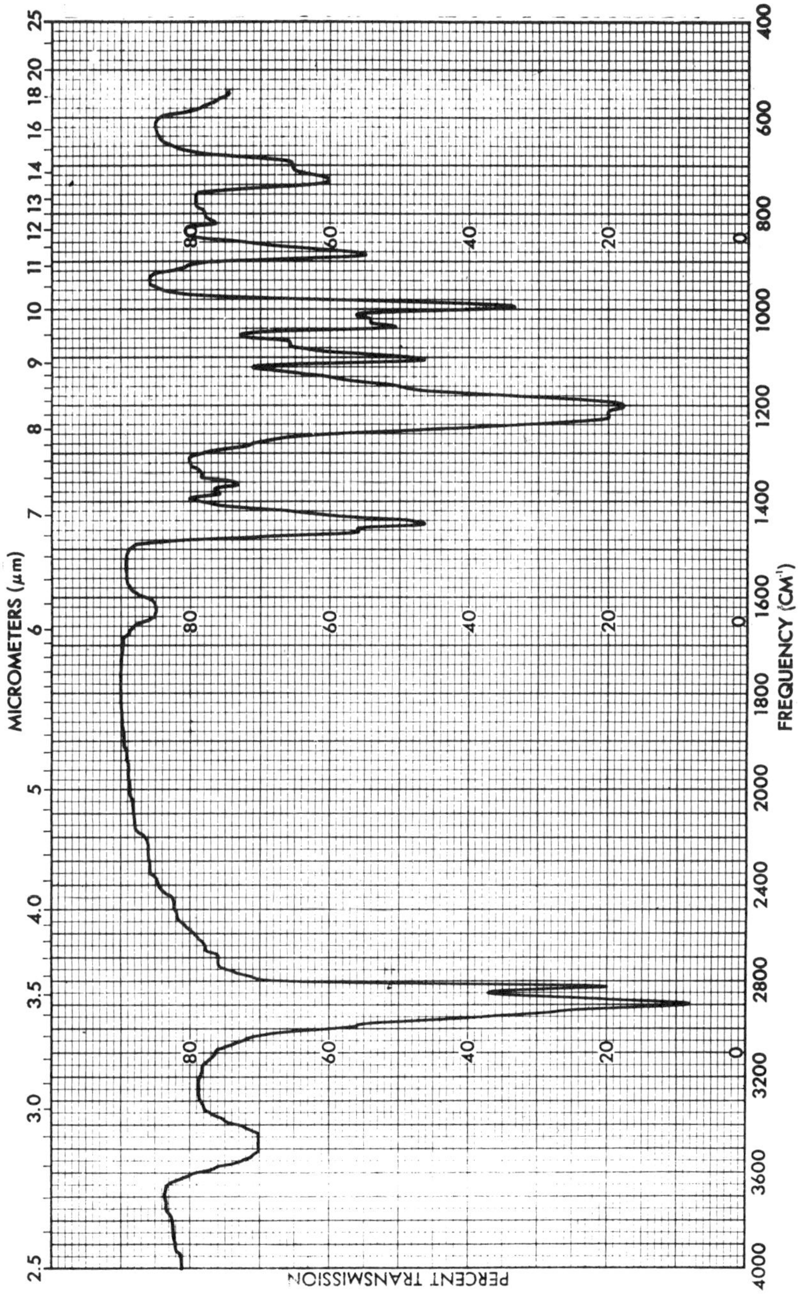

FIG. 4.65. Cetyl Morphilinum Ethosulfate.

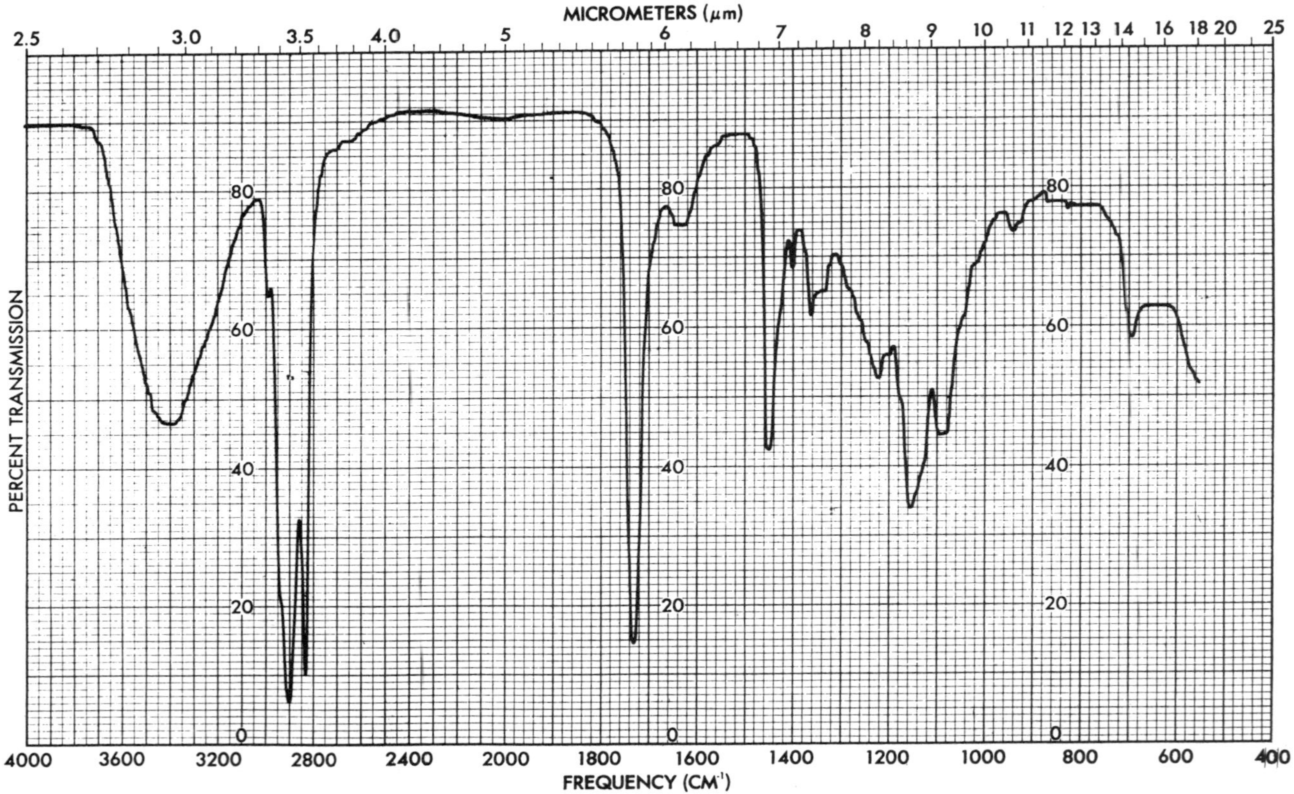

FIG. 4.66. Commercial size lubricant.

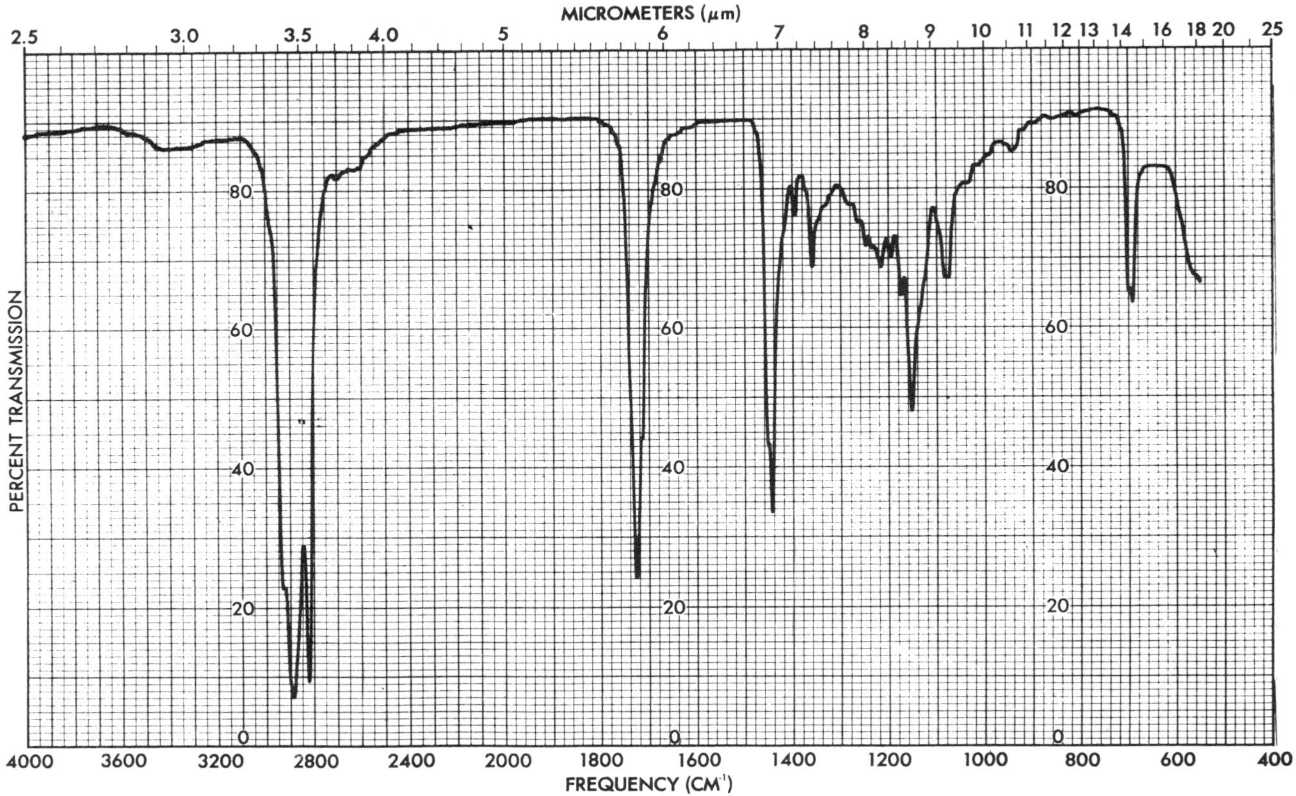

FIG. 4.67. Commercial kettle wax.

CHAPTER FIVE

PREPARATION OF FIBERS AND YARNS FOR FURTHER PROCESSING

INTRODUCTION

The most critical point in textile processing and the one that sets the stage for problems that can arise subsequently, is termed *preparation*.

Preparation is that process in which grège goods, whether yarn or fabric, knitted or woven, natural or synthetic, and still in a sized, oiled, unscoured, unbleached, undyed, or unfinished state, are made ready for further processing operations. Style and customer wishes will determine whether the goods are to be prepared for bleaching, dyeing, printing, or finishing. Fibers do not have to be woven or knitted, or even spun into yarn, to be prepared, as filament fibers can be handled directly on cones or metal tubes.

In the fiber manufacturing process, as the individual fibers are being transformed into yarn or being processed as lengthy continuous filaments and wound on cones, many various chemical entities are used to aid in the processing and to increase processing efficiency. If these additives, which help in the fiber processing, were left on the yarn, uniformity of printing, dyeing, and finishing would be very difficult to achieve. These fiber finishes and/or oils, waxes, polymers, lubricants, antistats, starches, and tints, if not properly removed prior to dyeing or finishing, can be the cause of blotchiness or unevenness during the dyeing, finishing, or printing process. In the case of synthetics that are heavily oiled prior to or during knitting, dye spots and speckiness later attributed to poor dyeing practices or faulty carrier emulsions in reality may be caused by the affinity of the unremoved oils in grège manufacture. Improper size removal is the major cause of unlevelness encountered in thermosol or range-dyed goods. Thus it becomes necessary to have an operation that minimizes the variables that occur in these processes. Apparently, many factors occur during preparation over which plant personnel and specialty chemical vendors have no apparent control, making a ready market for the myriad number of products currently being used as aids in the achieving of satisfactory preparation. The fact that these products continue to proliferate the market is an indication that minute differences in composition of matter do, in fact, cause substantial differences in the outcome of a process. In this particular case, experience of the compounder and prior arts, coupled with a knowledge of the customer's process, make the supplier a valuable part of the preparation team. The knowledge of what to compound, and the minimum investment necessary to purchase mixing equipment, have led to the common name of "hoe and barrel," or "soapy water" chemicals.

PROBLEM IN PREPARATION

Some specific problems in preparation that can affect later stages of processing are as follows:

1. Fiber and yarn preparation is the first area to feel the effects of a cost reduction program.
2. Shortcuts taken usually cause incomplete processing.
3. In continuous preparation ranges, insufficient water flow will create insufficient size removal, which, in turn, creates blotchiness during dyeing.
4. Insufficient rinsing of preparation detergents can carry over and create foam in jet dyeing.
5. Concentrated chemicals can cause fiber degradation.
6. In fibers to receive fire-retardant properties, residual silicones or fats may lower rating.

7. Where an oil repellent is to be applied, contamination by silicones can lower oil ratings if not properly removed.
8. In the case of water repellents, residual surfactants can cause rewetting and render treatment worthless.
9. Improper batch size makes reproducibility of performance difficult.
10. In the dyeing of acrylic blends, the wrong choice of preparation surfactant will cause staining later during blend dyeing or can retard single-component dyeing.
11. Unevenly heat-treated polyester, in singeing or desize operations, will cause uneven dye takeup during dyeing.
12. Wrinkles that set in polyamide fibers, unlike polyester fibers, cannot be reset. Polyamide fibers, once set, cannot be changed.
13. In the dyeing of silk, fibers that have been improperly degummed during preparation will take up less dye and render shade reproducibility difficult.

Preparation consists of several different operations, the main functions of which are the removal of size. This is accomplished by the use of detergents, auxiliaries, and alkali; therefore, an understanding of surfactant chemistry is a help in understanding preparation procedures.

Even though there are many different surfactant compounds in common usage and it would be virtually impossible to identify and discuss each of them, there are several basic family groups. We discuss these in detail by reviewing their manufacturing procedures, the patent literature relating to their manufacture, and their end use. It is necessary to realize that the compounds offered to the preparation department are rarely, if ever, a single-component product, but are, in reality, a two-, three-, or four-component blend. We also consider theoretical or possible combinations of these basic family groups as they could conceivably be found in a commercial preparation department and examine typical infrared spectra of major compounds and blends.

CHAPTER SIX

SURFACTANT AUXILIARIES AND THEIR CHEMISTRY

IONIC NATURE OF SURFACTANTS

Surfactants used in preparation are usually either anionic or nonionic but occasionally may be amphoteric or cationic. As a general rule, they are usually blends of anionics and nonionics. Table 6.1 lists the ionic charge for each of these classes.

Cationic surfactants are rarely used in preparation unless applied as a lubricant or antistat at the end of the scour cycle on goods that are to be dyed but not bleached. They may also be used on acrylics where a residual softener or retarder is applied prior to dyeing. Technical literature recommends the use of a small quantity of cationic surfactant as an aid in weight reduction in the caustic treatment of polyester.

Chemical specialties that utilize surfactants for use in preparation are (1) penetrants and wetting agents, (2) solvent and nonsolvent scouring compounds, (3) lubricants and antistats, and (4) oil and wax emulsifiers.

ANIONIC SURFACTANTS

Advantages of anionic surfactants are good oil emulsification, good dye dispersibility, excellent wetting speeds, and excellent rewetting speeds. Disad-

TABLE 6.1. Electrical Charges of Surfactant Classes

Class	Charge
Anionic	Negative (−)
Nonionic	Neutral (0)
Cationic	Positive (+)
Amphoteric	Positive and negative (+ and −)

vantages may be incompatibility with cationics, calcium and magnesium sensitivity on the part of sulfates, high foam level, and carry-over from various processing steps that can affect dye yield in basic or cationic dyeing. Anionic-type surfactants are used primarily as wetters, emulsifiers, dispersants, scouring agents, and lubricants.

Common anionic-type surfactants are composed chemically of sulfates, sulfamates, sulfonates, sulfosuccinates, phosphate esters, methyl taurines, carboxy methylates, and metallic and amine soaps.

Aliphatic Sulfur-Containing Surfactants

Sulfated, sulfonated esters, oils, and phenolic derivatives are the basic anionic surfactants used in preparation. One of the original synthetic auxiliaries manufactured for use by the textile industry was sulfated castor oil, which at that time was known as "turkey red oil." On today's market we have available the sulfur derivatives of products such as castor oil, oleic acid, peanut oil, and olive oil, as well as the alcoholic esters of these oils, as typified by butyl or propyl oleate or butyl castor. Literature on the sulfation of oils is plentiful.[70] Even the patent literature gives exhaustive procedures for the preparation of these compounds.[71]

The method by which the sulfur group is introduced into the molecule will depend on the choice of reagent used. The following are SO_3H-producing intermediates:[72]

Sulfuric acid (H_2SO_4)
Oleum [$HO—(SO_3)_xH$, where $x = 1–4$]
Sulfur trioxide (SO_3)
Chlorosulfonic acid ($ClSO_2OH$)

Sulfamic acid (NH_2SO_3H)

Mixtures of chlorosulfonic acid ($ClSO_2OH$) and sulfur trioxide (SO_3)

Mixtures of dimethyl sulfate and sulfur trioxide

Dimethyl formamide and trimethylamine complexed with sulfur trioxide

Structures of typical chemical compounds that have an —SO_3H group are:

Sulfation—results in the formation of a sulfur–oxygen bond

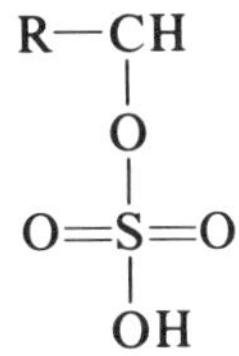

Sulfonation—results in the formation of a sulfur–carbon bond

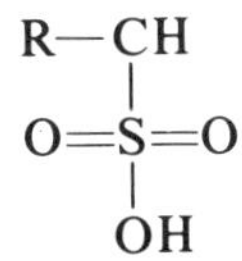

Sulfamation—results in the formation of a sulfur–nitrogen bond

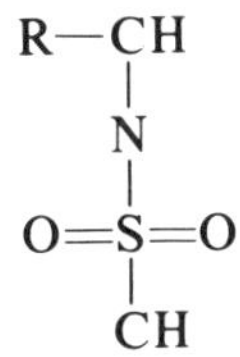

The following compounds when treated with —SO_3H-producing radicals, are used by the textile industry as surfactants, wetting agents, or lubricants:

Vegetable oils	Alkyl benzenes
Red oil	Diphenyl oxide
Castor oil	Fatty alcohol
Olive oil	Branched chain alcohols
Peanut oil	Linear alcohol
Tall oil	Fatty esters (butyl oleate)
Cottonseed oil	Naphthalene formaldehyde condensates
Safflowerseed oil	
Mineral oil	Lignin

TABLE 6.2. Normal Absorption Frequencies

Group	Type	Wavelength λ in Microns, 10^{-4} cm	Frequency ν in Reciprocal Centimeters, cm^{-1}
NH, NH_2	Amine	2.8	3500
OH	Hydroxyl	2.8–3.0	3300
H—F		2.59	3958
N—H	(In amides and amines)	2.9	3400
N—H	(In imines)	2.95	3375
C—H	Acetylenic	3.0	3350
C—H	Ethylenic	3.30	3020
C—H	Aromatic	3.25	3080
C—H	Aliphatic	3.46	2890
CH_2	Methylene	3.4	2940
H—Cl		3.46	2886
CH_3	Methyl	3.5	2870
S—H		3.87	2585
H—Br		3.91	2559
P—H		4.26	2350
Si—H		4.42	2260
C≡N		4.45	2250
Se—H		4.40	2230
H—I		4.48	2231
C≡C		4.50	2220
C=N	Iso-	4.65	2150
C=O	In anhydrides	5.63	1775
		5.48	1825
C=O	In esters	5.48	1825
COO	Ester or carboxyl	5.8	1675
C=O	In acids	5.85	1710
C=O	In ketones	5.90	1695
$\overset{O}{\overset{\parallel}{\text{R-C}}}\ \text{N}$	Amide	6.1–6.3	1640–1560
C=C	Ethylenic	6.10	1640
C=N		6.10	1640
C=C	Aromatic	6.27	1595
C=C	Conjugated	6.30	1590
$—NO_2$	Nitro group	6.3	1565
COONa		6.5	1540
$(CH_2)_n$	Methylene	6.8, 13.8	1475, 775
$C=CH_2$		7.09, 10.15	1410, 985
CH_3	Methyl	7.22	1385
CF_3	Trifluoromethyl	7.09	1380
$C—CH_3$		7.3	1420

TABLE 6.2. *(Continued)*

Group	Type	Wavelength λ in Microns, 10^{-4} cm	Frequency ν in Reciprocal Centimeters, cm^{-1}
CH_2		8.0	1240
S, S=O, SO_3H, SO_4	Sulfonic	8.2, 9.4, 10.4	1225, 1070, 960
C—F		8.3	1200
C—O		8.5	1150
C—O—C	Ether	9.0	1060
C—N		9.0	1100
	Aromatic	6.2–6.5	1600
		6.6–6.7	1500
COH		9.4	1060
SO_3H	Sulfonic	9.8	1010
C—C		10.0	1000
SO_3H	Sulfonic	10.8	920
C—Cl	Chloride	13.3	750
$(CH_2)_n$		13.8	720

These sulfur derivatives have excellent alkaline compatibility and are excellent emulsifiers. They are, however, subject to attack by hard-water ions such as calcium or magnesium, with which they form insoluble salts. For this reason, they are usually formulated with a nonionic surfactant, a chelate, or a buffer, such as a phosphate, to protect them from metallic hardness formations.

The propyl, butyl, and amyl esters of both castor oil and oleic acid are commonly used as raw materials for sulfation.[73] Table 6.2 lists representative functional groups and their normal absorption frequencies.

Infrared curves showing functional structures and wavelength frequencies of various sulfonated oils and esters are shown in Figures 6.1 through 6.5.

The key infrared bands for the sulfonic groups are 8.1–8.2, 9.4, 10.6, and 13.8 μm. An absence of a peak at 6.5 μm would seem to indicate, in this particular sample, that the end COOH groups have not been saponified as would be the case in a disodium salt.

The two major products of the reaction of sulfuric acid with fatty compounds are sulfates and sulfonates.

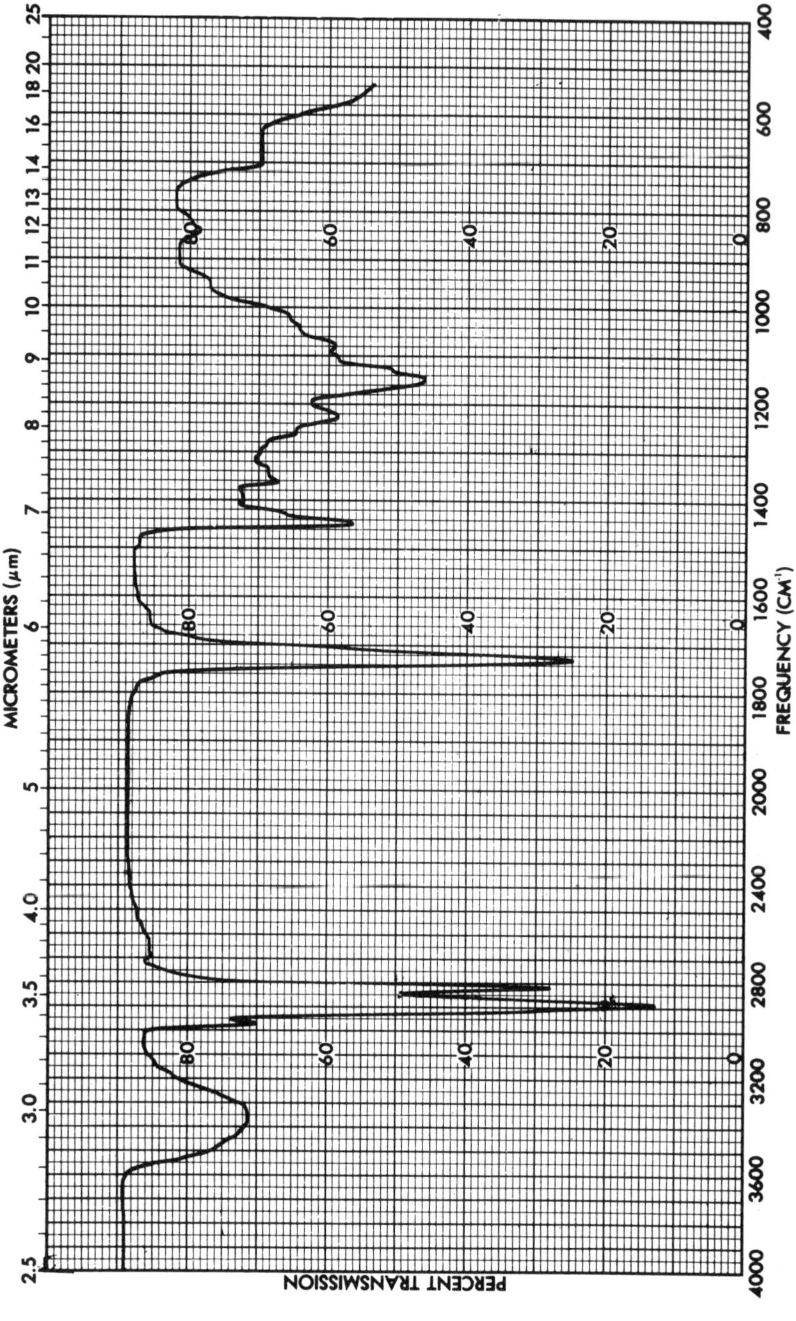

FIG. 6.1. Infrared spectrum of castor oil.

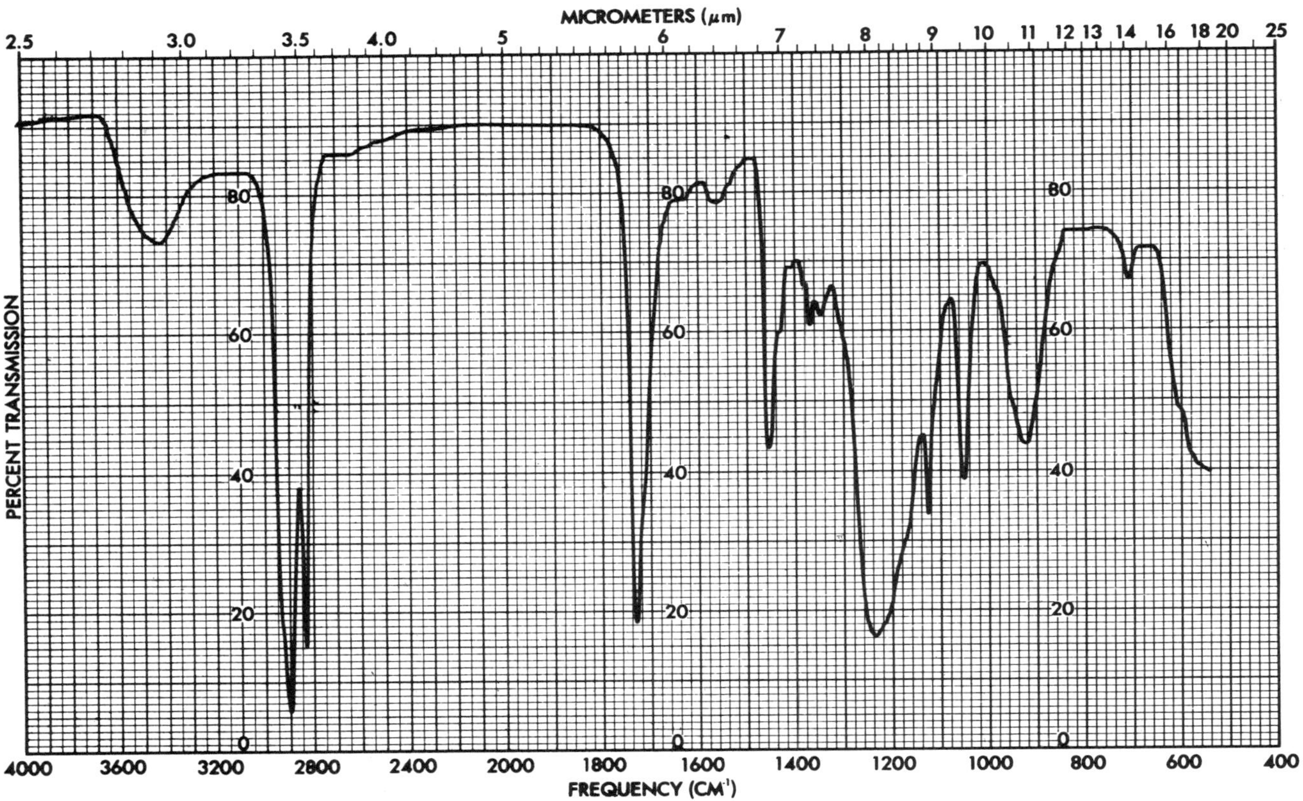

FIG. 6.2. Infrared spectrum of butyl castor oil.

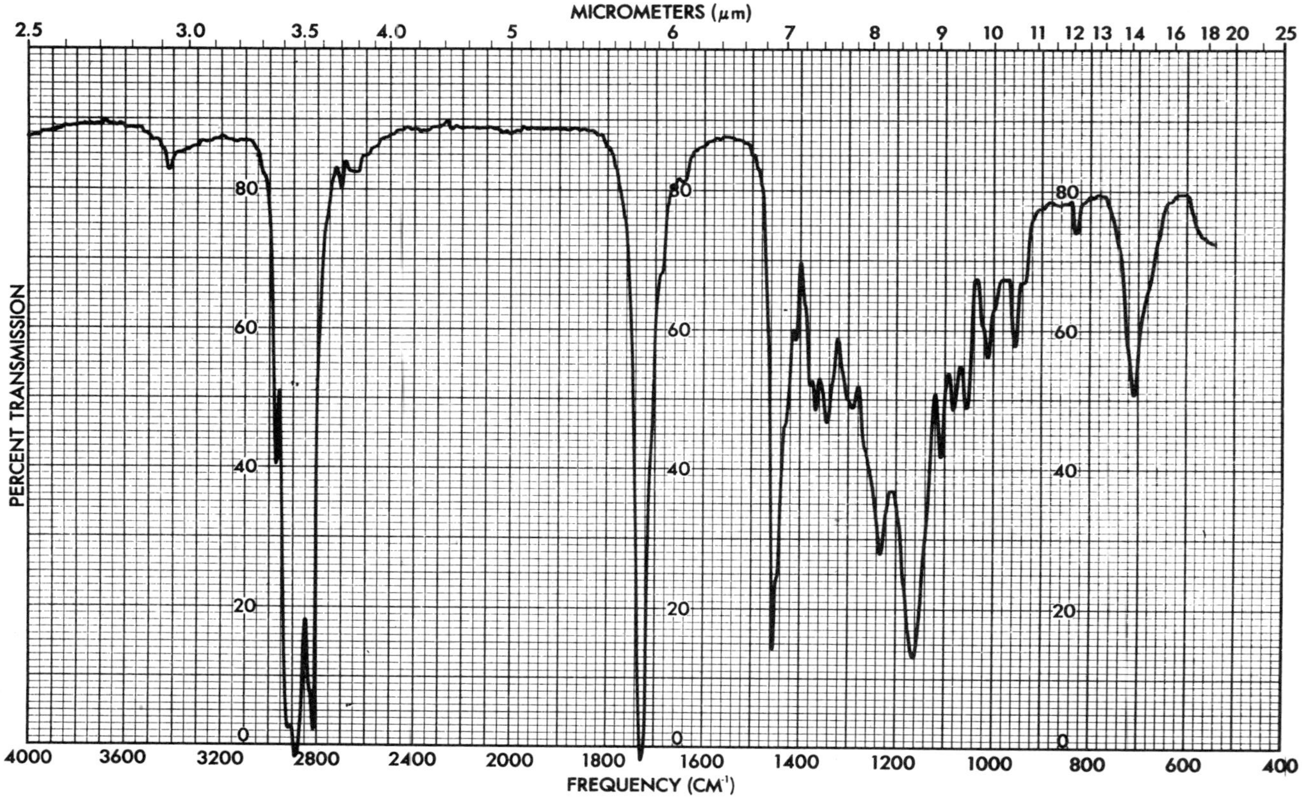

FIG. 6.3. Infrared spectrum of butyl oleate.

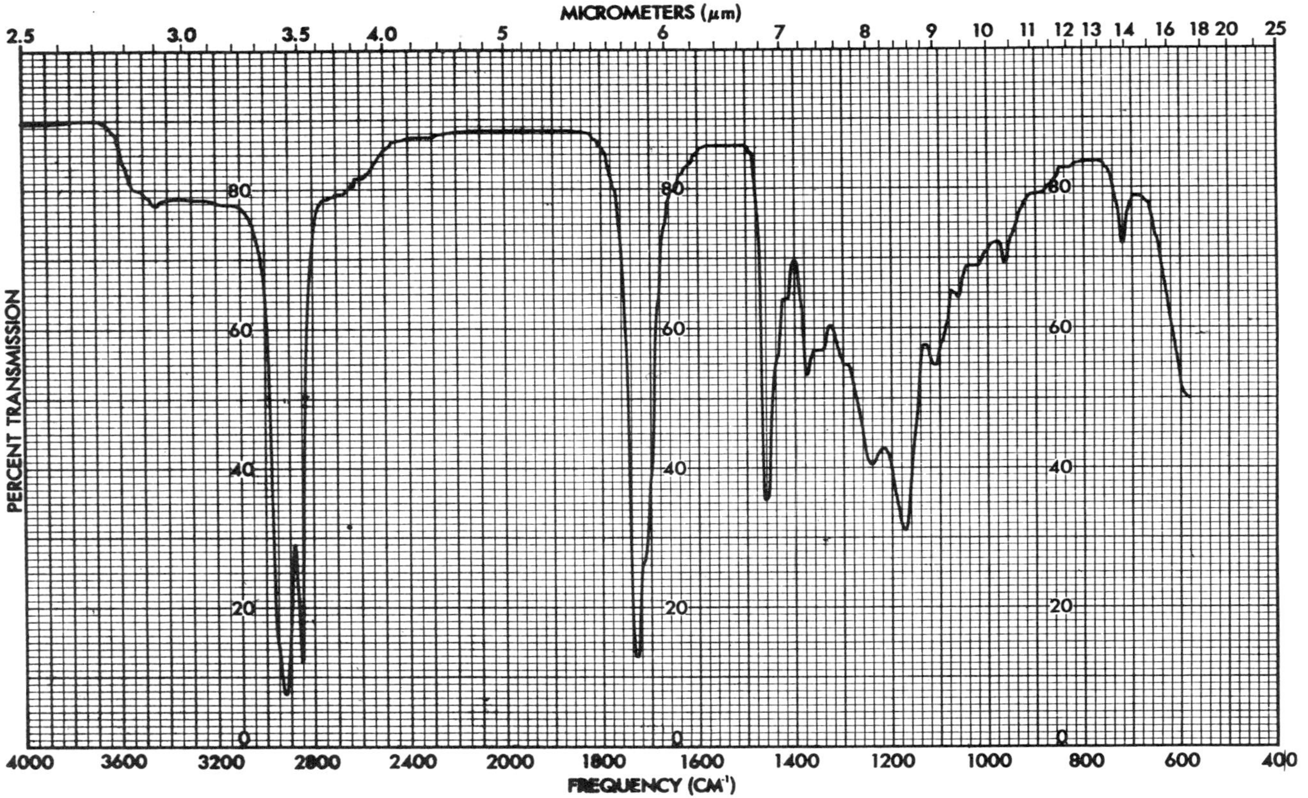

FIG. 6.4. Infrared spectrum of sulfated butyl oleate.

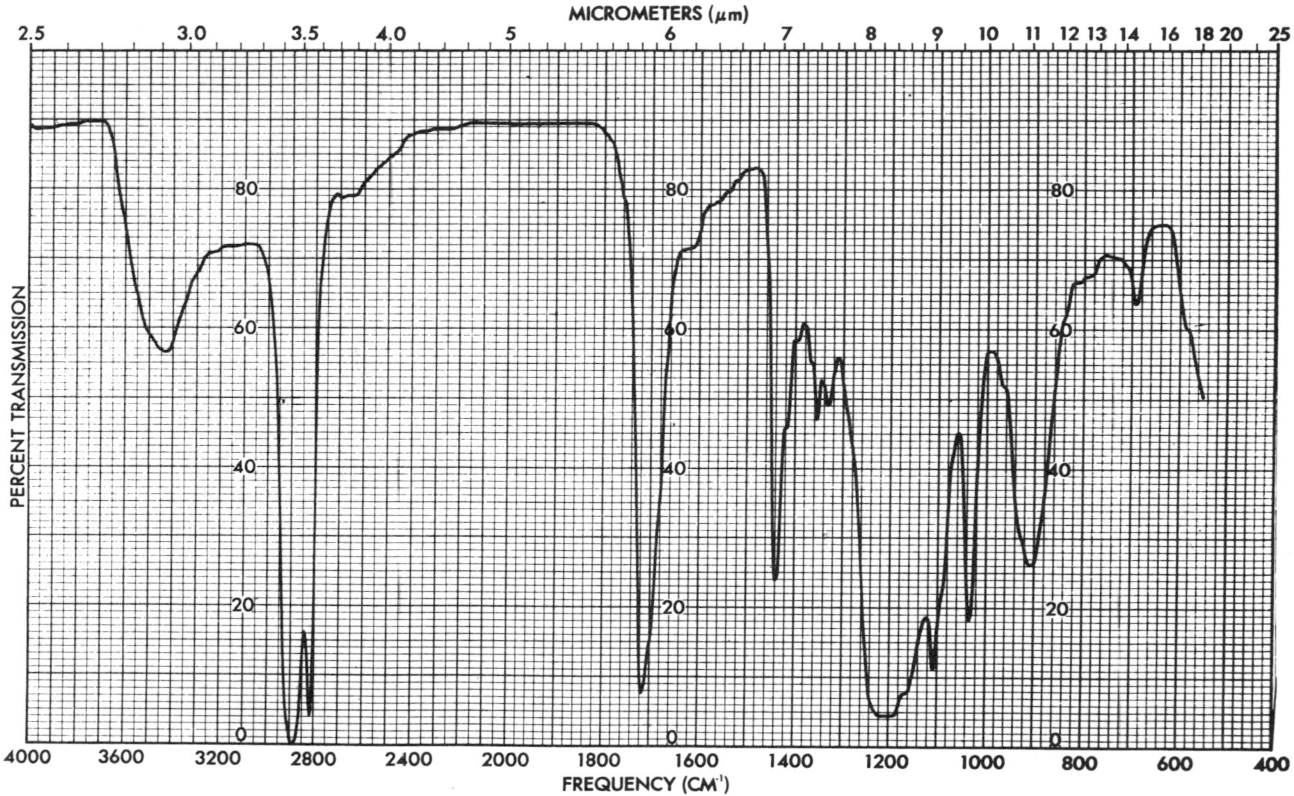

FIG. 6.5. Infrared spectrum of sulfated glycerol trioleate.

With experience and good reference samples, one may learn to recognize the difference between a sulfate and a sulfonate by observing the shift in peaks at 7.9, 9.1–9.3, 10.0, and 11.9–12.0 μm.

Sulfated butyl and propyl oleate have shown good commercial acceptance and are used most commonly as high-temperature dyeing assistants, fiber lubricants, and wetting agents.

A look at the structure of propyl alcohol shows a CH_2 chain, a terminal CH_3, and a hydroxyl group—infrared spectra confirm these functional groups:

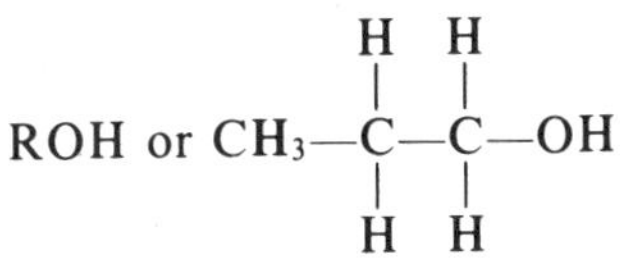

Propyl alcohol

When propyl alcohol is reacted with oleic acid, propyl oleate is formed:

$$ROH + RCOOH \rightarrow RCOOR + H_2O \text{ or}$$
$$CH_3CH_2\text{—}CH_2OH + CH_3(CH_2)_7\text{—}CH{=}CH\text{—}(CH_2)_7\text{—}COOH \rightarrow$$
$$CH_3CH_2\text{—}CH_2OOC\text{—}(CH_2)_7CH{=}CH(CH_2)_7CH_3$$

Propyl oleate

When sulfuric acid is reacted with this ester, the sulfuric radical attaches at a secondary position:

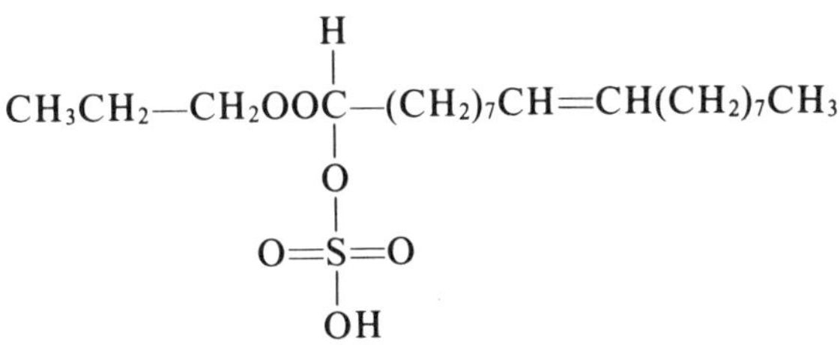

An interesting note on the spectra of the *N*-propyl oleate ester is that sulfonic groups (8.2, 8.6, and 10.4μm) can show up in the raw ester prior to sulfation. This is probably due to the presence of paratoluene sulfonic acid, used as a catalyst in the esterification reaction.

TABLE 6.3. Infrared Bands for Propyl Oleate Ester

Wavelength, μm	Group	Type
3.4	CH_2	Methylene
3.5	CH_3	Methyl
5.8	C=O	Ester
6.5	COOH	Free fatty acid
6.9	$(CH_2)_n$	Methylene
7.3	CH_3	Terminal
10.4	CH=CH	Unsaturation
	COOH	Carboxyl

Infrared bands for propyl oleate ester, sulfate propyl oleate ester, and sulfated oleic acid are listed in Tables 6.3 through 6.5. Infrared spectra of propyl oleate, sulfated propyl oleate, and sulfated oleic acid are shown in Figs. 6.6 through 6.8.

When oleic acid of the structure RCOOH or $CH_3(CH_2)_7{=}CH(CH_2)_7COOH$ is reacted with sulfuric acid (H_2SO_4), most peaks remain as they were in the spectra of the original compound, but with a major change in the 8–14-μm range.

TABLE 6.4. Infrared Bands for Sulfate Propyl Oleate Ester

Wavelength, μm	Group	Type
3.4	CH_2	Methylene
3.5	CH_3	Methyl
5.8	C=O	Ester
6.5	COO	Carboxyl
6.9	$(CH_2)_n$	Methylene
7.3	CH_3	Terminal
8.2	S=O	
8.6		Sulfonic
9.1		
10.0	SO	
10.4		Vibration
11.2		

TABLE 6.5. Infrared Bands for Sulfated Oleic Acid

Wavelength, μm	Group	Type
3.3	OH	COOH
3.4	CH_2	Methylene chain
3.5	CH_3	Methyl group
5.8	C=O	
6.8	$(CH_2)_n$	
8.0		Sulfonic groups
8.15		Characteristic of secondary sulfate
9.4		
10.6		
13.8	$(CH_2)_n$	

Butyl Oleate

A commercial procedure for the production of the sulfate ester of normal butyl or propyl oleate might be as follows:

Sulfated Ester: 1000-lb Batch[161]

1. Charge: Butyl oleate 490 lb
2. Cool: To 15°C
3. Add: Sulfuric acid, 98% 196 lb
   (add at a rate not to exceed 35°C; use cooling as necessary)
4. Hold: At 25–30°C for 6 hours
5. Add: Crushed ice 118 lb
   (agitate well)
6. Allow to stand until batch splits.
7. Draw off aqueous layer on bottom.
8. Discharge to neutralizing tank and discard.
9. Charge to acid mix:
   Caustic soda, 20% 196 lb
10. Hold temperature at 30–35°C.
11. When pH has stabilized, adjust to a pH of 5.0 + 0.2.
12. Allow to stand overnight.
13. Discard any aqueous layer that forms on bottom, and drum.

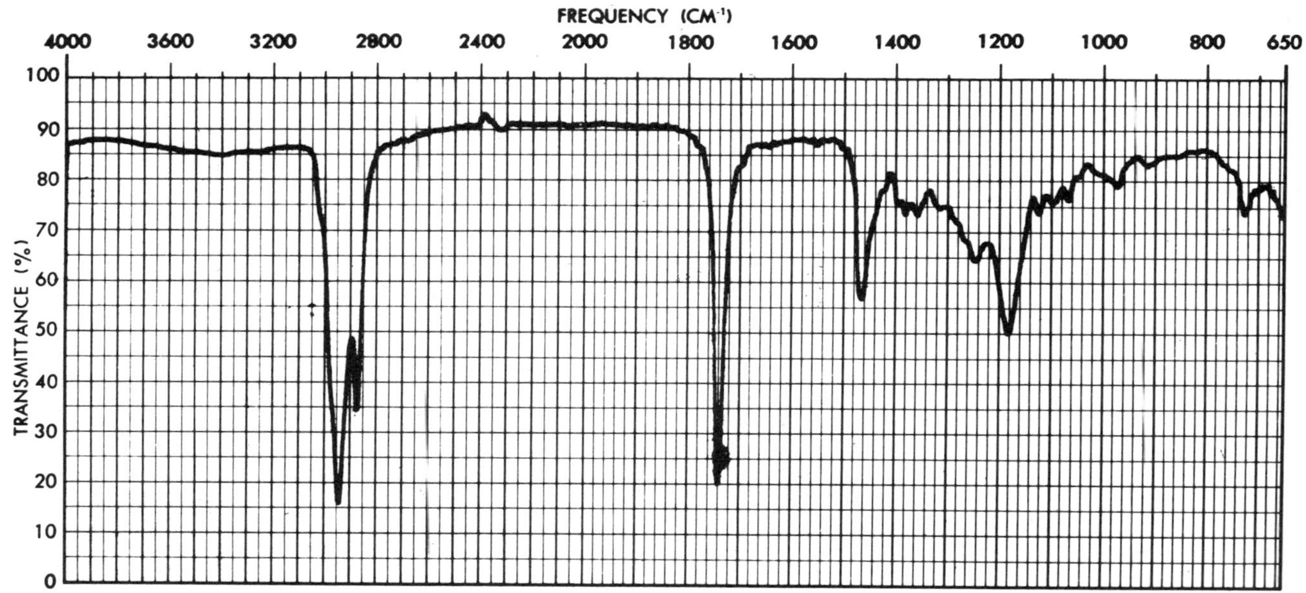

FIG. 6.6. Infrared spectrum of propyl oleate.

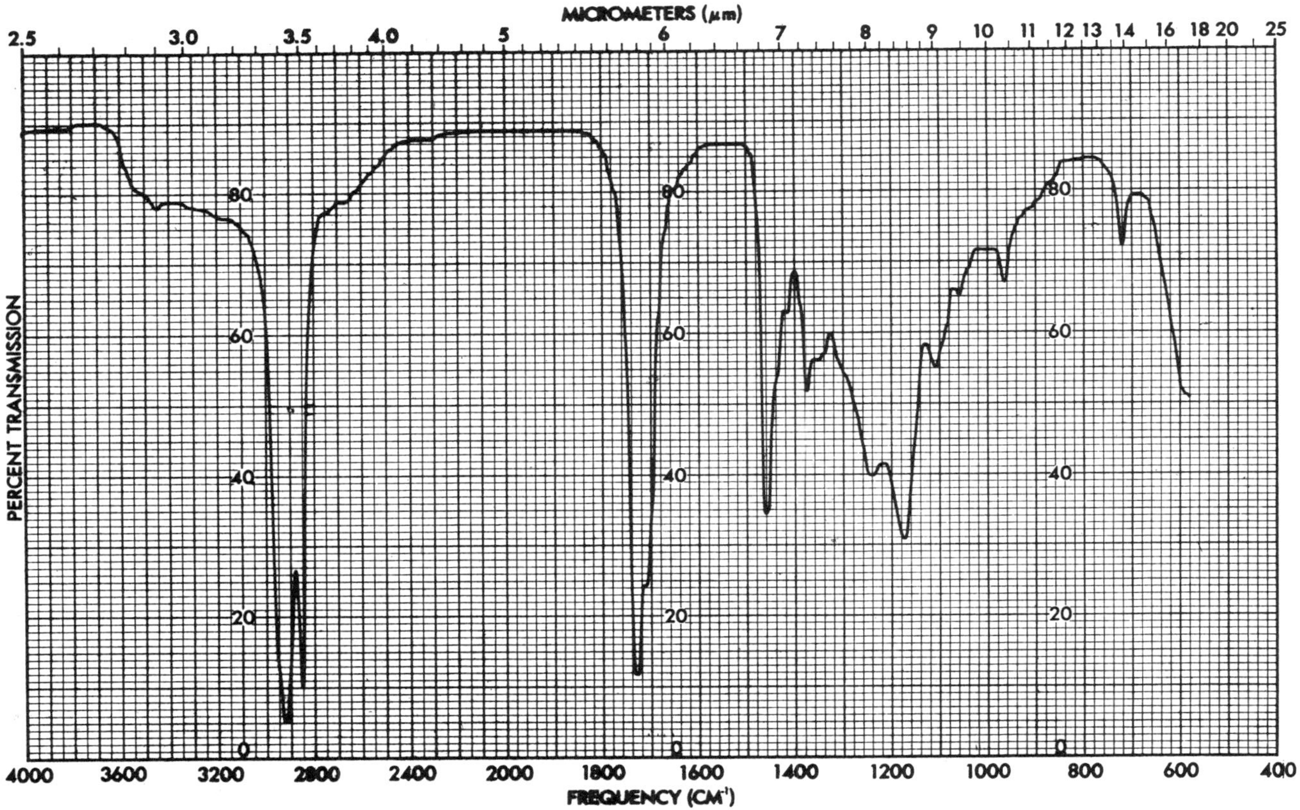

FIG. 6.7. Infrared spectrum of sulfated propyl oleate.

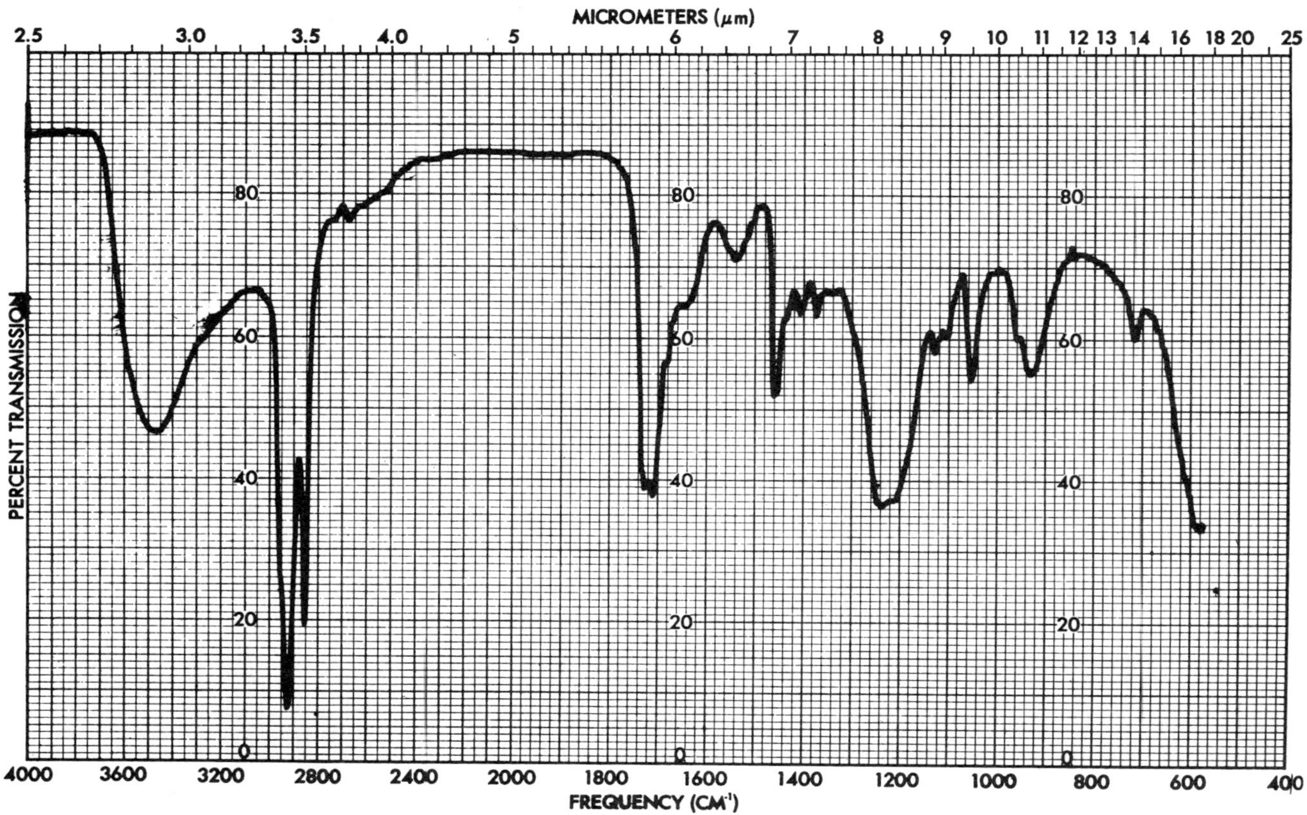

FIG. 6.8. Infrared spectrum of sulfated oleic acid.

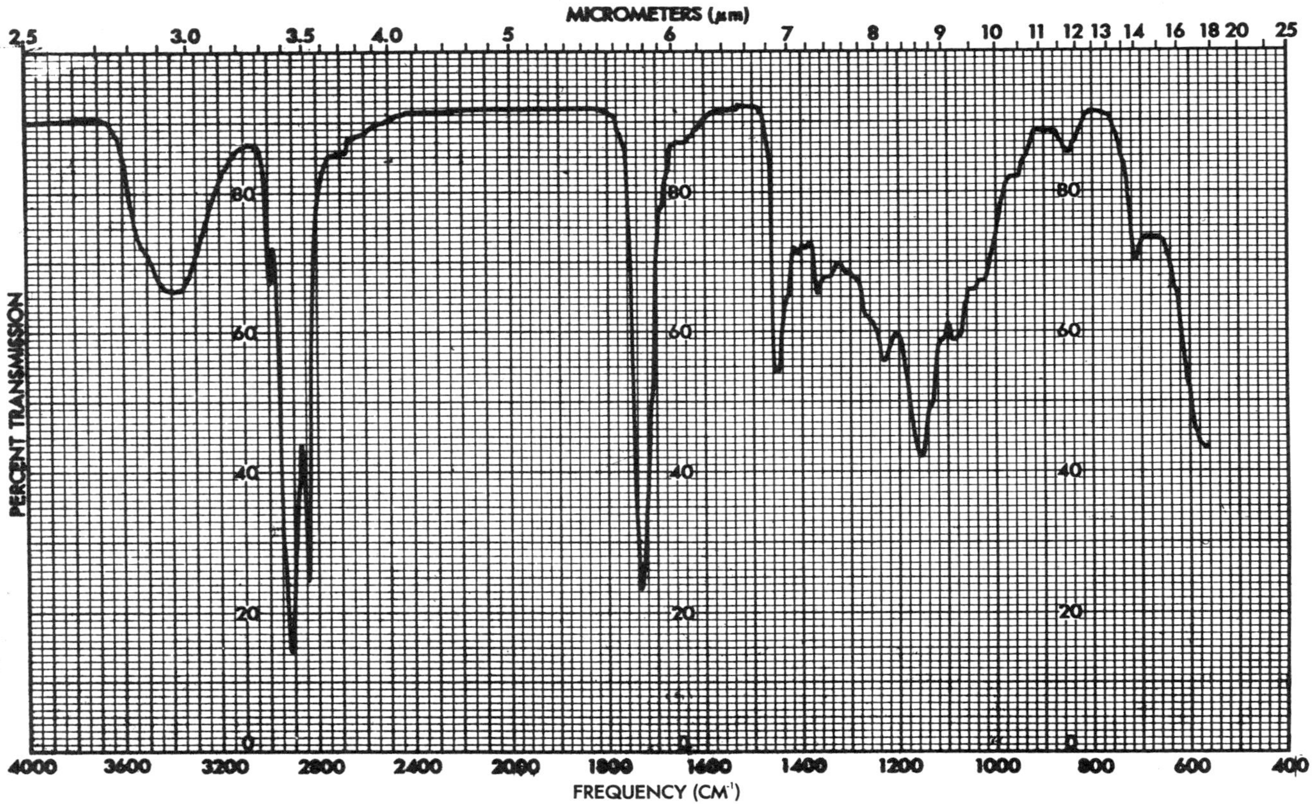

FIG. 6.9. Infrared spectrum of sulfated castor oil.

Sulfated Castor Oil: 1000-lb Batch (Courtesy of Caschem, Inc.)

1. Charge: Castor oil 650 lb
2. Add: Sulfuric acid, 98% 152 lb
   (at a temperature not above 30–35°C)
3. Continue agitation for 4 hours
4. While the oil is being sulfated, prepare neutralizing bath consisting of:

Water	700 lb
Flake caustic	90 lb

5. When sulfation is complete, adjust cooling on neutralization kettle and add acid at such a rate that temperature is not allowed to rise above 55°C.
6. Adjust pH with caustic soda, 20%, until pH is neutral to methyl red end point.
7. Allow to stand overnight.
8. Draw off aqueous layer on bottom.
9. Adjust pH to desired range.
10. Adjust water content to desired level.

Shown in Fig. 6.9 is infrared spectra of sulfated castor oil.

Sulfonation of Tall Oil (Courtesy of Arizona Chemical Company)[162]

1. Charge: Tall oil fatty acid 1000 parts
   Sulfuric acid, 98% 250 parts
   (at a temperature below 30°C, while maintaining viscosity for good agitation)
2. Acid charge should be 25% on the weight of the tall oil, and added as rapidly as possible to allow removal of the exotherm while batch is still fluid. This will shorten the total addition time for the acid. Acid charging time should be approximately 1 hour.
3. Continue reaction time after acid charge for 1 additional hour, maintaining a temperature below 30°C.

4. Prepare neutralizing bath consisting of a mixture of 30° baume sodium hydroxide and water, (percent sodium hydroxide determined by titration) approximately equivalent to twice the volume of the sulfonated oil. Use good agitation.
5. Alkali concentration of bath is then adjusted to 0.5% as KOH.
6. Temperature is raised to 65°C and batch is allowed to settle overnight, during which time the temperature is allowed to drop to 40°C. Let batch split.
7. The aqueous layer is drawn off, and the oil is again washed with a volume equivalent to that removed, using concentrated brine. At a temperature of 65°C batch is again allowed to stand overnight. The next day the salt solution is drawn off.
8. Water and combined alkali are then determined and the necessary adjustments made to bring the oil within specifications.

Sulfonation of Oleyl Alcohol (U.S. Patent 2,114,042)[74]

1. Charge to lead-lined sulfonator, with agitation and cooling on:

Commercial-grade oleyl Alcohol	1170 lb
Inert solvent (ethyl alcohol)	200 lb
Sulfuric acid, 98%	750 lb

2. Charge acid at rate to maintain temperature at 100°F, and when sulfonation step is complete, agitate for 30 additional minutes.
3. While acid and alcohol are reacting, prepare neutralization bath consisting of:

Water	500 gal
Sodium hydroxide, 50%	760 lb

4. Add acid–alcohol mixture at such a rate that temperature does not exceed 90–100°F. Control pH with addition of additional alkali as necessary so that pH of mix stays constant at 7.5–8.0.
5. Maintain final pH at 7.5–8.0.

Sulfated Lauryl Alcohol (U.S. Patent 2,114,042)[74]

1. Per 1000-lb batch, charge to glass-lined reactor capable of being sparged with nitrogen and scrub tank available to neutralize free hydrochloric acid:

Lauryl alcohol	200 lb

2. Cool to 20°C or lower
   Add: Chlorosulfonic acid 43 lb
   (at a rate to maintain a temperature of 30°C)
   (Reaction is highly exothermic and evolves hydrochloric acid as a by-product. Therefore, scrubbing of acid fumes must be provided for.)
3. When exotherm has subsided and temperature has stabilized once again at 20°C, add chlorosulfonic acid, 80 lb, maintaining a temperature of 20°C.
4. When all chlorosulfonic has been charged, continue agitation for 1 hour.
5. Prepare neutralization bath of:

Crushed ice	100 lb
Water	500 lb
Caustic, 50%	100 lb

6. Charge acid mix to neutralization bath with full cooling as rapidly as possible.

Sulfonated Amine Condensate (U.S. Patent 1,932,180)[75]

$$RCONHC_2H_4OSO_3Na$$

1. Charge to reactor equipped with receiver–collector:

Coconut fatty acid	420 lb
Monoethanolamine	134 lb

2. Condense under nitrogen sparge with distillate collected in receiver until acid number determined in alcohol is 5 or below.

3. Water evolved and excess monoethanolamine yield approximately 40 lb of distillate.
4. On cooling, melting point of base compound is 60–65°C. When base has reached 70°C or begins to thicken, add:

Inert solvent (e.g., alcohol)	30 gal

5. Cool to lowest possible temperature at which mix is still drummable.
6. Charge to lead-lined sulfonator:

Sulfuric acid, 98%	750 lb

7. Slowly add, with vigorous agitation, the coconut monoamide, 750 lb.
8. Maintain a temperature of 60–65°C, allowing wax to melt and react.
9. When all wax has entered acid bath, allow reaction to continue for 1 hour.
10. While mix is reacting, prepare neutralization bath of:

Water	500 gal
Caustic soda, 50%	816 lb

11. When sulfonation is complete, add acid base to alkali bath with cooling and agitation. When temperature has stabilized and base is neutralized, allow air to escape from mix, and drum.

Infrared spectra formed at various stages during the sulfonation and neutralization procedure of a coconut monoethanolamide are shown in Figures 6.10 through 6.12.

Sulfated Tallow: 1000-lb Batch

1. Charge: Sulfation-grade tallow 400 lb
   (to reactor)
2. Heat to 50°C or until completely melted.
3. With agitator on, and cooling water on, add the acid at a rate that will allow temperature to hold at approximately 40°C. Cooling will need to be adjusted so that tallow will not solidify around coils if sulfonator has internal cooler.
4. Continue sulfation for 30 minutes after all acid has been added.

5. While sulfation is proceeding, prepare the neutralizing bath by charging:

Water	200 lb
Caustic soda, 50%	50 lb

6. When sulfation is complete, transfer the tallow–acid mix as rapidly as possible into the neutralizing bath, holding temperature to 55°C.
7. Neutralize to methyl red end point, and allow to stand overnight.
8. Draw off aqueous layer, and recharge sufficient water to adjust solids.
9. Adjust pH 8.5–9.0 with caustic soda, 20%, and draw at a temperature sufficient that material thickens in container upon drumming.

The infrared spectrum of sulfated tallow is shown in Fig. 6.13.

Sulfonation for Oleic Acid-Mineral Oil Combination: 1000-lb Batch[161]

1. Charge: Oleic acid 180 parts
   White oil 360 parts
2. Adjust temperature to 20°C
3. Add: Sulfuric acid, 98% 75 parts
   (at a rate that temperature holds at 20°C)
4. Addition time for acid should not exceed 1 hour.
5. Acid mixture is then washed with cold water at 10–15°C.
6. Wash water used should be approximately equal to the combined weight of the oleic acid and sulfuric acid.
7. Allow batch to split. (Three layers will be formed.) The bottom layer is the aqueous layer and should be drained. The middle layer consists essentially of sulfonated acids and should be drawn off and neutralized to a methyl orange end point using 50% sodium hydroxide. (Do not overneutralize as excess alkali will yield a product that has poor water solubility. During the washing and neutralization, temperature should be maintained below 30°C.) The upper layer will contain the white oil and any unsulfonated oleic acid and may be used for the next batch.

Infrared spectrum for the sulfated oleic acid–mineral oil combination is shown in Fig. 6.14.

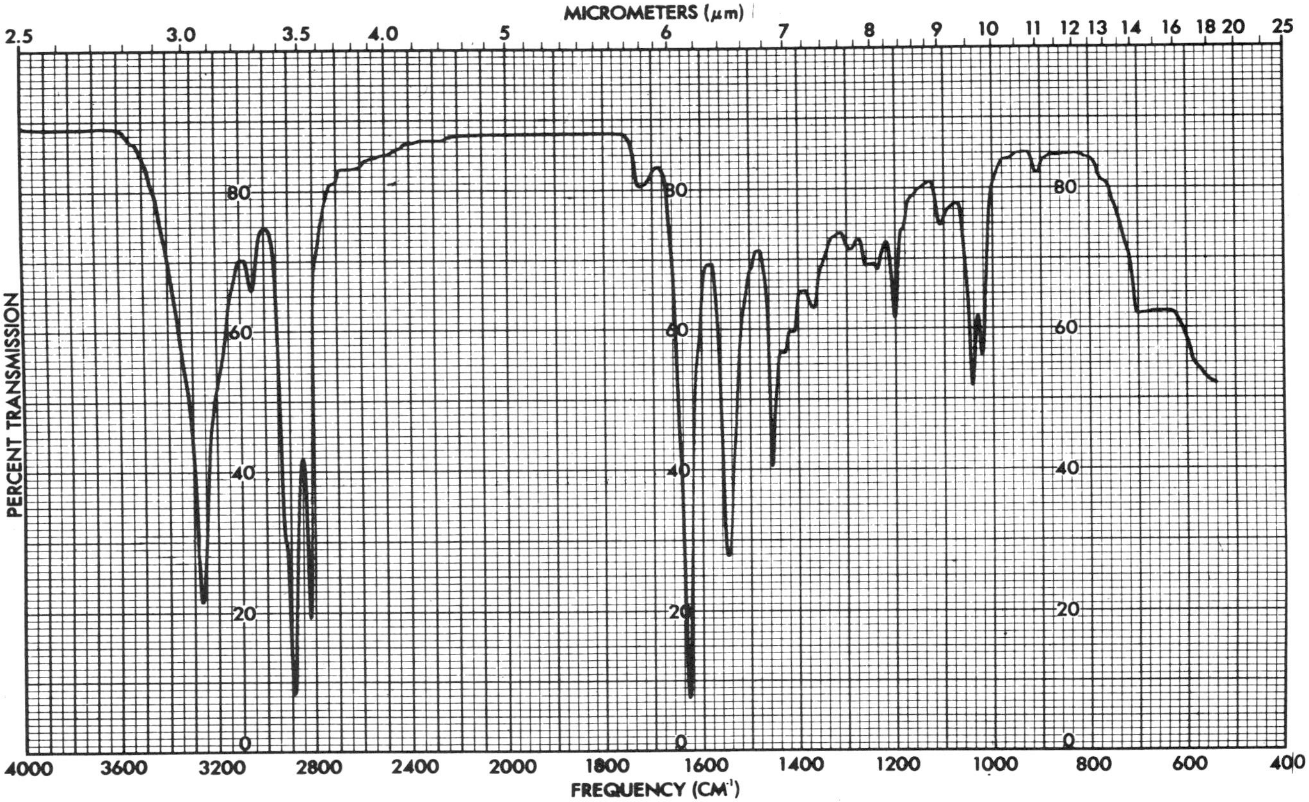

FIG. 6.10. Infrared spectrum of coconut monoethanolamide.

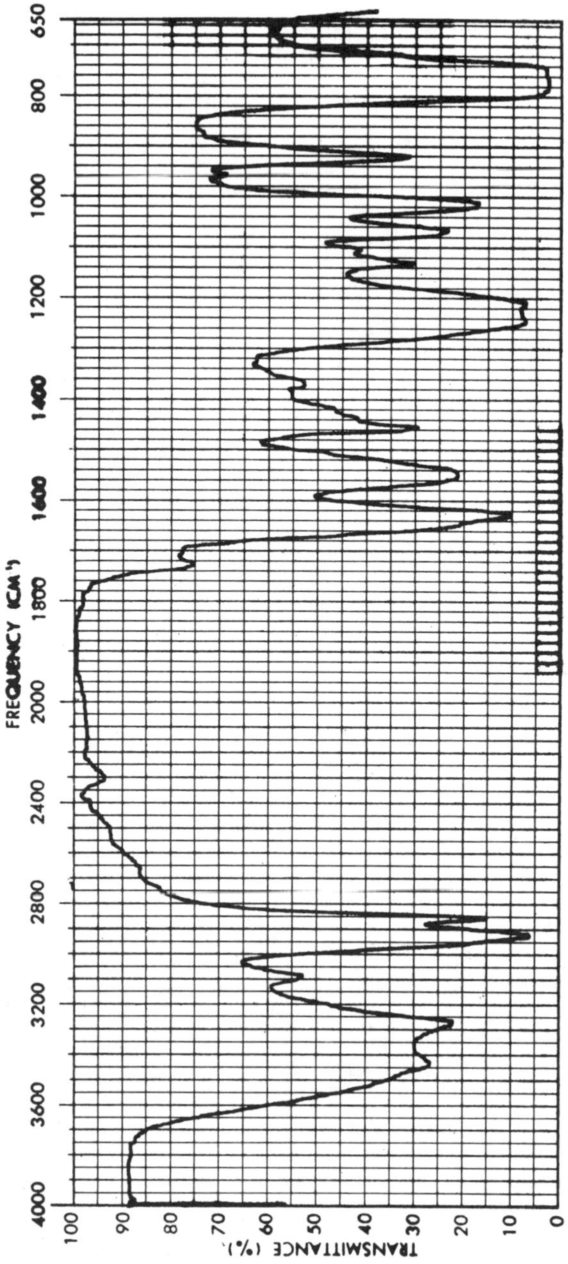

FIG. 6.11. Infrared spectrum of coconut monoethanolamide sulfonate.

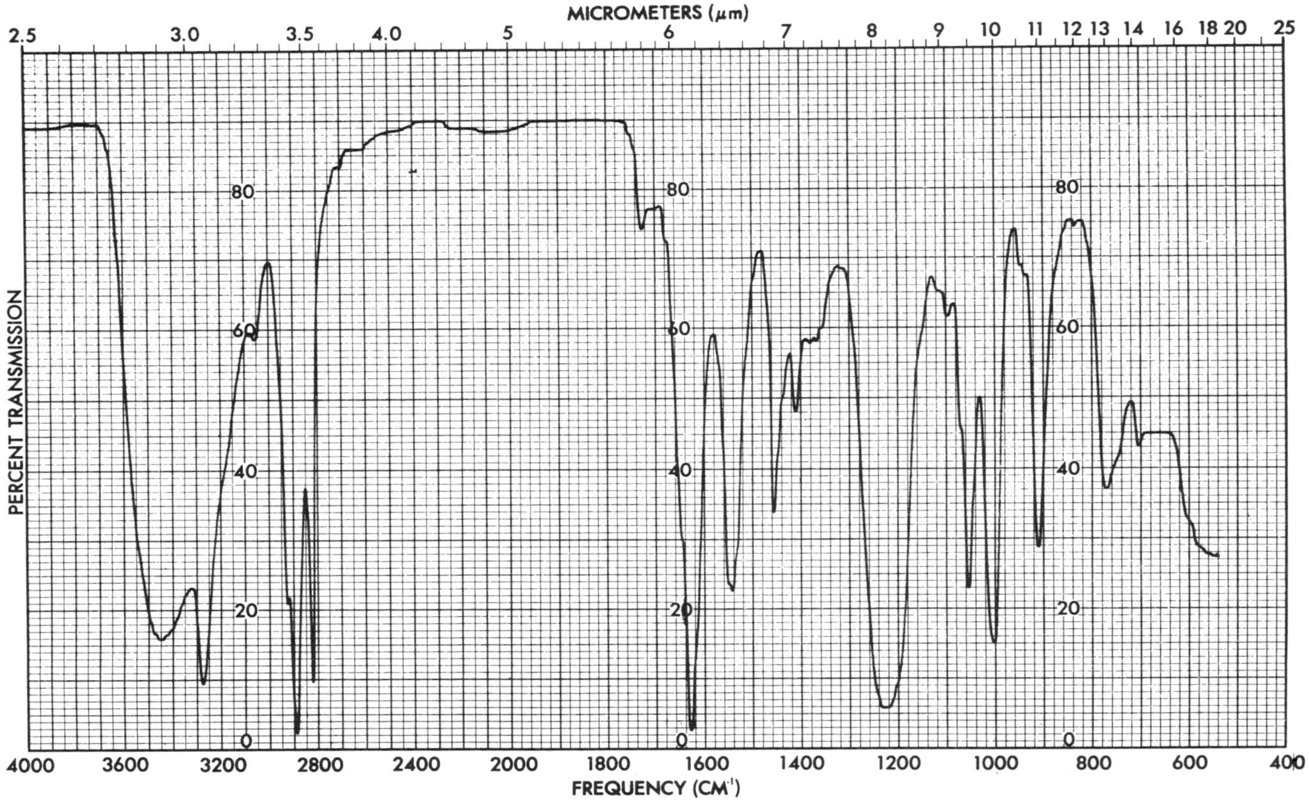

FIG. 6.12. Infrared spectrum of coconut monoethanolamide sulfonate–sodium salt.

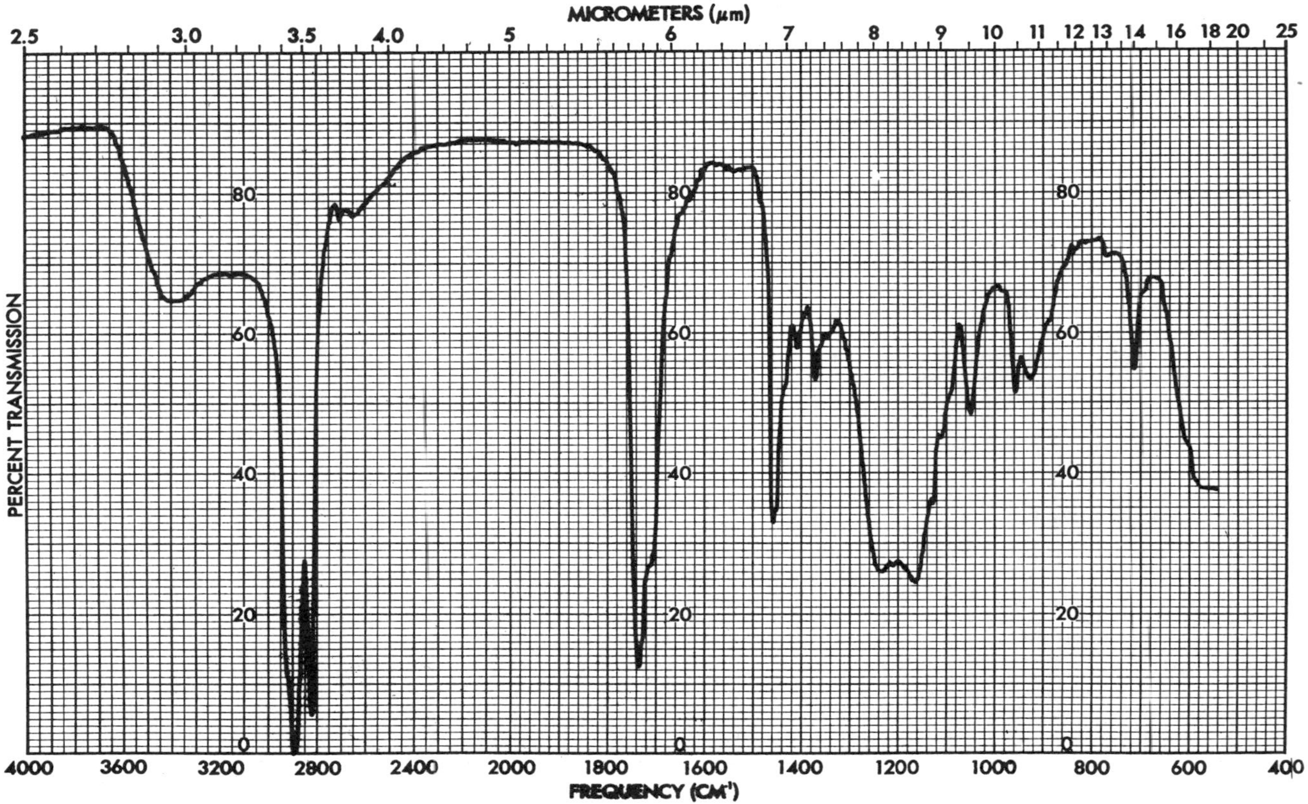

FIG. 6.13. Infrared spectrum of sulfated tallow.

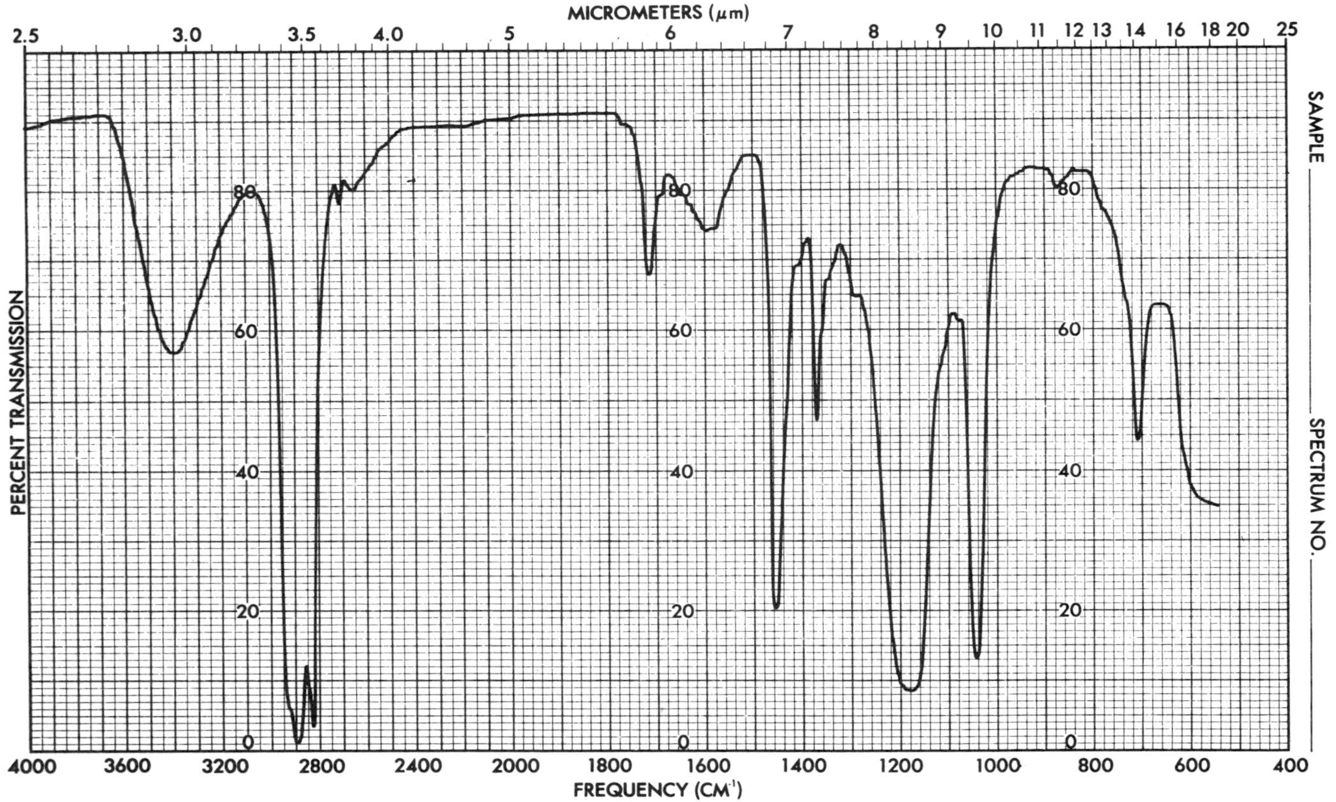

FIG. 6.14. Infrared spectrum of sulfated oleic acid–mineral oil.

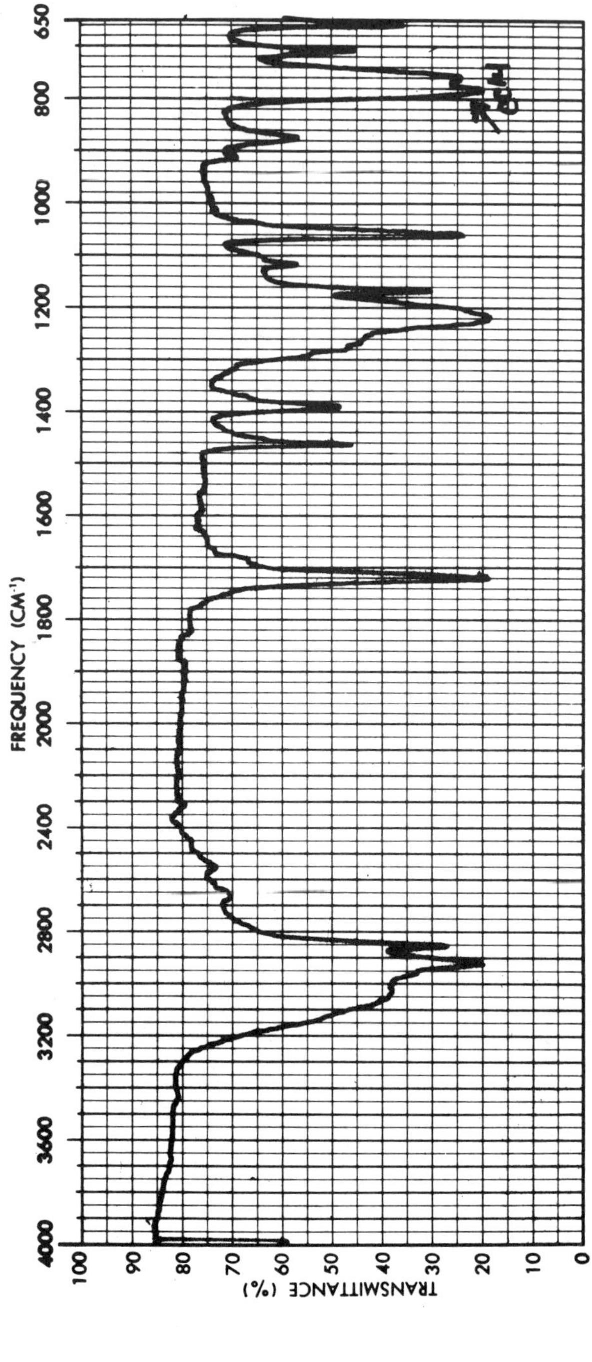

FIG. 6.15. Infrared spectrum of α-sulfa ester.

α-Sulfo Esters of Fatty Acids

These esters are covered in great detail by the following patents: U.S. Patents 2,806,044 (Weil, 1957), 2,844,606 (Bistline, 1958), 3,128,294 (Stirton, 1964), 3,219,584 (Stirton, 1965), 2,915,473 (Stirton, 1959), and 2,867,586 (Weil, 1959). These products are of interest because of their extreme biodegradability and potential economics. The U.S. Department of Agriculture, Eastern Utilization Research Center in Philadelphia, has excellent literature available on these products, including a bibliography of patents and published articles. The structure for α-sulfo esters is represented by:

$$RCH—(SO_3Na)CO_2R'$$

An infrared spectrum is shown in Fig. 6.15.

Aromatic Sulfonates

Alkylbenzene Sulfonates

All the previous sulfur derivatives have been of linear or aliphatic structure. There are, however, aromatic compounds that are reacted with SO_3H groups and whose functions and use are similar to those of the linear compounds. Typical of these are the alkylbenzene sulfonates, which are covered under U.S. Patent 2,133,282 and are discussed in detail in references such as Schwartz[76] and Trotman.[77] Alkylbenzene sulfonates are well known to the trade and are widely used as rapid wetting agents, mild dye levelers, and mild scouring agents, and their amine salts are useful as solvent and oil emulsifiers. Alkylbenzene sulfonates, although exhibiting excellent wetting properties, are hampered by their rewetting characteristics. They have extremely high foaming properties and are difficult to defoam. As they are one of the more inexpensive surfactant bases, they are found both by themselves and in combination with other products. At the present time, there are two forms of alkylbenzene sulfonates currently available to the marketplace. One has the structure as represented in structure **1** (below) and is known as linear alkylate sulfonate (LAS). The other, structure **2,** which is designated as "hard" or branched chain alkylate sulfonate, is known as ABS. The linear form is considered to be the more biodegradable, as reported by R. L. Huddleston and A. M. Nielsen.[78]

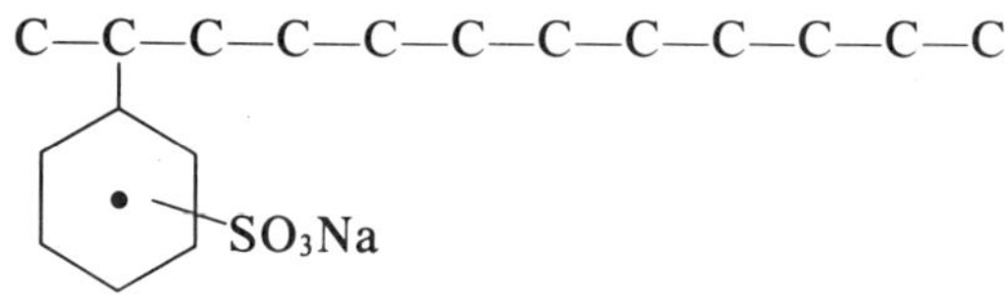

LAS
1

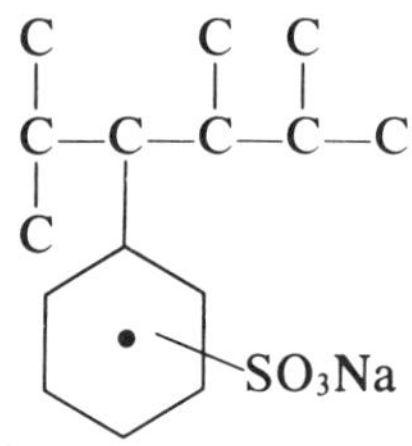

ABS
2

Infrared spectra of the potassium, triethanolamine, and isopropylamine salts are shown in Figs. 6.16 through 6.20.

The use of amine salts is covered in U.S. Patent 2,673,215, and U.S. Patent 3,030,410 covers the use of calcium salts as an emulsifier.

Sulfonated Naphthalene Formaldehyde Condensates

A second aromatic compound that finds wide use is naphthalene condensed with formaldehyde and then reacted with SO_3H. It is covered in detail by Schwartz.[79] These types of products are useful as wetting and dispersing agents and when properly synthesized, are relatively low foamers. They find their greatest use as dye dispersants for dispersed dyes. A theoretical structure is:

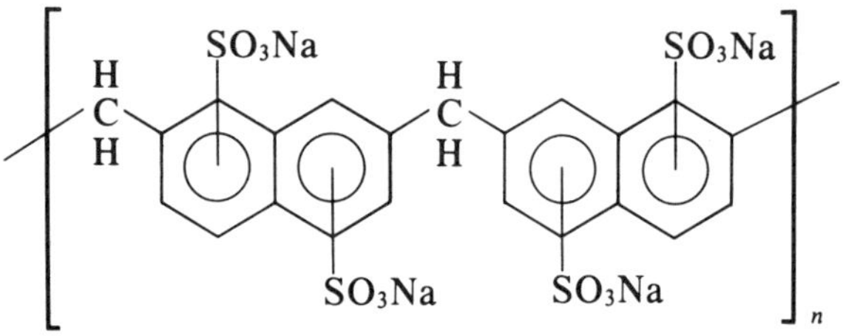

An infrared spectrum for this compound is shown in Fig. 6.21.

Sulfonates of Diaryl Compounds

A third type of sulfonate, which is characteristic of aryl sulfonates, is that characterized by diphenyl oxide. Substituted diphenyl sulfonates are covered by U.S. Patent 2,081,876 to E. I. du Pont and U.S. Patents 2,202,686 and 2,320,846 to Hercules Powder Company, du Pont also has an additional patent, 2,315,951, in this field.

A typical structure for diaryl sulfonate is:

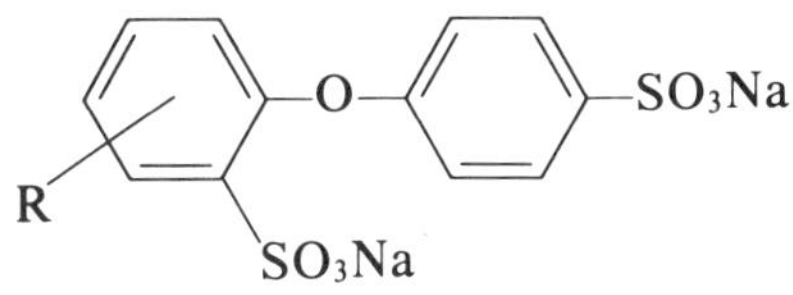

Diphenyl oxide sulfonate

An infrared spectrum for this compound is shown in Fig. 6.22.

Sulfates of Alkylphenols

A fourth type of aromatic sulfonate is that represented by U.S. Patents 1,987,228; 2,008,017; 2,008,032; 2,075,018; and 2,143,759. French Patent 847,549 also refers to these types of structures. An excellent discussion on procedures for sulfation of ethylene oxide alkylphenol derivatives is presented in technical service bulletin no. 10 by Jefferson Chemical Company, Inc.[80] The structure for this type of compound is:

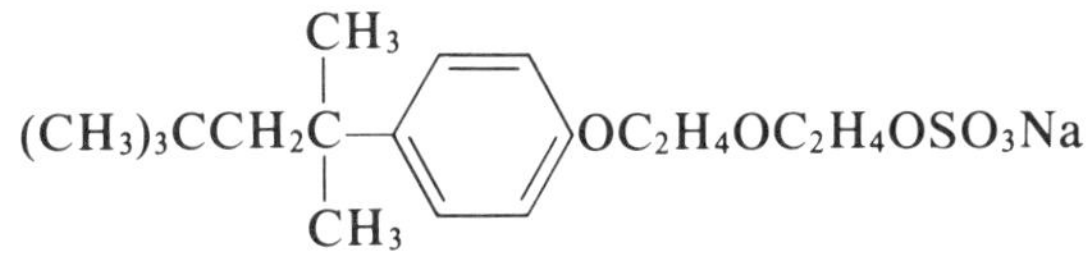

Alkyl phenol ethoxylate sulfate[81]

Infrared spectra showing the vibrational difference between an alkylphenol ethoxylate sulfate and a linear alcohol ethoxylate are shown in Figs. 6.23 and 6.24.

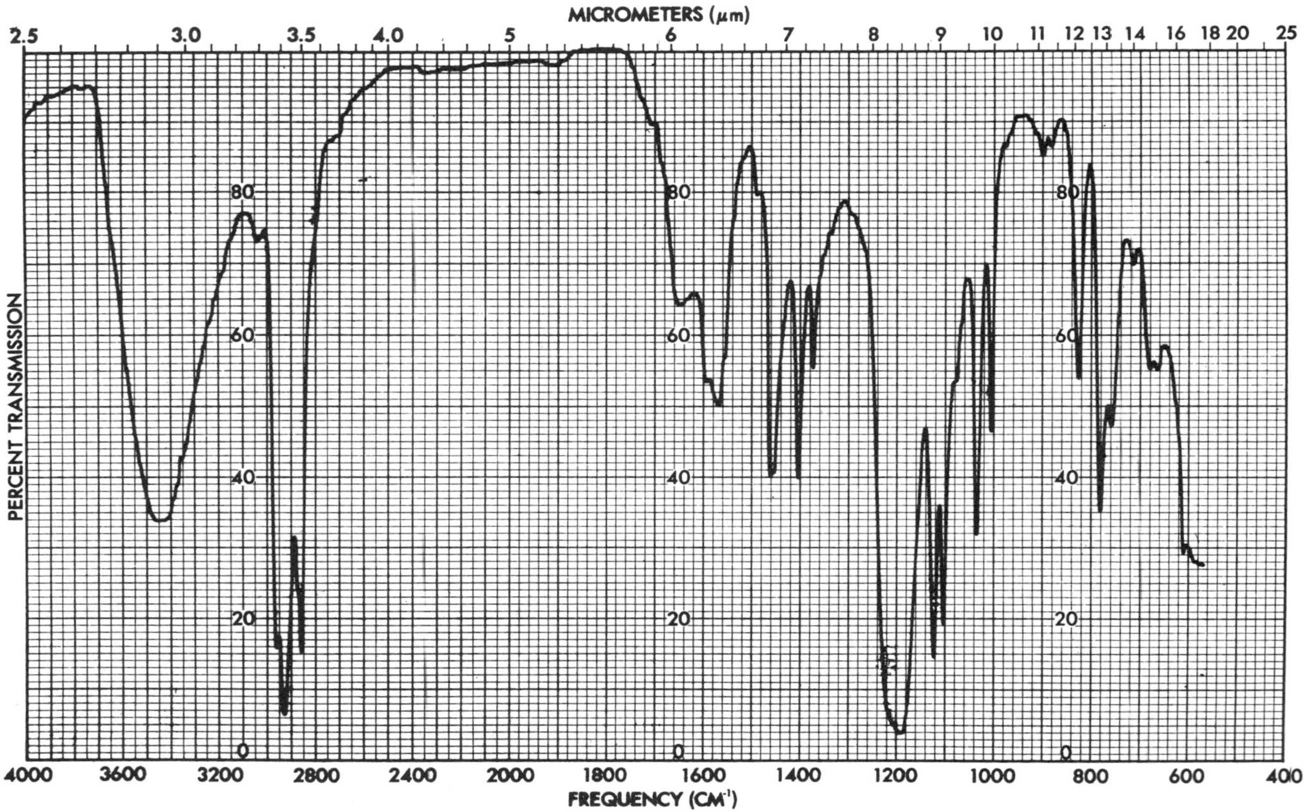

FIG. 6.16. Infrared spectrum of DDBSA–potassium salt.

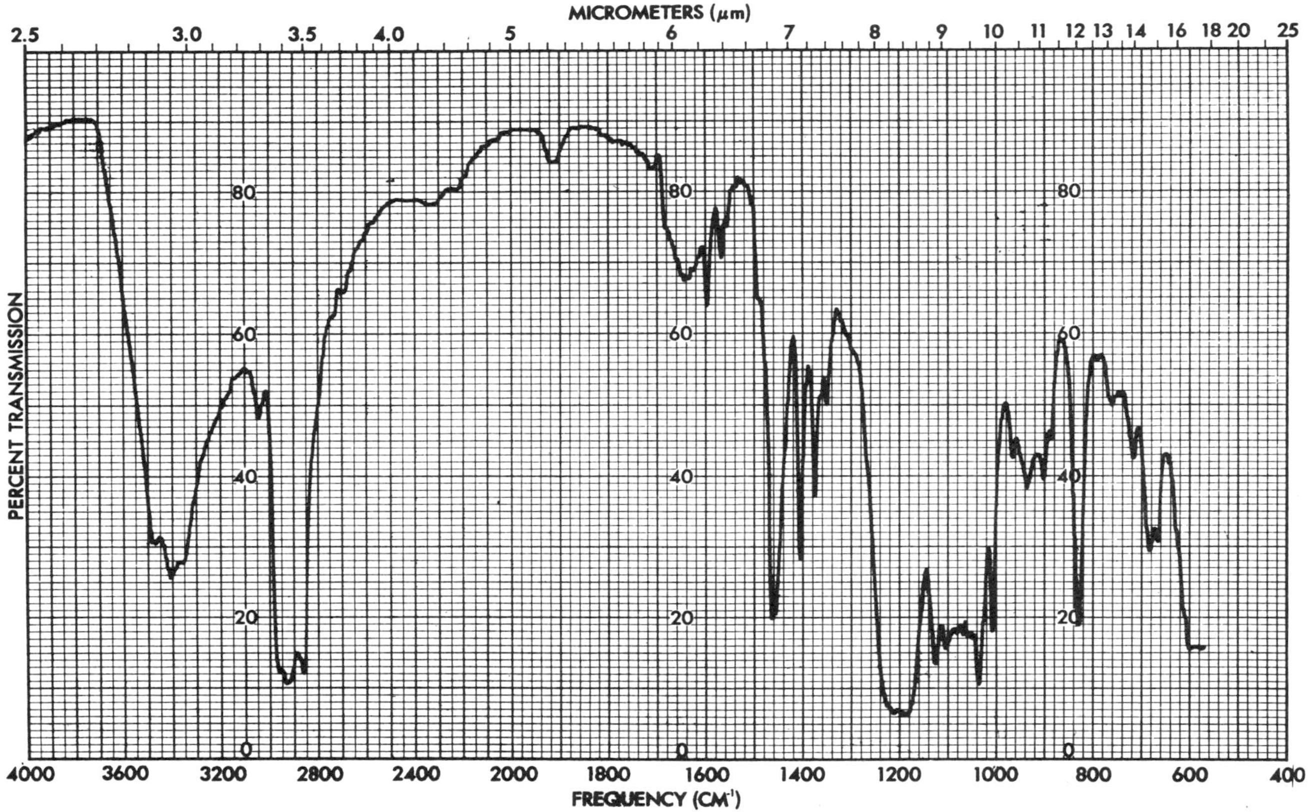

FIG. 6.17. Infrared spectrum of DDBSA–triethanolamine salt.

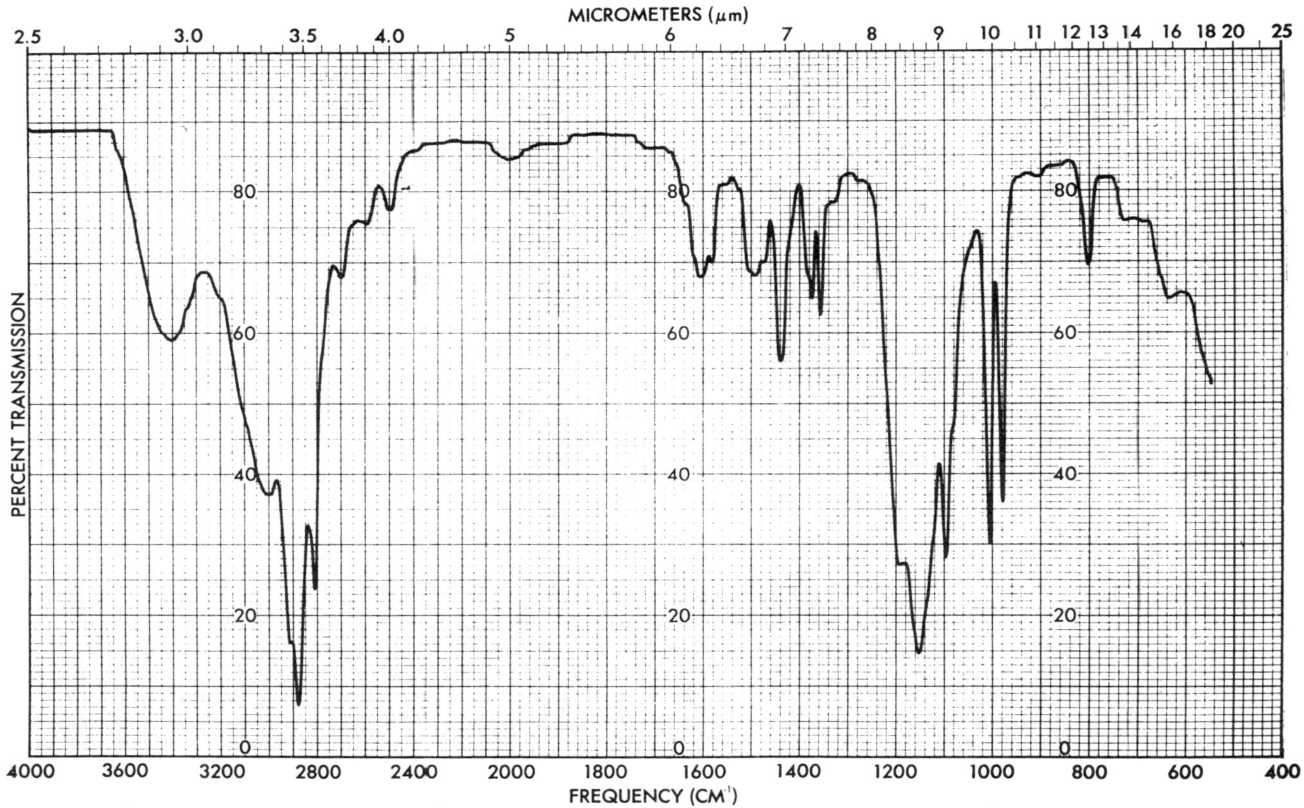

FIG. 6.18. Infrared spectrum of DDBSA–isopropylamine salt.

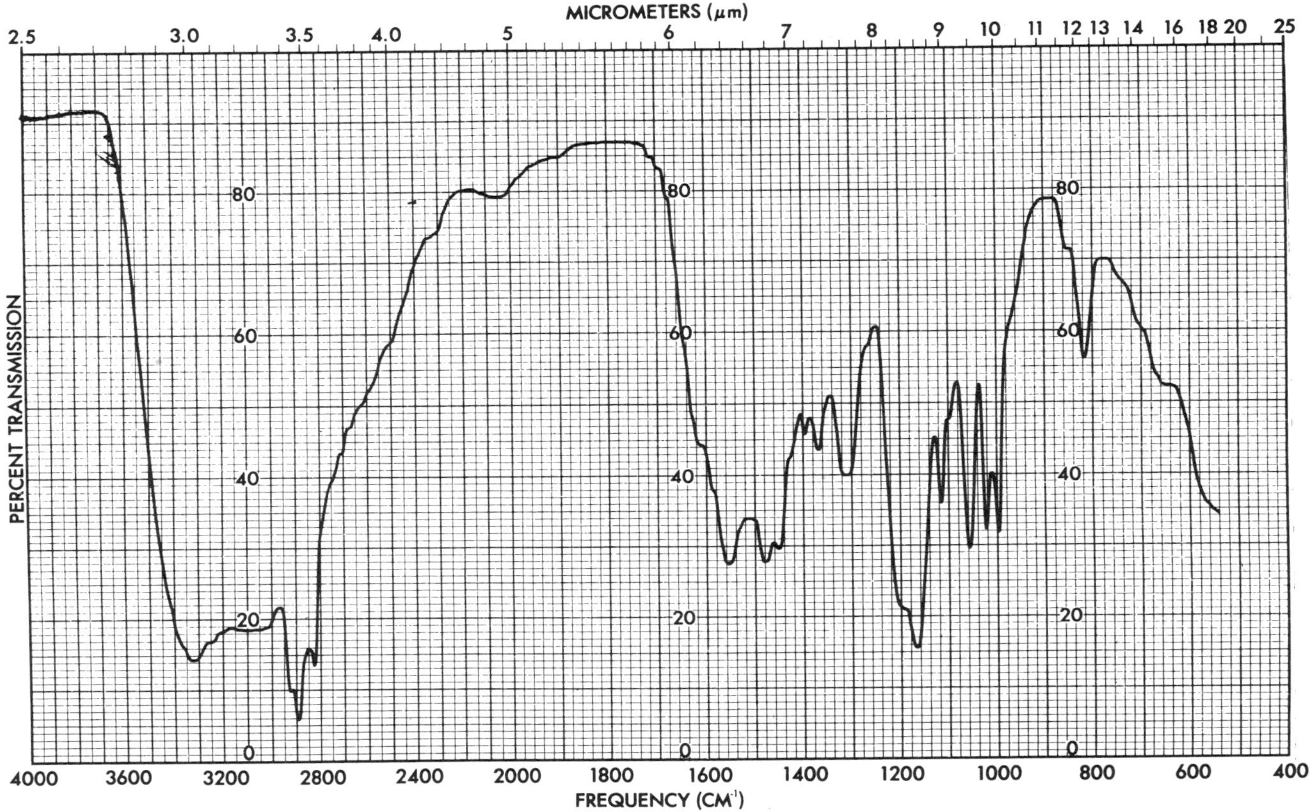

FIG. 6.19. Infrared spectrum of DDBSA–monoethanolamine salt.

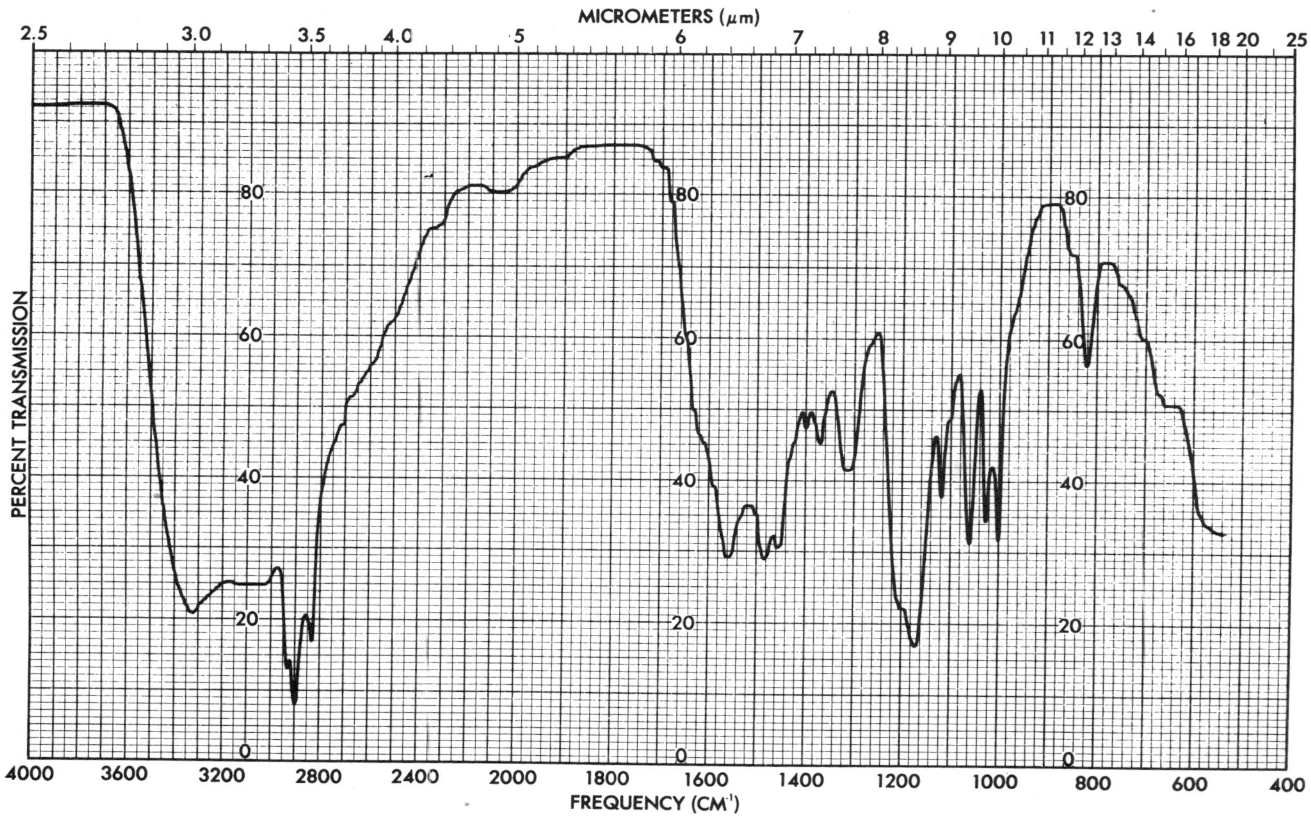

FIG. 6.20. Infrared spectrum of DDBSA–morpholine salt.

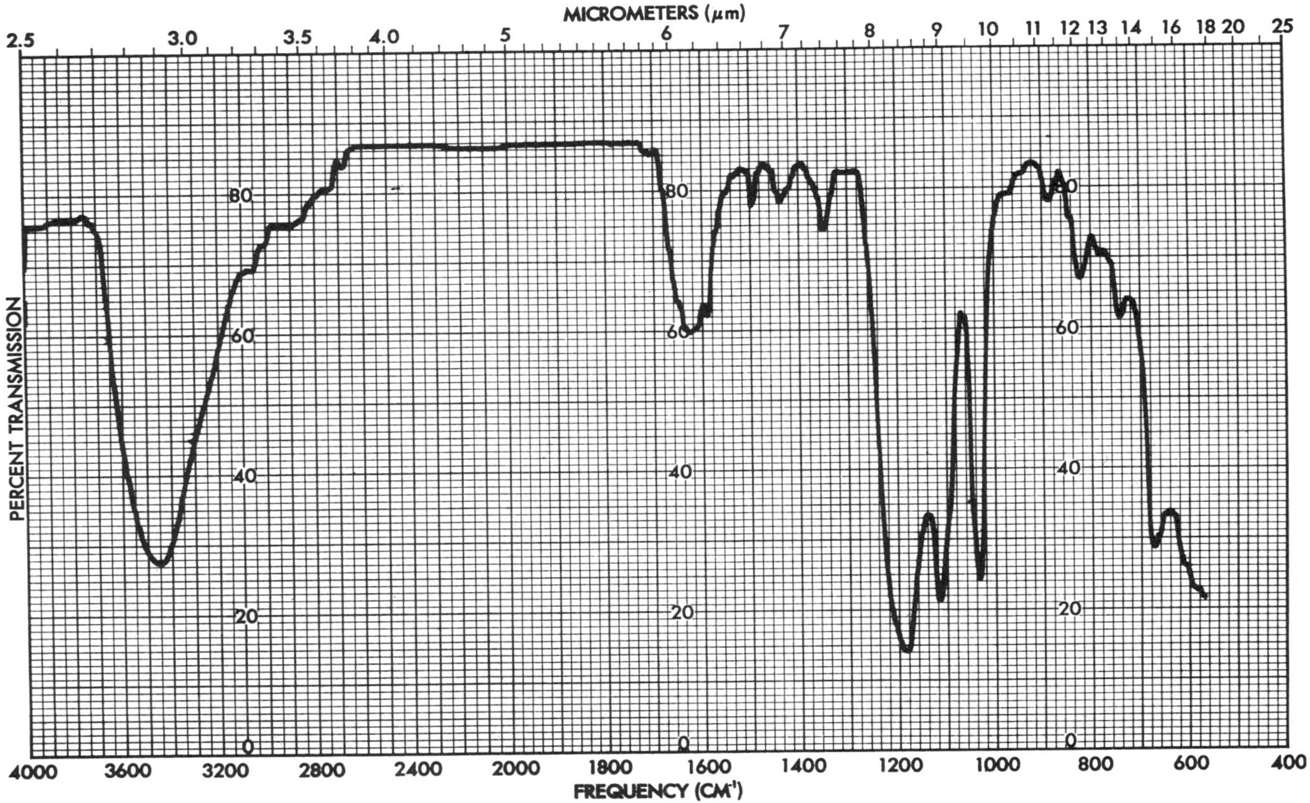

FIG. 6.21. Infrared spectrum of sulfonated naphthalene formaldehyde condensate.

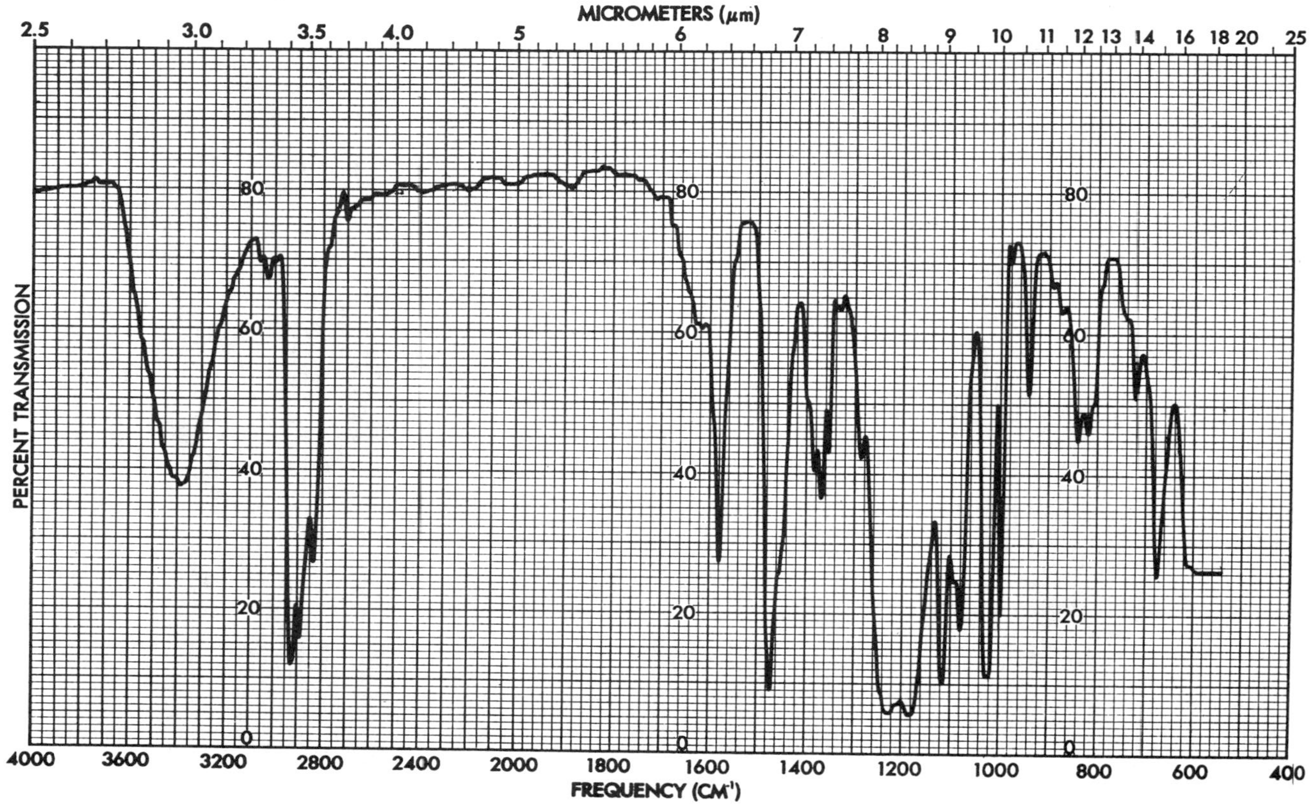

FIG. 6.22. Infrared spectrum of diaryl sulfonate.

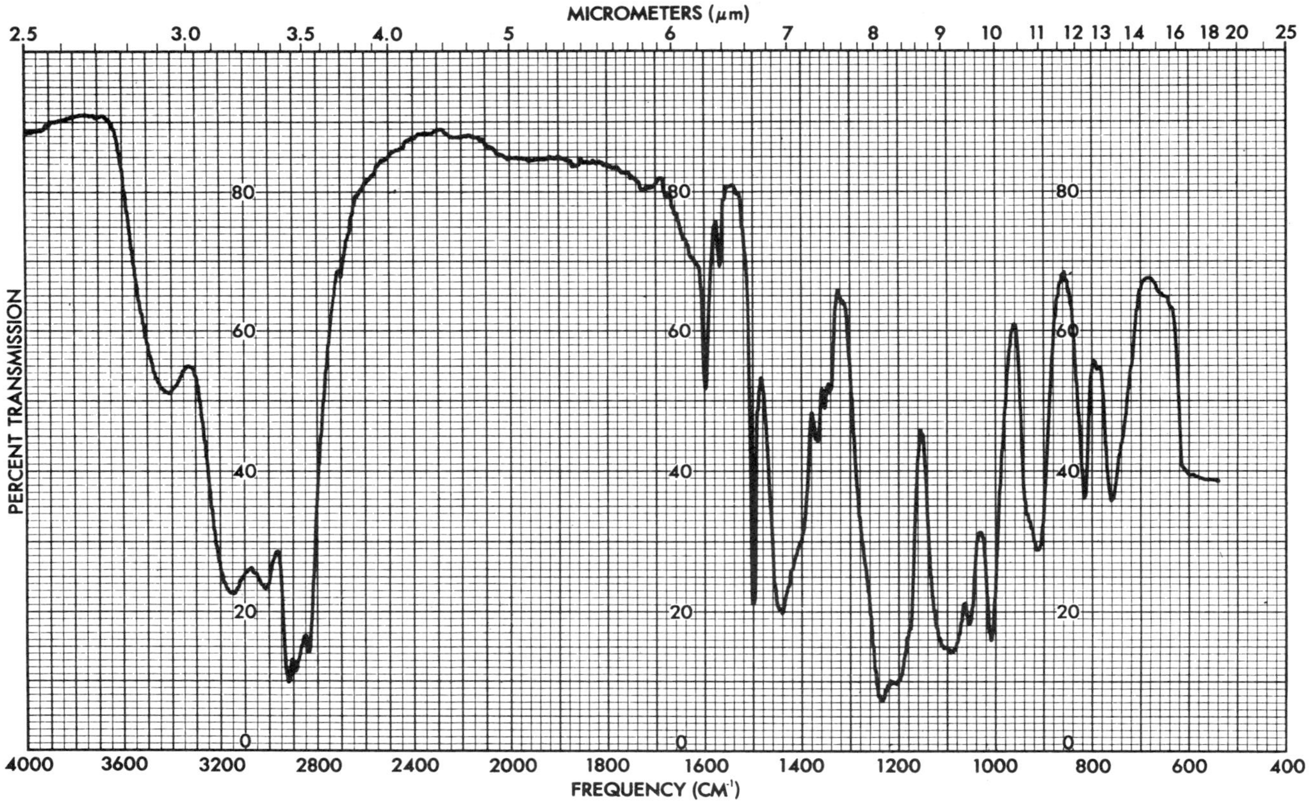

FIG. 6.23. Infrared spectrum of alkylphenol ethoxylate sulfate.

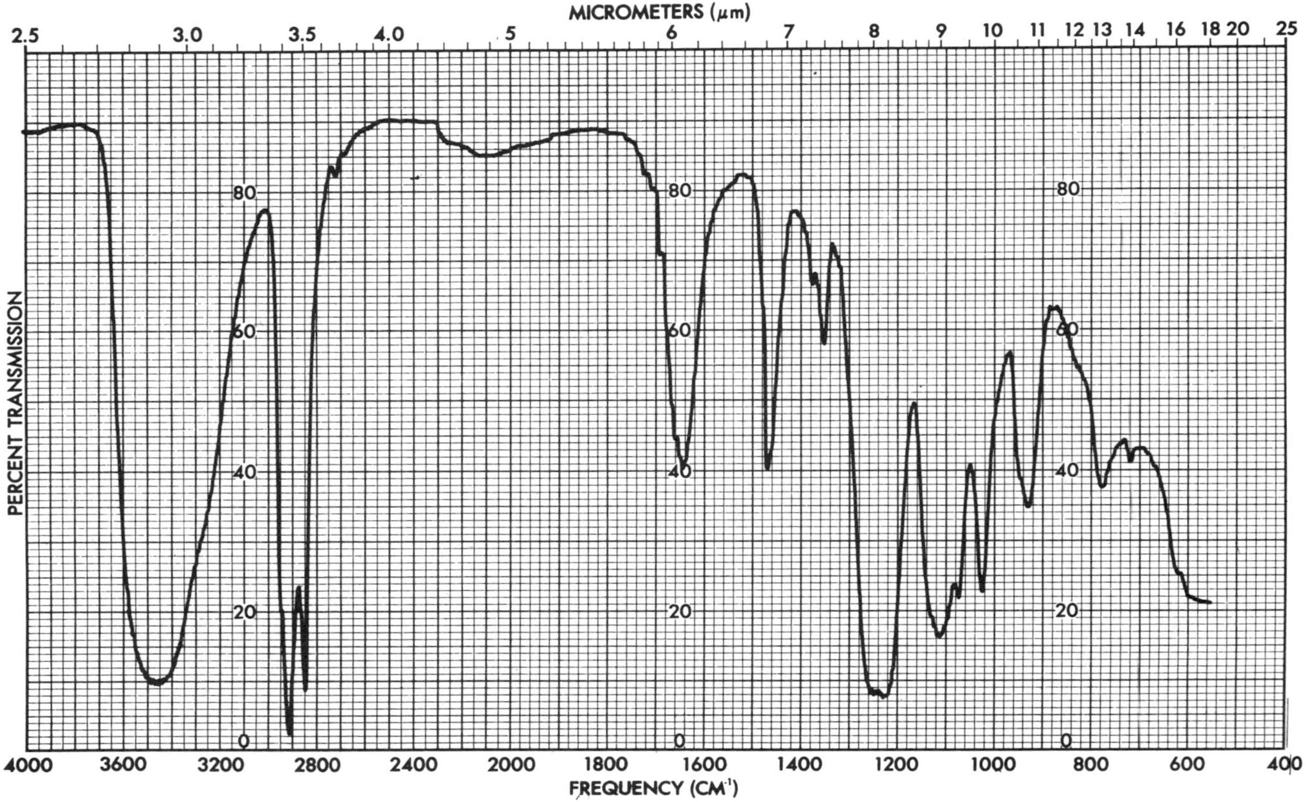

FIG. 6.24. Infrared spectrum of tridecyl (EO) sulfate.

Sulfonates of Succinate Esters

A novel class of surfactants known for their high foam characteristics and their rapid wetting, coupled with their excellent rewetting properties, are esters of alkane dicarboxy sulfonates. These are covered under U.S. Patents 2,441,341, 2,028,091, and 2,176,423; British Patent 446,568; and French Patent 776,495. Additional references in the patent literature are U.S. Patents 2,316,234 to Allied Chemical and 2,181,087 and 2,265,944 to American Cyanamid.

Sulfosuccinate derivatives are extremely fast wetters at low temperatures and low concentrations. They do not have very good alkaline stability unless compounded with other products. They are very useful in scouring or dyeing, where rapid penetration of the fiber is desirable, and where absorbency at later stages of processing is desirable. As a rule, they should not be used where goods are to be later treated with a water repellent, as they have a tendency to remain in the fabric and will create rewetting. Sulfosuccinic esters are produced by the reaction of sodium bisulfite on an ester of an alcohol and a difunctional acid; their infrared spectra show characteristic vibrational frequencies. The structure for sulfosuccinates is:

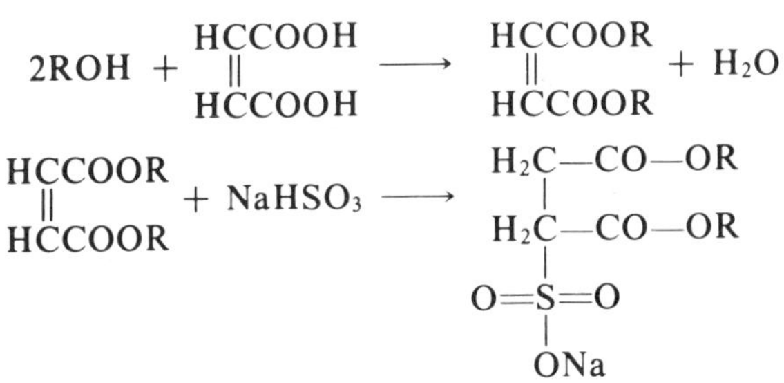

Sulfosuccinate

The infrared spectra are shown in Figs. 6.25 and 6.26 and the characteristic vibrational frequencies, in Table 6.6.

The CSO_3Na linkage shows the esters to be sulfonates; therefore, the points of reference are C═O at 5.8, SO_3 at 8.0 and 8.2, SO_3 at 9.5 and 9.6, and C—O—C at 8.6.

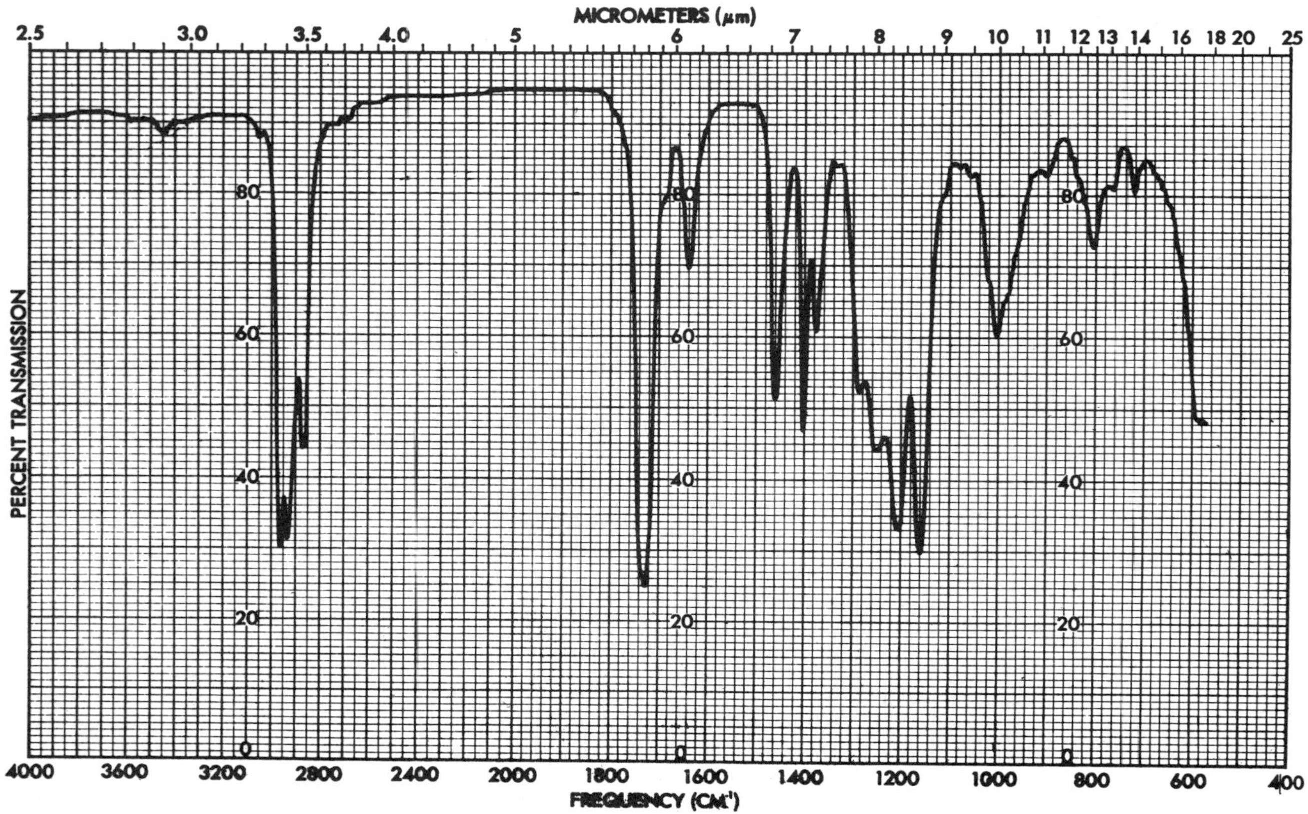

FIG. 6.25. Infrared spectrum of di(2-ethylhexyl) maleate.

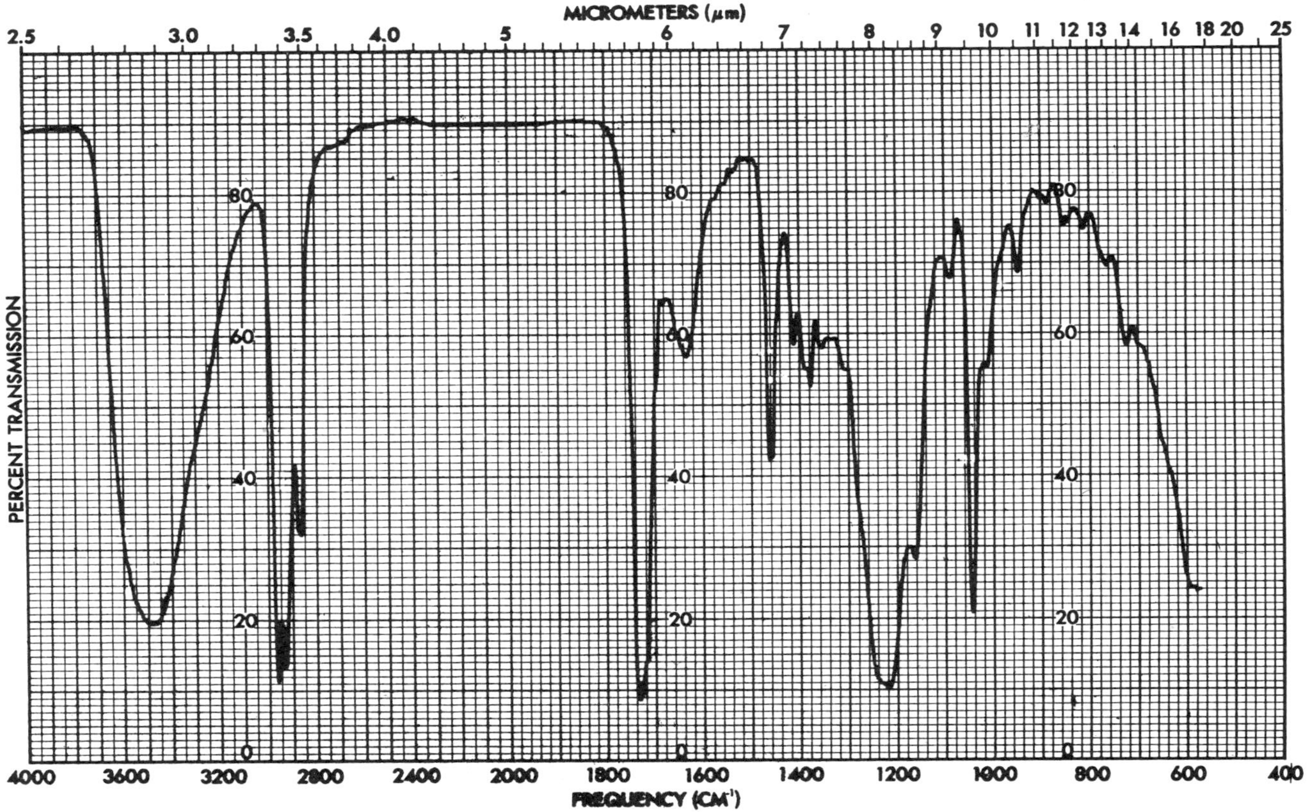

FIG. 6.26. Infrared spectrum of di(2-ethylhexyl) sulfosuccinate.

TABLE 6.6. Sulfosuccinate Ester Sulfonate

Wavelength, μm	Group	Type
2.8	OH	Hydroxyl
3.4	CH_2	Methylene
3.5	CH_3	Methyl
5.8	C=O	Ester
6.2	COO	Ester
6.8	CH_2	Methylene
7.2	COO	Ester
8.0, 8.2	SO_3H	Sulfonic
8.6	C—O—C	Carbon to oxygen
9.5, 9.6	SO_3H	Sulfonic

A typical manufacturing procedure for the synthesis of a sulfosuccinate-type surfactant as determined from the patent literature would be as follows:

Step 1. Manufacture of Ester

1. Charge to reaction vessel:

2-Ethyl-1-hexanol, MW 130.22	631 lb
Maleic anhydride, MW 98.06	216 lb
Xylene	400 lb
Paratoluene sulfonic acid	1 lb

2. Close kettle and set up for azeotropic distillation.
3. When temperature reaches 135°C, azeotrope water into separator and hold at reflux azeotrope for 5 hours.
4. Titrate for percent maleic, and when acidity is at 0.2–0.4%, strip solvent and excess alcohol. Final temperature at end of stripping procedure should be approximately 160°C. Use vacuum and nitrogen sparge for good color.

Step 2. Sulfonation Procedure

1. Charge to reactor:

Water	200 lb
Sodium bisulfite, MW 104.07	200 lb

Isopropyl alcohol	100 lb
Liquid caustic, 20%	16 lb

2. Adjust pH to 7.0–7.5 and hold for 30 minutes.
3. Add the above ester (MW 322), 660 lb. (heel from old batch for catalyst, 40 lb.)
4. Take to reflux for 3 hours or until bisulfite is less than 0.5%.
5. If color is not satisfactory, add 2–4 lb of hydrogen peroxide and bleach until color meets specifications.
6. Cool and drum.

Phosphate Esters

Complex organic phosphate esters of alcohols, nonionic surfactants, and combinations of the two, with ethylene oxide (EO) adduct-type products, are widely used in the textile industry. They are, perhaps, among the most popular compounds used in open-width processing. They are also useful in the dispersion of dispersed dyes in package-dyed yarns and under pressure conditions. These surfactants are formed by the reaction of a phosphorus-containing compound such as polyphosphoric acid[82] or P_2O_5[83] with an alcohol or an intermediate containing an alcoholic hydroxyl. The acid ester produced by the reaction of the phosphorus donor is then neutralized, depending on end use, with either a metal hydroxide or an amine. The phosphate esters so produced have excellent compatibility with other surfactants and by themselves are good emulsifiers. Patent literature covers this reaction under U.S. Patents 3,235,627 to Rohm and Haas and 3,004,056 and 3,004,057 to GAF. Additional patents referring to phosphation procedures are German Patent 696,317 to I. G. Farben and U.S. Patents 2,027,785, 2,052,029, and 2,177,983 to Benjamin Harris. United States Patent 2,137,792 to Victor Chemical outlines surfactants characterized by high wetting power combined with very low foaming tendencies.

The choice of phosphating agents is dictated by the performance properties desired. As phosphorus compounds have three reactive sites and as mono-, di-, and triesters can be formed, end use dictates which compound to use. Polyphosphoric acid is useful in preparing esters that have good solubility in concentrated caustic solutions. Phosphorus pentoxide (P_2O_5) is used to prepare products that contain both mono- and diacid esters. Polyphosphoric acid produces primarily monoesters. Composition and properties of the finished product can be varied by changing the mole ratio of the reactants.[84] Figure 6.27 shows the effect on mono–diester content produced by three different phosphorylating agents—concentrated phosphoric acid, polyphosphoric acid, and P_2O_5.

Figure 6.28 shows the effect on mono–diester content by changing the mole ratio of the phosphorylation agent.

Figure 6.29 shows the effect on mono–diester content when a mixture of ethoxylated nonyl phenol plus *n*-decyl alcohol is reacted with different phosphorylation agents.

Phosphate esters are controlled by the phosphorus atom of the phosphate radical and are analagous to sulfate radicals in that the phosphorus atom is

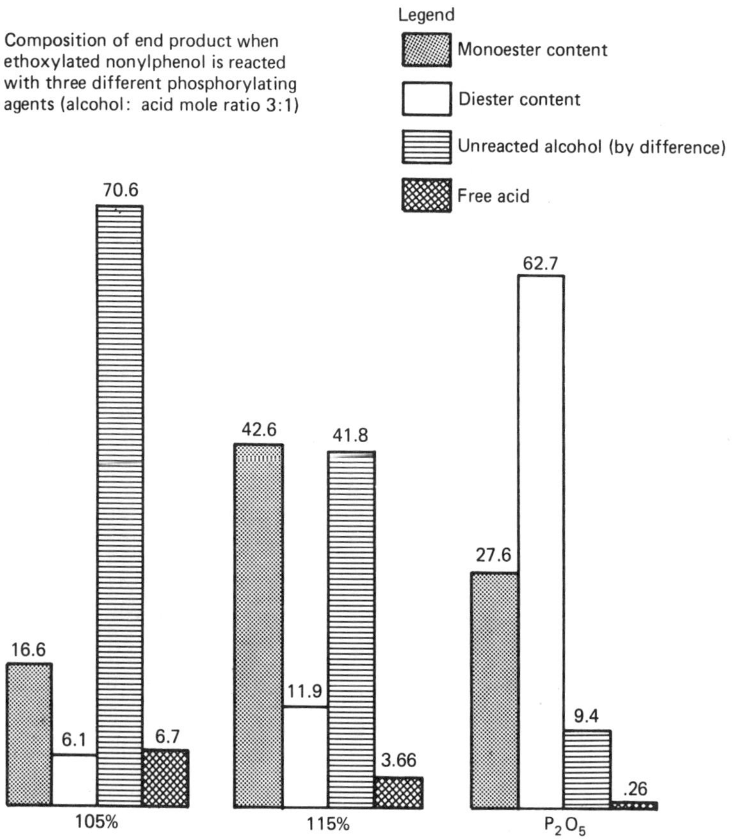

FIG. 6.27. Mono–diester content. Numbers on top of bars indicate percentage of component by weight. Courtesy of FMC Corporation.

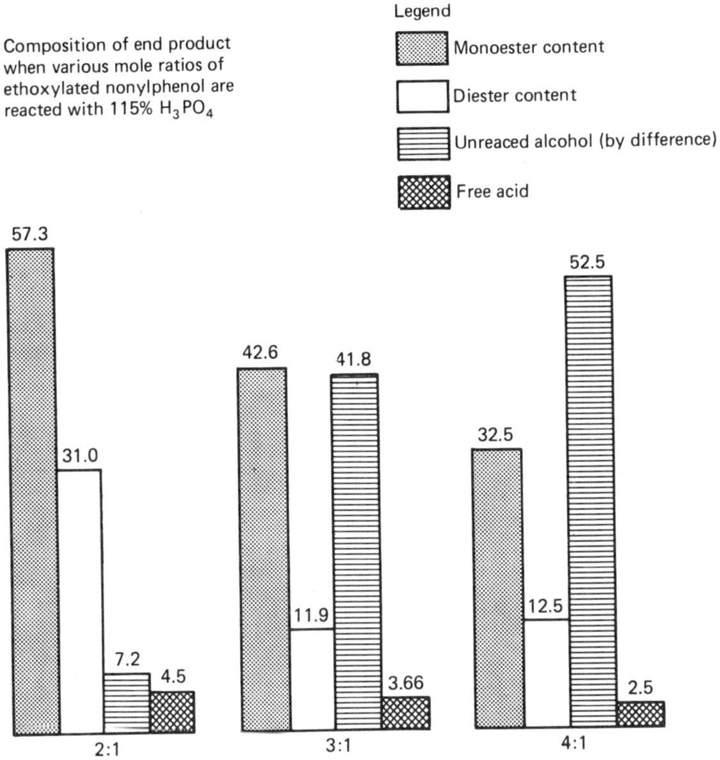

FIG. 6.28. Mole ratio of alcohol to H_3PO_4 (115%). Numbers on top of bars indicate percentage of component by weight. Courtesy of FMC Corporation.

separated from the rest of the surfactant molecule by an oxygen atom. Structures for a mono- and a diester are:

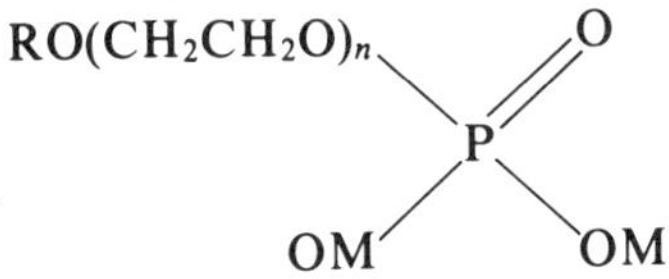

Monoester

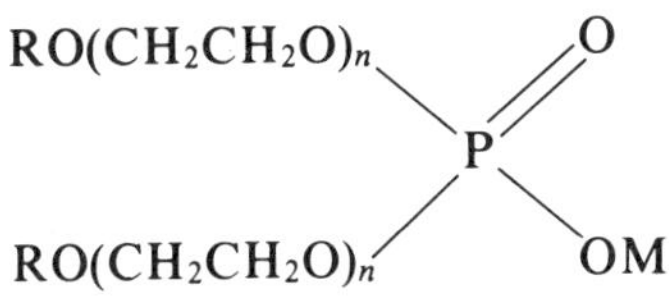

Diester

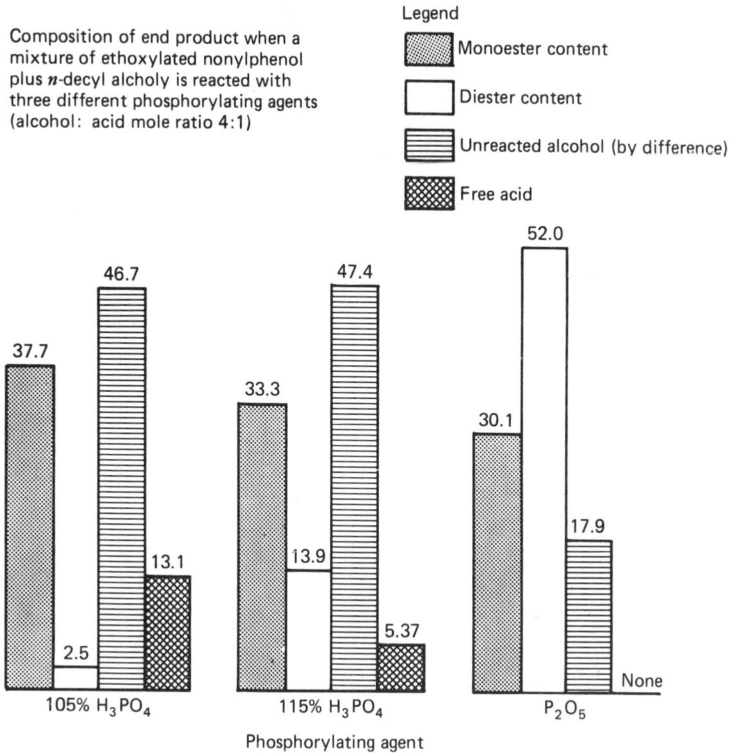

FIG. 6.29. Mixture of (EO) nonyl phenol + *n*-decyl alcohol. Numbers on top of bars indicate percentage of component by weight. Courtesy of FMC Corporation.

where R = alkyl or alkylaryl radical, n = average number of moles of ethylene oxide reacted with one mole of hydrophobe, and M = H, Na, etc. Infrared spectra for the free acid, potassium, and triethanolamine salts of a phosphated nonyl phenol that has been ethoxylated are shown in Figs. 6.30 through 6.32.

Procedures for Manufacture of Phosphate Esters

Procedure 1. Phosphation with Polyphosphoric Acid (115%)[85]

1. Charge to reactor:

Tergitol® 15-S-9[86] MW 600.0	910 lb
Polyphosphoric acid MW 170.6 (use 3:1 ratio of hydroxyl to phosphate)	90 lb

2. Adjust temperature of Tergitol® to 40–50°C.
3. Add polyphosphoric acid at a rate that will permit good mixing and does not allow temperature to exceed 60°C.
   Note: In cold weather polyphosphoric has a tendency to thicken. It may be necessary to allow polyphosphoric to warm in separate add tank or by approved method until viscosity is at an acceptable level.
4. When all polyphosphoric acid is in, allow reaction to continue for 1–1½ hours.
5. Cool and draw into specified containers.

Procedure 2. Phosphation with Polyphosphoric Acid (115%)[87]

1. Charge to reactor:

Tridecyl alcohol–MW 332.0 3 moles of EO	64.00%
Polyphosphoric acid	40.00%

2. Charge tridecyl alcohol to reactor, and add polyphosphoric acid with sufficient agitation to maintain dispersion.
3. When all polyphosphoric has been added, raise temperature to 160°F and hold for 5 hours.
4. Cool to 120°F and check for solubility and caustic stability.
5. A 2% solution when neutralized to a pH of 7.0 should be clear in water and stable to a 5% caustic solution.
6. If product does not pass these tests, phosphation is incomplete, and heating should be resumed, with the addition of polyphosphoric acid in 10-lb increments until a satisfactory product is obtained.

Procedure 3. Neutralization of Phosphates

Once phosphation in either of the above procedures has reached a satisfactory end point, the material is usually neutralized as follows:

1. Charge to reactor:

Water	21.00%
Isopropyl alcohol	5.00%
Butyl cellosolve	5.00%
Caustic potash, 45%	15.00%
Phosphate ester	40.00%
Caustic potash, 45% (to adjust pH)	QS

2. Adjust pH to 7.0–7.5.
3. Check water content and adjust solids to desired level.
4. Commercial products range in activity from 27.5 to 40% solids.
5. If a high-solids liquid product is necessary, the use of triethanolamine in place of the caustic potash may be used.
6. Viscosity of finished product may be adjusted by varying the ratios of isopropyl alcohol and butyl cellosolve.

Procedure 4. Phosphation of Isooctyl Alcohol with P_2O_5[88]

1. Charge to reactor:

Isooctyl alcohol	41.00%	2000 lb
With agitation, add phosphorus pentoxide (P_2O_5)	15.00%	740 lb

2. Add at a rate so that temperature reaches 35°C optimum.
3. When all P_2O_5 has been added, allow temperature to rise to 50°C and hold for 2 hours.
4. Cool to 25°C.
5. Dissolve in separate neutralization tank:

Flake caustic potash	15.00%	725 lb
Water	20.00%	949 lb

6. Cooling to 30°C, add:

Isopropyl alcohol	10.00%	480 lb

7. With cooling on, add phosphate acid from above step, maintaining a temperature of 30–50°C.
8. On completion of acid addition, allow to stir 30 minutes. Adjust pH to 7.5 ± 0.5 as necessary.
9. Adjust solids content with water to specification.
10. Commercial products usually contain 30–40% solids.

United States Patent 3,380,927 discusses coesters of branched-chain alcohols and ethylene oxide adducts of nonyl phenol. These products are typified in the trade and were made popular by Dexter Chemical Corporation, holder of this patent, and sold commercially under their trade name "Barisol."

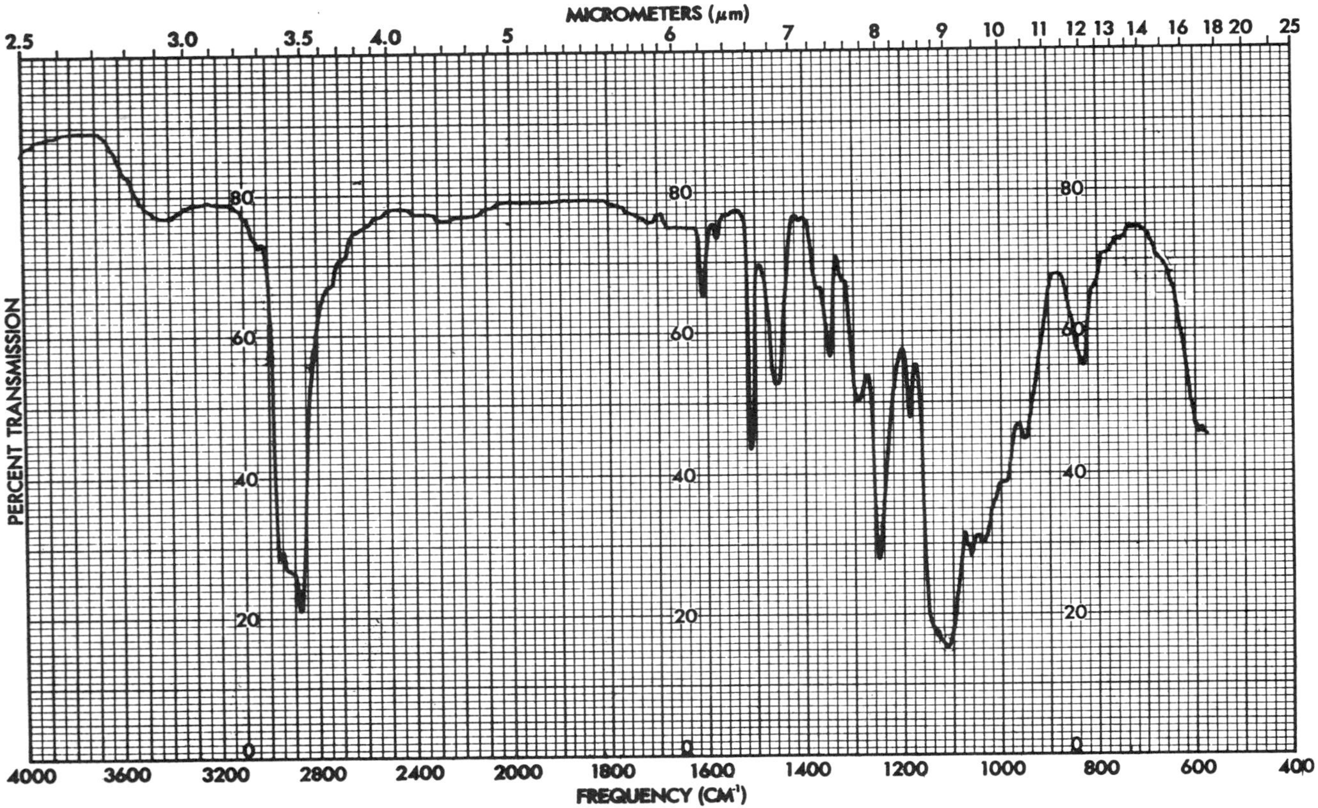

FIG. 6.30. Infrared spectrum of phosphated 9-mole (EO) nonyl phenol.

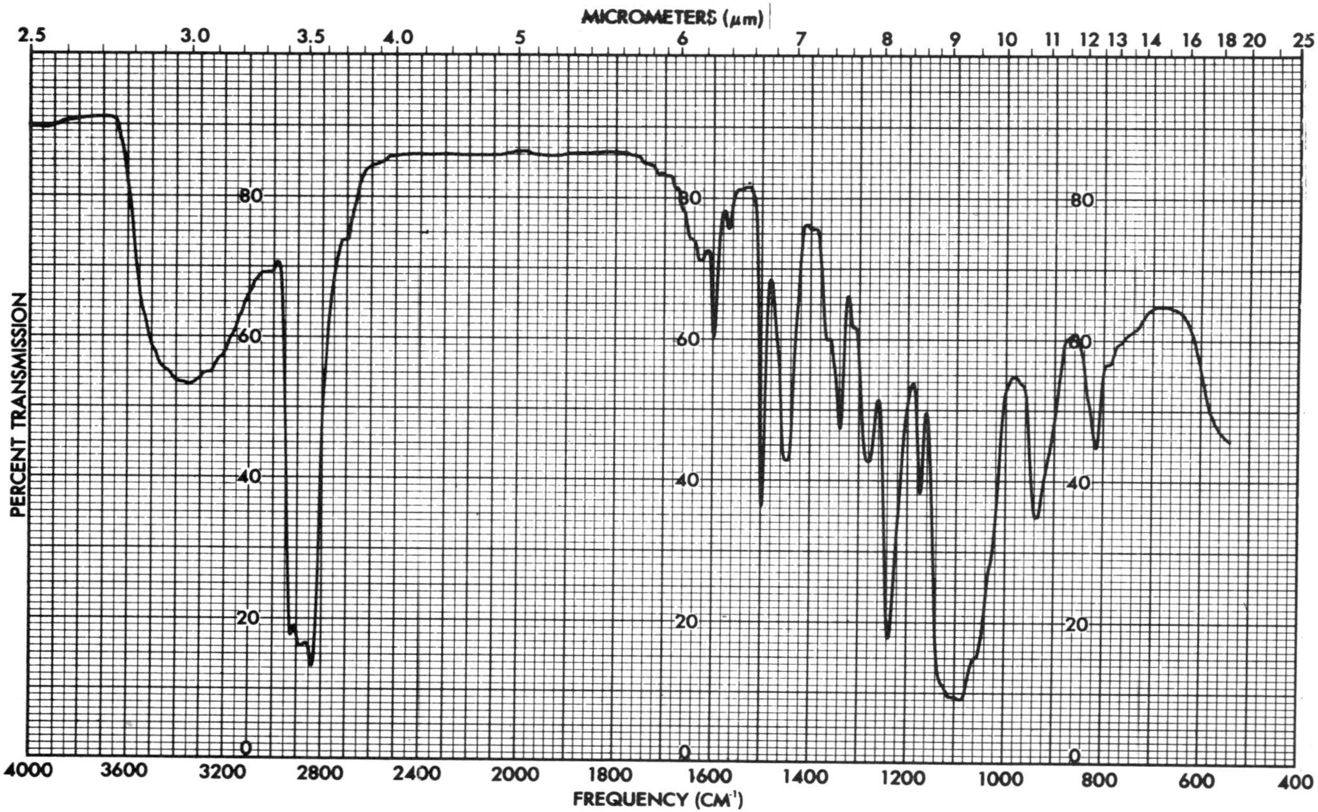

FIG. 6.31. Infrared spectrum of potassium salt of 9-mole (EO) nonyl phenol phosphate.

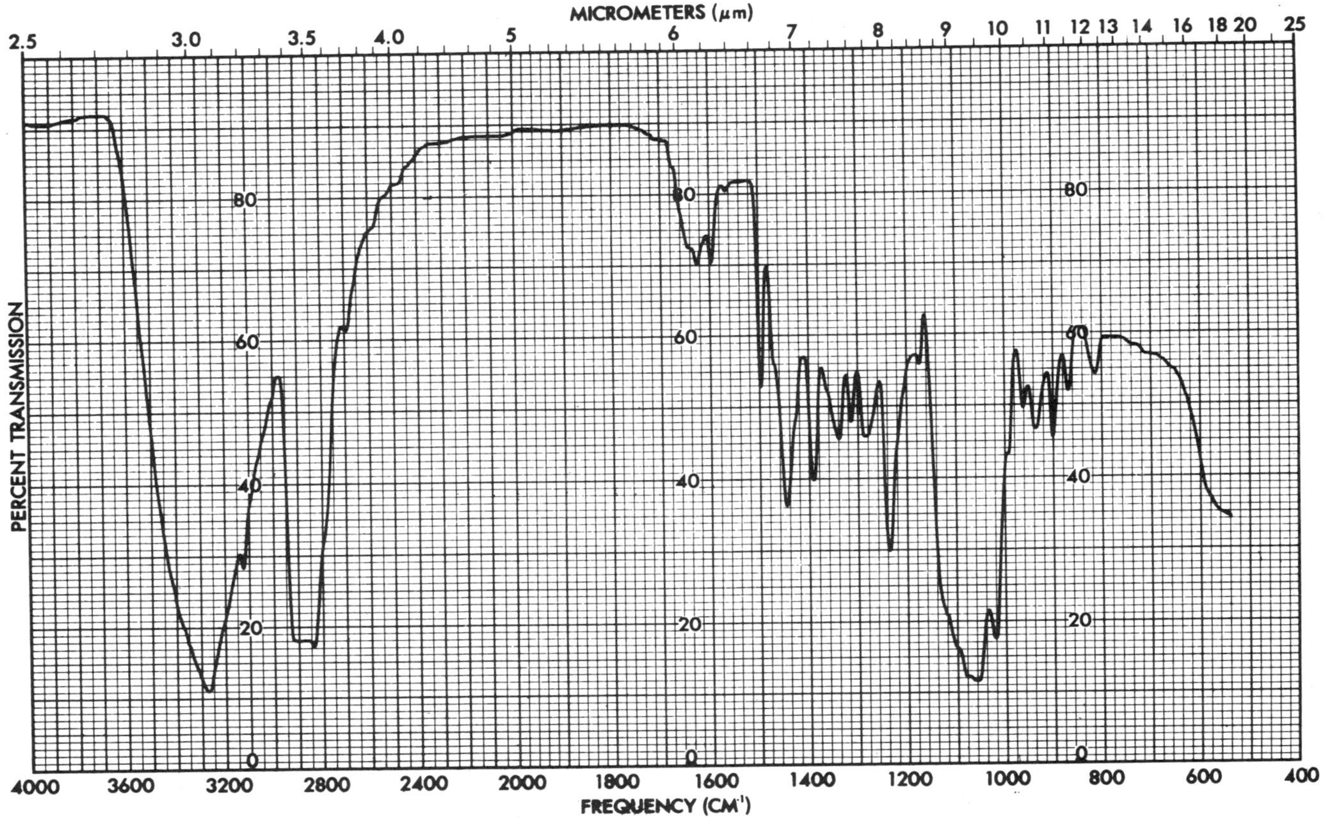

FIG. 6.32. Infrared spectrum of triethanolamine salt of 9-mole (EO) nonyl phenol phosphate.

United States Patent 4,135,878, to Bishop and Moon, discloses an emulsifier–solvent scour composition that is stable at elevated temperatures when incorporated into aqueous caustic media for the treatment of textile materials. Phosphate esters are a key ingredient to these compositions.

Polyether Acetic Acids (Carboxymethylates)

Carboxymethylates have found widespread use in preparation and bleaching under the trade name of "Sandopan,"[89] Their manufacture is covered under U.S. Patent 2,623,900. Additional patent references are U.S. Patents 2,252,401; 2,493,126; 2,438,092; 2,490,109; 2,480,817; 2,474,175; and 2,458,741. An interesting note is cited in Patent 2,623,900, which uses as a reference Patent 554,938 to Lederer dated February 18, 1896. A further discussion of aspartic acid derivatives *N*-methylglycine and acetylglycine may be found in German Patent 546,942, U.S. Patent 1,973,860, and U.S. Patent 2,215,367 to I. G. Farben. There is also an excellent discussion of these compounds in *Surface Active Agents* by Schwartz and Perry, pp. 35–36. The structure for sodium alkyl alkoxy carboxylate is:

$$R{-}O{-}\left[CH_2{-}CH_2{-}O\right]_n{-}CH_2{-}CH_2{-}CO_2Na$$

Sodium alkyl alkoxy carboxylate

Carboxylate surfactants have the unique property of being anionic while possessing the favorable characteristics of nonionic surfactants. This property is brought about by a terminal carboxyl group similar to those found in fatty acid soaps. However, because carboxylates are based on a nonionic surfactant, they do not form insoluble salts with calcium and magnesium ions, unlike soaps, which *do* form these salts. Carboxylates have excellent oil solubility and emulsification properties also attributed to their carboxyl group and exhibit good resistance to alkali mediums.

N-Methyltaurines, Sulfoalkylamines, and Sulfoalkyl Esters

N-Methyltaurines were the first real attempt by synthetic detergent manufacturers to apply the knowledge of chemistry to the problems encountered with natural derived soaps. As it was well known that tertiary nitrogen compounds are less hydrolizable[90] under alkaline conditions, the introduction of a series of surfactants based on this technology was readily accepted because of the good detergency, satisfactory wetting qualities, and stability to alkali of these compounds.

United States Patents 1,881,172 (October 1933) and 1,923,180 (October 1934) describe the manufacture of *N*-acyl-*N*-methyltaurine. Kastens and Ayo describe in detail commercial size processes for manufacturing these compounds.[91] An excellent discussion on *N*-acyl-*N*-methyltaurine is given by Shapiro in *Fatty Acids and Their Industrial Applications.*[92] Schwartz and Perry, in *Surface Active Agents*, also discuss these compounds in detail.[93] The structure for a methyltaurine is:

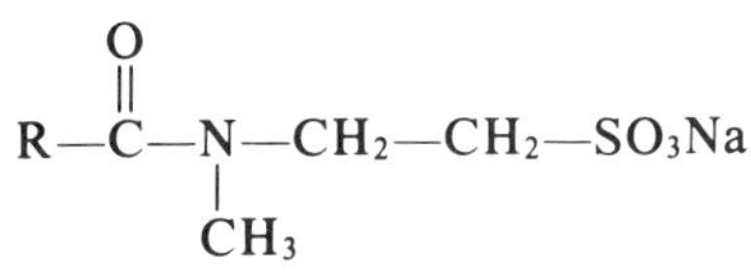

N-Methyltaurine

The GAF Corporation manufactures a line of taurine derivatives known as Igepon® surfactants.[94] The structures of their sulfoalkyl amides and their sulfoalkyl esters are:[95]

Sulfoalkyl amide

Sulfoalkyl ester

Figure 6.33 is a reproduction of charts furnished by GAF showing the chemical breakdown of their Igepon® line of surfactants.

Alkoyl Sarcosinates and Benzimidazoles

G. F. Longman, in *The Analysis of Detergents and Detergent Products*,[96] discusses two "prewar products" that found commercial importance as anionic surfactants. They were alkoyl sarcosinates and benzimidazole sulfonates. The sarcosinates were covered under U.S. Patent 3,079,980, and their structure is shown below. Sarcosinates are generally prepared by the reaction of a

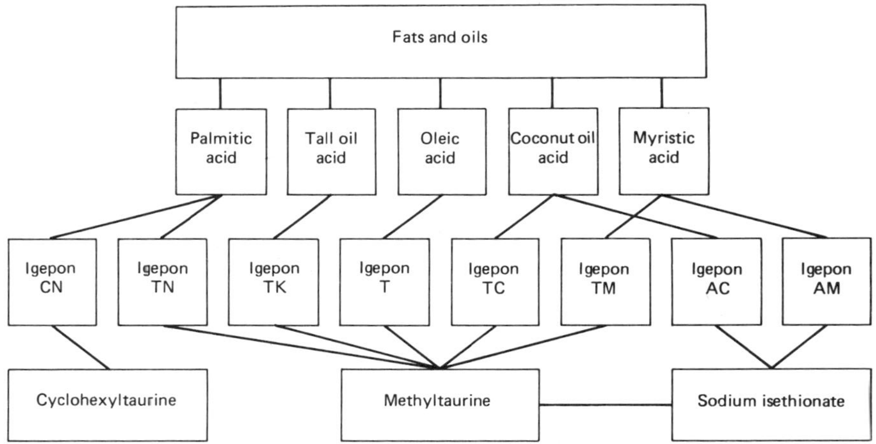

Igepon Surfactant Type	Structure	Name	CTFA Designation
AC	$R-\overset{\overset{O}{\parallel}}{C}-O-CH_2-CH_2-SO_3Na$	Coconut fatty acid, 2-sulfoethylester, sodium salt	Sodium cocoyl isethionate
T	$C_{17}H_{33}-\overset{\overset{O}{\parallel}}{C}-\overset{\overset{CH_3}{\vert}}{N}-CH_2-CH_2-SO_3Na$	*N*-Methyl-*N*-oleoyltaurine, sodium salt	Sodium methyl oleoyl taurate

TC	$R-C(=O)-N(CH_3)-CH_2-CH_2-SO_3Na$	*N*(coconut oil acyl)-*N*-methyltaurine, sodium salt	Sodium methyl cocoyl taurate
TK	$C_{18}H_{33}-C(=O)-N(CH_3)-CH_2-CH_2-SO_3Na$	*N*-methyl-*N*-(tall oil acyl)-taurine, sodium salt	
CN	$C_{16}H_{31}-C(=O)-N(C_6H_{11})-CH_2-CH_2-SO_3Na$	*N*-cyclohexyl-*N*-palmitoyl-taurine, sodium salt	
TN	$C_{15}H_{31}-C(=O)-N(CH_3)-CH_2-CH_2-SO_3Na$	*N*-methyl-*N*-palmitoyl-taurine, sodium salt	
AM	$C_{13}H_{25}-C(=O)-O-CH_2-CH_2-SO_3Na$	Myristic acid-2-sulfoethyl ester, sodium salt	
TM	$C_{13}H_{25}-C(=O)-N(CH_3)-CH_2-CH_2-SO_3Na$	*N*-Methyl-*N*-myristoyl taurine, sodium salt	

FIG. 6.33. Chart of Igepon® surfactant structures.

fatty acid chloride and an amino acid such as sodium sarcosinate. Surfactants of this type exhibit many of the properties of soap but are less susceptible to hardness ions. Their structure bears a strong resemblance to the carboxylates in current usage today.

$$C_{11}H_{23}CON(CH_3)COONa$$

Alkoyl sarcosinate

Benzimidazole sulfonates are formed by the reaction of a fatty acid chloride and *ortho*-phenylenediamine with the resulting benzimidazole sulfonated with $—SO_3H$ groups and then neutralized with sodium hydroxide. The literature reports that these products have many properties in common with alkylaryl sulfonates, and the presence of the linear chain renders them biodegradable.[99] The structure of a benzimidazole sulfonate is:

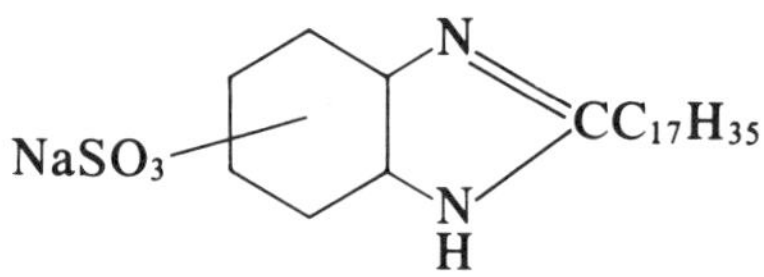

Benzimidazole sulfonate

Fatty Acid Soaps

The final anionic fatty acid surfactant commonly used in preparation is fatty acid soap. These compounds date back to antiquity and are even reported in Roman legend of some 3000 years ago. According to legend, fat from burnt offerings trickled through wood ashes on Sapho Hill outside Rome. The alkali of the wood ashes changed the fatty acids from the burnt offerings into soap that was absorbed into the clay of the hillside. People who sat on this clay noticed that their clothes were easier to clean and attributed their cleanliness to factors other than chemistry.

The mechanism by which a soap is formed is:

$$RCOOH + NaOH \longrightarrow RC(=O)ONa + H_2O$$

Soap structure

Soaps are formed by the replacement of the terminal hydroxyl group of a fatty carboxylate with a metal salt or amine. When this compound comes into contact with dirt or oil, a very thin film of soap forms around the dirt and oil gathers together in small droplets that lift the oil to the surface. The organic end of the soap lifts the oil from the fabric, allowing the water to creep in from underneath. Soap lost favor in postwar years because of its poor stability to hard-water ions. However, with the advent of synthetic surfactants, soaps compounded with phosphates and sequestering agents can once again be used as a part of the preparation process. Inasmuch as soap imparts lubricity to cellulosic fibers, it can function as a lubricant in scouring and preparation compounds.

Several factors to be considered in the choice of which soap to use are those that determine the physical properties of the soap. The properties of the soap are controlled by the molecular weight of the fatty acid used to manufacture the soap, but since fatty acids are not pure compounds, but in reality mixtures, properties vary over a given range. Any fatty acid that is present in a soap not in the form of its salt should be regarded as undesirable because the nonsaponifiable matter is less effective in terms of its detergent property than is the soap.[98] A second factor to consider is the titer, or the melting point, of the fatty acid used to manufacture the soap. The higher the titer, the higher the temperature necessary to solubilize the soap. Titers of common fatty acids are listed in Table 6.7.

A third factor of interest is the degree of unsaturation in the fatty acid. The presence of one double bond as in oleic acid is not undesirable, as it aids in the production of liquid soaps. However, a high degree of unsaturation renders the soap susceptible to atmospheric oxidation, which can produce a rancid odor that may be absorbed by the fabric. Oxidation can also bring about polymerization of the fatty acid with the formation of insoluble yellow residues that are difficult, if not impossible, to remove.[100]

Fabrics held in storage under high humidity are subject to oxidation attack. If goods are dyed and/or finished, unsaturated fatty acids remaining in

TABLE 6.7. Titers of Common Fatty Acids[99]

Castor oil	3°C
Coconut oil	29–30°C
Olive oil	25°C
Cottonseed oil	34–35°C
Tallow	40–45°C

the fabric can fade or discolor substrates. If synthetic fabrics such as nylon, in the greige state, contain spin finishes with unsaturated soap emulsifiers, yellowing of the fabric can occur, which can be whitened only by a harsh bleach such as peracetic acid.

United States Patents 3,376,327; 3,469,245; and 3,475,336 discuss detergent modified formulations. Figures 6.34 and 6.35 show infrared spectra of potassium soap of mixed coconut and oleic fatty acids and the triethanolamine soap of the same fatty acids, respectively.

In the infrared spectra shown in Figs. 6.34 and 6.35, certain vibrational frequencies are characteristic of fatty acid and soaps. Table 6.8 lists in tabular form the wavelength frequencies for the fatty acid and soap structures.

An examination of the soap spectra shows a change from the starting raw material. The broad, sloping peak at 3.0–3.5 μm is now more distinct, showing the transition of the OH group of the COOH. This is confirmed by the loss of the COO peak at 5.8 and the rise of a new peak at 6.4 (COONa). Peaks at 6.8–7.0 are also characteristic of COOH salts.[101]

TABLE 6.8. Wavelength Frequencies for Fatty Acid and Soap

Wavelength, μm	Group	Type
	Fatty Acid	
3.3	OH	COOH
3.4	CH_2	Methylene chain
3.5	CH_3	Methyl group
3.7	OH	COOH
5.8	COO	COOH
6.8	CH_2	$(CH_2)_n$
7.0	COOH	Carboxyl group
7.7	CH_2	Methylene chain
10.5	COH	COOH
13.8	$(CH_2)_n$	Methylene chain
	Fatty Acid Soap	
3.0	COOH	
3.4	CH_2	
3.5	CH_3	
6.4, 7.0		Characteristic of COOH salts
6.9	$(CH_2)_n$	
12.8, 13.2	CCl_4	Solvent used to smear sample

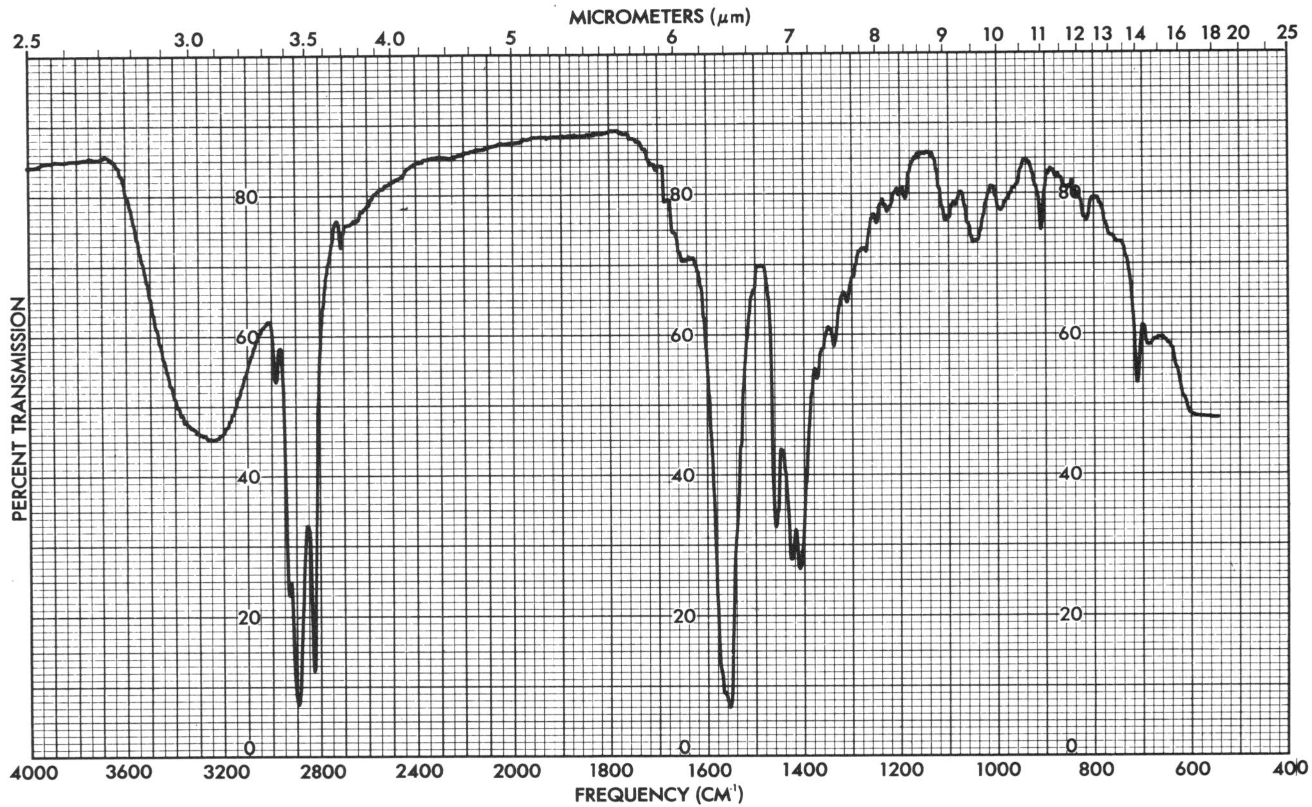

FIG. 6.34. Infrared spectrum of potassium soap of coconut and oleic fatty acids.

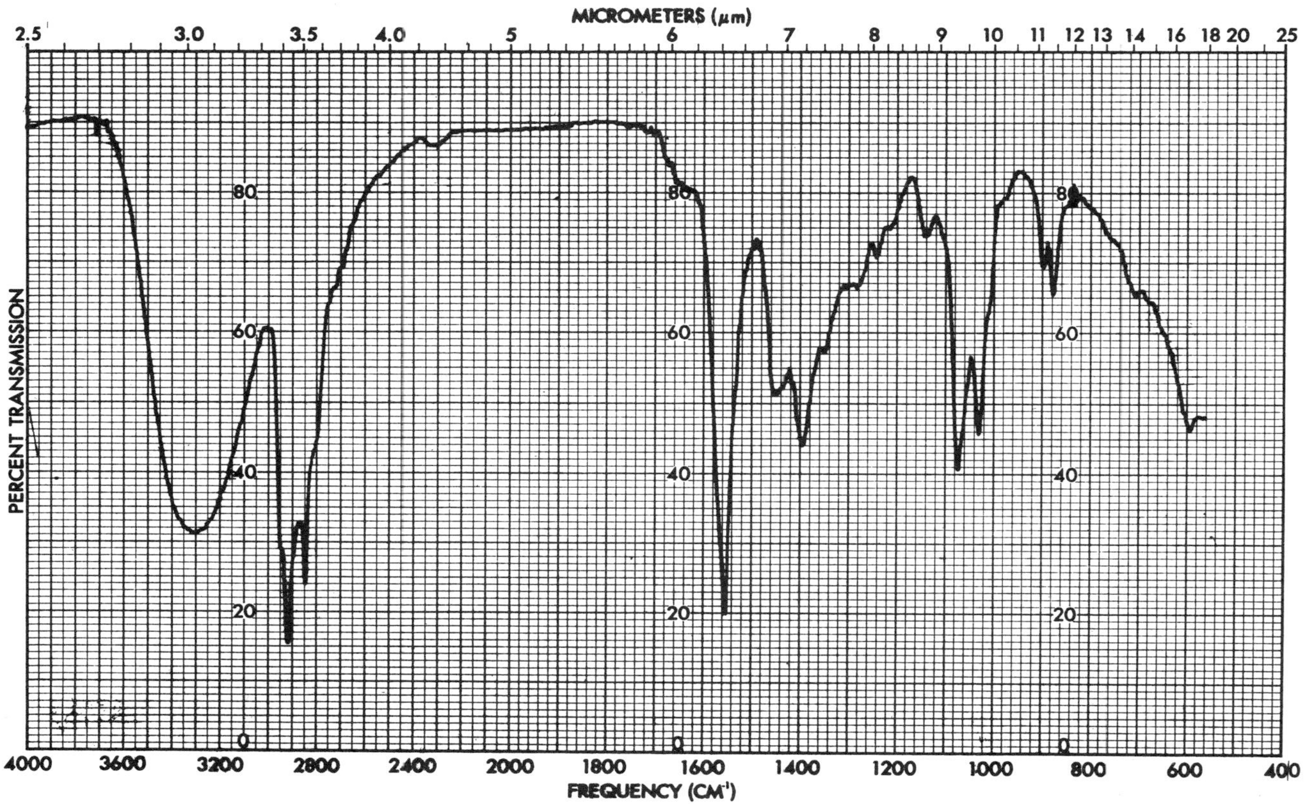

FIG. 6.35. Infrared spectrum of triethanolamine soap of coconut and oleic fatty acids.

Procedures for Production of Fatty Acid Soaps

Procedure 1. Coconut Fatty Acid-Potash Soap (Courtesy of Emery Industries, Inc., TSB-103-B.)

1. Charge to soap kettle:

Water	62.50%
Flake caustic potash	7.90%
Distilled coconut fatty acid	29.60%
Water	QS

2. Dissolve the caustic in the necessary amount of water, heating to 120–130°F.
3. Add the melted coconut fatty acid in a slow, steady stream, with agitation.
4. Slow agitation should be applied until saponification is complete.
5. Reaction temperature should reach 150–160°F.
6. Depending on the temperature of the water in the initial charge, heat as necessary to maintain 150–160°F until saponification is complete.
7. Adjust pH to specification, and allow to stand and settle.
8. Decant or filter soap.
9. Coconut fatty acid soaps are characterized by their clarity and profuse sudsing. A trace of sequestering agent (0.1–0.5%) is normally added as a safety factor against hardness ions in the water. Alcohol may be substituted for a portion of the water. The alcohol will improve clarity and increase the resistance of the soap to low-temperature gel formation but will reduce foaming.[102]

Procedure 2. Oleic Acid-Coconut Fatty Acid Soap (Courtesy of Emery Industries, Inc., TSB-103-B.)

1. Charge to soap kettle:

Water	62.70%
Flake potash (100%)	7.10%
Oleic acid	15.10%
Coconut fatty acid	15.10%

2. Charge water and potash and heat to 140°F.
3. Charge oils to alkali–water mix that will exotherm to 180–200°F.
4. Apply cooling if necessary to control exotherm.
5. Allow some head room in kettle in case mixture should boil.
6. When exotherm subsides, continue applying heat until free alkalinity is 2–3 mg/g.
7. Cool and drum.
8. In the event that it is necessary to use water of questionable hardness or fatty acids containing metal contaminants, the addition of 0.1–0.5% sequestering agent in initial charge of water may be desirable.

Procedure 3. Rosin-Potash Soap (Courtesy of Hercules Powder Company.)[103]

1. Charge to soap kettle:

Water	26.30%
Caustic potash, 45%	15.00%
Soap-grade rosin	58.40%

2. Charge water and caustic potash and heat to 90°C.
3. Add rosin with agitation until rosin lumps have melted.
4. Hold temperature at 90°C with agitation until acid value reaches 30 mg/g KOH.
5. Dilute with water to product specifications.

Additional Formulations for Fatty Acid Soaps[104]

1.	Tall oil	220 lb
	Water	180 lb
	Potassium hydroxide, 45%	91 lb
2.	Olive oil	100 lb
	Flake potash, 45%	47 lb
	Water	90 lb
	Alcohol	20 lb

3.		
3.	Cottonseed oil	408 lb
	Coconut oil	102 lb
	Potassium hydroxide, 45%	242 lb
	Water	120 lb

As can be seen from the above processes, soaps are easy to prepare and are inexpensive to use. They should be considered as an integral part of any anionic preparation formulation.

NONIONIC-TYPE SURFACTANTS

Nonionic-type surfactants have an electrically neutral charge. They are the second major surfactant class used in preparation and are manufactured in many different forms, typical of which are those listed in Table 6.9.

Advantages of Nonionics. Excellent compatibility with all classes of surfactant; Good wetters and rewetters; good emulsifiers; excellent oil solubility; good components of oil emulsifiers.

Disadvantages of Nonionics. Normally are high foamers; exhibit temperature sensitivity; have low alkali tolerance; show low electrolyte tolerance; can cause harshening of cellulosic fibers; residual surfactant can cause dye spotting and/or crocking of disperse dyes.

Major Uses. Used primarily as rapid wetters, coemulsifiers, and lubricants; may be used as compatibilizers, anticoprecipitants, and components in carrier emulsion systems.

TABLE 6.9. Nonionic Surfactants That Bear an Electrically Neutral Charge

EO–PO Ethers	Sorbitan ester–EO
EO–PO Esters	Sorbitol esters
Fatty esters	Sorbitol ester–EO
Amphoterics (at isoelectric point)	Emulsified fatty acids
	EO fatty acid
Thioethers	PO fatty acid
Sortiban esters	Fatty alkanolamides

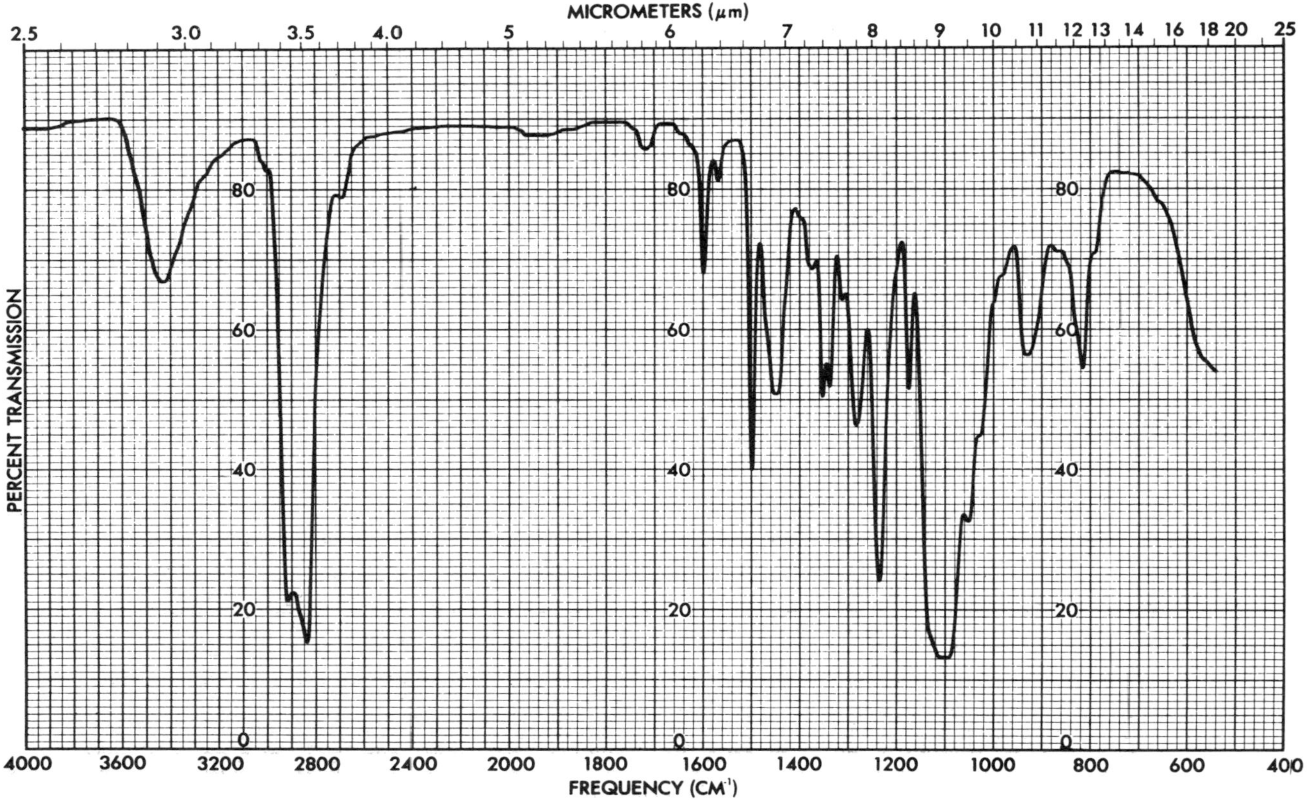

FIG. 6.36. Infrared spectrum of $(EO)_9$ octyl phenol.

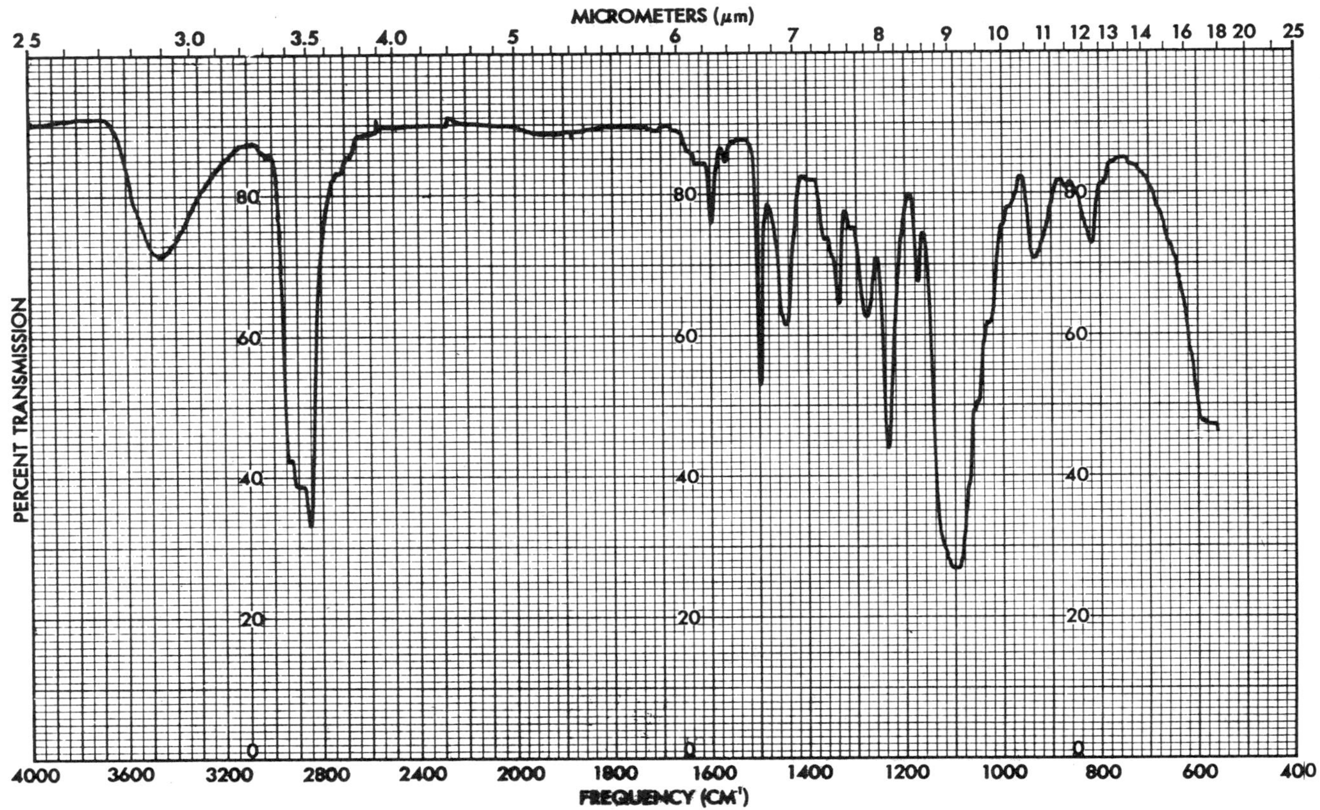

FIG. 6.37. Infrared spectrum of $(EO)_9$ nonyl phenol.

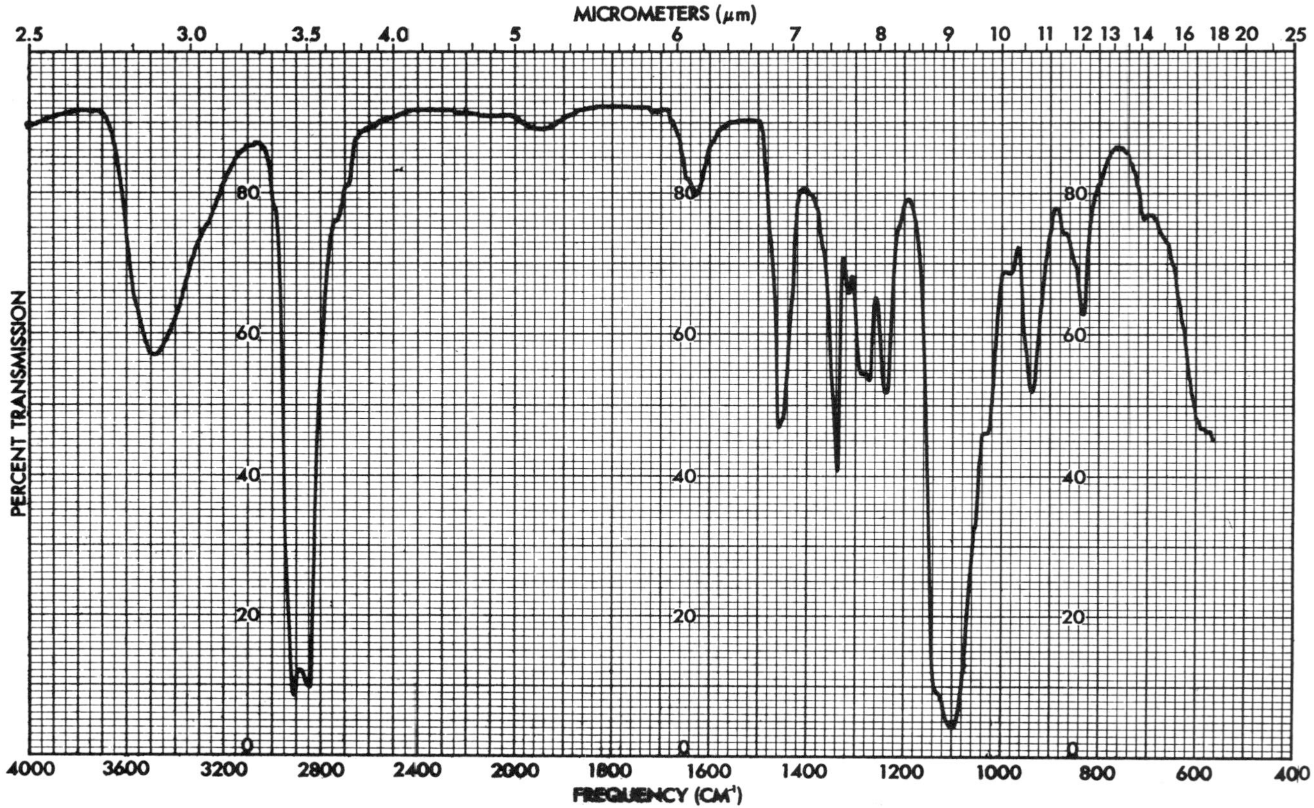

FIG. 6.38. Infrared of $(EO)_{20}$ oleyl alcohol.

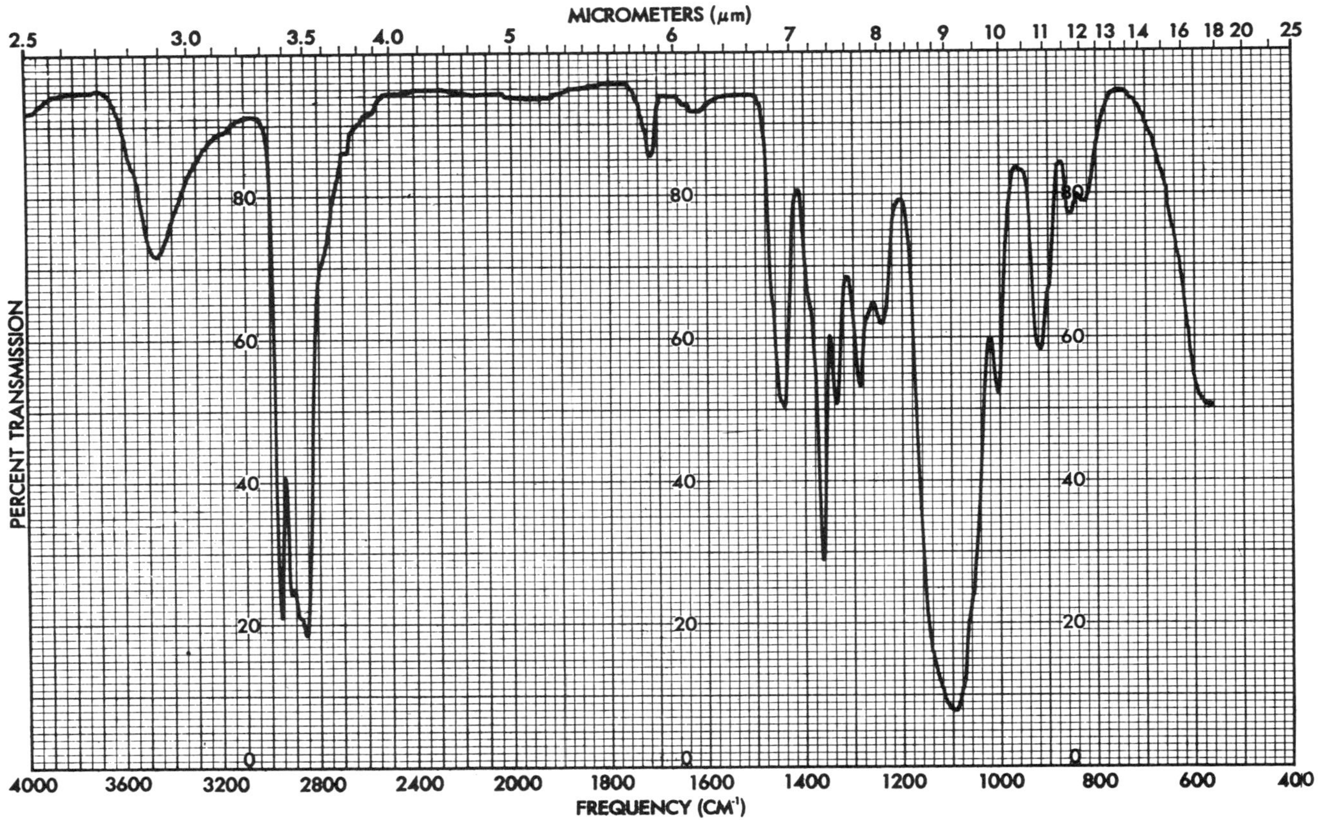

FIG. 6.39. Infrared spectrum of EO–PO copolymer.

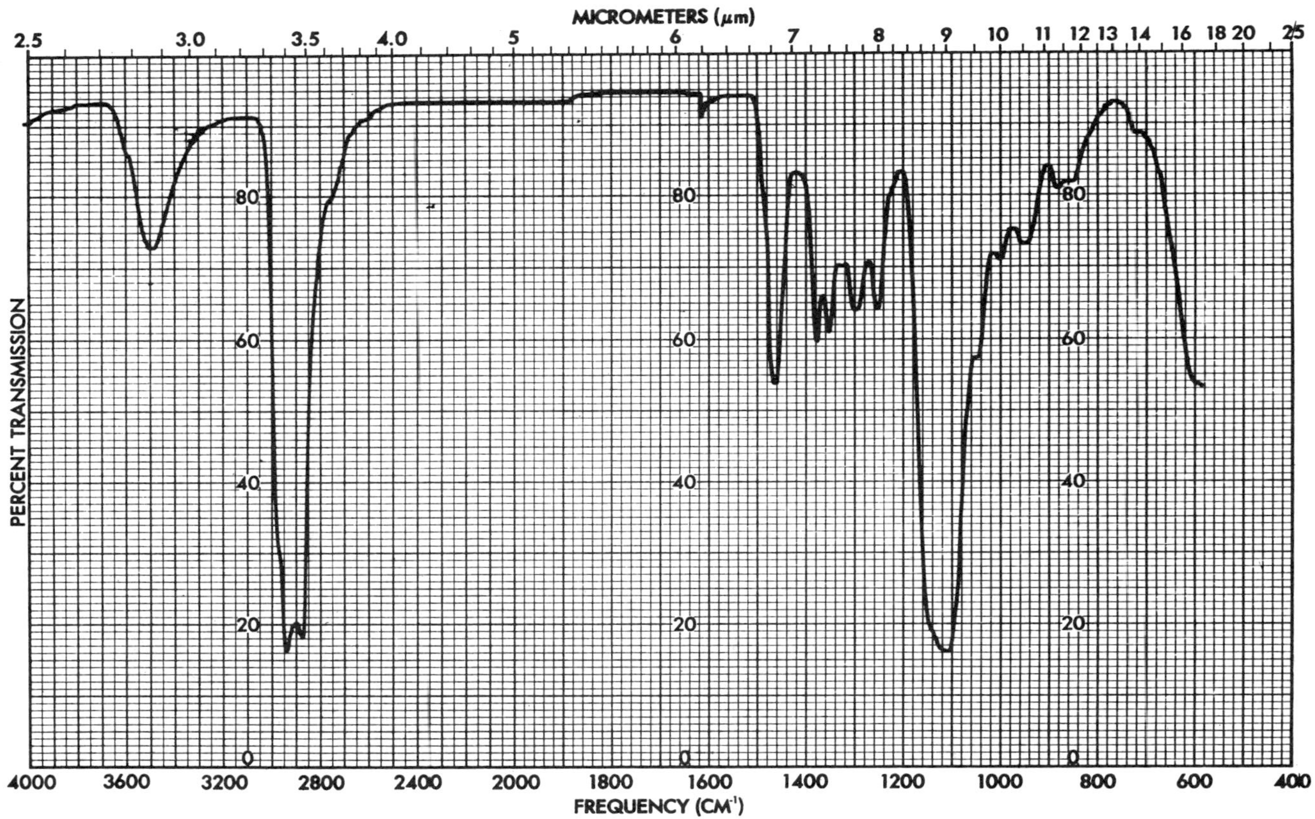

FIG. 6.40. Infrared spectrum of capped EO–PO copolymer.

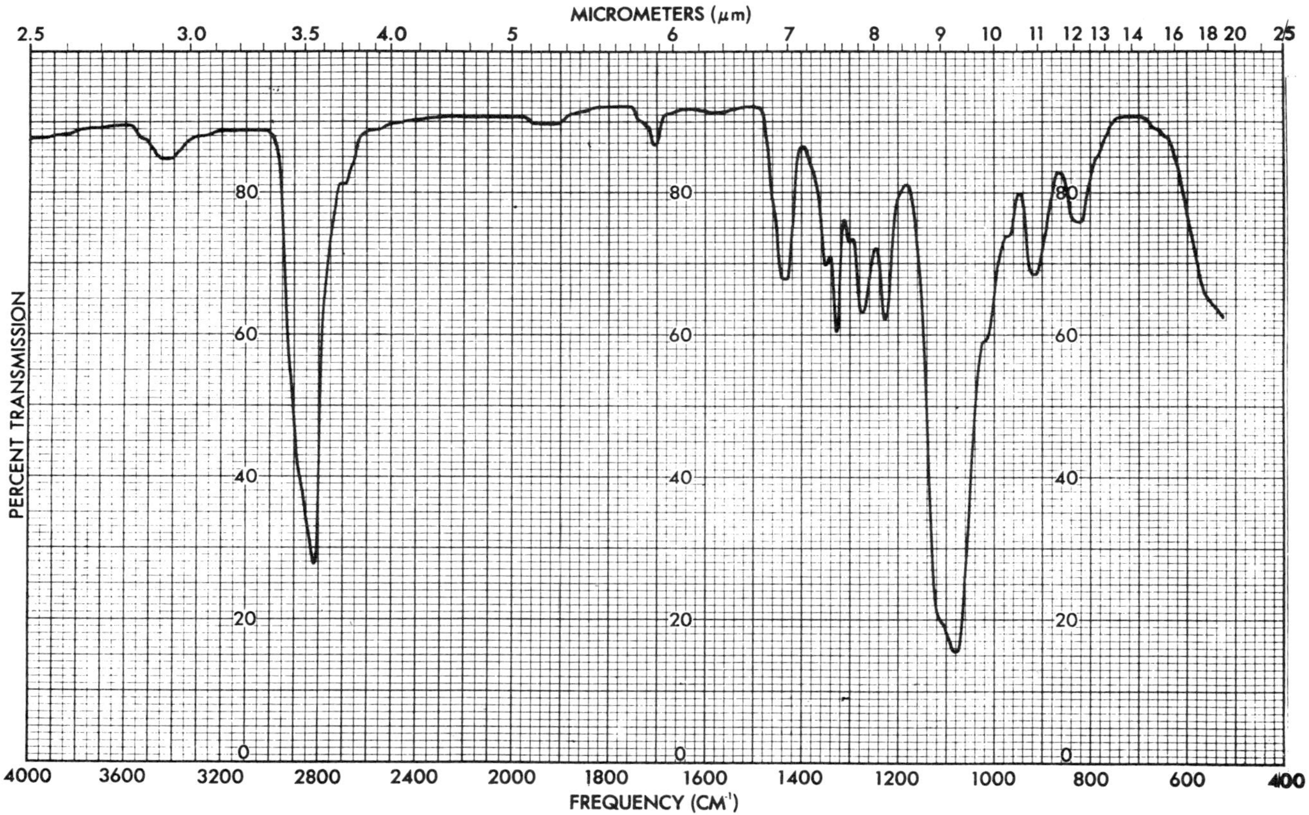

FIG. 6.41. Infrared spectrum of long-chain nonyl phenol EO–PO copolymer.

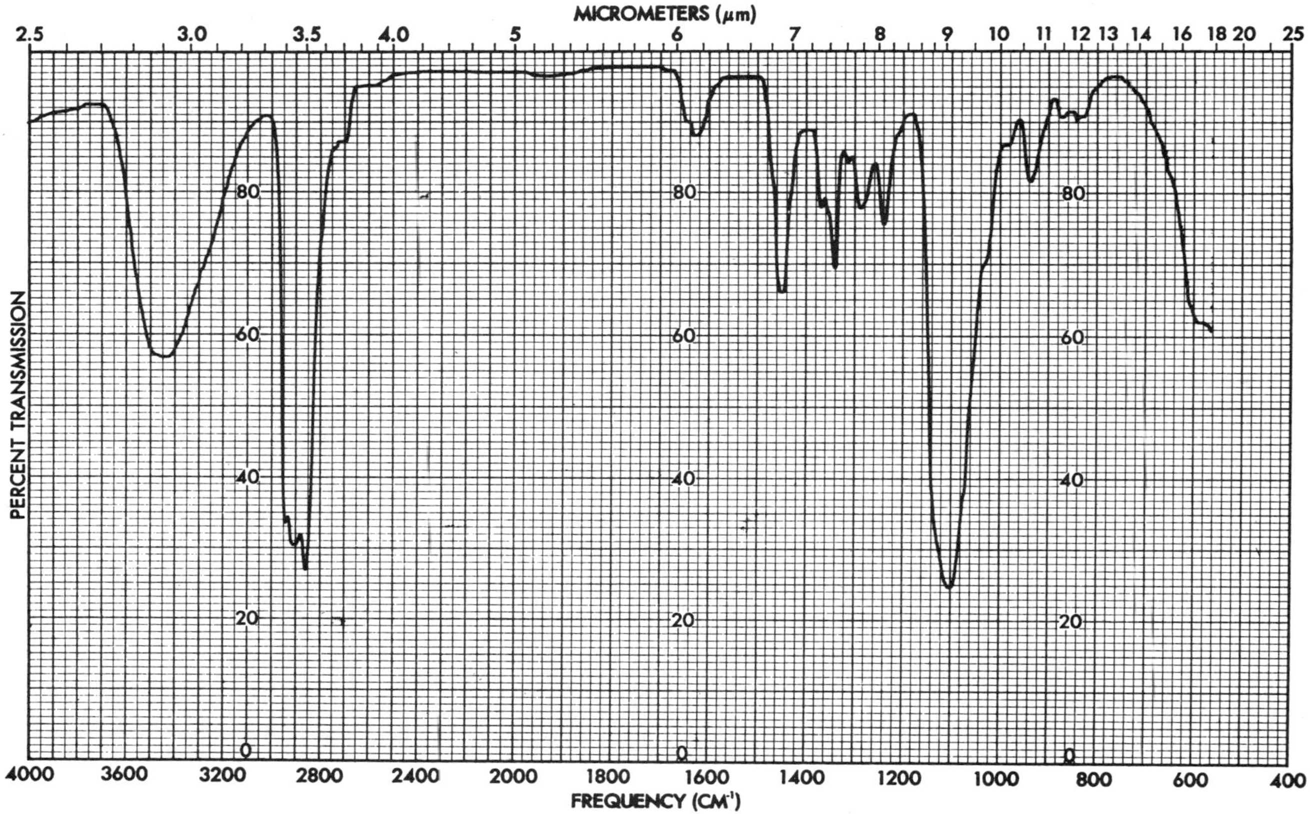

FIG. 6.42. Infrared spectrum of branched-chain alcohol–EO.

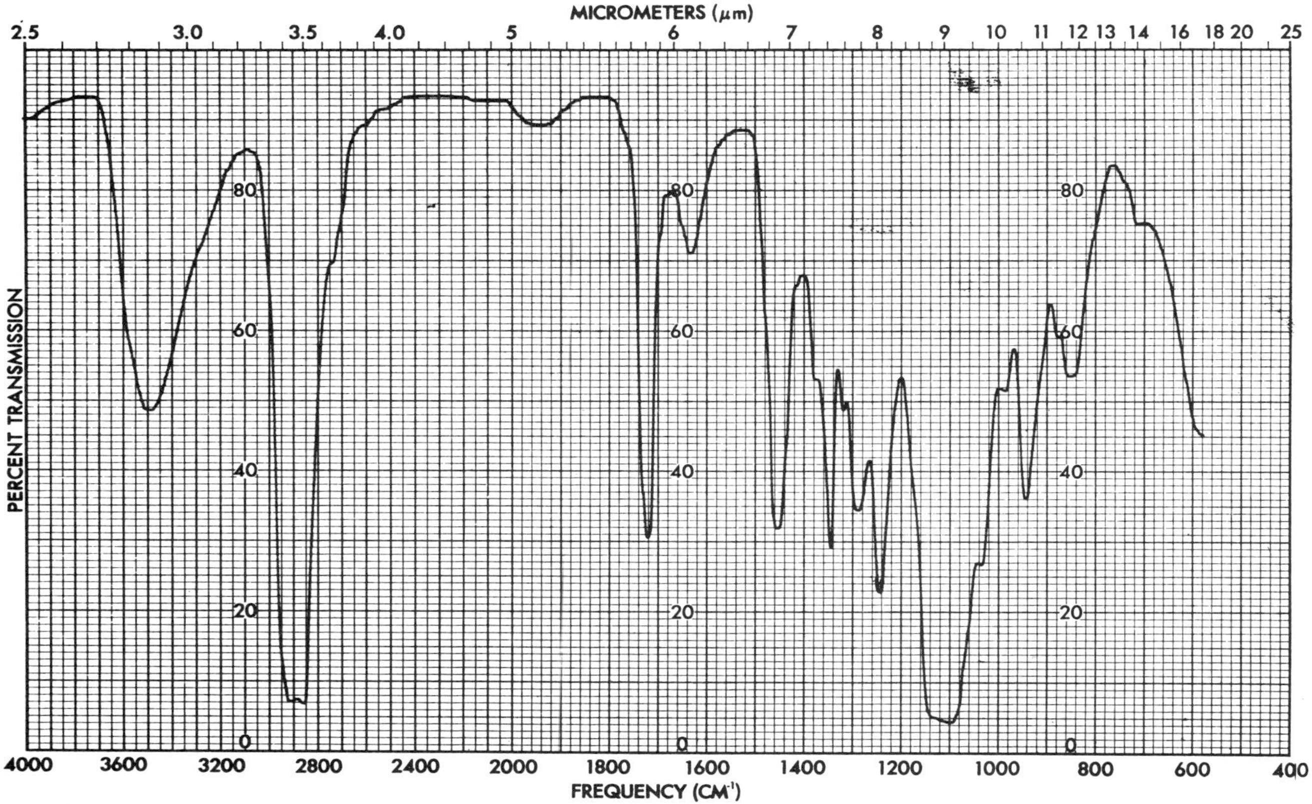

FIG. 6.43. Infrared spectrum of $(EO)_{10}$ tall oil fatty acid.

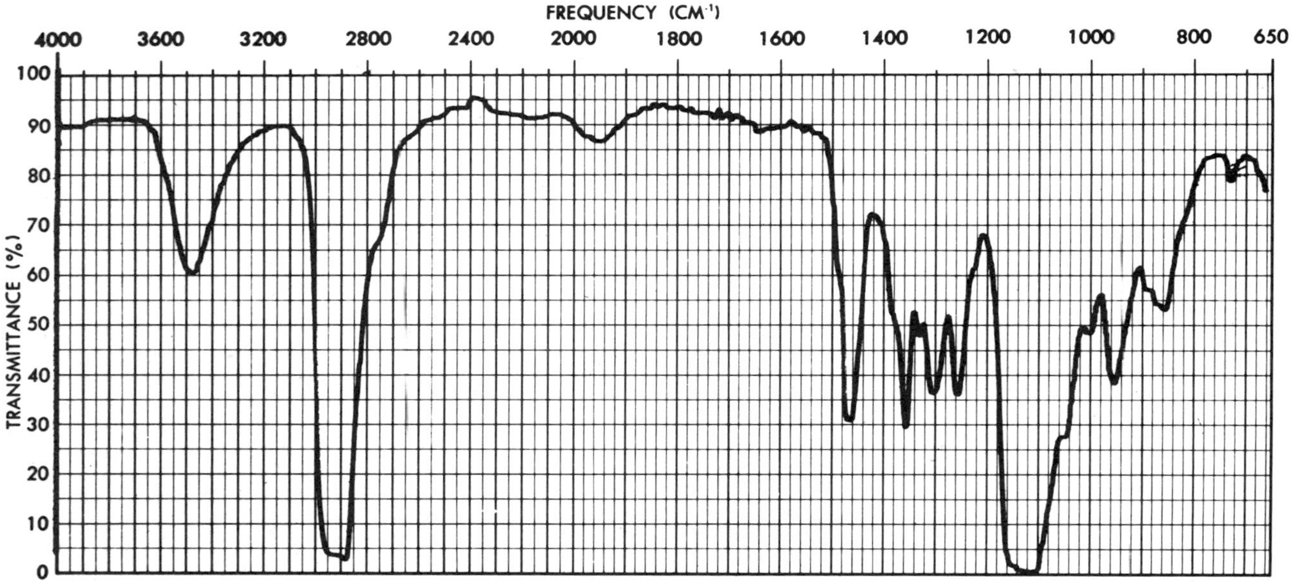

FIG. 6.44. Infrared spectrum of $(EO)_{12}$ secondary alcohol.

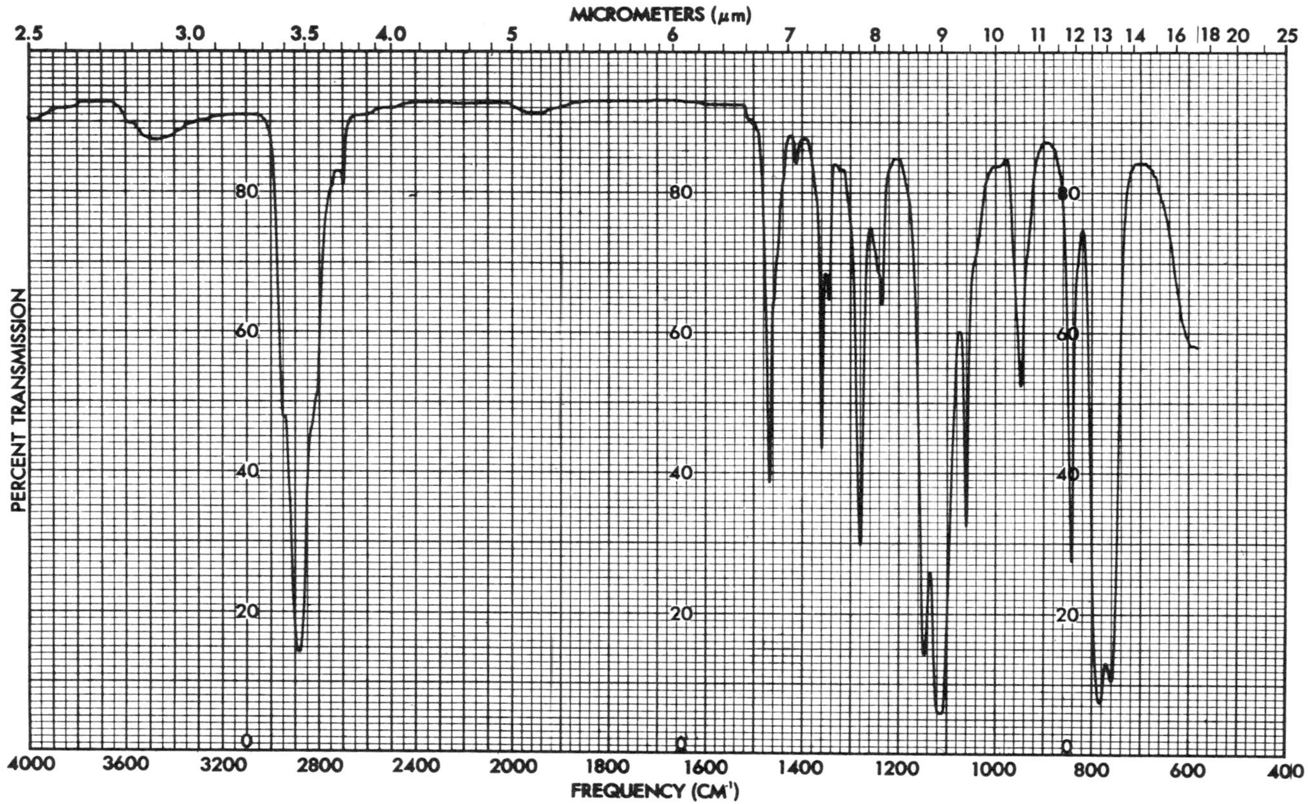

FIG. 6.45. Infrared spectrum of $(EO)_{50}$ nonyl phenol.

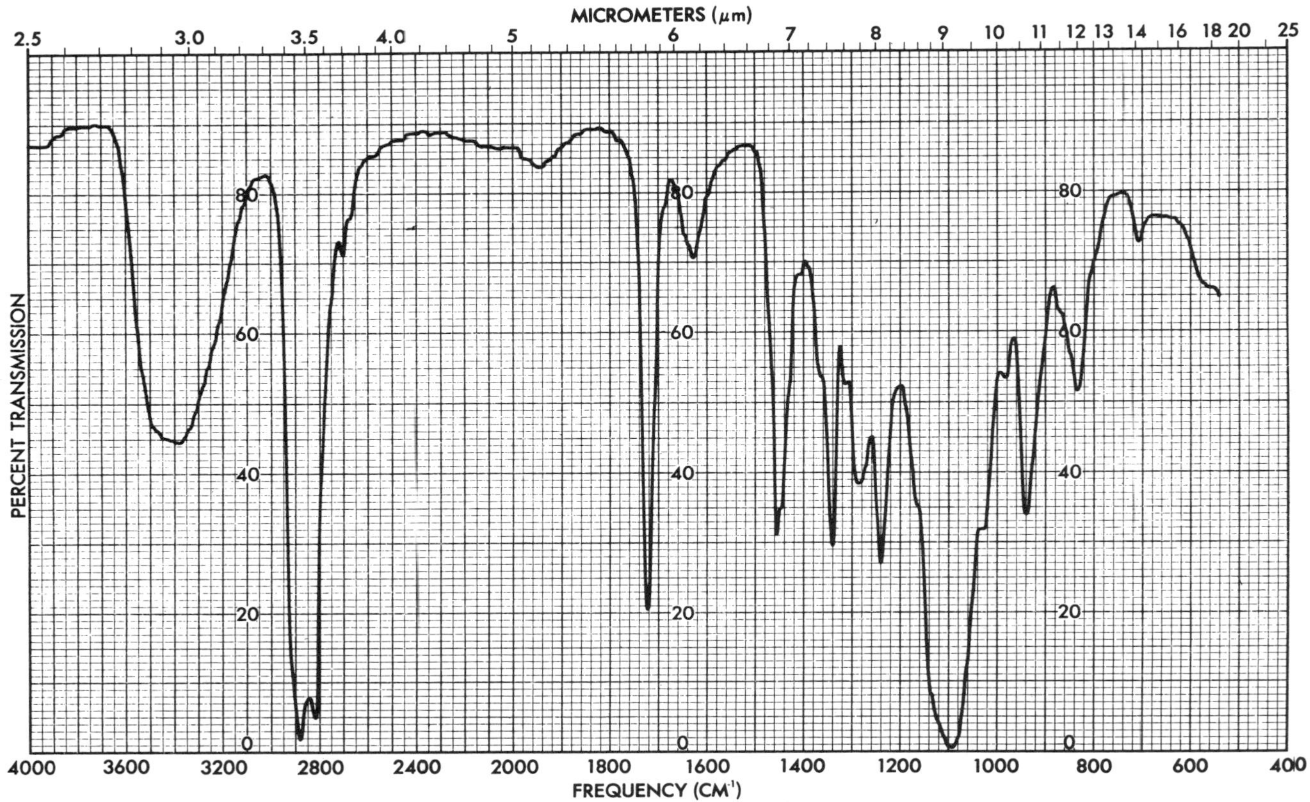

FIG. 6.46. Infrared spectrum of $(EO)_{40}$ castor oil.

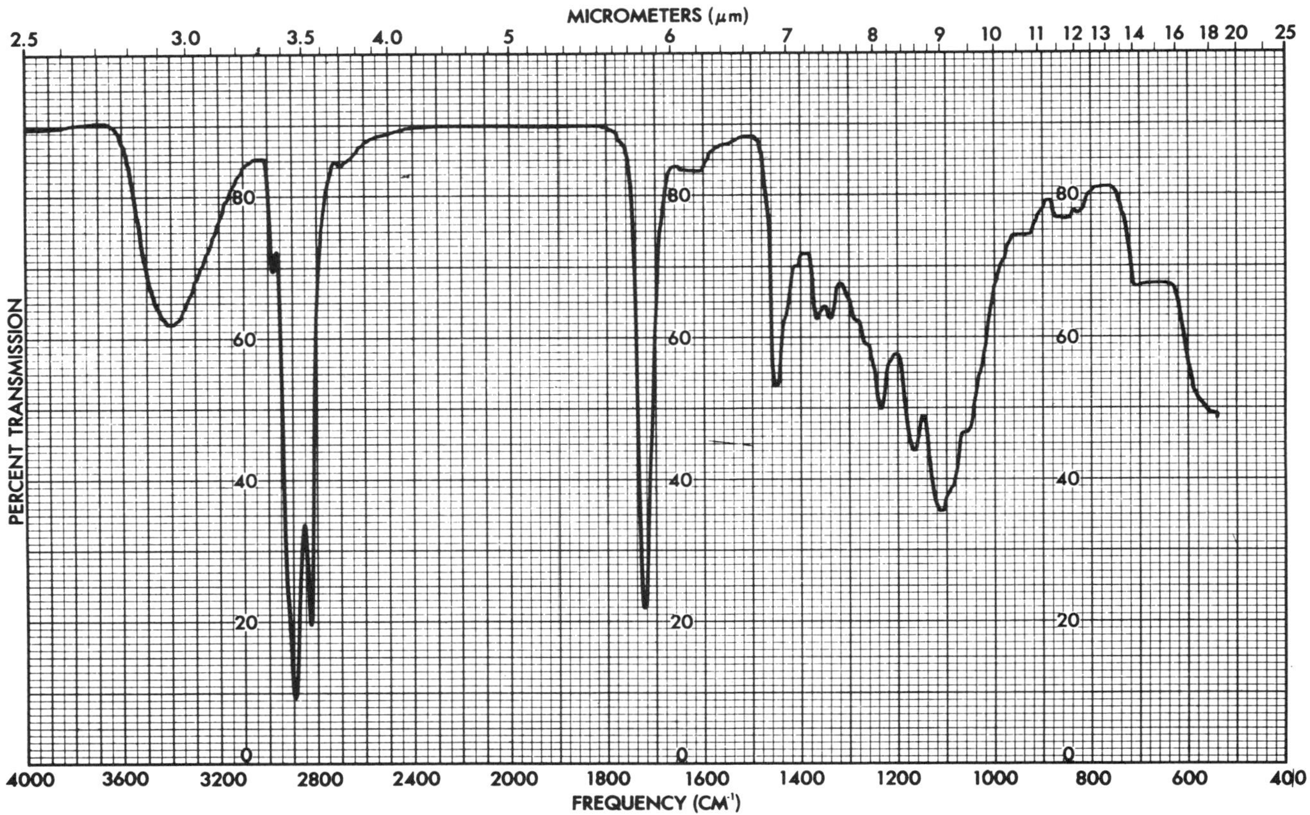

FIG. 6.47. Infrared spectrum of $(EO)_5$ castor oil.

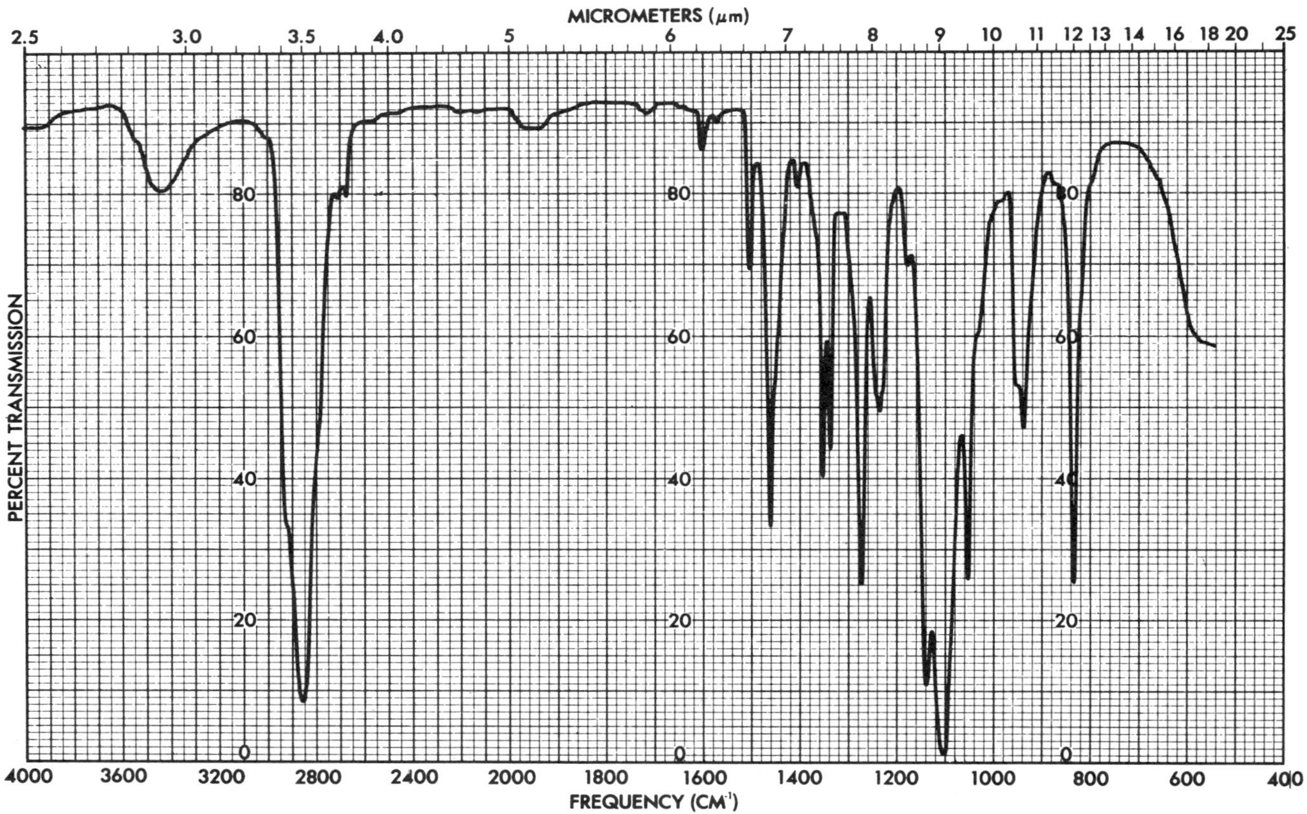

FIG. 6.48. Infrared spectrum of $(EO)_{30}$ nonyl phenol.

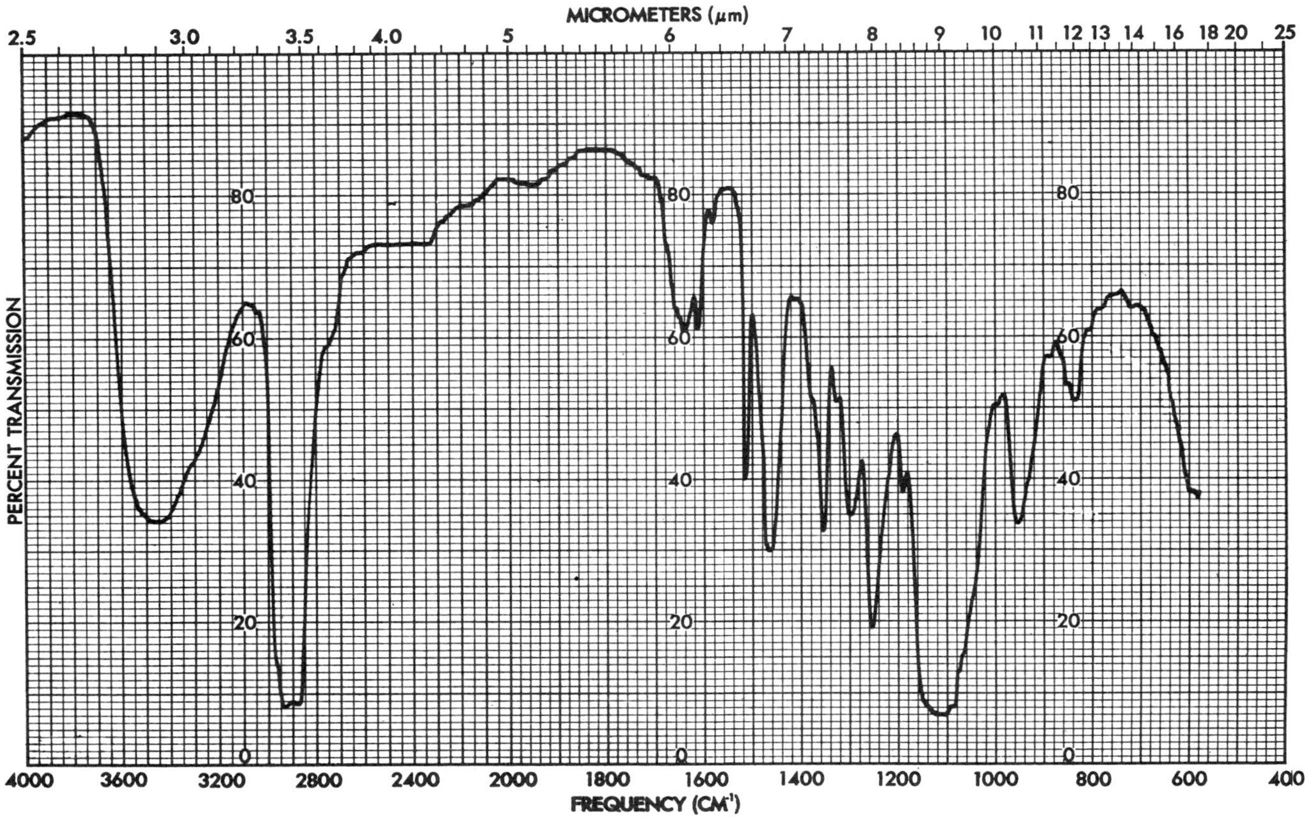

FIG. 6.49. Infrared spectrum of $(EO)_{30}$ nonyl phenol/$(EO)_{20}$ oleyl alcohol.

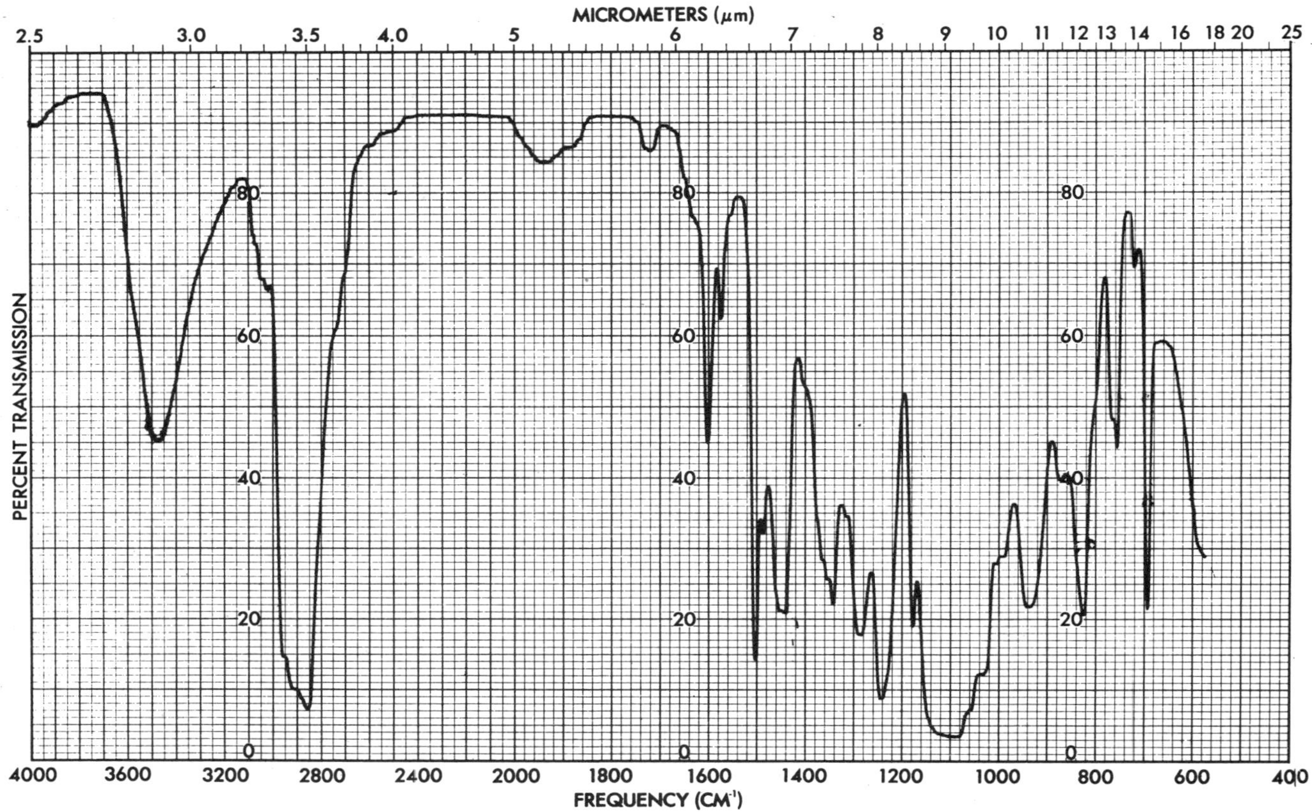

FIG. 6.50. Infrared spectrum of (EO$_{10}$ *p*-cumyl phenol.

Nonionic detergents, as used by the textile industry, are usually the ethylene or propylene oxide condensates of primary or secondary alcohols. They are also manufactured from ethylene and propylene oxide derivatives of octyl or nonyl phenol, dodecyl mercaptans, sorbitols and sorbitans, and fatty acids. By varying the amount of ethylene or propylene oxide added to the parent compound, one can vary the water–oil solubility characteristics of the final product. Ethylene oxide–propylene oxide derivatives may be prepared from any compound containing a labile hydrogen that is attached to either a nitrogen or an oxygen atom, such as amines, vegetable oils, fatty acids, alcohols, or phenols. The end use of the product determines the amount of ethylene or propylene oxide used, with ethylene oxide being more water soluble and propylene oxide more oil soluble. Low-EO-content products have a tendency to be more oil soluble, whereas higher-EO-content products have a tendency to be more water soluble.

Ethylene oxide compounds can be used alone but also work very well in combination with phosphate esters, sulfates, and soaps. With the exception of a few products that are made from mixtures of ethylene, propylene, and butylene oxide and used as defoamers, there are no commercially available butylene oxide-containing surfactants in common use by the textile industry today. Pictured in Figs. 6.36–6.50 are infrared spectra of miscellaneous ethylene oxide–propylene oxide derivatives.

Interpretation of Infrared Spectra of Nonionic Surfactants

Absorption wavelengths for nonionic surfactants are given in Tables 6.10 through 6.17.

TABLE 6.10. Absorption Wavelengths for Primary Alcohols

Wavelength, μm	Group	Type
2.8	OH	Alcohol
3.3	CH=CH	Unsaturated methylene
3.4	CH_2	Methylene
3.5	CH_3	Methyl
6.8	$(CH_2)_n$	Methylene
7.3	CH_3	Terminal
9.5	C—O	Alcohol

Ethoxylated Primary Alcohol (German Patent 623,482 to I. G. Farben)

One of the most simple ethylene oxide reactions is that of a primary alcohol with ethylene oxide:

$$ROH + n(\underset{\diagdown\,O\,\diagup}{CH_2{-}CH_2})_n \longrightarrow R{-}O{-}(C_2H_4O)_nH$$

where n = moles of ethylene oxide.

An ethoxylated oleyl alcohol is formed by the reaction:

$$CH_3(CH_2)_7CH{=}CH(CH_2)_7CH_2OH + n(\underset{\diagdown\,O\,\diagup}{CH_2{-}CH_2}) \longrightarrow$$

$$CH_3(CH_2)_7{-}CH{=}CH(CH_2)_7CH_2O(C_2H_4O)_nH$$

The most important identification characteristics are the hydroxyl at 2.8 μm, the CH_2 bands at 7.4 and 7.8 μm, the C—O—C band at 9.0 μm, and the C—OH band at 10.5 and 11.8 μm. Unbranched long-chain aliphatics have peaks at 13.8 μm.

Since the CH_3 methyl group concentration is low, the vibration at 7.3 μm is only a weak shoulder, and the unsaturated characteristic of the oleyl group can be seen as a shoulder at 3.3 μm.

TABLE 6.11. Absorption Wavelengths for Ethoxylated Primary Alcohol

Wavelength, μm	Group	Type
2.8	OH	EO alcohol
3.3	CH=CH	Unsaturated methylene
3.4	CH_2	Methylene
3.5	CH_3	Methyl
6.9	$(CH_2)_n$	Methylene
7.4	$(CH_2)_n$	CH_2 of EO
9.0	C—O—C	Ether
10.5	C—OH	Alcohol
11.8	CH_2	CH_2OH

Ethoxylated primary alcohols are used basically as detergents, cleaning agents, and dye leveling compounds, whereas secondary or highly branched chains are used to make wetting agents.

Ethoxylated Secondary Alcohols

In a highly branched chain such as isotridecyl alcohol or trimethylnonyl alcohol, the band at 13.9 μm does not appear. The band at 7.25 μm is more pronounced as a result of the higher CH_3 methyl group concentration. The presence of an isopropyl group shows up as a distinct band at 7.3 μm. The structure of a secondary alcohol ethoxylate is very similar to that of the primary alcohol since the molecular arrangement is similar:

$$\begin{array}{c} CH_3{-}(CH_2)_n{-}\underset{\displaystyle O(CH_2CH_2O)_y{-}H}{\underset{|}{CH}}{-}(CH_2)_m{-}CH_3 \end{array}$$

Polyethylene Glycol Esters (U.S. Patent 2,275,494 to Harry Bennett)

To confer solubility to fatty residues, polyethylene glycol polymers are reacted with them to form mono- or diesters. An example of this type of ester is polyethylene glycol 400 monooleate, which has a structure of:

$$RCOOR \text{ or } CH_3(CH_2)_7CH{=}CH(CH_2)_7COO(C_2H_4O)_nH$$

TABLE 6.12. Absorption Wavelengths for Ethoxylated Secondary Alcohol

Wavelength, μm	Group	Type
2.8	OH	Hydroxyl
3.4	CH_2	Methylene
3.5	CH_3	Methyl
6.8	$(CH_2)_n$	Methylene
7.3	CH_3	Terminal
7.4		
7.8	CH_2	EO
8.0		alcohol
9.0	C—O—C	Ether
10.5	C—OH	Of alcohol

TABLE 6.13. Absorption Wavelengths for Oleic Acid

Wavelength, μm	Group	Type
3.3	OH	COOH
3.4	CH_2	Methylene
3.5	CH_3	Methyl
3.7	OH	COOH
5.8	C=O	COOH
6.8	CH_2	$(CH_2)_n$
7.0	COOH	Carboxyl
7.7	CH_2	Methylene
10.5	COH	COOH
13.8	$(CH_2)_n$	Methylene

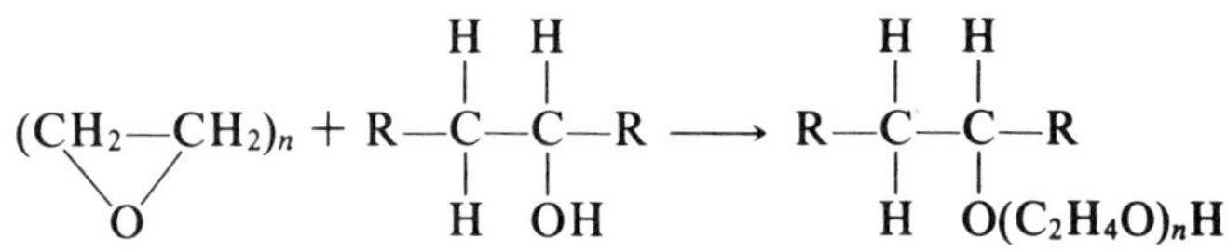

The structure of PEG-400 is characteristic of ethylene oxide polymers:

$$HO—(C_2H_4O)_nH$$

The major points of reference are the absence of a CH_3 group at 3.5 μm and peaks at 6.9, 7.4, 8.0, and 9.0 μm.

TABLE 6.14. Absorptions Wavelengths for Polyethylene Glycol 400

Wavelength, μm	Group	Type
2.8	OH	Hydroxyl
3.4	CH_2	Methylene
6.9	$(CH_2)_n$	Methylene
7.4, 7.8, 8.0	CH_2	EO CH_2
9.0	C—O—C	Ether
10.5	C—OH	Terminal OH

TABLE 6.15. Absorption Wavelengths for Polyethylene Glycol 400 Monooleate

Wavelength, μm	Group	Type
2.8	OH	OH of glycol
3.4	CH_2	Methylene
3.5	CH_3	Of COOH
5.8	COO	Ester
6.9	$(CH_2)_n$	Methylene
7.4 }		
7.6 }	CH_2	EO CH_2
8.0 }		
9.0	C—O—C	Ether linkage
10.5	C—OH	Terminal C—OH
11.5	CH_2	Of glycol

Ethoxylated Phenols (U.S. Patent 2,148,103 to Rohm & Haas)

Two major phenolic derivatives widely used by the textile industry are:

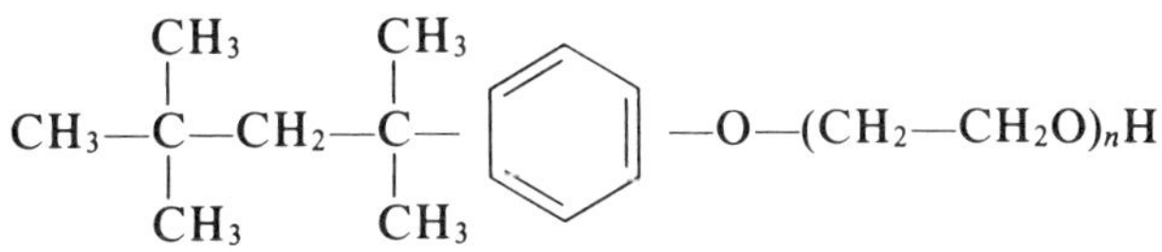

Ethoxylated octyl phenol

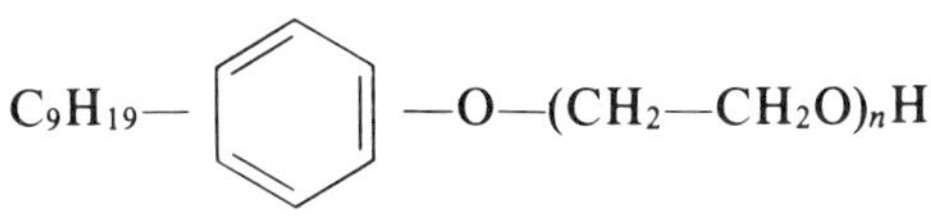

Ethoxylated nonyl phenol

In ethylene oxide phenol derivatives, the aromatic ring shows weakly at 6.7 μm and is very important in distinguishing this class of nonionic surfactants from ethoxylated alcohols. Another point of reference is at 6.1 μm, which indicates CH=CH of a ring compound. Note the apparent reversal of the normal CH_2/CH_3 bands at 3.4–3.5 μm as the increase in CH_3 content is shown at 3.5 and 7.3μm. The peak at 3.4 μm also gives a hint of aromatic structure, which is confirmed by the peak at 6.7 μm.

TABLE 6.16. Absorption Wavelengths for Ethoxylated Nonyl Phenol

Wavelength, μm	Group	Type
2.8	OH	EO
3.4	CH_3	Aromatic
3.5	CH_3	Methyl
6.2	C=C	Aromatic unsaturated
6.7	CH_2	Benzene ring
6.9	$(CH_2)_n$	Methylene
7.3	CH_3	Methyl
7.5 7.8	$(CH_2)_n$	Methylene
8.0	$AROCH_2$	Aromatic ether
8.1 8.6	R	Substituent on aromatic nucleus
9.0	C—O—C	Ether linkage
10.5	C—OH	Terminal COH
12.1	AR	1,4-Substituted benezene

TABLE 6.17. Absorption Wavelengths for Thioether

Wavelength, μm	Group	Type
2.8	OH	Hydroxyl
3.4	CH_2	Methylene
3.5	CH_2	Methyl
3.8–3.9	SH	Sulfur–hydrogen bond
6.8	$(CH_2)_n$	Methylene
7.3	CH_3	Terminal methyl
7.4 7.8 8.0	CH_2	CH from EO
9.0	SO COS	Sulfur–oxygen Carbon–oxygen
10.5	C—S—C	Thioether
11.5	COH	Terminal COH
12.5	C—S	Mercaptan

Ethoxylated Thioethers (French Patent 780,144 to I. G. Farben)

Ethoxylated thioethers are alkoxylates composed of both sulfur and oxide molecules. These compounds are thiol analogs similar to alcohol alkoxylates. They perform similarly to the alcohol or alkylaryl structures but show two outstanding properties: (1) they have an exceptionally high potential for carbon soil detergency; and (2) they are readily biodegradable. Thioethers, when viewed by infrared spectrosophy, are very difficult to distinguish from ethoxylated alcohols or phenols in the 2.5–15-μm range. However, by simply heating a sample, one may readily discern the characteristic mercaptan or hydrogen sulfide odor. Literature shows the structure for ethoxylated thioether as:

$$CH_3\text{—}S\text{—}C_2H_4\text{—}(OCH_2\text{—}CH_2)H$$

Note the peaks at 6.9, 7.3, 7.4, 7.8, and 8.0 μm denoting ethylene oxide. The peak at 9.0 μm is the thioether band. Thioethers also show a characteristic peak at 14.5 μm.

Other Typical Nonionic Structures

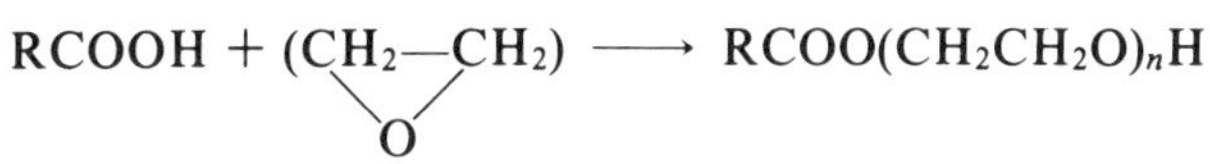

EO plus fatty acid (British Patent 563,725 to British Nylon Spinners)

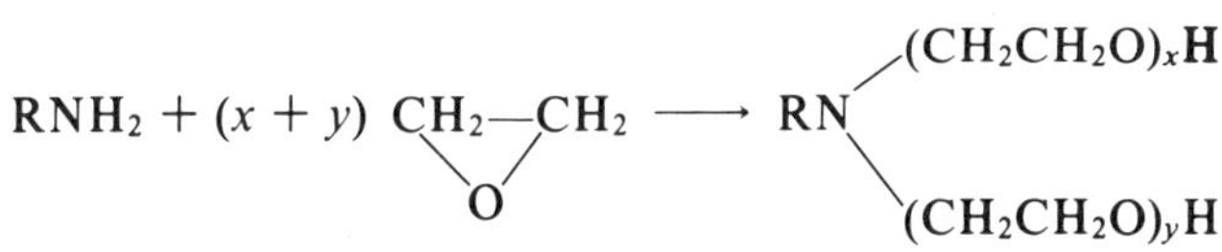

EO plus amine (British Patent 420,518 to ICI)

$$RCOO(CH_2CHCH_3O)H$$

Propylene oxide derivative (U.S. Patent 2,174,760 to I. G. Farben)

The structure of sorbitan ester (U.S. Patents 3,018,281 and 3,021,324) is:

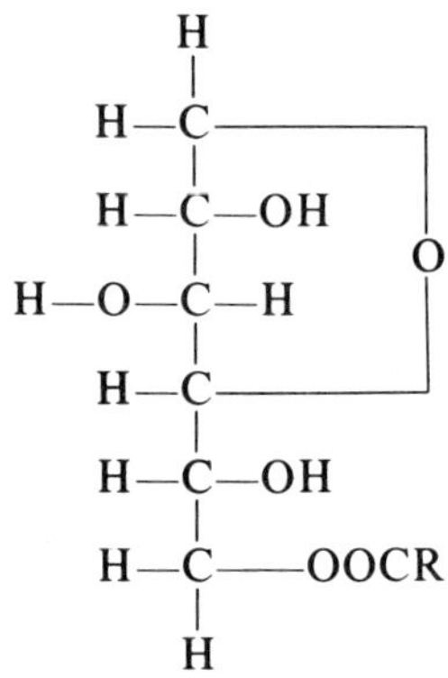

Sorbitan ester

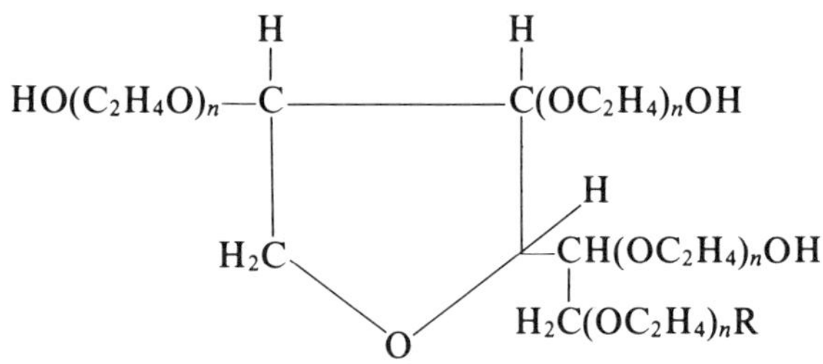

Ethoxylated sorbitan ester (U.S. Patents 2,322,820; 2,322,821; and 2,380,166 to Atlas)

HLB System

In describing nonionic molecules, one terms the fatty portion the ***hydrophobe*** and the alklylene oxide polymer, the ***hydrophile***. These Greek words mean "water hating" and "water loving," respectively. It is the ratio of hydrophobe to hydrophile (hydrophilic–lipophilic balance) that determines the solubility of the nonionic surfactant. As a rule of thumb, the more ethylene oxide (EO) in a molecule, the more water soluble it becomes. Conversely, the more propylene oxide a molecule has, the more oil soluble it becomes. To control foam, a mixture of propylene oxide and ethylene oxide are used simultaneously.[105] Table 6.18 lists common hydrophobes and hydrophiles.

TABLE 6.18. Common Hydrophobes and Hydrophiles

Hydrophobe		Hydrophile	
Dodecyl	$(C_{12}H_{25})$	Polyoxyethylene	$(C_2H_4O)_n$
Tetradecyl	$(C_{14}H_{29})$	Polyoxypropylene	$(C_3H_6O)_n$
Hexadecyl	$(C_{16}H_{33})$	Sulfate	(SO_4^{2-})
Octadecyl	$(C_{18}H_{37})$	Sulfonate	(SO_3^{2-})
Heptadecenyl	$(C_{17}H_{33})$	Phosphate	(PO_4^{2-})
		Carboxyl	(COOH)
		Aminosulfate	(NH_2HSO_4)
		Substituted ammonium	$(N^+H_3R, N^+H_2R_2, N^+HR_3, NR_4)$

Cloud Point

A peculiar phenomenon that is characteristic of nonionic surfactants and is a direct indication of the solubility of the surfactant in water is "cloud point." As the temperature of the bath increases, water solubility of nonionic surfactants decreases inversely. That temperature at which the nonionic surfactant becomes insoluble and relatively ineffective is a characteristic specific to a given hydrophobe–hydrophile combination and is characterized by cloud point. Nonionic surfactants emulsify oily residues by being absorbed into the compounds and emulsifying them and lifting the oily compounds from the fabric. When the cloud point is exceeded, nonionics lose the ability to emulsify the oils. It has been postulated that nonionics are most effective within 5° of their cloud point. The exception to this rule is a nonionic that is coreacted, or capped, with propylene oxide. Because of the propylene oxide, the surfactant appears to continue to function above its cloud point. It is felt that these surfactants emulsify the residues by being absorbed into them, and remaining with them, through later stages of processing. When the temperature is once again lowered below the cloud point of the surfactants, these surfactants become soluble again and remove the oily residues into the processing liquor.

One method of overcoming the problem of cloud point is to blend the nonionic with an anionic such as a soap, a sulfonate, or a phosphate. Table 6.19 shows the cloud points of several typical nonionic surfactants.

TABLE 6.19. Typical Cloud Points of Nonionic Surfactants[a]

	1% Solution
Primary alcohol	
3 moles EO	<0
5 moles EO	<0
7 moles EO	50°C
9 moles EO	60°C
12 moles EO	90°C
Secondary alcohol	
3 moles EO	<0
5 moles EO	<0
7 moles EO	37°C
9 moles EO	60°C
12 moles EO	90°C
Nonyl phenol	
4 moles EO	<0
7 moles EO	22°C
9 moles EO	54°C
10.5 moles EO	63°C
Blocked copolymers	
Sample A	40°C
Sample B	21°C

[a] Courtesy of Union Carbide Corporation.

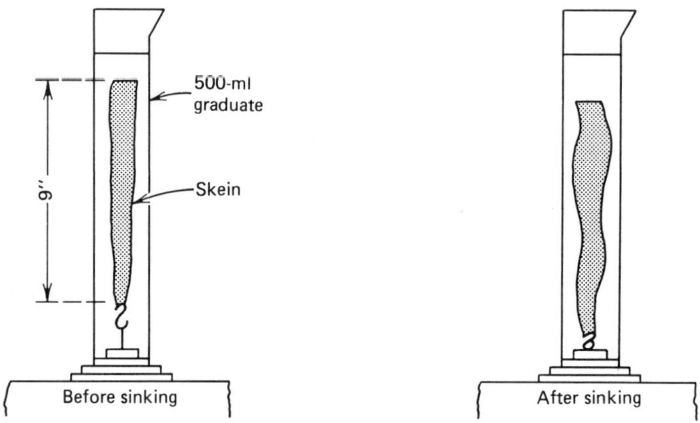

FIG. 6.51. Apparatus for running Draves wetting test. Courtesy of Union Carbide Corporation.

TABLE 6.20. Draves Wetting Time in Seconds at 25°C of Typical Nonionic Surfactants[a]

Product	Concentration 0.05	0.10	0.20
Primary alcohol (EO 7)	28	12	5
Primary alcohol (EO 7)	25	11	5
Secondary alcohol (EO 7)	18	6	2
Primary alcohol (EO 9)	28	15	8
Primary alcohol (EO 9)	26	11	5
Nonyl phenol (EO 9.5)	23	9	2
Secondary alcohol (EO 9)	20	7	3
Primary alcohol (EO 12)	68	38	21
Primary alcohol (EO 12)	33	23	14
Secondary alcohol (EO 12)	31	11	5

[a]Courtesy of Union Carbide Corporation.

Wetting Speeds

Nonionic surfactants are excellent wetting agents in commercial practice. Figure 6.51 shows the technique for evaluating wetting speed by the standard Draves wetting test.[106] Table 6.20 is a representation of wetting speeds at three different concentrations at equal temperatures. Wetting and scouring are normally associated as being synonymous; however, this unfortunately is not true. Experimental tests indicate that the best wetting agents are usually nonionics with low ethylene oxide content, whereas the best scouring compounds generally have a long-chain hydrophile and hydrophobe.

Table 6.21 shows Draves wetting time at elevated temperatures, and Table 6.22 shows Draves wetting time in hard water.

Amine Condensates

A second class of nonionic surfactant important in the preparation process is that termed *alkanolamide*, or those compounds known to the trade as *amine condensates.* These products are manufactured by one of two processes, the first of which produces the Kritchevsky-type liquid products derived from the reaction of a fatty acid or fatty acid derivative with an ethanolamine at a 1 : 2 ratio at elevated temperatures. Kritchevsky-type alkanolamides contain

TABLE 6.21. Draves Wetting Time in Seconds at 62.2° C[a]

Product	Concentration 0.05	0.10	0.20
Primary alcohol (EO 7)	41	28	20
Primary alcohol (EO 7)	25	17	10
Secondary alcohol (EO 7)	19	8	4
Primary alcohol (EO 9)	17	7	3
Primary alcohol (EO 9)	15	7	4
Nonyl phenol (EO 9.5)	17	6	2
Secondary alcohol (EO 9)	21	11	6
Primary alcohol (EO 12)	23	11	5
Primary alcohol (EO 12)	21	7	3
Secondary alcohol (EO 12)	35	11	3

[a] Courtesy of Union Carbide Corporation.

TABLE 6.22. Draves Wetting Time in Seconds at 25°C in Hard Water (300 ppm)[107]

Product	Concentration 0.05	0.10	0.20
Primary alcohol (EO 7)	36	12	4
Primary alcohol (EO 7)	28	9	4
Secondary alcohol (EO 7)	21	6	2
Primary alcohol (EO 9)	27	16	12
Primary alcohol (EO 9)	30	12	5
Nonyl phenol (EO 9.5)	19	7	2
Secondary alcohol (EO 9)	23	6	2
Primary alcohol (EO 12)	100+	51	24.4
Primary alcohol (EO 12)	51	27	14
Secondary alcohol (EO 12)	35	12	5

60–70% amide content plus varying percentages of amine esters, diesters, and piperazine derivatives that are formed by side reactions. There also remains an excess of alkanolamine available for further reaction, which renders these type products water soluble.

The second type of alkanolamide is the superamide, which is prepared by the reaction of a fatty acid ester and an alkanolamine in a 1 : 1 mole ratio. These products are generally solids and have an alkanolamide content of 90% or better. Superamides, because of their purity and their lack of excess alkanolamine content, have poor water solubility and are usually used with small quantities of additional anionic or nonionic surfactant that acts as a solubilizer for the superamides.

Amine condensates, although generally considered to be nonionic, are, as a result of their amine linkage, partially substantive to textile fibers. Therefore, they make excellent lubricants and are extremely efficient as soaping-off compounds because of their antiredeposition properties. With the proper choice of fatty acid alkanolamine content, they are useful in producing anti-crack mark–anticrease agents for use in the beck dyeing of acetates, nylons, and wools. Amine condensates, along with amine soaps, are the primary constituents of old-time wool scours.

The generic structure reported for the reaction of a carboxylic acid and diethanolamine is:

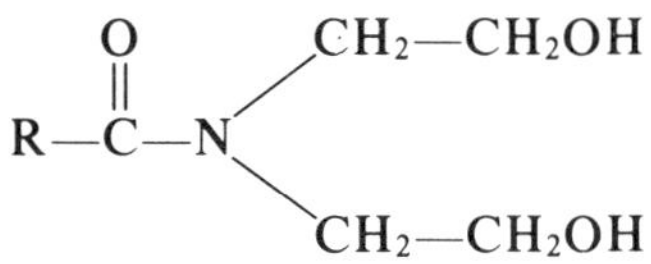

Amine condensate

Patent literature on the synthesis of amine condensates is extensive. The first patent was issued to Wolf Kritchevsky in 1937 when he was issued U.S. Patents 2,098,212 and 2,096,749. In his process, Kritchevsky used 1 mole of fat, fatty glyceride, fatty ester and 2 moles of diethanolamine. This reaction was carried out under atmospheric conditions at temperatures of 100–300°C.

In 1949 Edwin M. Meade was issued U.S. Patent 2,464,094 for the preparation of amides that used an alkoxide catalyst that allowed them to be synthesized under atmospheric conditions at temperatures of approximately 100°C.

In 1958 Guiliana C. Tesoro was issued U.S. Patent 2,844,609, in which sodium methoxide was used as a catalyst, and the reaction was run under a reduced pressure of 30–60 ml at a temperature of 55–75°C.

In 1958 Jack W. Schurman patented a continuous process under U.S. Patent 2,863,888, which was followed in 1962 by Robert Ernst with U.S. Patent 3,024,260 and John W. Lohr with U.S. Patent 3,040,075 with various modifications of the above processes. Additional references to other modifications and procedures are listed in the patent literature references at the end of this chapter.

Figures 6.52 through 6.57 show the effects of processing variables on amide content (courtesy of Dow Chemical Company.).[108]

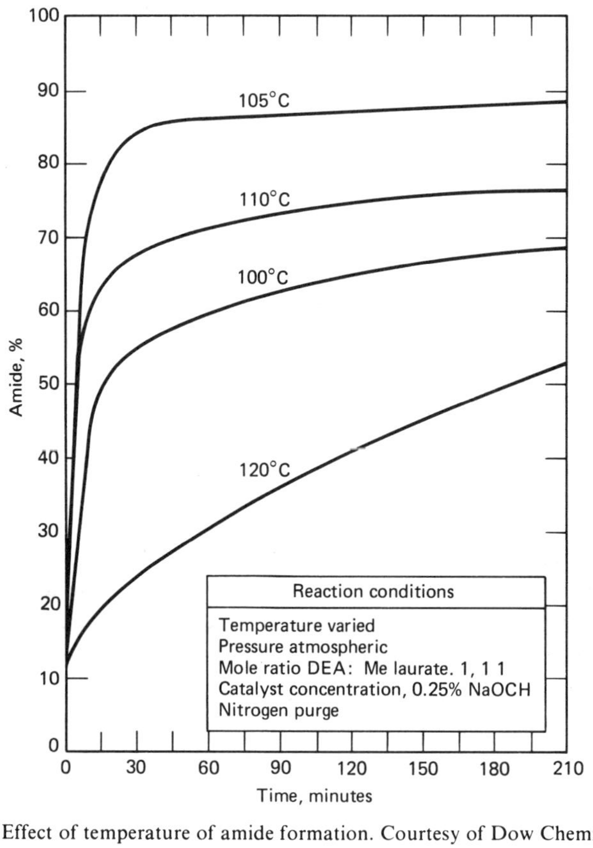

FIG. 6.52. Effect of temperature of amide formation. Courtesy of Dow Chemical Company.

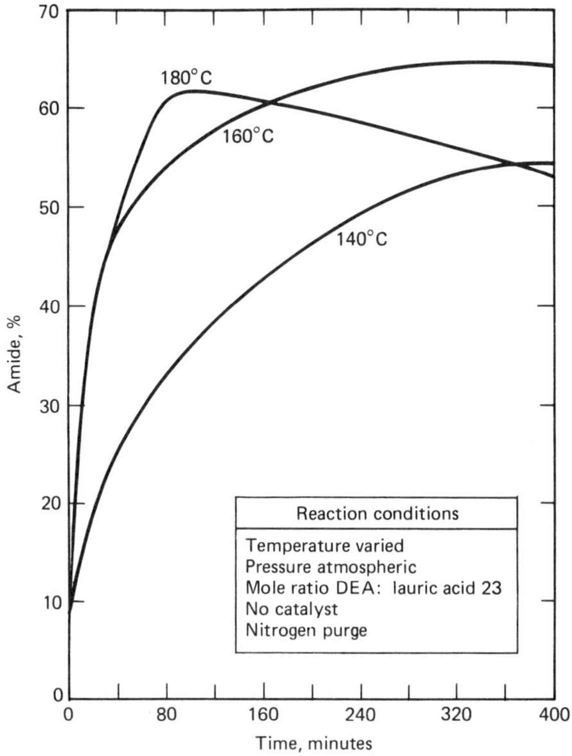

FIG. 6.53. Effect of temperature on amide formation. Courtesy of Dow Chemical Company.

Reaction of Monoethanolamine and Lauric Acid

The structure for this reaction is:

$$RNH_2 + RCOOH \longrightarrow R-\overset{\displaystyle O}{\overset{\|}{C}}-\underset{\displaystyle H}{\underset{|}{N}}-R + H_2O$$

$$\begin{matrix} H \\ \quad\diagdown \\ \quad\; N-CH_2-CH_2OH + CH_3(CH_2)_{10}COOH \longrightarrow \\ \quad\diagup \\ H \end{matrix}$$

Monoethanolamine + Lauric acid

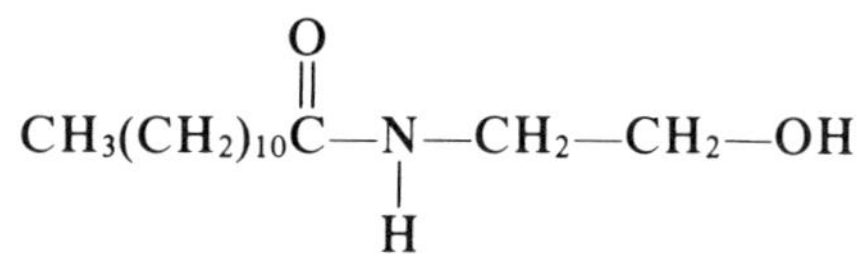

Amide

Absorptivity data for coconut fatty acid and coconut monoethanolamine amide are presented in Tables 6.23 and 6.24, respectively, and those for diethanolamine and oleic diethanolamine amide, in Tables 6.25 and 6.26, respectively.

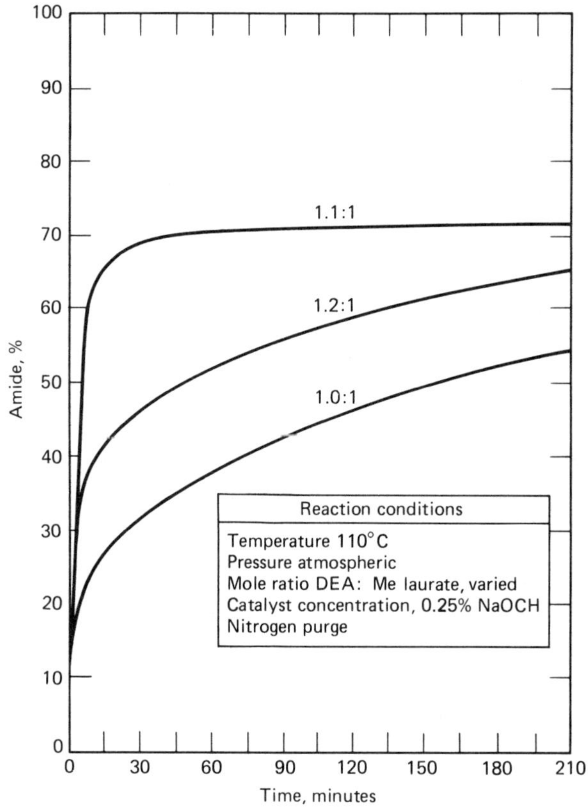

FIG. 6.54. Effect of mole ratio of reactants on amide formation. Courtesy of Dow Chemical Company.

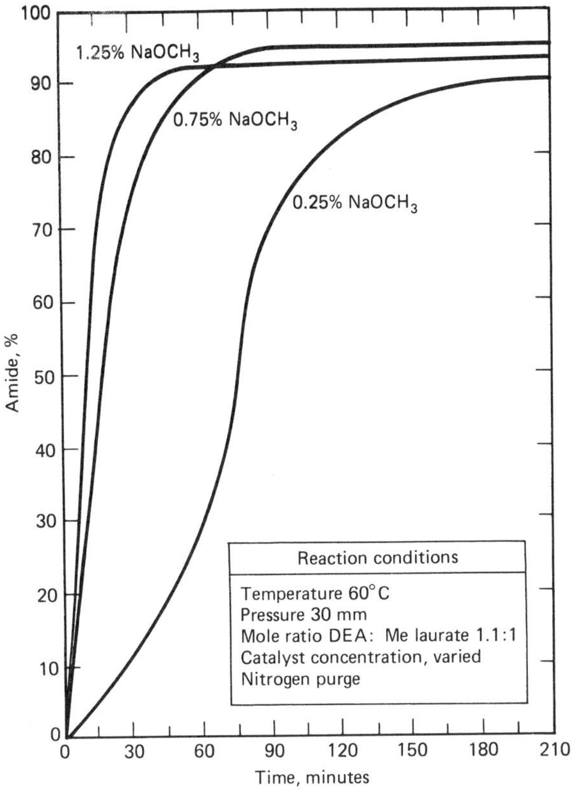

FIG. 6.55. Effect of amount of catalyst on amide formation at reduced temperature. Courtesy of Dow Chemical Company.

Infrared Spectra of Nonionic Amide Surfactants

Figures 6.58 through 6.69 show the infrared spectra of various fatty acid-alkanolamine reaction products at various mole ratios.

From the curves shown in Figs. 6.58 through 6.69, note the presence of large quantities in amine ester as shown by the ester carbonyl absorption at 5.8 μm and the 1,4 diethanol piperazine characterized by the absorption at 8.7 μm. Amide carbonyl absorption can be seen at 6.15 μm and carboxyl soap absorptions, at 6.5 μm.

These spectra show that vibrations of alkanolamides are fairly specific and generally fairly definitive in nature. Reactions and vibrational frequencies of major wavelength absorption bands are outlined in the following subsections.

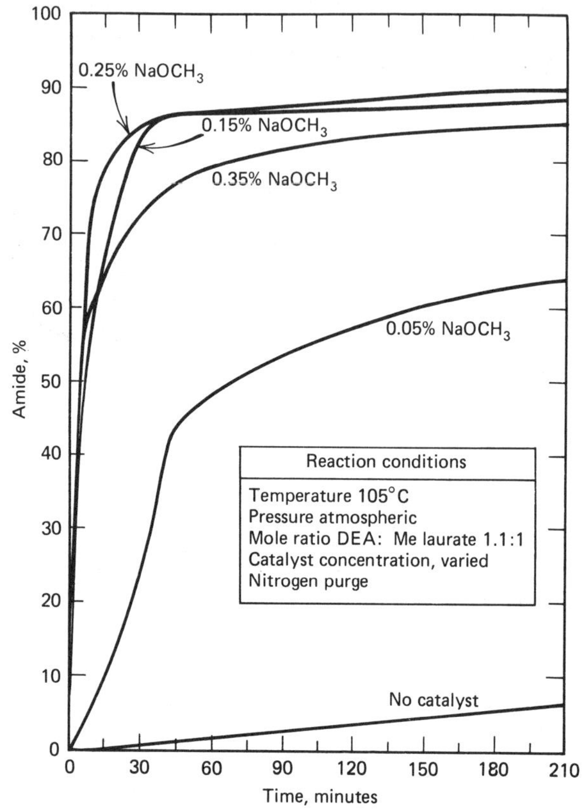

FIG. 6.56. Effect of sodium methylate catalyst on amide formation. Courtesy of Dow Chemical Company.

TABLE 6.23. Absorption Wavelengths for Coconut Fatty Acid

Wavelength, μm	Group	Type
3.4	CH_2	Methylene
3.5	CH_3	Methyl
5.8	C=O	Ester of COOH
6.5	COO	Free fatty acid
6.8	$(CH_2)_n$	Methylene
7.3	CH_3	Terminal methyl
8.0	$(CH_2)_n$	Methylene
8.6	COO^-	Of COOH
9.1	COO^-	Of COOH

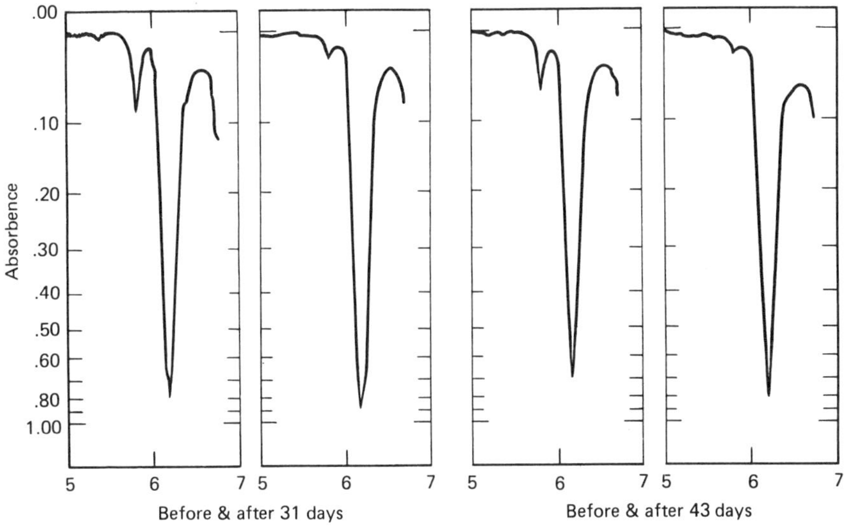

Reaction Conditions	Left-Hand Pair	Right-Hand Pair
Temperature	105°C	110°C
Pressure	Atmospheric	Atmospheric
Mole ratio, DEA: Me laurate	1.1: 1	1.2: 1
Catalyst concentration, $NaOCH_3$	0.15%	0.25%
Nitrogen purge	Yes	Yes

FIG. 6.57. Effect of prolonged standing on ester and amide formation. Courtesy of Dow Chemical Company.

TABLE 6.24. Absorption Wavelengths for Coconut Monoethanolamine Amide

Wavelength, μm	Group	Type
3.05	NH_2	Amine
3.25	NH	Amide
3.4	CH_2	Methylene
3.5	CH_3	Methyl
6.1	$R\overset{\overset{O}{\|}}{C}N$	Secondary amide
6.5	COO^-	Free fatty acid–soap
6.8	$(CH_2)_n$	Methylene
9.4	CH_2OH^-	C—O, Alcohol

TABLE 6.25. Absorption Wavelengths for Diethanolamine

Wavelength, μm	Group	Type
2.8	NH	Of amine OH
3.0	OH	Of amine
3.4	CH_2	Methylene
6.2	OH	Of amine
6.9	$(CH_2)_n$	Methylene
7.4	$(CH_2)_n$	Methylene
8.9	NH	Secondary amine
9.5	CH_2OH	Terminal OH
10.6	C—OH	Of amine

TABLE 6.26. Absorption Wavelengths for Oleic Diethanolamine Amide

Wavelength, μm	Group	Type
3.0	OH	Of alkanolamine
3.4	CH_2	Methylene
3.5	CH_3	Methyl
5.8	C=O	Of COOH
	CH=CH	
6.2	RC=ON	Of tertiary amide
6.5	COOH	Of free fatty acid
7.4	$(CH_2)_n$	Methylene
8.5	C=O	Of ester
9.4, 9.5 }	CH_2OH	Of amine
10.6	C—OH	Terminal group
13.8	$(CH_2)_n$	Methylene

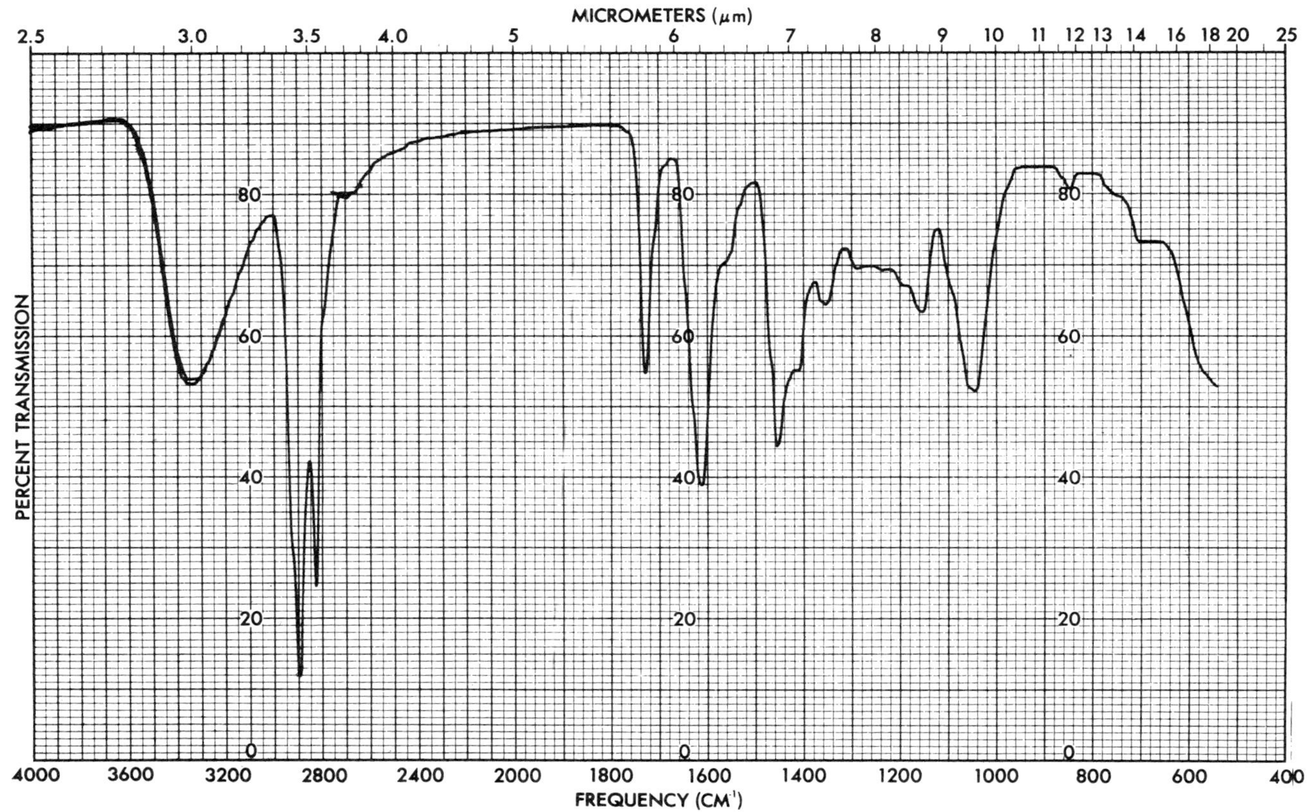

FIG. 6.58. Infrared spectrum of tall oil–DEA 1:2.

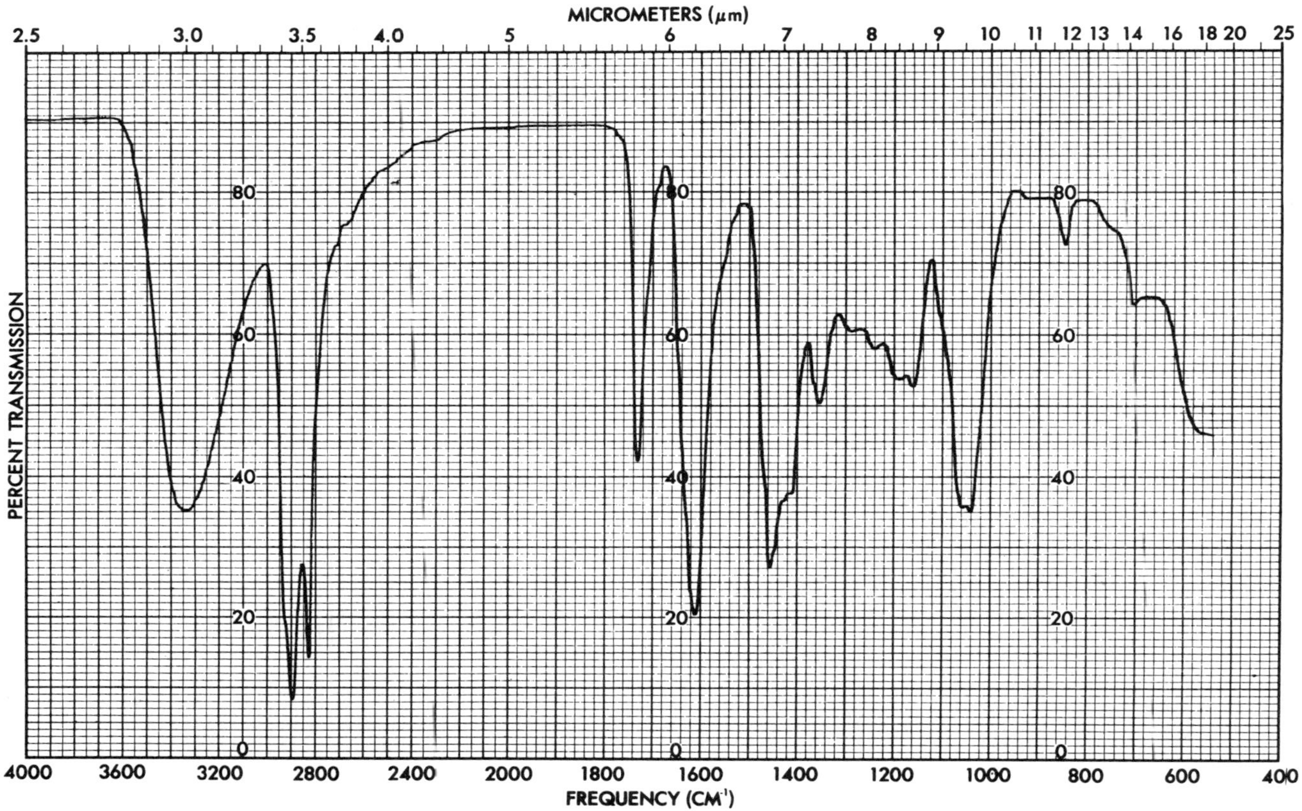

FIG. 6.59. Infrared spectrum of coconut fatty acid–DEA 1:1.

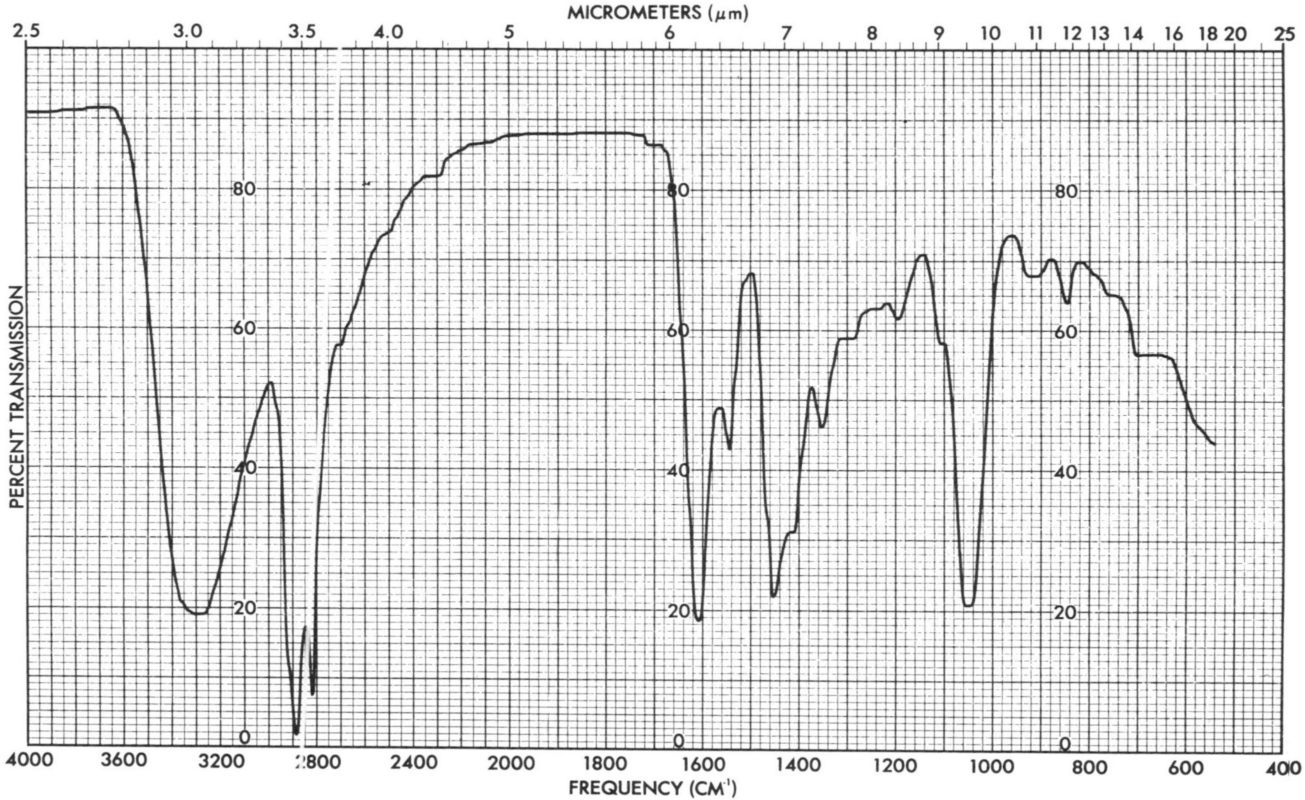

FIG. 6.60. Infrared spectrum of coconut fatty acid–DEA 1:2.

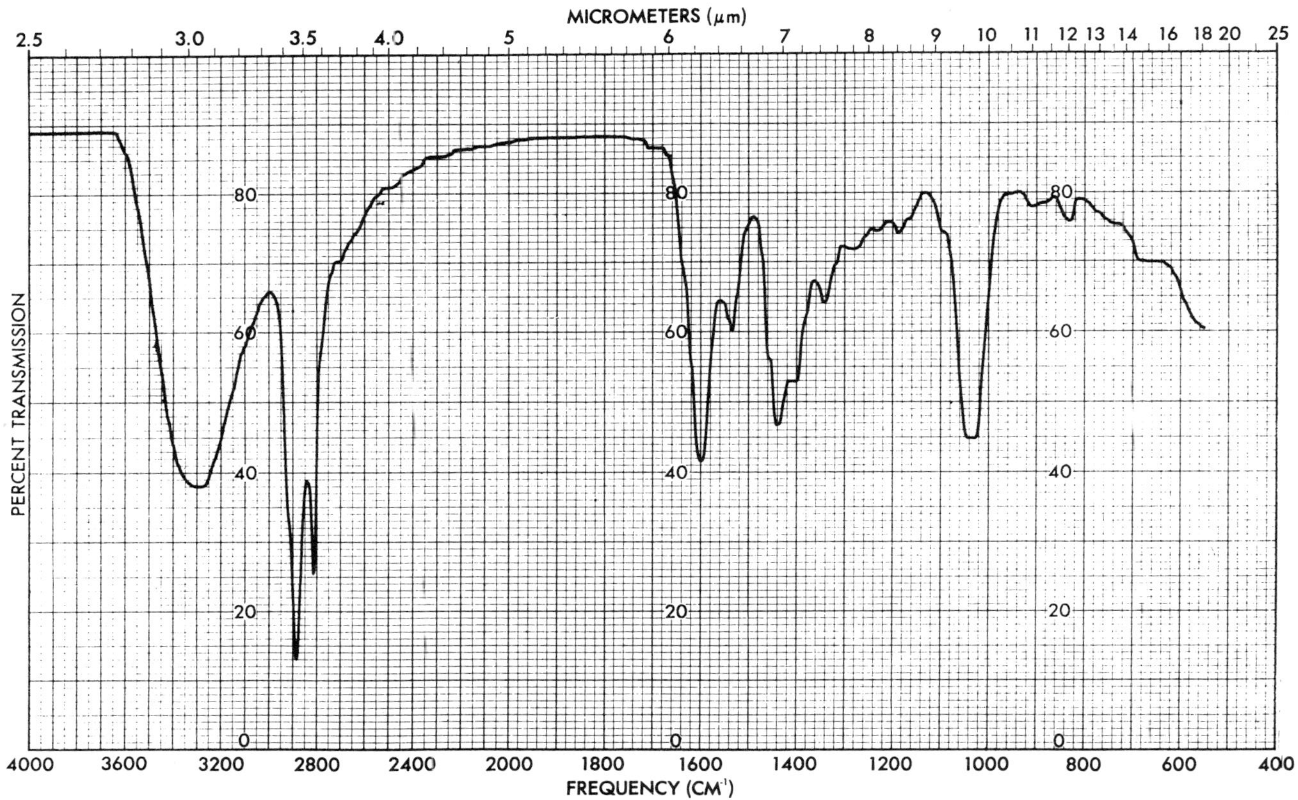

FIG. 6.61. Infrared spectrum of mixed fatty acid–DEA 1:2.

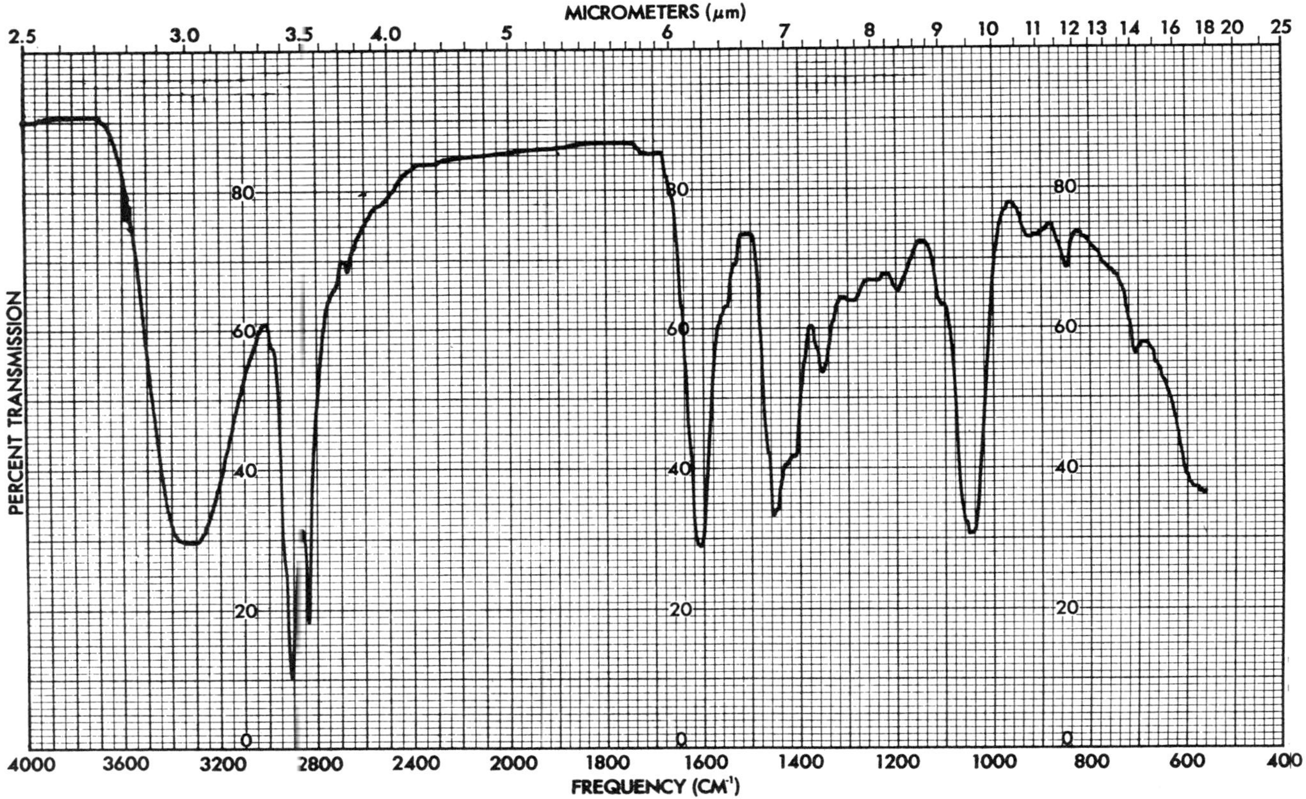

FIG. 6.62. Infrared spectrum of mixed fatty acid–DEA 1:2 ($NaOCH_3$ catalyst).

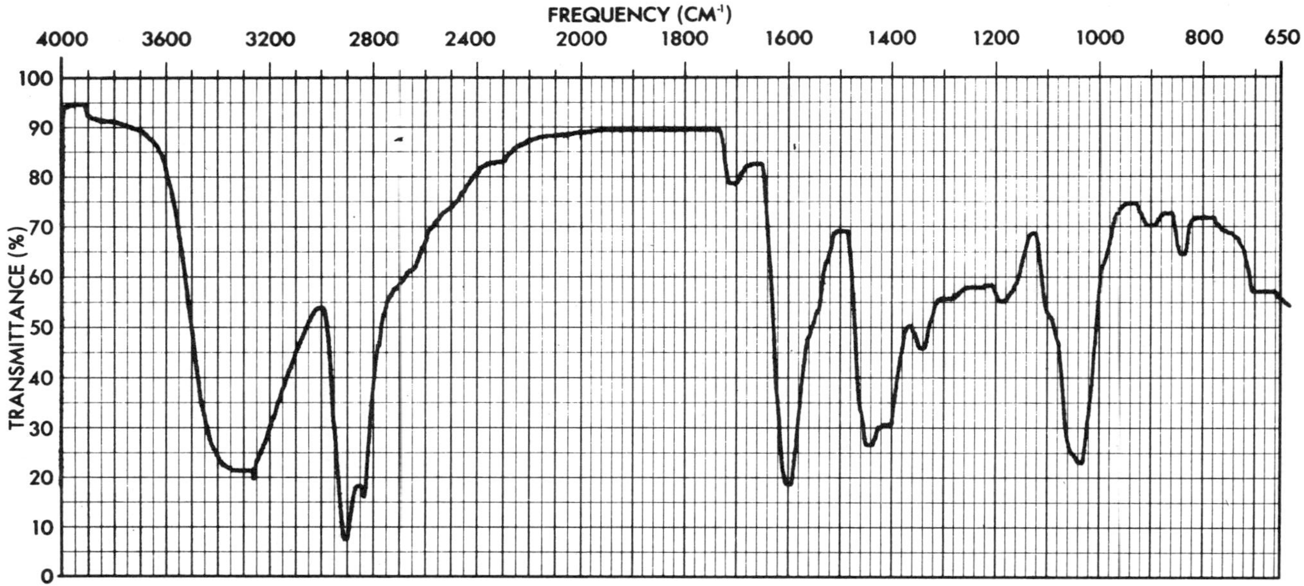

FIG. 6.63. Infrared spectrum of coconut oil–DEA 1:2 (KOH catalyst).

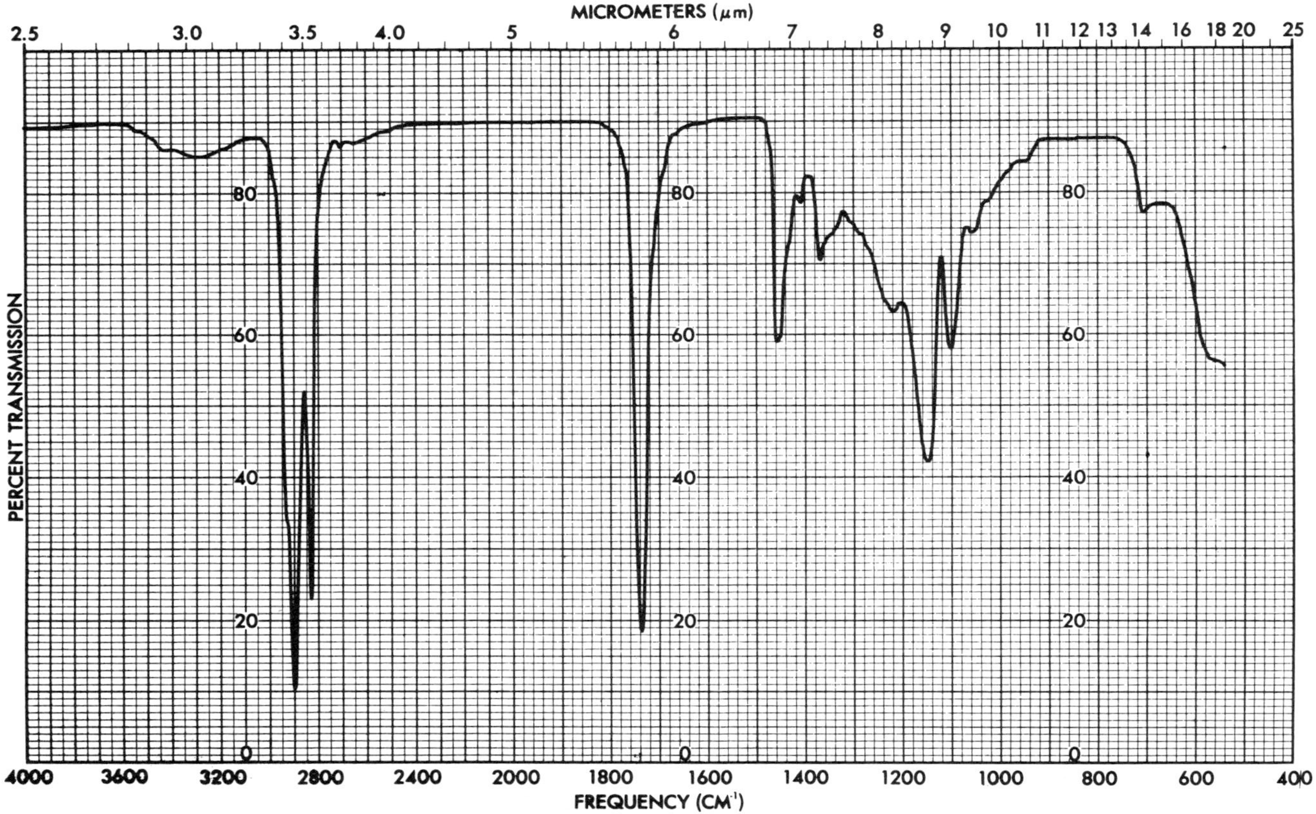

FIG. 6.64. Infrared spectrum of crude coconut oil.

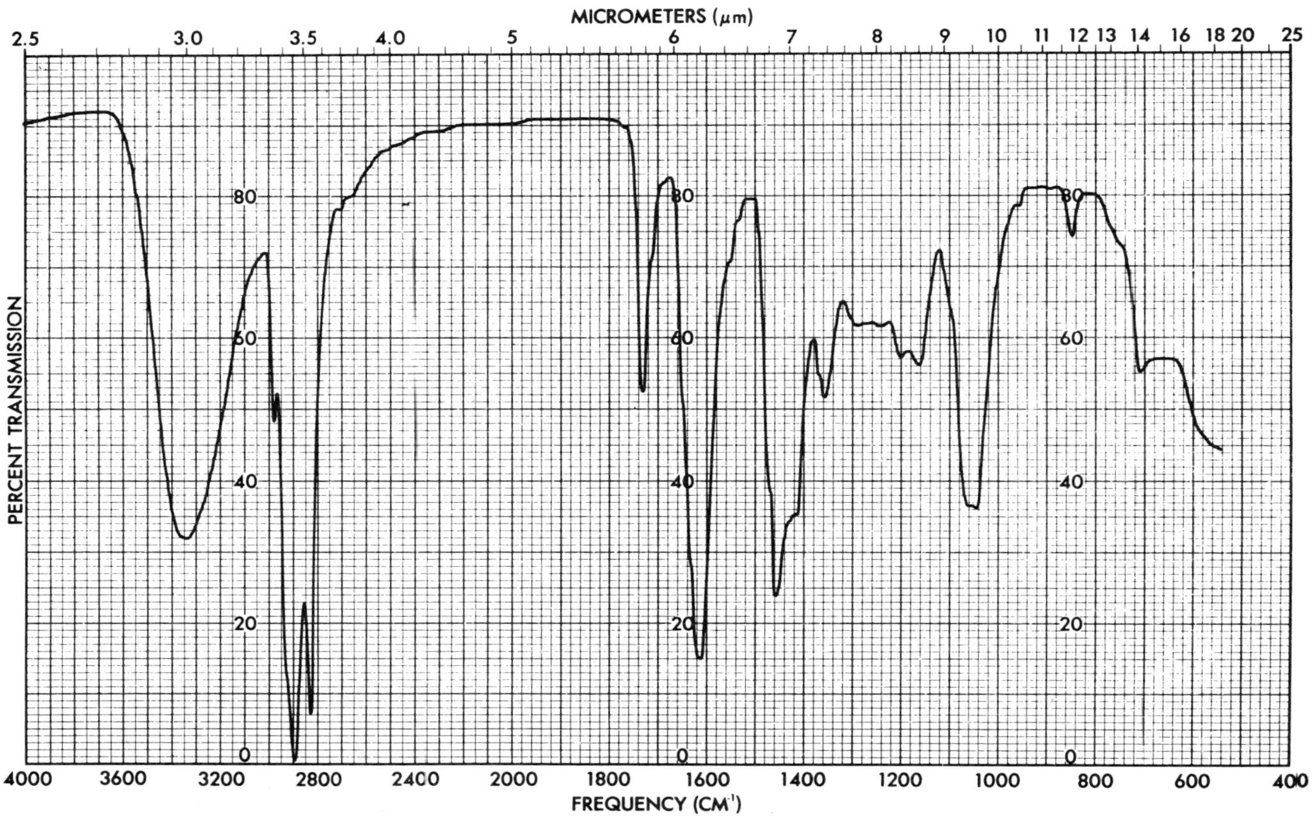

FIG. 6.65. Infrared spectrum of oleic acid–DEA 1:1.2.

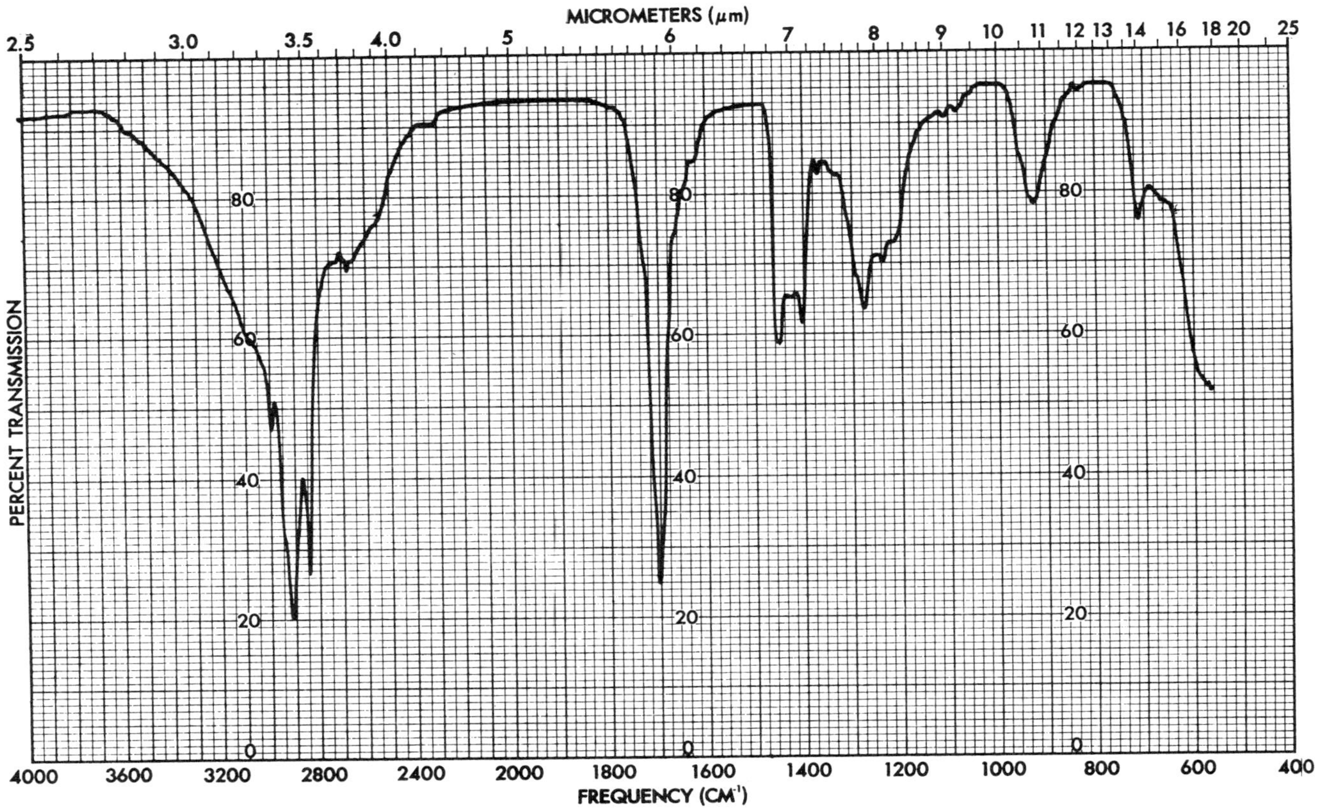

FIG. 6.66. Infrared spectrum of oleic acid.

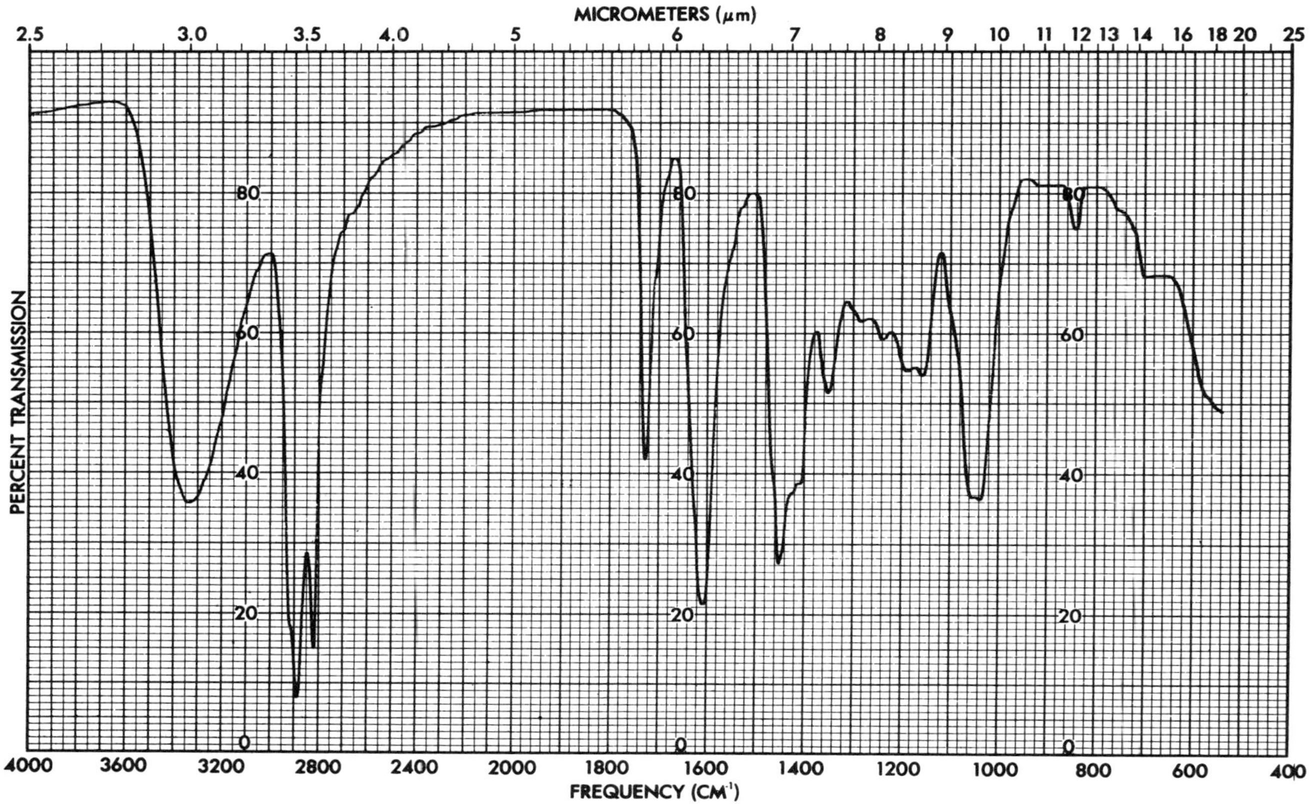

FIG. 6.67. Infrared spectrum of methyl laurate–DEA 1:1.

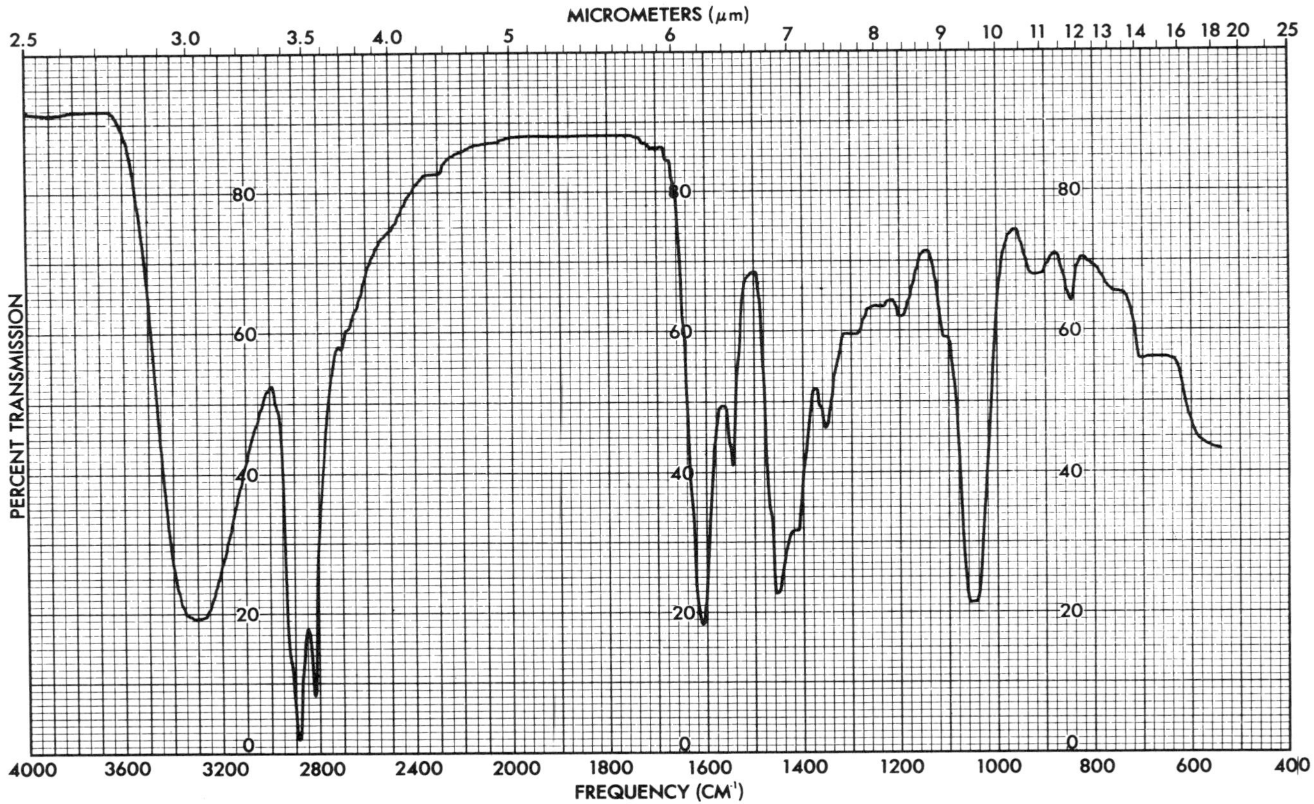

FIG. 6.68. Infrared spectrum of methyl laurate–DEA 1:2.

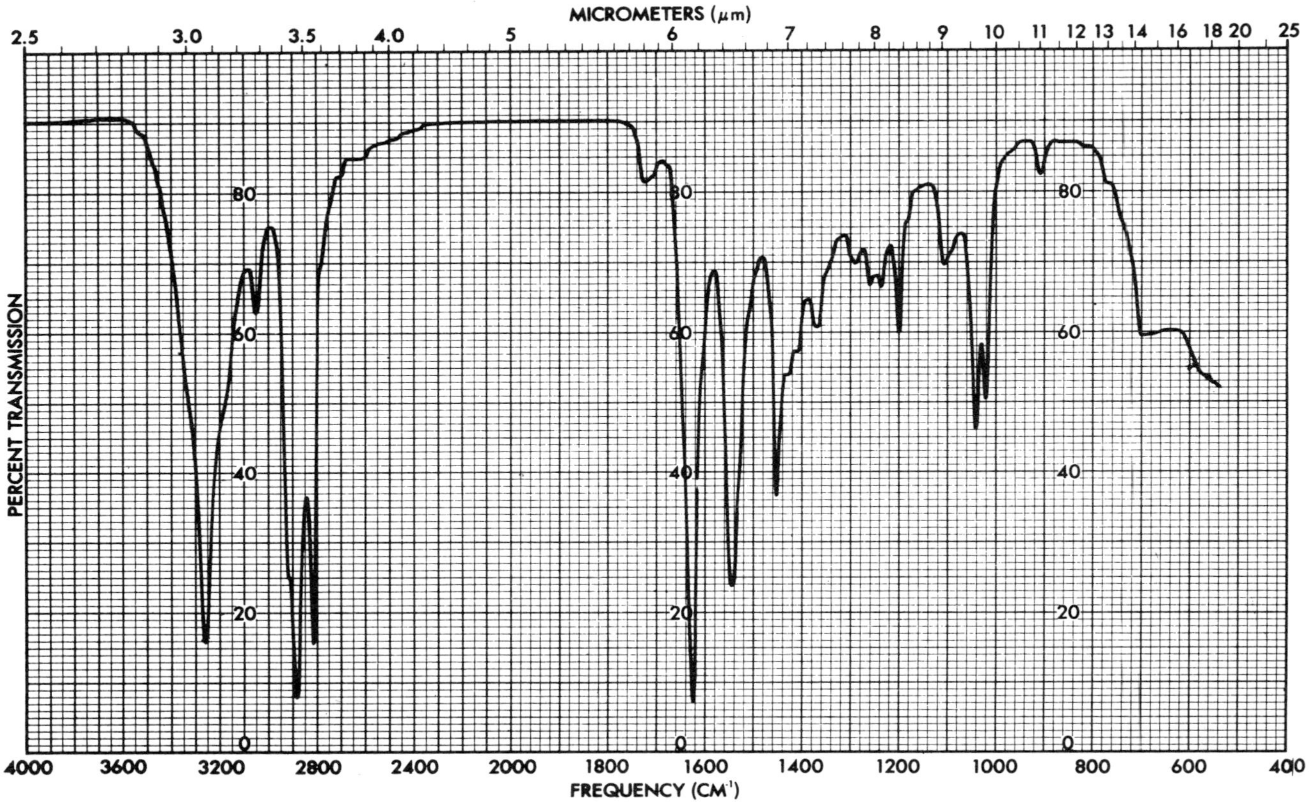

FIG. 6.69. Infrared spectrum of coconut fatty acid–MEA 1:1.1.

Reaction of Oleic Acid and Diethanolamine

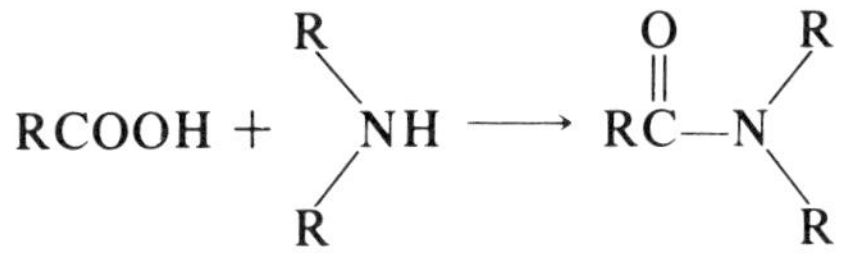

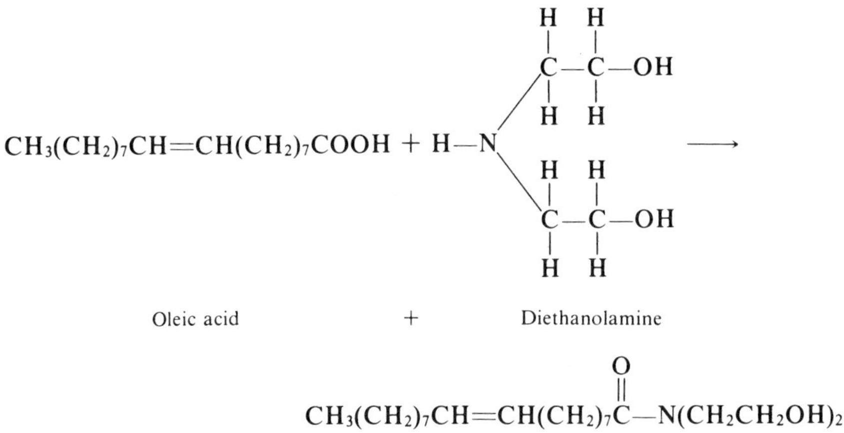

In diethanolamine, a typical secondary amine, the peaks at 2.8–3.0 are overlaps of both the NH and OH groups. A change to a grating-type instrument improves the spectra considerably. Thc peak at 8.9 indicates a secondary amine, and the peak at 9.4–9.6 is the CH_2OH of alkanolamine.

A further reaction that primary and secondary amides may undergo is that of methylolation. This reaction follows the outline shown below.

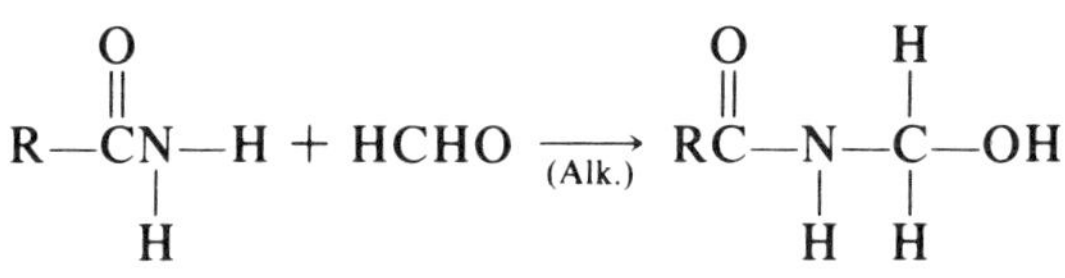

Primary amide methylolation

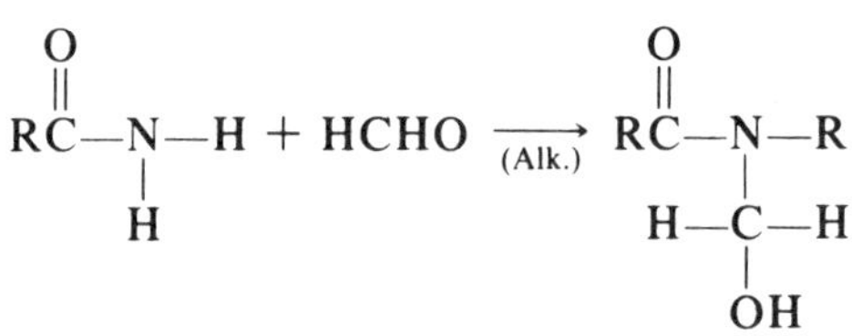

Secondary amide methylolation

A typical running procedure for the manufacture of coconut oil amide as outlined in U.S. Patent 2,844,604 is as follows.

Per 2850-lb Batch

1. Charge to kettle equipped with distillation receiver:

Refined coconut oil	1770 lb
Sodium methylate (25%)	2.4 lb
Potassium hydroxide, flake	30 lb

2. Close kettle; rig for distillation; apply nitrogen sparge.
3. Heat to 125°C, and add:

Diethanolamine	1200 lb

4. With steam full on jacket, cook at 125°C. Allowing water and excess amine to distill over. (As amine condensate is being manufactured from a glyceride, there will be little water evolved.)
5. Continue to cook until material conforms to standard specifications.

A typical manufacturing procedure for a coconut fatty acid–diethanolamide per U.S. Patent 2,096,749 is as follows.

Per 1000-lb Batch

1. Charge kettle with 505 lb of diethanolamine and start heat. Apply nitrogen sparge.
2. Heat until mixture reaches 165°F. Add premelted coconut fatty acid, 488 lb, and continue to apply heat until temperature of 135°C is achieved. Nitrogen sparge and addition of vacuum to aid in the removal of the water of reaction will aid in the maintaining of a light-colored product.
3. Although this reaction is not as viscous as that with monoethanolamine, agitation should be kept at a minimum during the preliminary soap phase and should not be applied until the basic reaction has been completed. Acid value determination should be run at periodic intervals and should be stopped as soon as the value reaches 30 mg/g KOH.
4. Apply cooling, and drum at lowest possible temperature consistent with viscosity of material.

A running procedure for the manufacture of a mixed fatty acid–diethanolamine condensate per U.S. Patent 2,464,094 is as follows.

Per 3000-lb Batch

1. Charge to reactor equipped with agitator, nitrogen sparge and receiver:

   Mixed fatty acids 1550 lb

2. Heat until melted, and add:

   Sodium methylate 5 lb

3. Begin addition of diethanolamine and continue at a rate at which mix will be fluid and begin to evolve water.
4. As mixture passes through the soap phase, it will be very viscous and difficult to stir. Slow agitation is preferred as it will minimize foaming.
5. Temperature should stabilize at 125–135°C.
6. Cook with heat at this temperature under nitrogen blanket, applying vacuum, if necessary, to pull over water. (Excessive vacuum will also distill short-chain fatty acids and some diethanolamine.)
7. When acid value reaches 20–25 mg/g KOH, temperature will begin to rise. At this point, apply cooling, and drum at temperature consistent with viscosity.

Ethylene oxide derivatives and amine condensates form the two major nonionic surfactants used in preparation. Additional references to their use and manufacture are found in the material at the end of the book.

CATIONIC-TYPE SURFACTANTS

Cationic-type surfactants find very limited use in preparation. These compounds are used, however, in the finishing of textile fibers; therefore, their use and synthesis are discussed in greater detail in the sections on dyeing and finishing.

The major exception to this is where cationic compounds are used as accelerators in the weight reduction of polyester fibers by caustic treatment. This is covered in detail in a paper by Tony Dimaria of Tanatex Corporation presented to the Palmetto AATCC Sectional Meeting that outlines in great depth the effect of various cationic compounds on weight reduction of polyester during caustic treatment.[109]

AMPHOTERIC-TYPE SURFACTANTS

Amphoteric surfactants are those surfactants that possess both an anionic and a cationic charge. At that point where the electrocharge is equal, the isoelectric point, a zwitterion is formed, which is both positive and negative. Depending on pH, a compound which is amphoteric will either be anionic, nonionic, or cationic.[110] In acidic solutions, these compounds behave as if they were cationic by the reaction of the donor proton with the unshared electric pair. This linkage occurs through coordinate covalent bonding. The acquired proton imparts a positive charge to the molecule. In alkaline media, a carboxylic salt is formed and the resulting compound acquires a negative charge. Table 6.27 serves to explain how the shift in ionic nature occurs with a fluctuation in pH.

In an alkaline medium, an anionic salt is formed; in an acid medium, a cationic amine salt is formed; and at the isoelectric point, specific for each compound, a zwitterion is formed that is both anionic and cationic at the same time.

Amphoterics have found limited usage in preparation, except in specialized situations requiring wide ranges of compatibility and for all practical purposes have been replaced by nonionics and phosphate esters. However, because of the many claims made for amphoterics and the confusion surrounding them, a general discussion of their advantages, disadvantages, major uses, and general structures is in order at this point.

Advantages. Amphoterics may be used in either alkaline or acid media, with either anionic or cationic surfactants. They offer an excellent degree

TABLE 6.27. Effect of pH on Ionic Nature

Structure	pH	Ionic Charge
(−) (+) Cl $RNH_2CH_2CH_2COOH$	Acidic	Cationic
(+) (−) $RNH_2CH_2CH_2COO$	Isoelectric range	Nonionic
(−) (+) $RNHCH_2CH_2COO$ Na	Alkaline	Anionic

of lubrication, corrosion inhibition, and wetting action and provide a protective colloid for silk and wool processing.

Disadvantages. Amphoterics are expensive. Some forms are not heat stable and cannot be used at elevated temperatures. They exhibit few advantages other than compatibility over less expensive processing aids.

Major uses. Amphoterics are used primarily in scouring and dyeing of protein fibers to prevent chafing, crack marks, and crow's feet.

Basic structures.

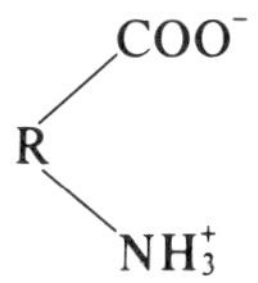

Generic

$\overset{(-)}{Cl}\ \overset{(+)}{R}NH_2CH_2CH_2COOH$

Amine salt

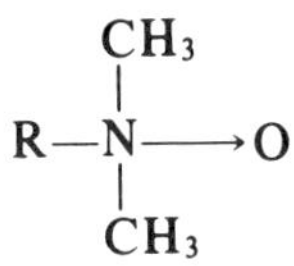

Dimethyl COCO amine oxide

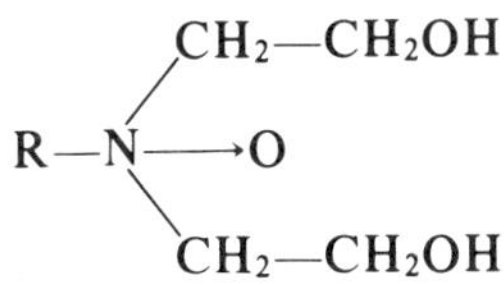

Bis-2-hydroxy ethyl COCO amine oxide

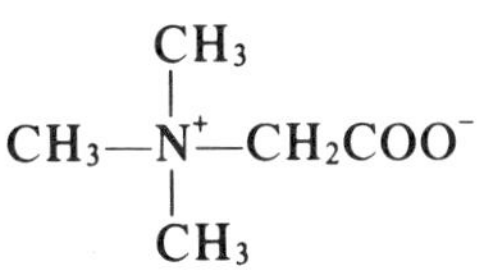

Betaine

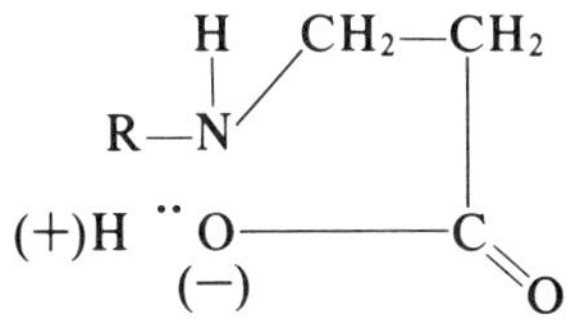

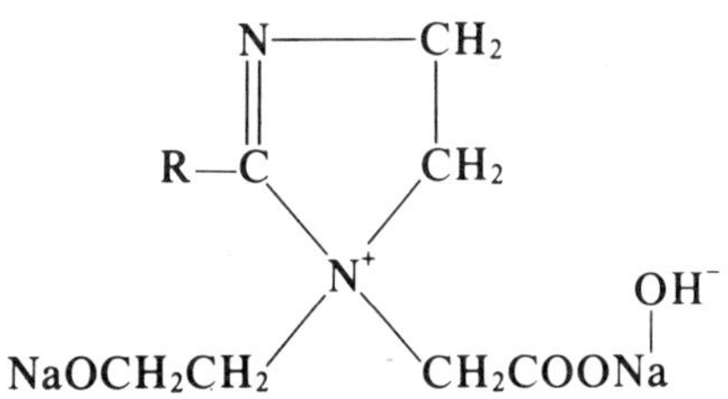

Imino derivative[111]

$CH_3(CH_2)_{17}NHCH_2CH_2\overset{O}{\overset{\|}{C}}O^-\ H^+$

N-Stearoyl β-amino propionic acid

The synthesis of amphoteric compounds is well documented in the literature.[112] Typical of the synthesis is the structure disclosed in U.S. Patent 2,206,249, in which a primary amine of structure RNH_2 is reacted with monochloroacetic acid of structure $ClCH_2COOH$ to yield a monocarboxylic acid of structure $RNHCH_2COOH$. If this carboxylic acid is reacted with a second mole of monochloroacetic acid, a dicarboxylic acid is formed whose structure is:

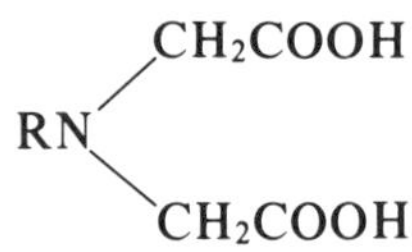

If a third mole of monochloroacetic acid is reacted with the dicarboxylate, an amphoteric surfactant is formed whose structure is:

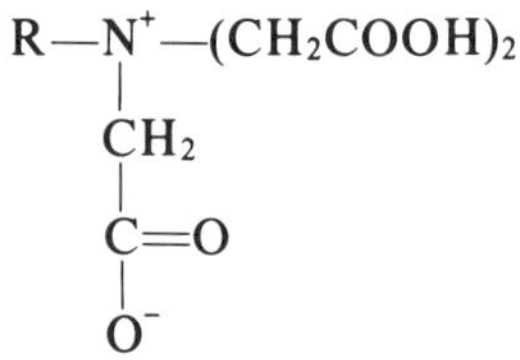

A second procedure for utilizing primary amines in the synthesis of amphoterics is that shown by the reaction of an aliphatic amine with the sodium salt of bromoethane sulfonate. This surfactant is synthesized as follows:

$$RNH_2 + BrC_2H_4SO_3Na \rightarrow RNH{-}C_2H_4SO_3H$$

A third procedure utilizes carboxymethyl fatty amines and cyano synthesis as outlined in British Patent 460,372 to I. G. Farben as follows:

$$RNH_2 + CH_2O + HCN \rightarrow RNHCH_2CN \xrightarrow{H_2O} RNHCH_2COOH$$

Under this same patent, a dicarboxylate is formed:

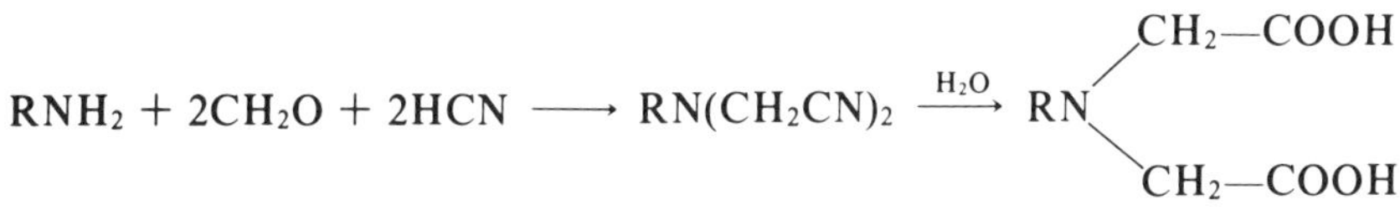

Additional references and structures for amphoteric surfactants are as follows:[113]

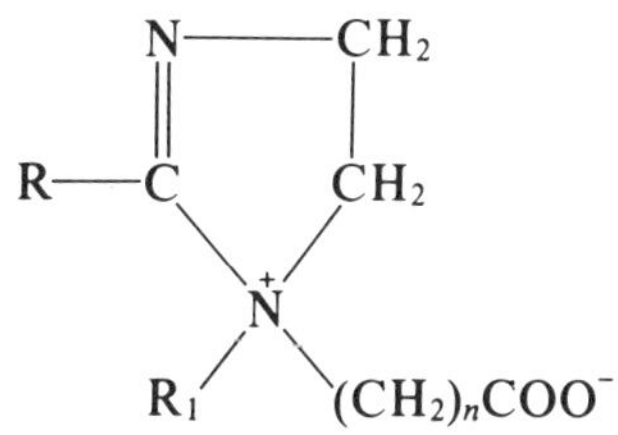

British Patent 1,186,411

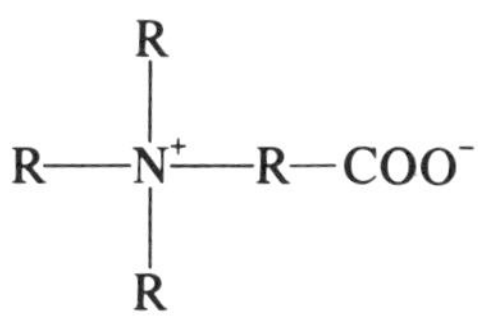

British Patent 1,185,111

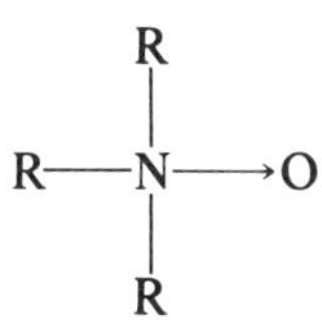

U.S. Patent 2,159,967

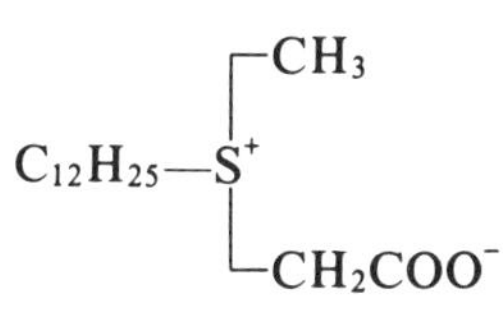

U.S. Patent 2,178,353

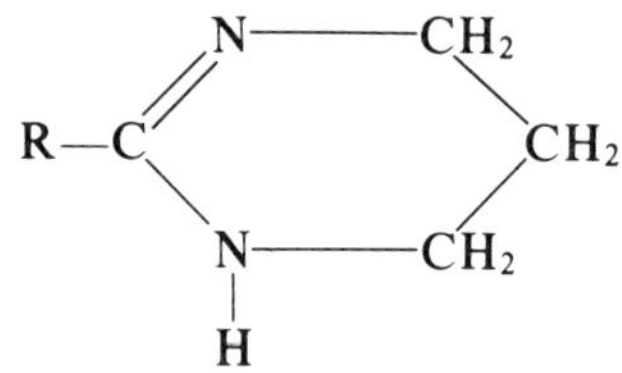

German Patent 700,731

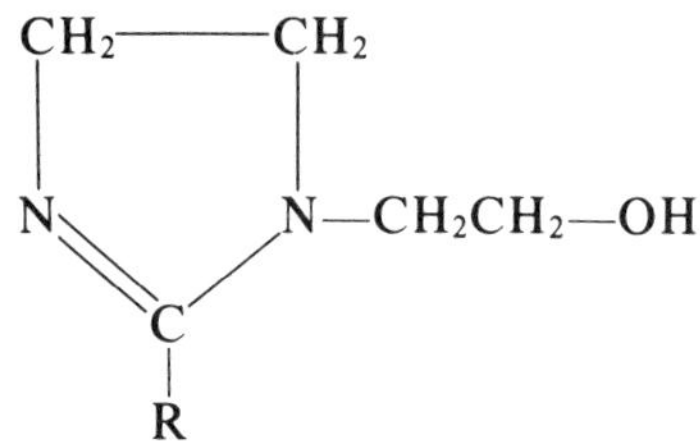

British Patent 985,321

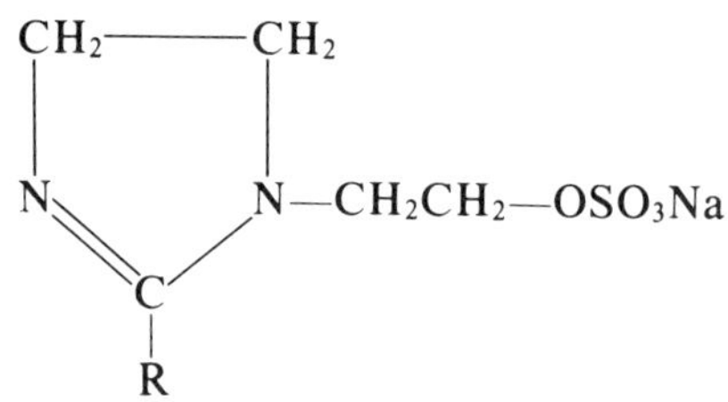

British Patent 985,321; U.S. Patent 2,329,086

Figure 6.70 shows the infrared spectra of sodium *N*-coco-β-aminopropionate, and Table 6.28 provides an interpretation of these spectra. This compound is produced by the condensation of a fatty primary amine with an acrylic monomer as disclosed in U.S. Patent 2,810,752 to General Mills:

$$RCOOH + NH_3 \rightarrow RCN + H_2 \rightarrow RCH_2NH_2$$
$$RCH_2NH_2 + \text{acrylic monomer} \rightarrow (RCH_2NH_2CH_2CH_2COOH)^+Cl^-$$

The key points of reference are the absence of a peak at 7.1 and the presence of a peak at 6.4.

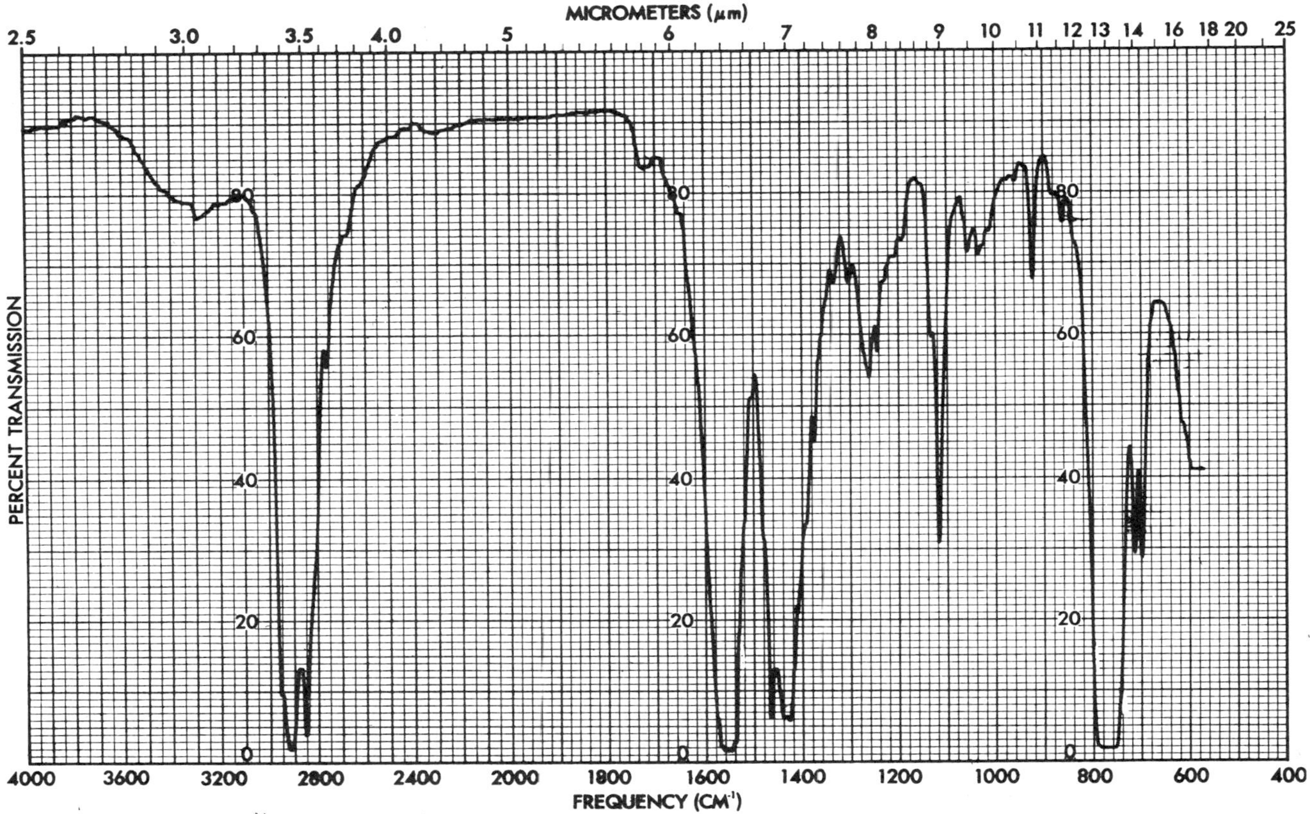

FIG. 6.70. Infrared spectrum of sodium *n*-coco-β-aminopropionate.

TABLE 6.28. Interpretation of Infrared Spectra of Sodium *N*-coco-β-Aminopropionate

Wavelength, μm	Group	Type
3.0	COOH	Carboxyl OH
3.4	COOH	Methylene
3.5	CH_3	Methyl
6.4	COONa	COOH salt
6.9	$(CH_2)_n$	Methylene
9.0	C—O	Of COOH
	NH	Of secondary amine

Manufacturing Procedure for Amine Oxide

A typical manufacturing procedure for the class of surfactants known as *amine oxides* is as follows:

Per 3000-lb Batch (U.S. Patent 2,043,965)

1. Charge to reactor:

Water	60.25%	1500 lb
Dimethyl coco amine	25.47%	654 lb
Silicone defoamer, 10%	00.20%	6 lb
Hydrogen peroxide, 35%	14.09%	362 lb

2. Heat water to 180°F, and add defoamer.
3. Charge total dimethyl coco amine.
4. Add one-fourth of hydrogen peroxide and heat to 180°F.
5. Add one-fourth of hydrogen peroxide and heat to 190°F.
6. Add one-eighth of hydrogen peroxide; exotherm will carry temperature of mix to 200°F or above.
7. When heat subsides, add one-eighth hydrogen peroxide; heat will again rise, and foaming may occur as temperature approaches reflux. (Point of exotherm occurs depending on size of batch at 60–75% of peroxide charge. When total of 75% of peroxide has been charged and exotherm has subsided, remaining peroxide may be added.)

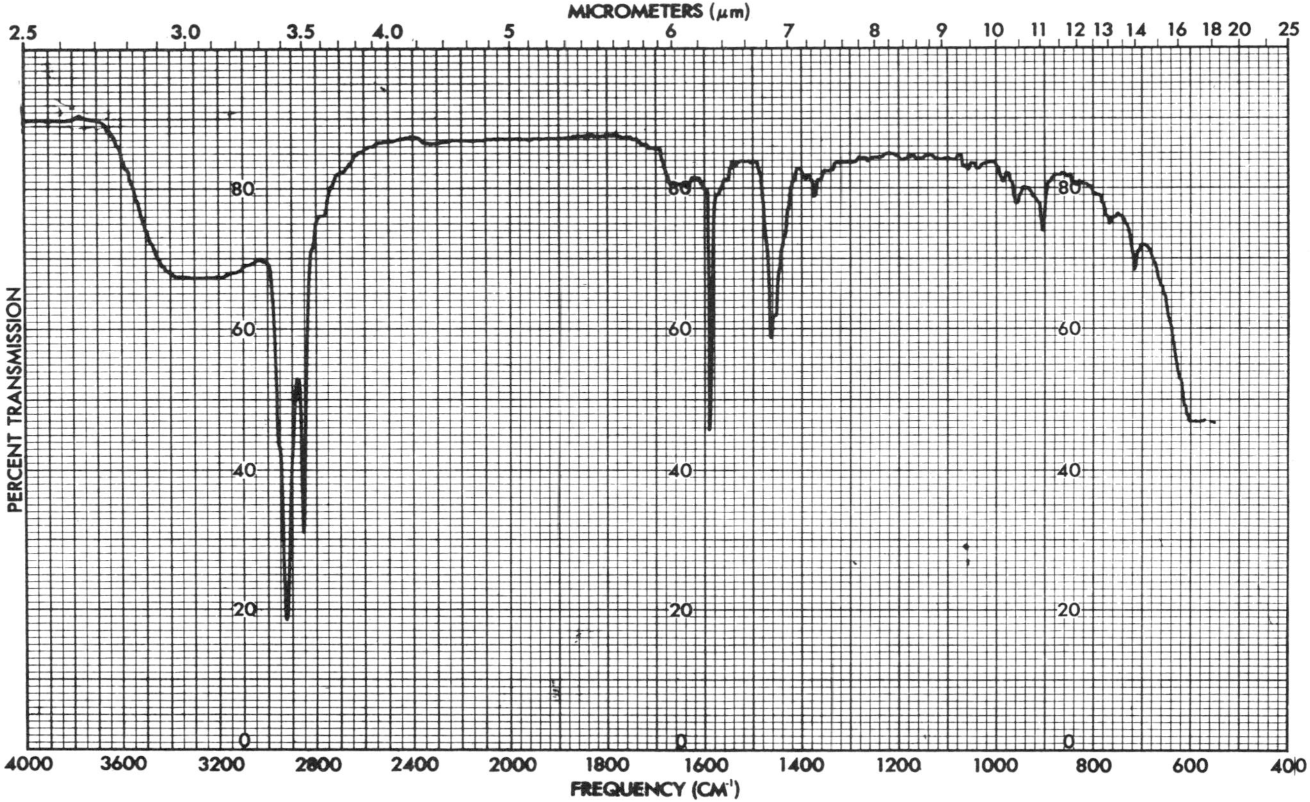

FIG. 6.71. Infrared spectrum of dimethyl cocoamine oxide.

8. When all peroxide has been charged, maintain reactor temperature at the boil under reflux until peroxide content drops to 1.2–2.0% free peroxide.
9. Foaming should subside and viscosity build as peroxide content drops.
10. Slow agitation is desirable, as excessive agitation will create uncontrollable foaming.
11. Final pH should be 6.0–6.5, and nonvolatile solids can be adjusted per specification.

Figure 6.71 shows the infrared spectra of dimethyl cocoamine oxide.

United States Patent 3,062,824 discloses the preparation of vinylpyridine *N*-oxides of the following general structure:

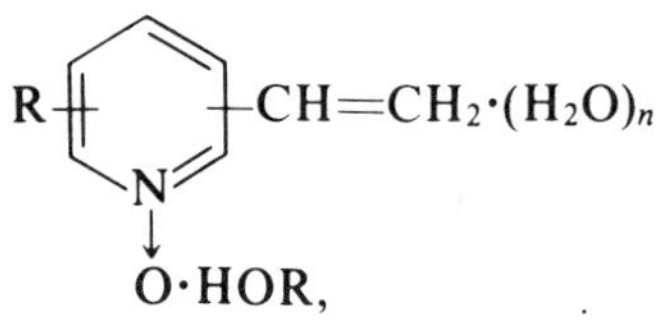

United States Patent 3,147,295 describes the preparation of dimethyldodecylarsine oxide according to the following reaction:

$$C_{12}H_{25}As(CH_3)_2 + H_2O_2 \longrightarrow C_{12}H_{25}\underset{\downarrow \atop O}{As}(CH_3)_2 + H_2O$$

Dimethyldodecylarsine oxide

Blends of Surfactants

Many of the preceding surfactants are blended together for use, and the finished compound is more efficient than the single component. When natural waxes and sizes are to be removed, anionics—or nonionics built with anionics—appear to be the preferred choice. This has been brought about by the necessity for high concentrations of alkali and elevated bath temperatures that can render some nonionic surfactants ineffective. The selection of surfactant blend, whether for batch or continuous, synthetic or natural, severe or mild, is generally based on the temperature of the scour, the degree of soiling, the percentage of oil or size, whether tint is present, and the degree of foaming that may be tolerated by the machinery in which scouring is taking place.

Probably nowhere are blended surfactants more commonly used than in preparation auxiliaries. This has been made necessary by the oils and sizes used as an outgrowth of the synthetic fiber boom. When used as the emulsifying component of water-insoluble solvents, excellent preparation and dyeing systems for use in aqueous processes can be formulated. In these formulations, low-mole ethoxylates, which are very hydrophobic, are the primary emulsifiers that make the solvents dispersible or soluble in the aqueous phase. Anionics, whether sulfate or phosphate derived, when neutralized with amines or alkanolamines become solvent and oil soluble and may be used in conjunction with low-mole ethoxylates. Commonly used in the trade are the triethanolamine or isopropylamine salts or dodecyl benzenesulfonic acid (DDBSA), alkyl phosphate esters, and carboxylic acids.[114]

Solvent and oil emulsifiers probably originated in early antiquity when vegetable oils such as olive oil or oleic acid were sulfonated. From this beginning, other natural oils were sulfonated and/or emulsified to form oil-solubilizing mediums. In today's specialty chemical industry, petroleum and synthetic compounds have found wide use in textile processing. Table 6.29 lists commonly used solvents.

The incorporation of emulsifier blends, which are both miscible in the solvent and dispersible in water, results in an excellent combination that is extremely effective in emulsifying the types of solvents listed in Table 6.29. For example, to emulsify a high-boiling mineral spirit-type solvent, a blend of propylene oxide-capped nonionics or thioether ethoxylates, coupled with a low-mole phosphate ester of a dinonyl phenol ethoxylate that has been neu-

TABLE 6.29. Commonly Used Solvents

Xylene	Tetrahydrofurfuryl alcohol
Butyl cellosolve	Isopropyl alcohol
Pine oil	Chlorinated benzenes
Varsol	Mineral oil
Kerosene	Cyclohexanol

tralized with an alkanolamine, is commonly used. This combination product provides good detergency and satisfactorily accomplishes the removal of processing oils, greases, and sizes.

The choice of surfactants to use as emulsifiers is endless, but the resultant blend must be one that will produce satisfactorily prepared goods. A simple blend of short-chain ethoxylate and long-chain ethoxylate of the same hydrophobe that produces a balanced HLB and is stable over a wide range of processing can sometimes be the best choice. On the other hand, it is sometimes necessary to incorporate a solvent processing aid and a complex mixture of ethylene oxide and fatty acid derivatives produced with inorganic compounds to provide optimum processing efficiency.

HLB Values

Examples of HLB values, the commonly designated unit for emulsification value, are listed in Table 6.30.

With the use of the procedure of dividing percent ethylene oxide by 5, the HLB values of common EO surfactants are as those shown in Table 6.31.

Biologically active surfactants, such as thioether derivatives, are becoming quite desirable in the wake of new, vigorous government regulations that limit the effluent composition of hydrocarbon, scour, and dye components. Pending legislation will render the use of many aromatic and chlorinated hydrocarbon ingredients now in common use unusable and cause them to be replaced with entities that are more environmentally acceptable. As government regulations cover chemical compounds from the cradle to the grave, interest in biologically active components and those that are nontoxic and nonpolluting will increase in interest to both manufacturing and management personnel.

TABLE 6.30. HLB Values of Emulsifiers

Emulsions–Emulsifiers	HLB
Water in oil	3–6
Oil in water	8–15
Wetting agents	7–9
Detergents	13–15
Solubilizers	15–18

TABLE 6.31. HLB Value of Common Surfactants[115]

Hydrophobe	Hydrophile, Moles EO	HLB
Linear alcohol	3	8.0
Linear alcohol	9	13.3
Linear alcohol	20	16.3
Linear alcohol	40	18.0
Nonyl phenol	4	8.9
Nonyl phenol	10	13.6
Nonyl phenol	40	17.8
Trimethyl nonanol	3	8.3
Trimethyl nonanol	10	14.1

Infrared Spectra of Commonly Used Solvents

Figures 6.72 through 6.92 show the infrared spectra of solvents commonly used by specialty chemists in the formulation of solvent surfactant blends.

Infrared Spectra of Commonly Used Surfactants and Surfactant Blends

Figures 6.93 through 6.136 show the infrared spectra of commonly used surfactants and surfactant blends as found in commercial practice.

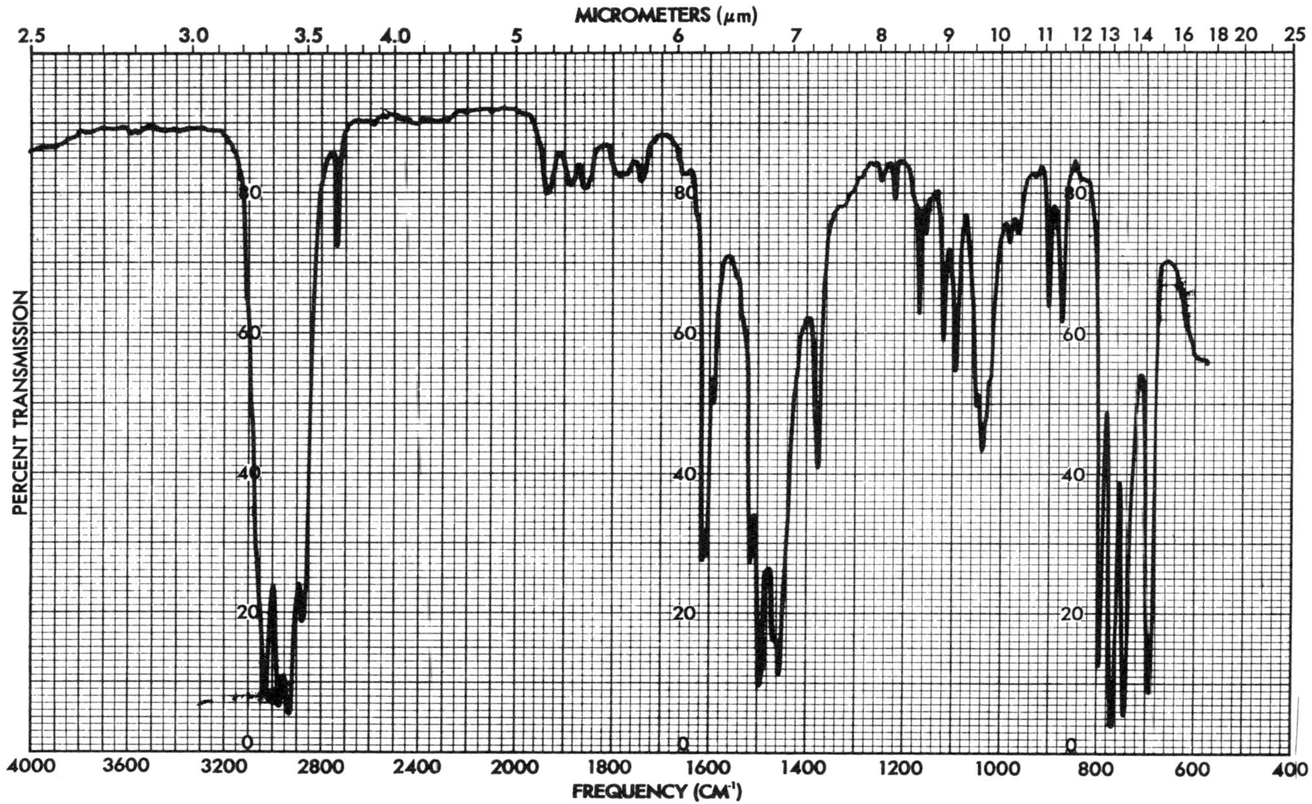

FIG. 6.72. Infrared spectrum of xylene.

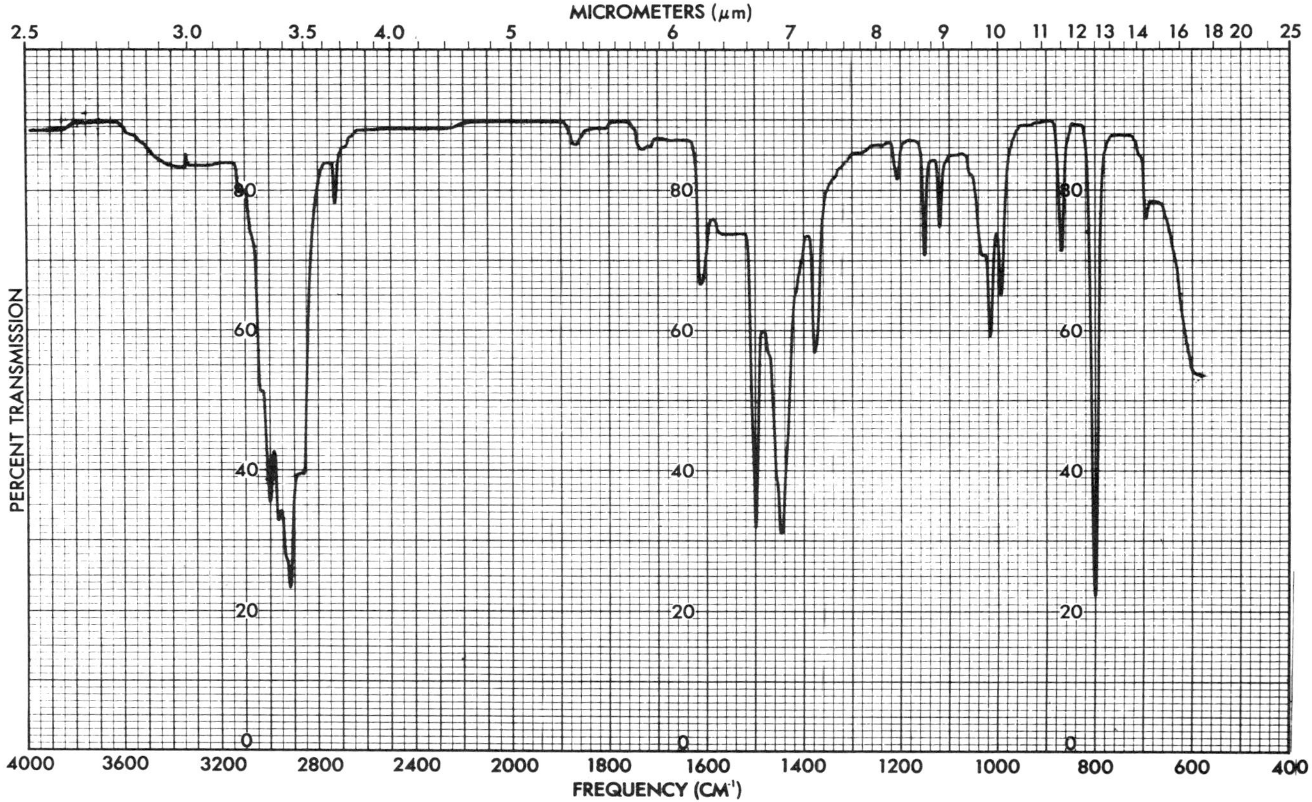

FIG. 6.73. Infrared spectrum of 1,2,4-trimethylbenzene.

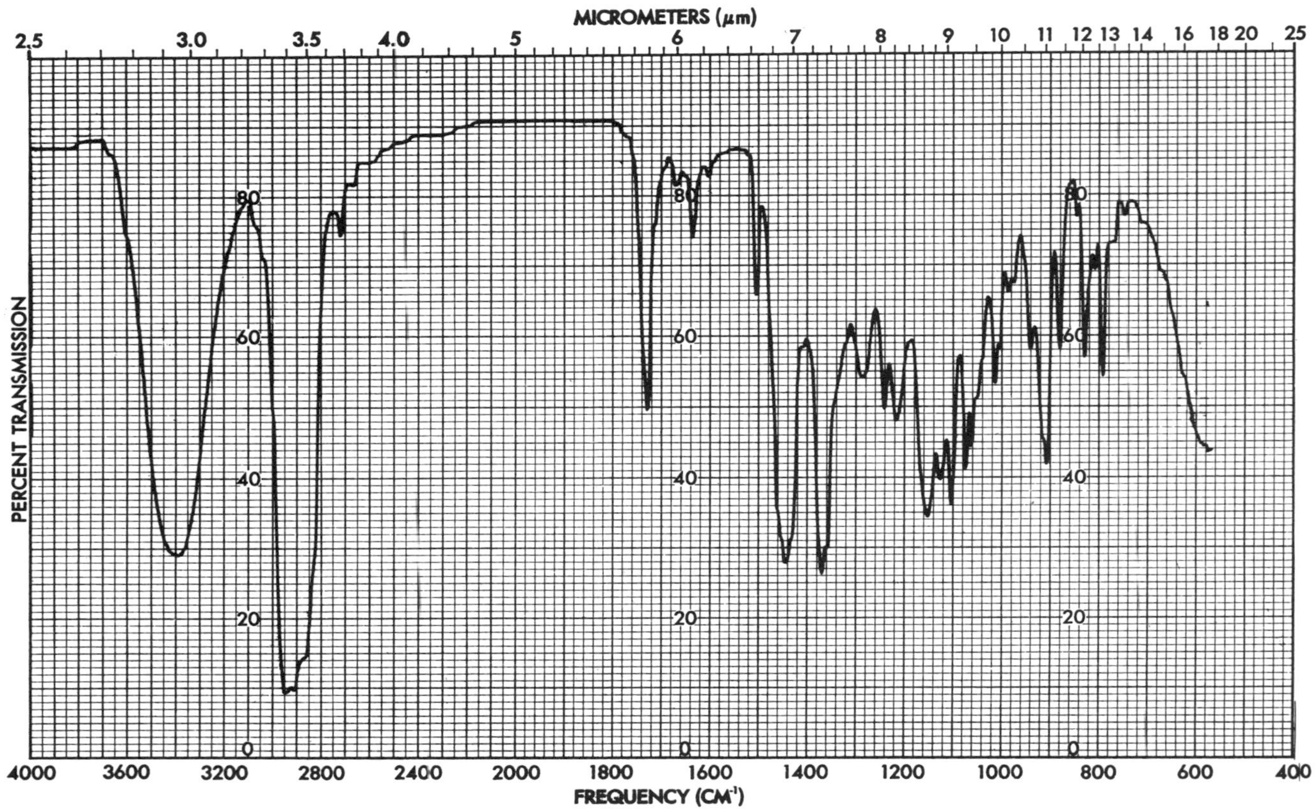

FIG. 6.74. Infrared spectrum of pine oil.

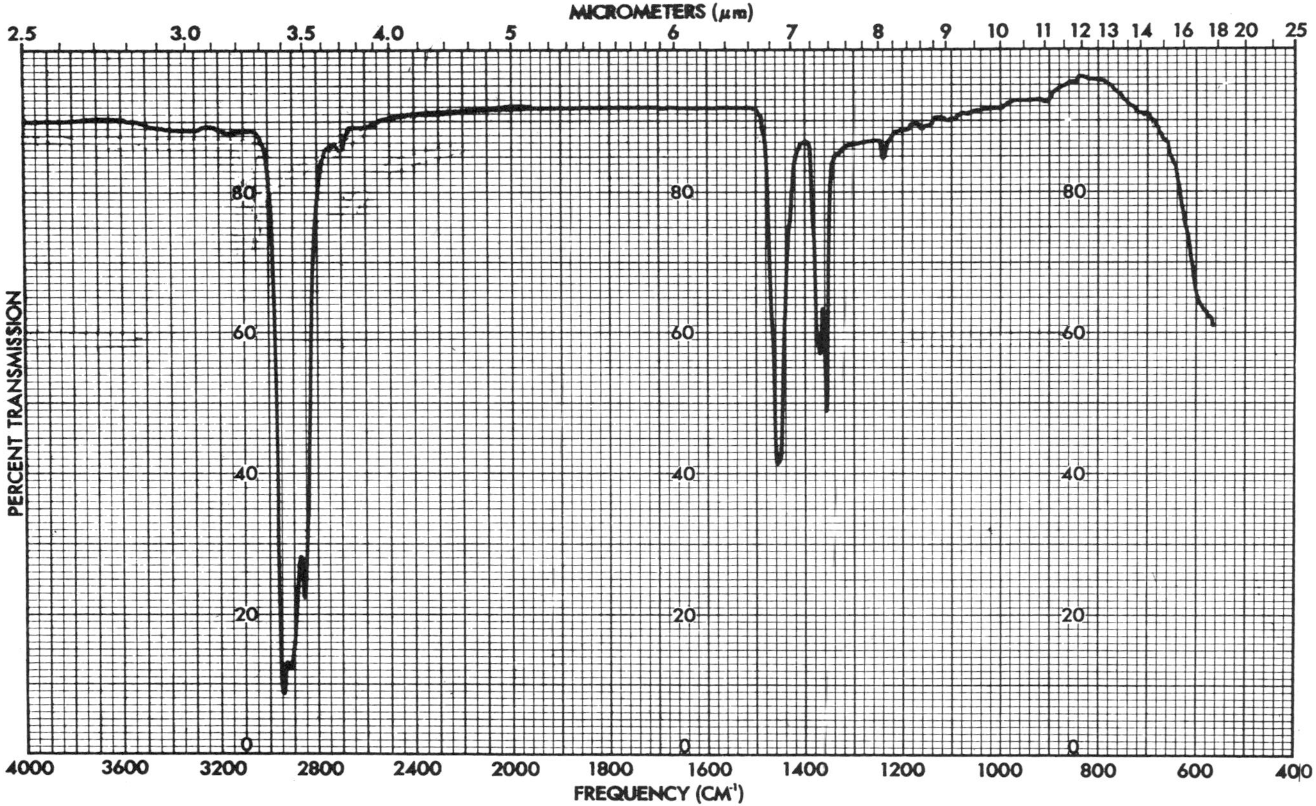

FIG. 6.75. Infrared spectrum of odorless mineral spirits.

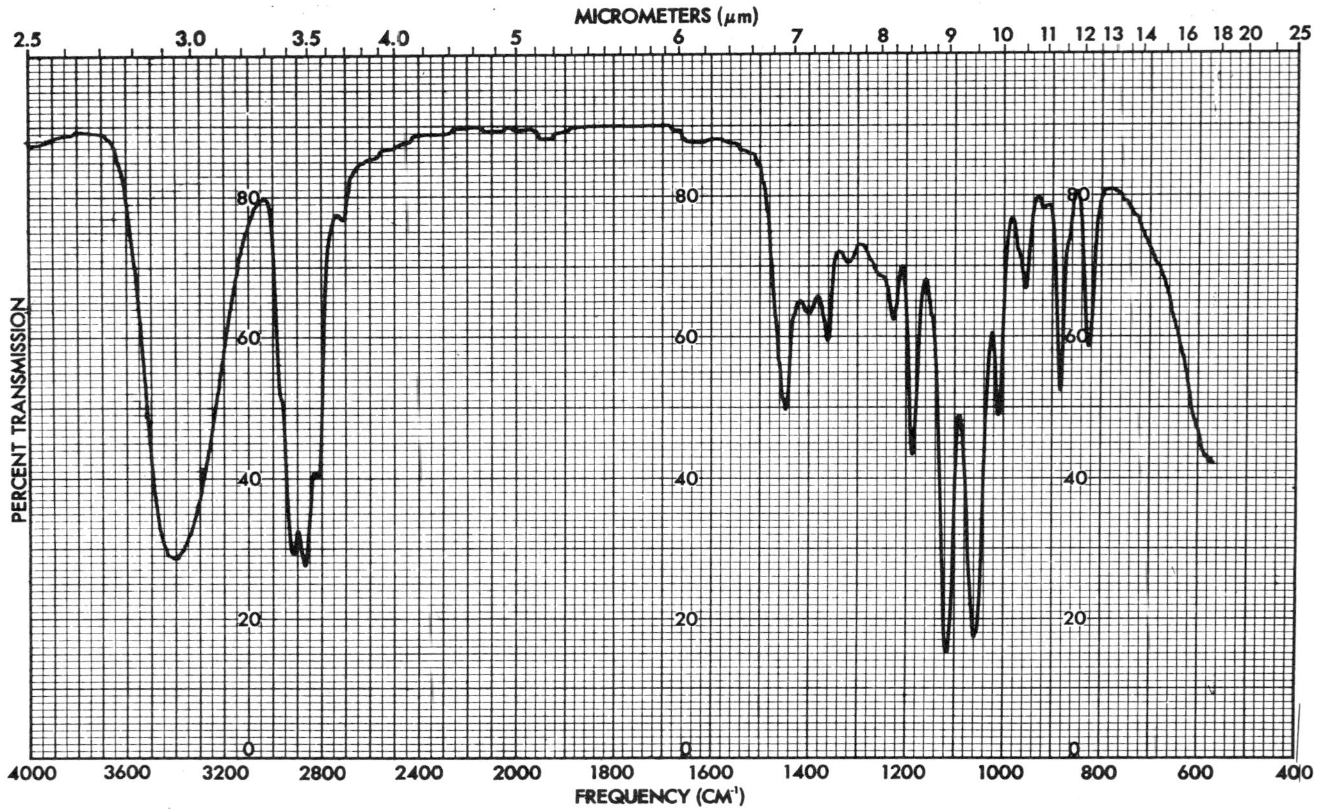

FIG. 6.76. Infrared spectrum lof ethylene glycol monomethyl ether.

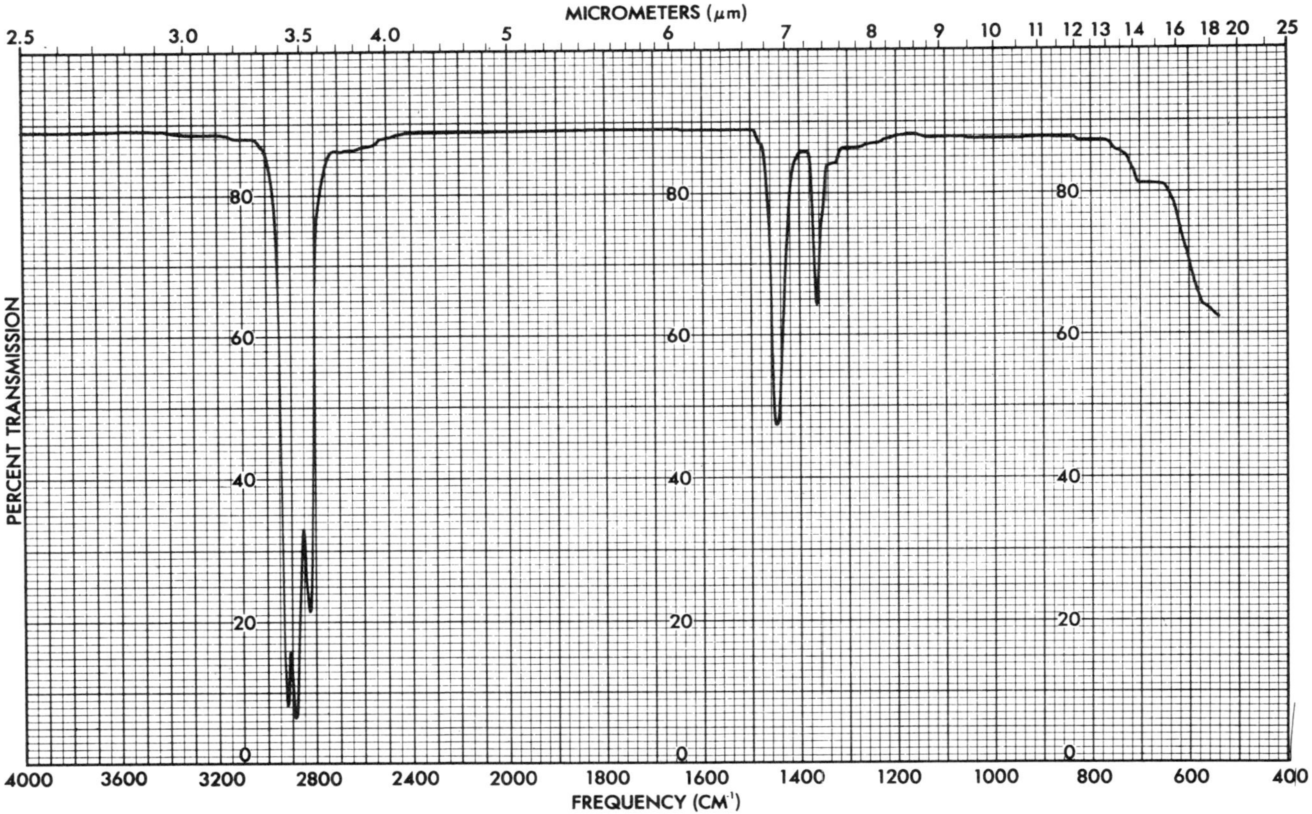

FIG. 6.77. Infrared spectrum of mineral spirits.

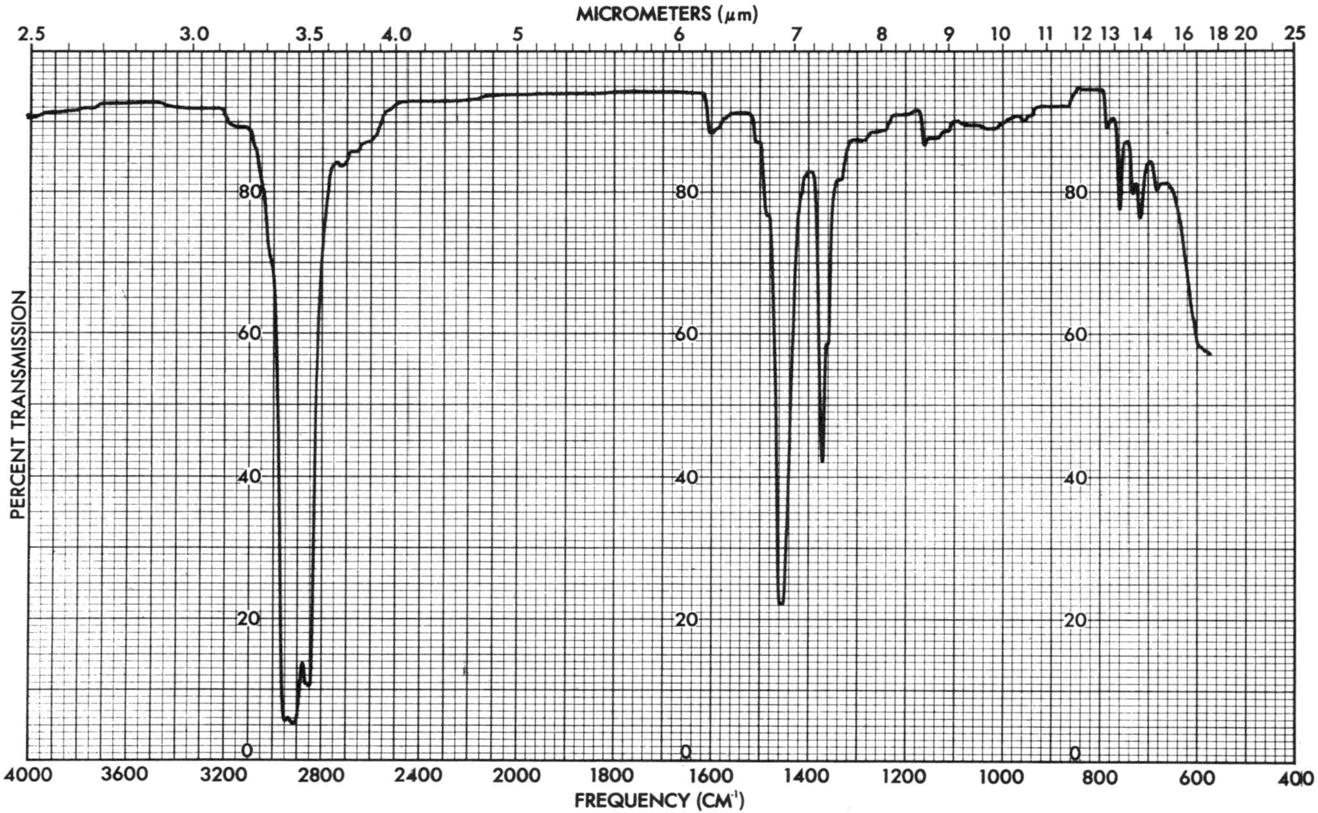

FIG. 6.78. Infrared spectrum of naphtha.

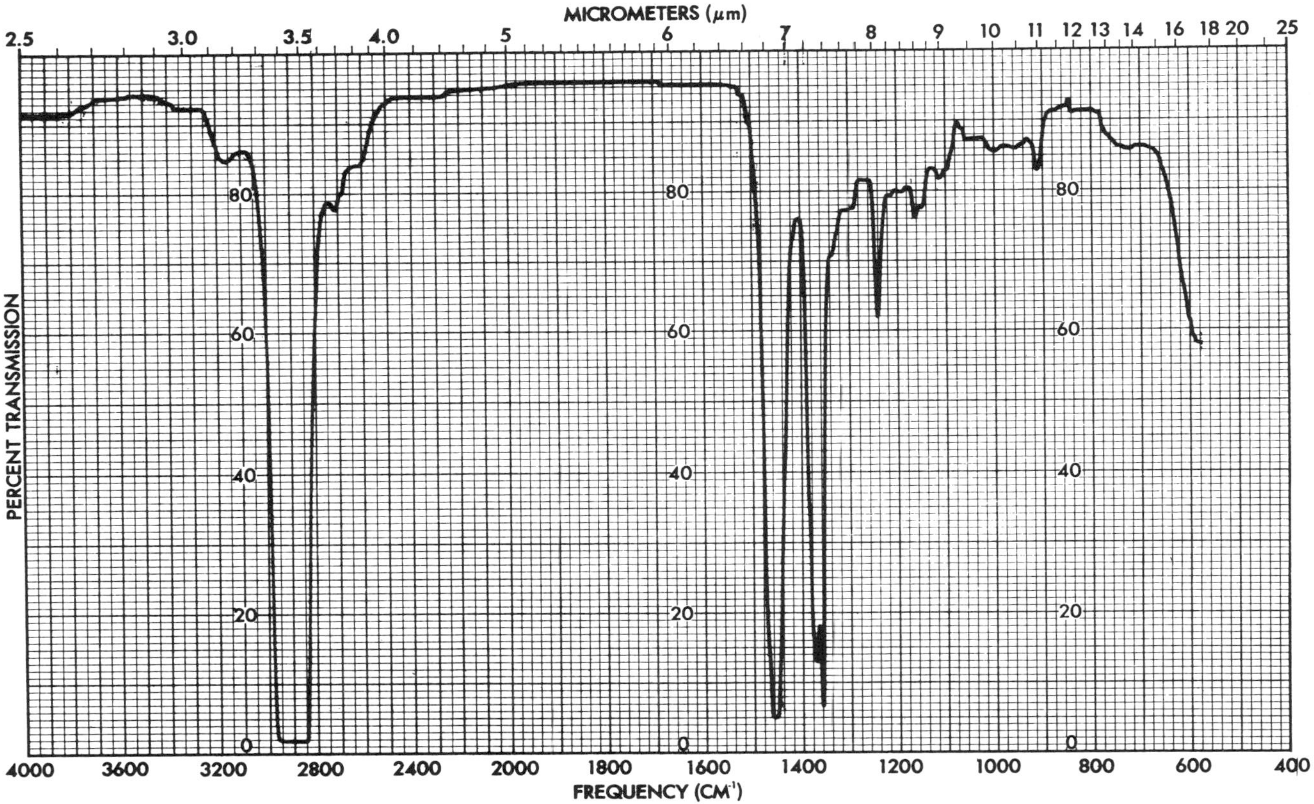

FIG. 6.79. Infrared spectrum of isoparaffinic hydrocarbon.

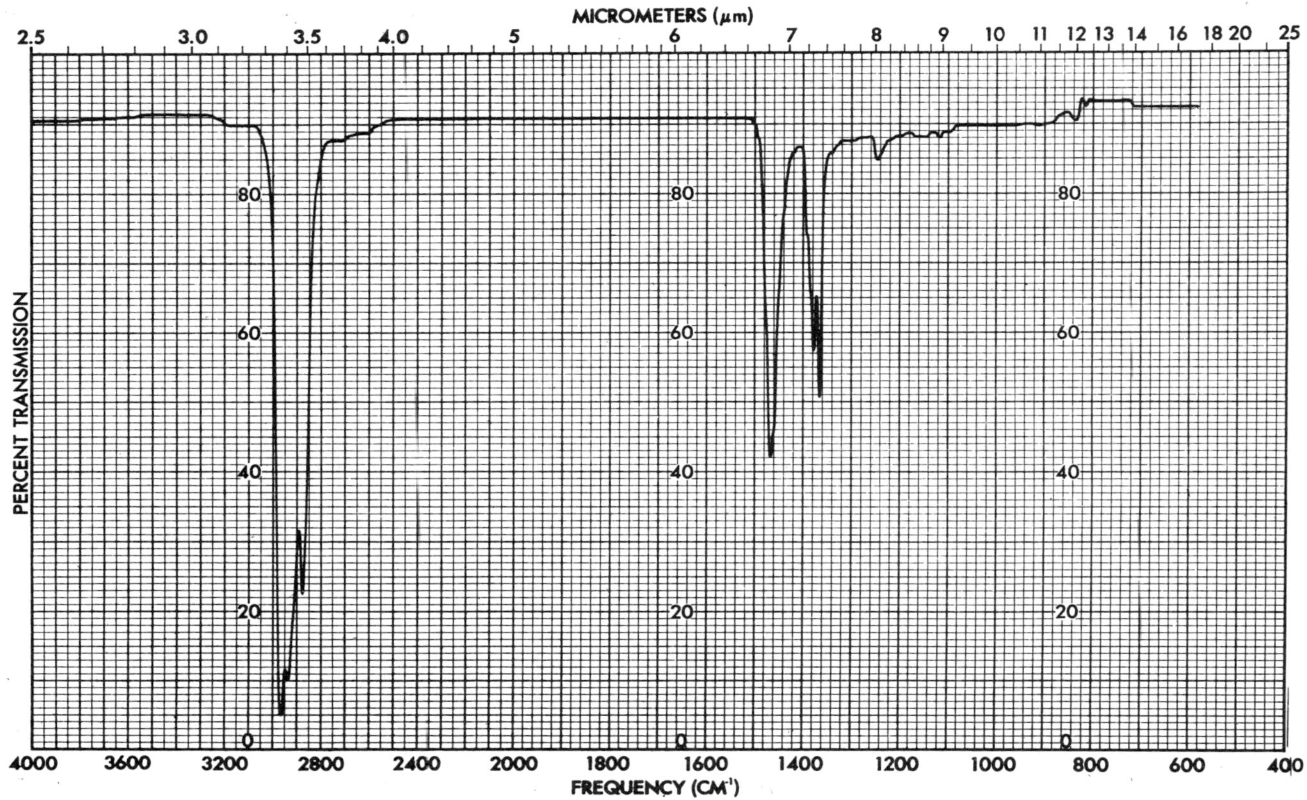

FIG. 6.80. Infrared spectrum of deodorized kerosene.

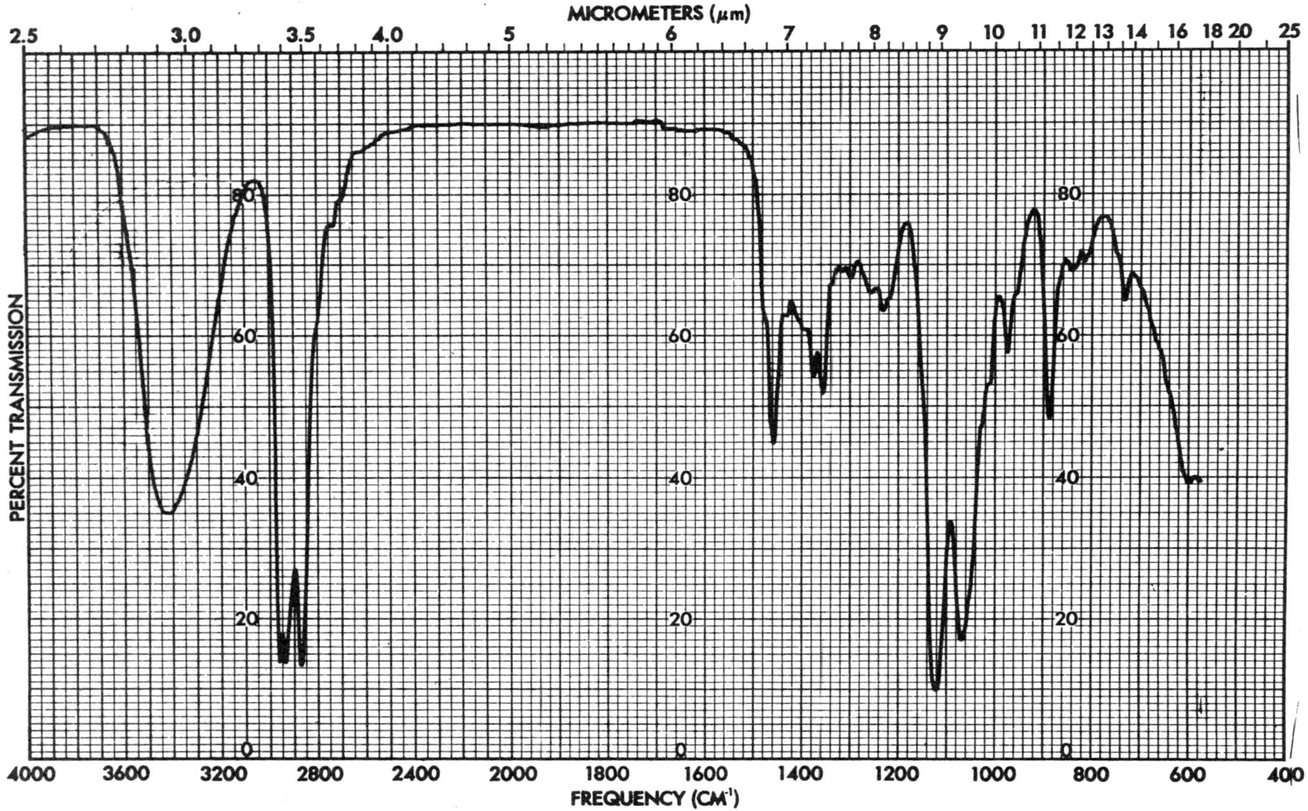

FIG. 6.81. Infrared spectrum of ethylene glycol monobutyl ether.

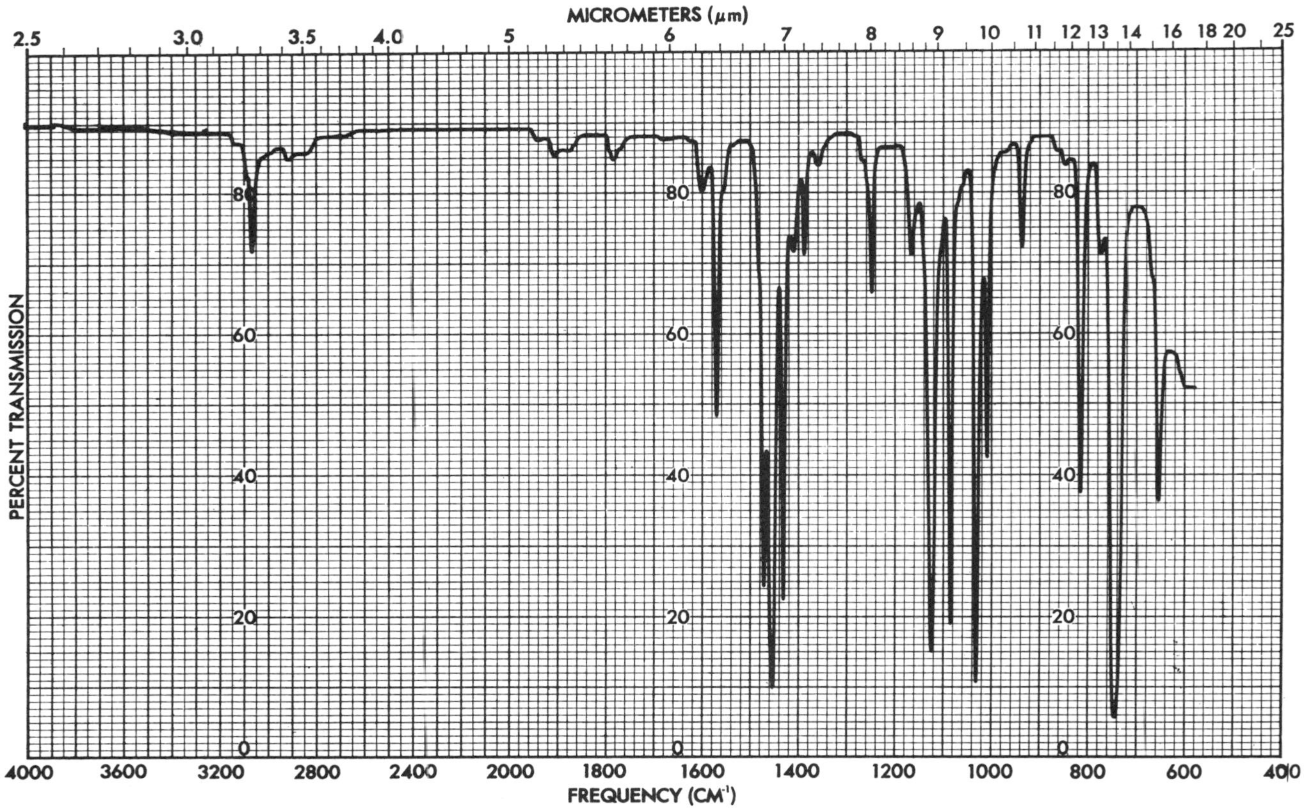

FIG. 6.82. Infrared spectrum of dichlorobenzene.

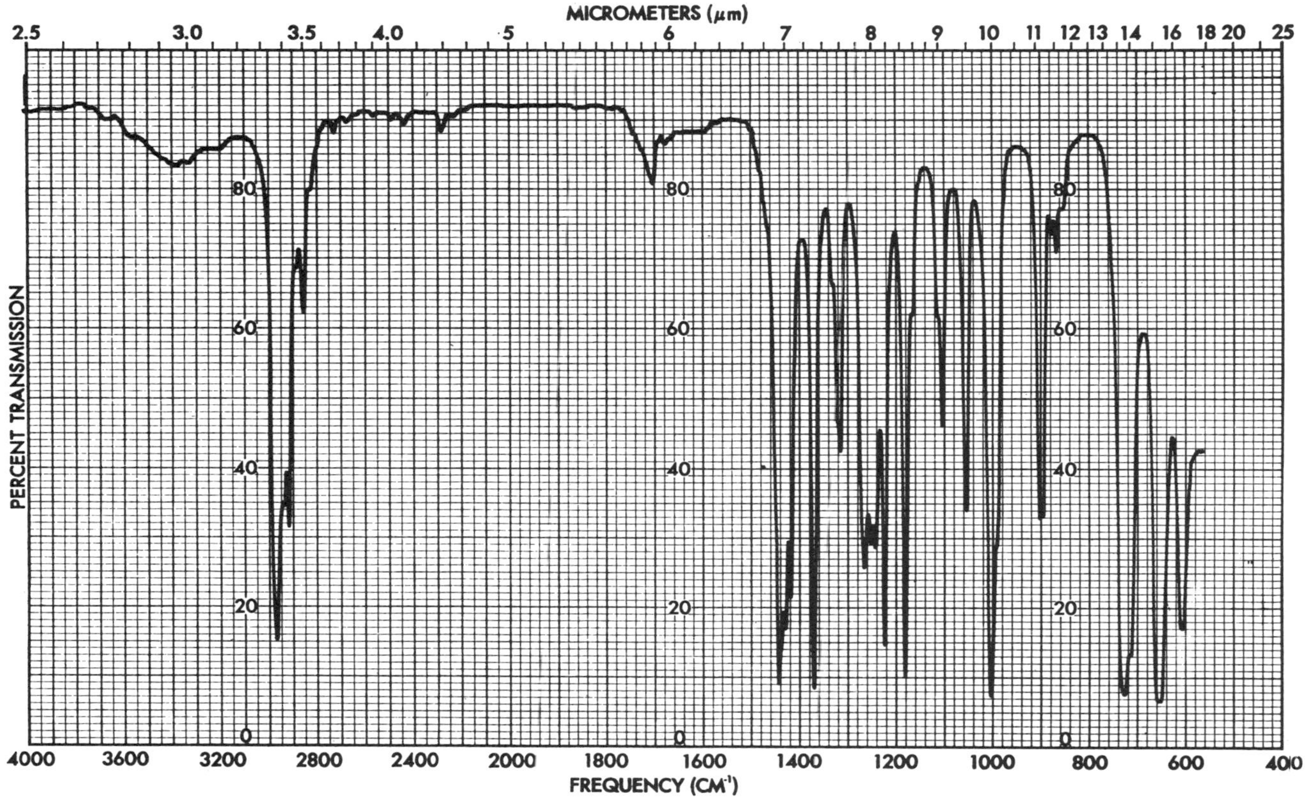

FIG. 6.83. Infrared spectrum of propylene dichloride.

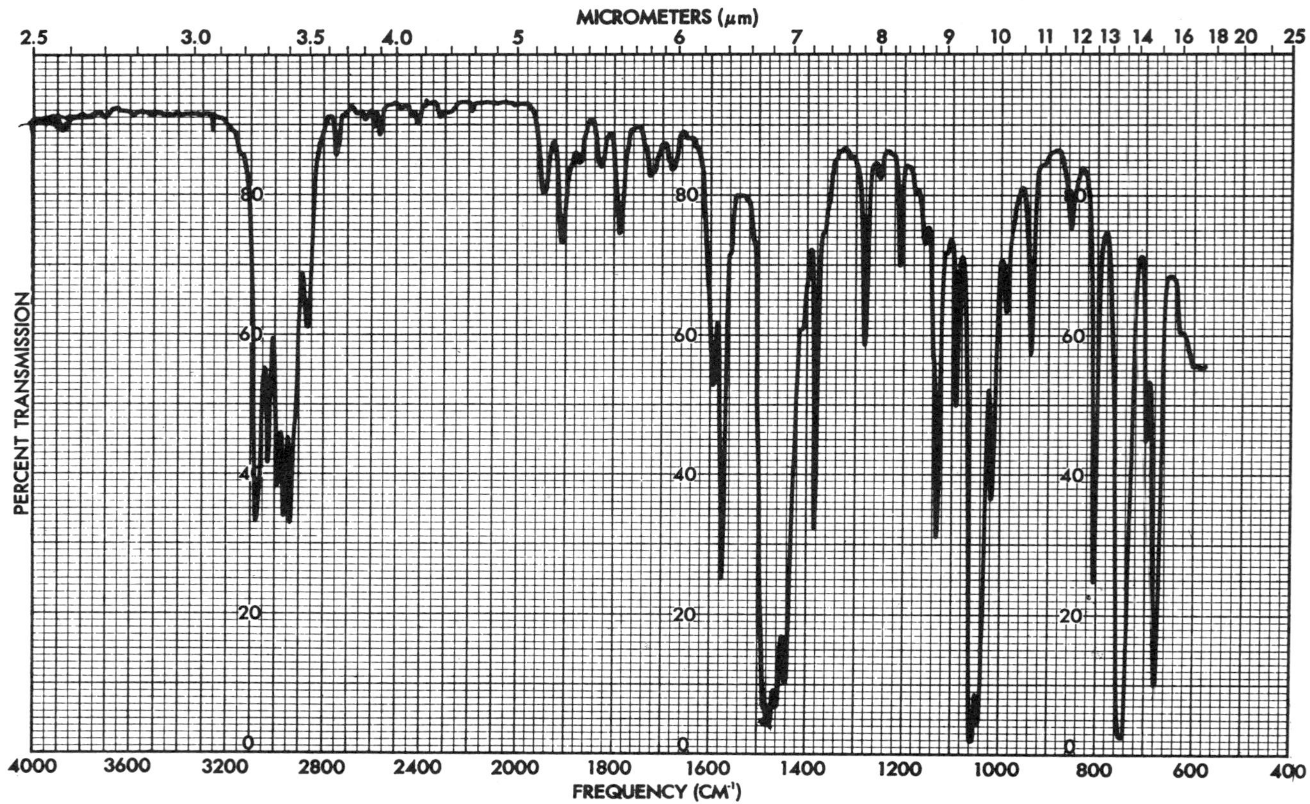

FIG. 6.84. Infrared spectrum of *o*-chlorotoluene.

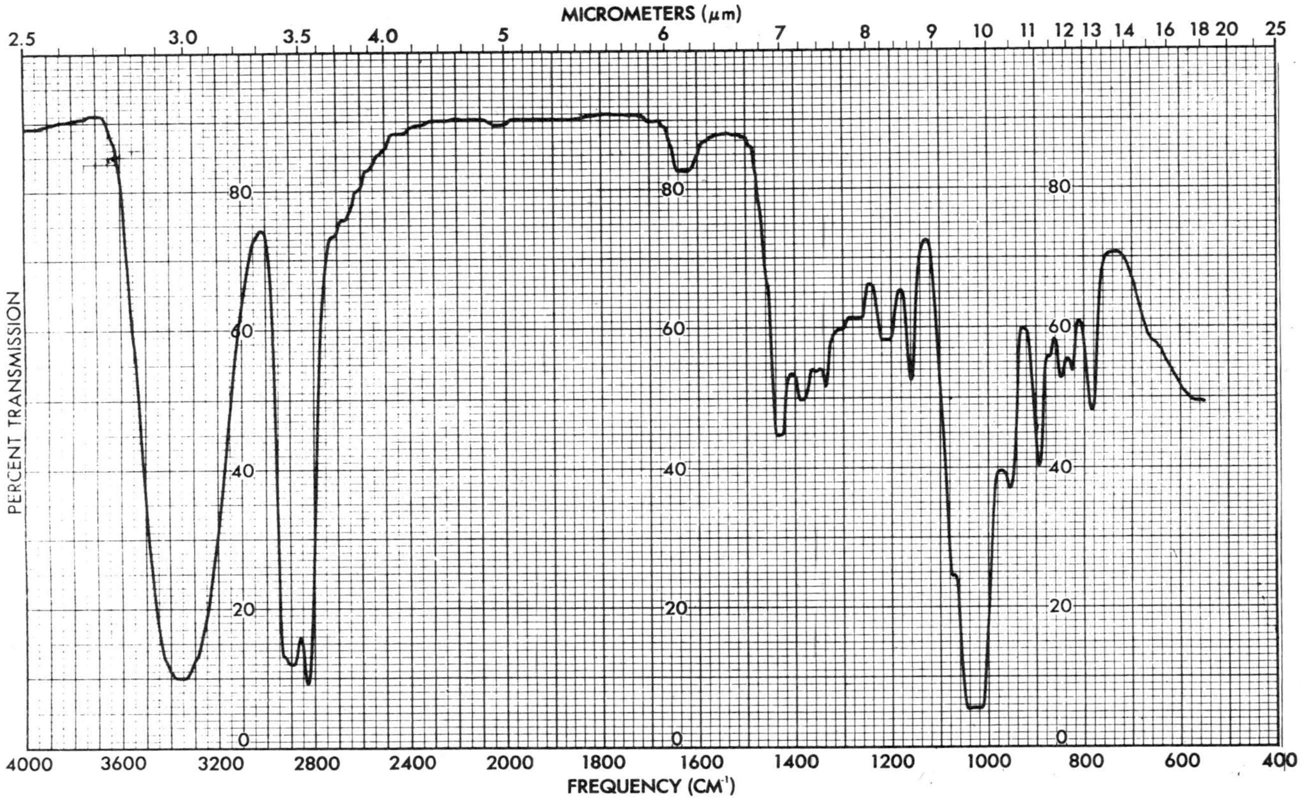

FIG. 6.85. Infrared spectrum of tetrahydrofurfuryl alcohol.

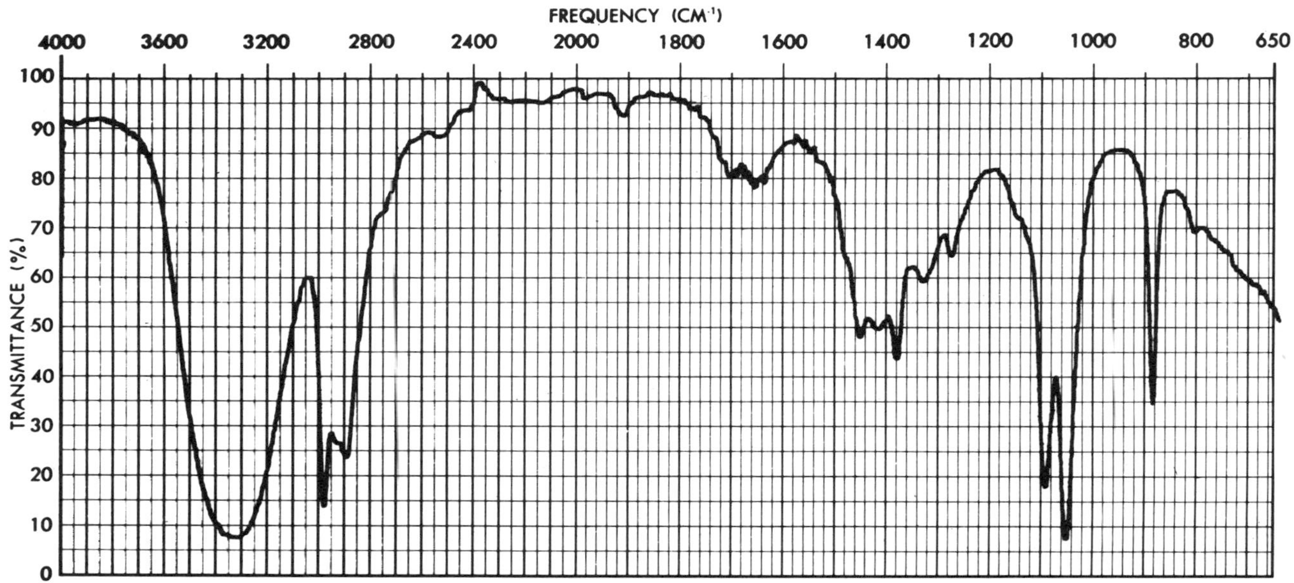

FIG. 6.86. Infrared spectrum of ethanol.

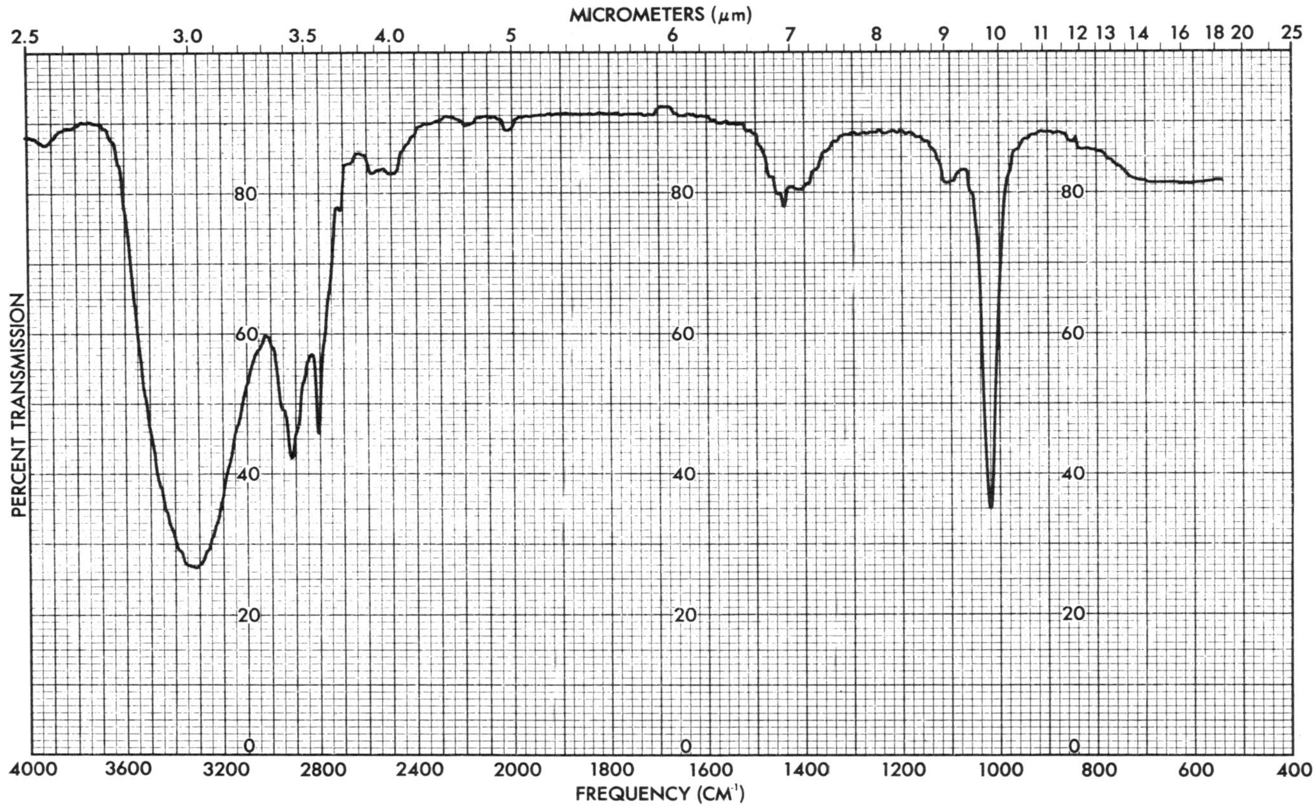

FIG. 6.87. Infrared spectrum of methanol.

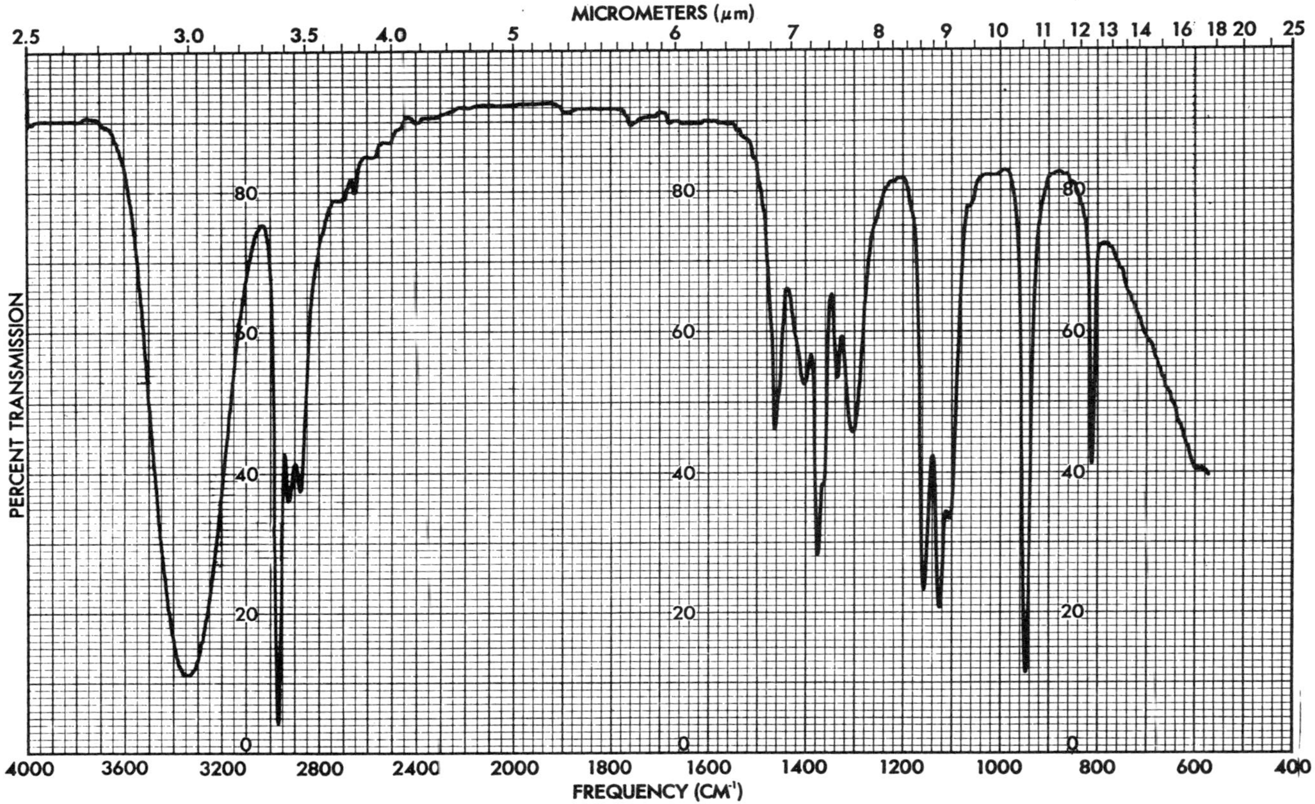

FIG. 6.88. Infrared spectrum of isopropanol.

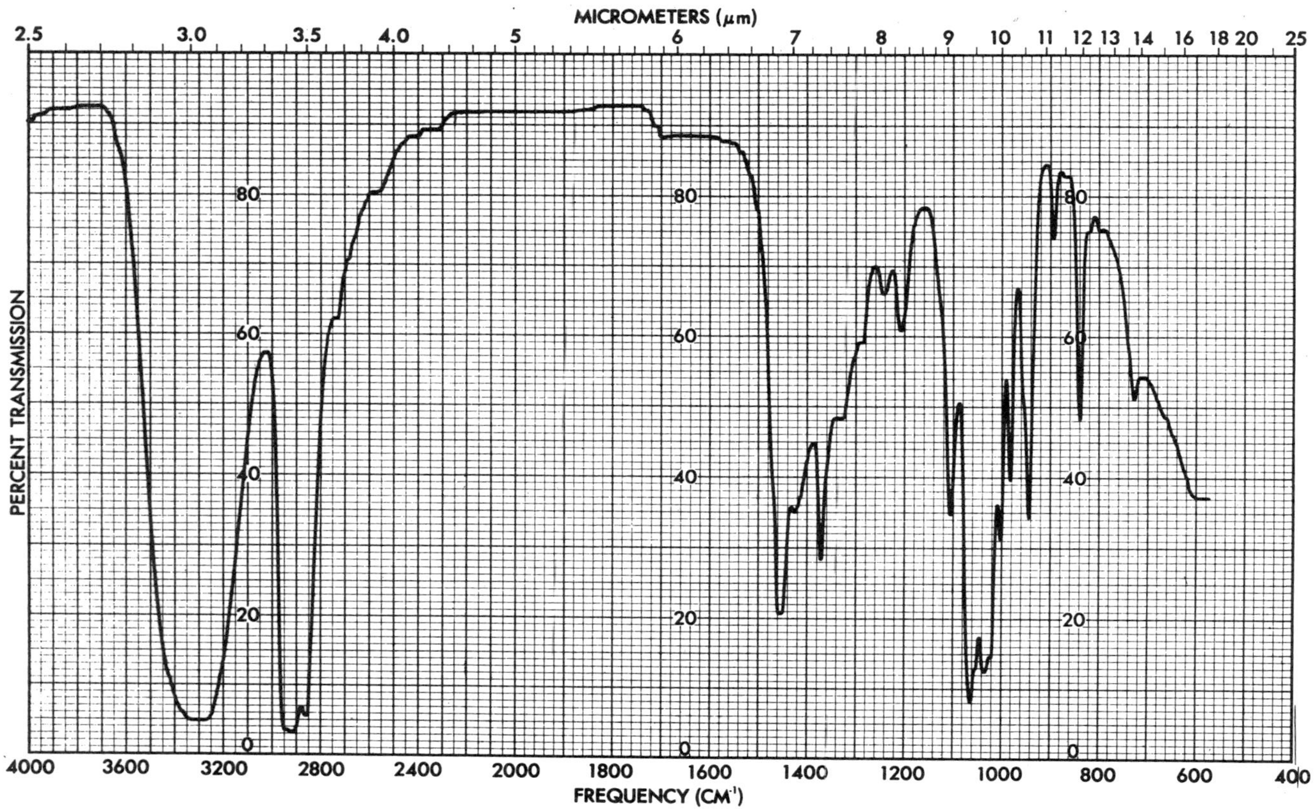

FIG. 6.89. Infrared spectrum of butanol.

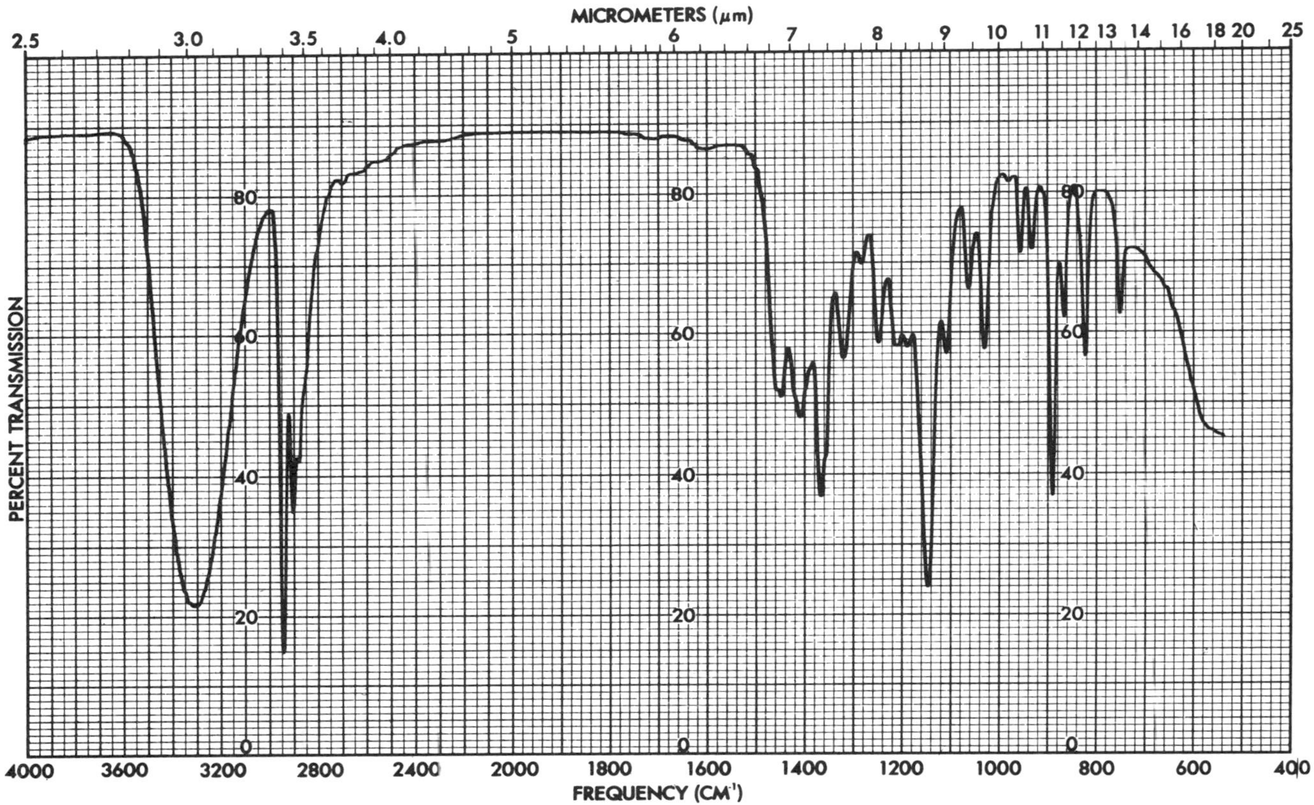

FIG. 6.90. Infrared spectrum of hexylene glycol.

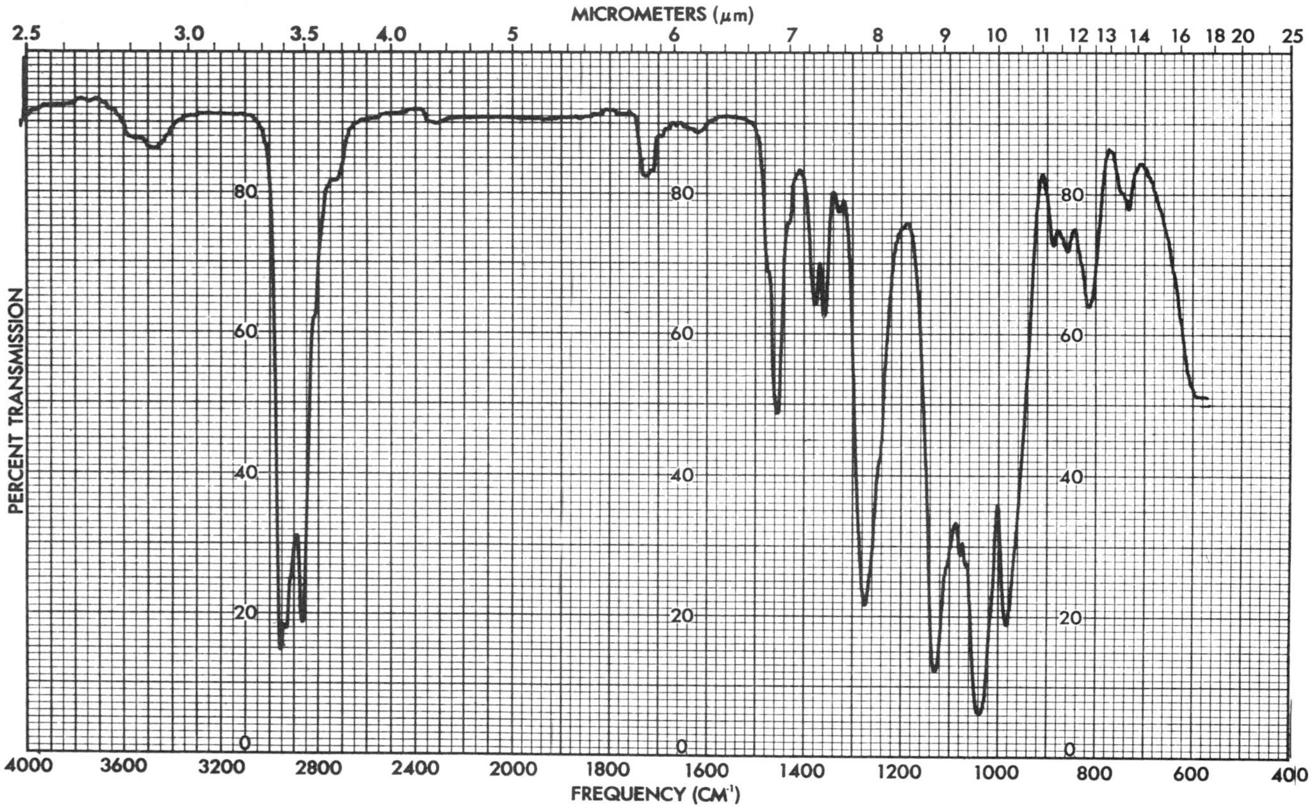

FIG. 6.91. Infrared spectrum of tributoxy ethoxy phosphate.

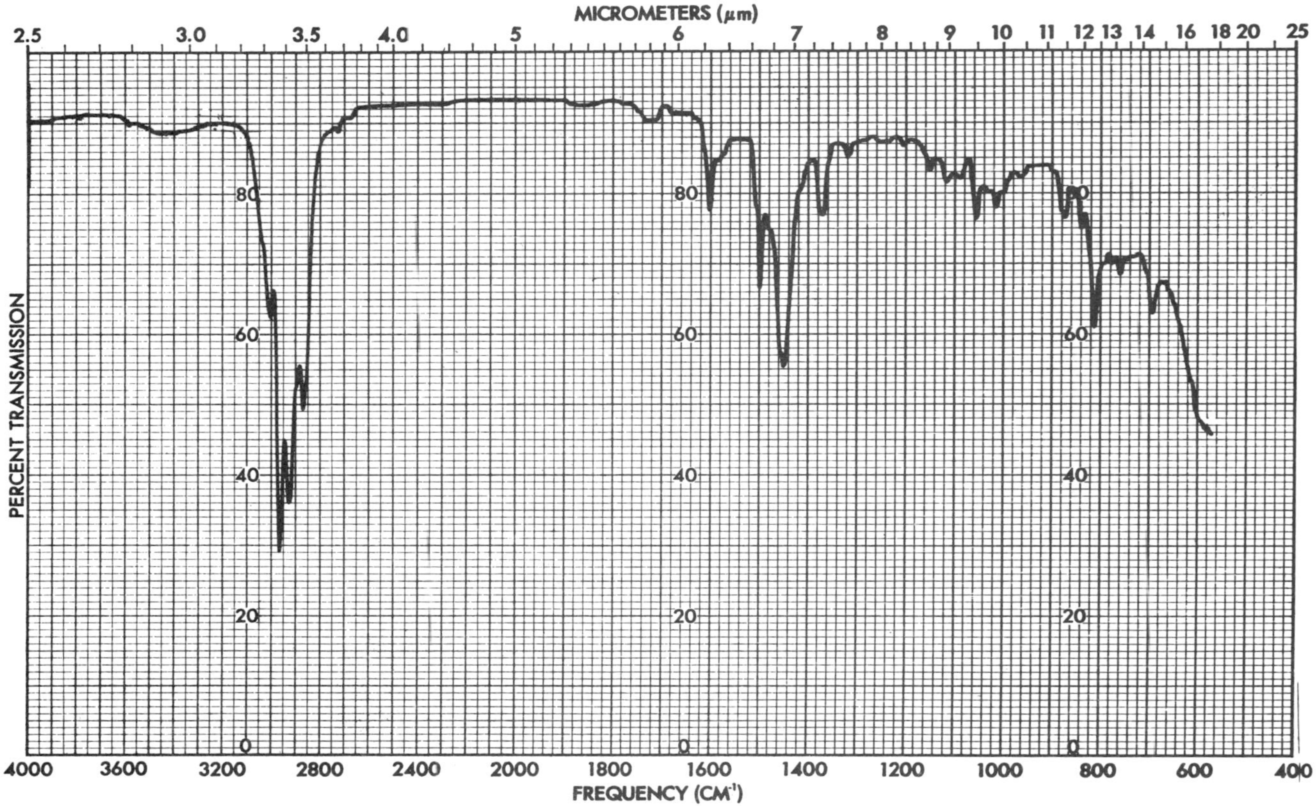

FIG. 6.92. Infrared spectrum of aromatic hydrocarbon blend.

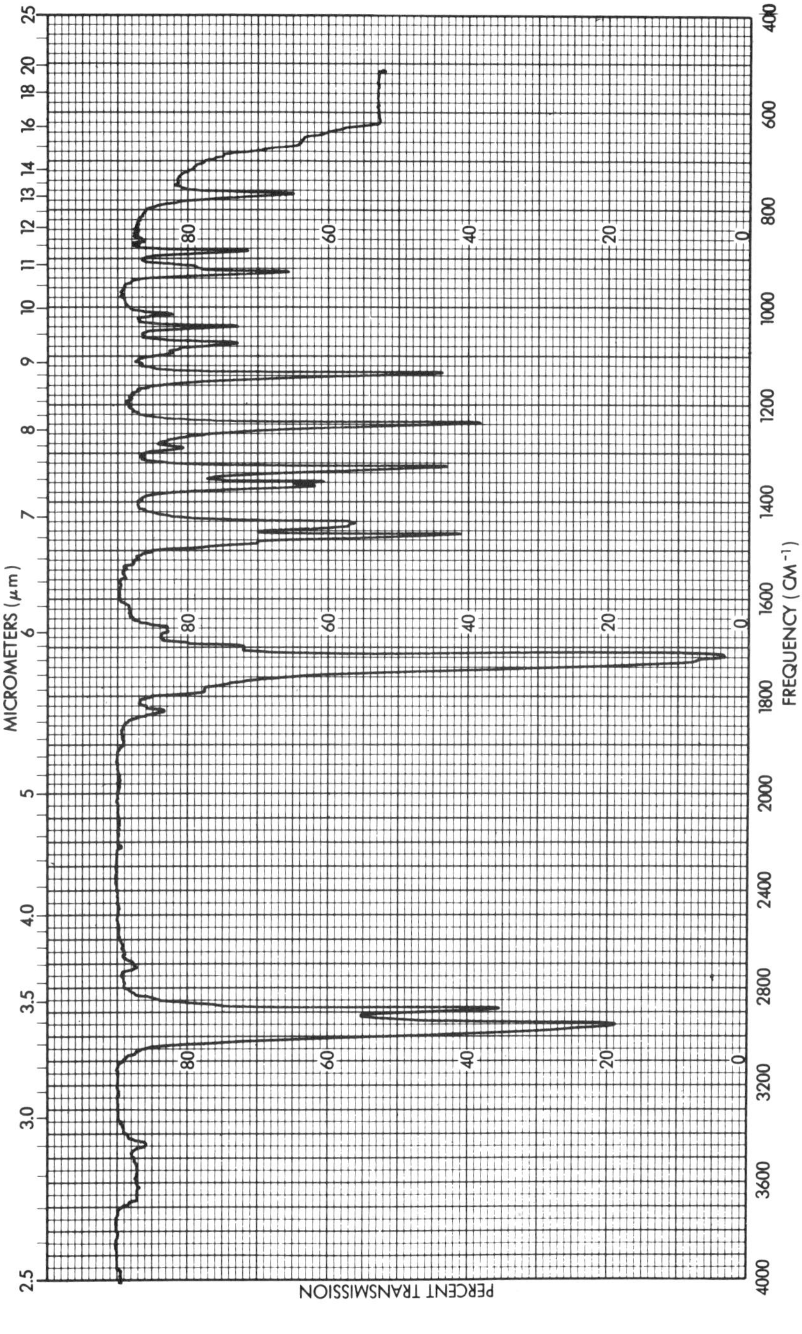

FIG. 6.93. Infrared spectrum of cyclohexanone.

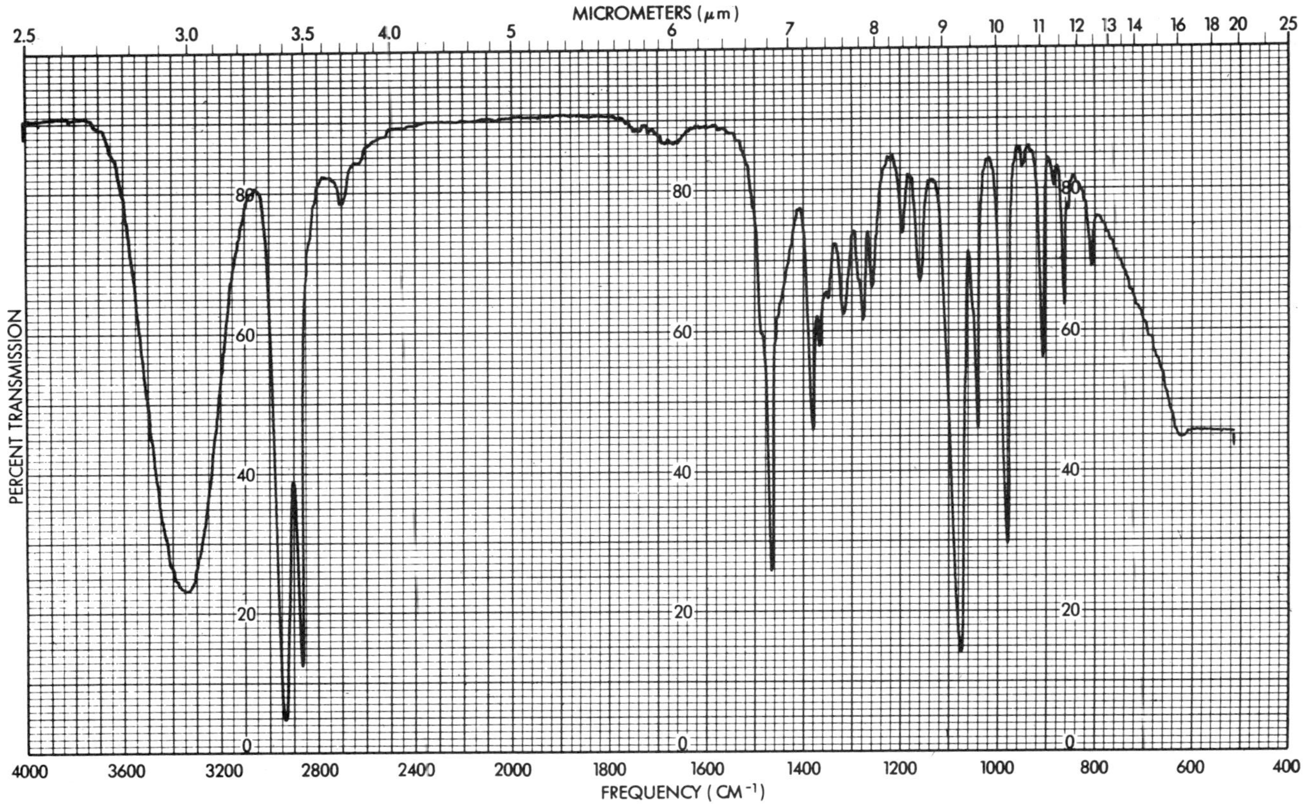

FIG. 6.94. Infrared spectrum of cyclohexanol.

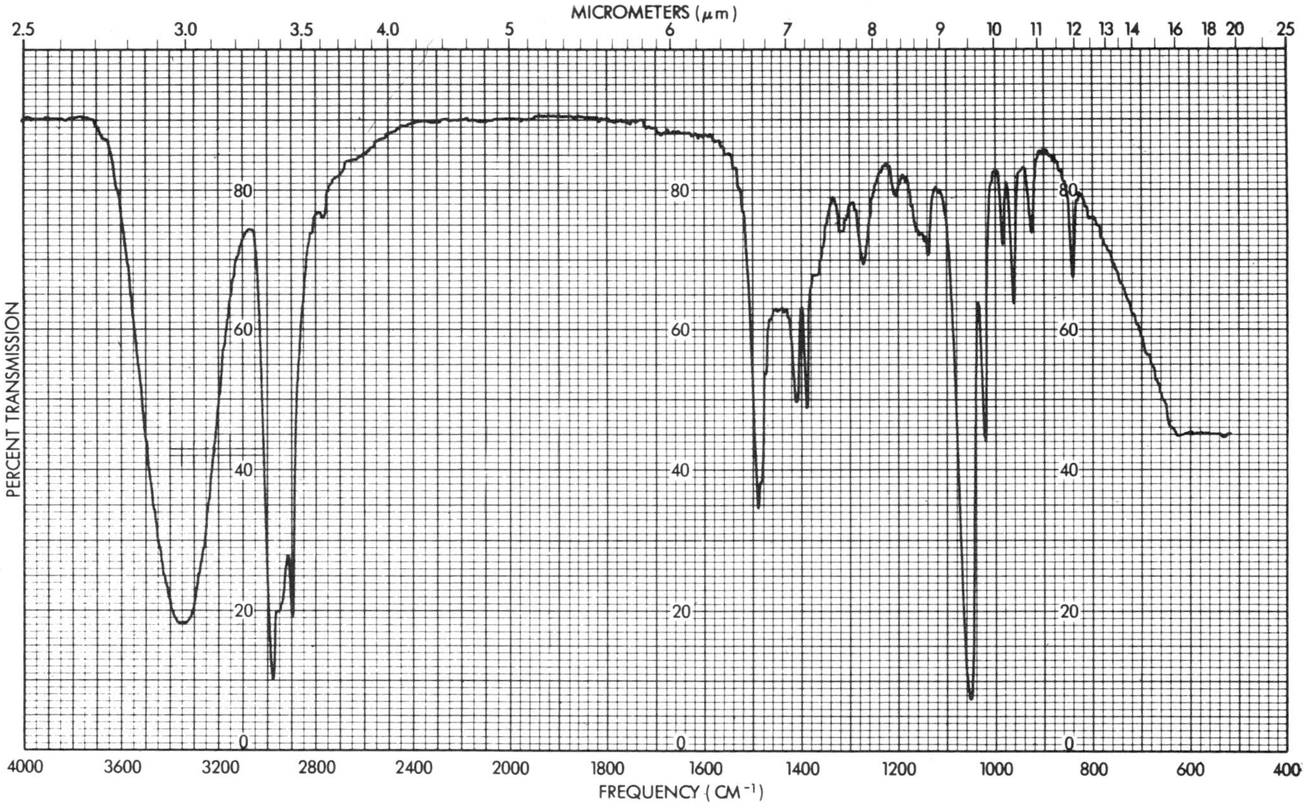

FIG. 6.95. Infrared spectrum of isobutanol.

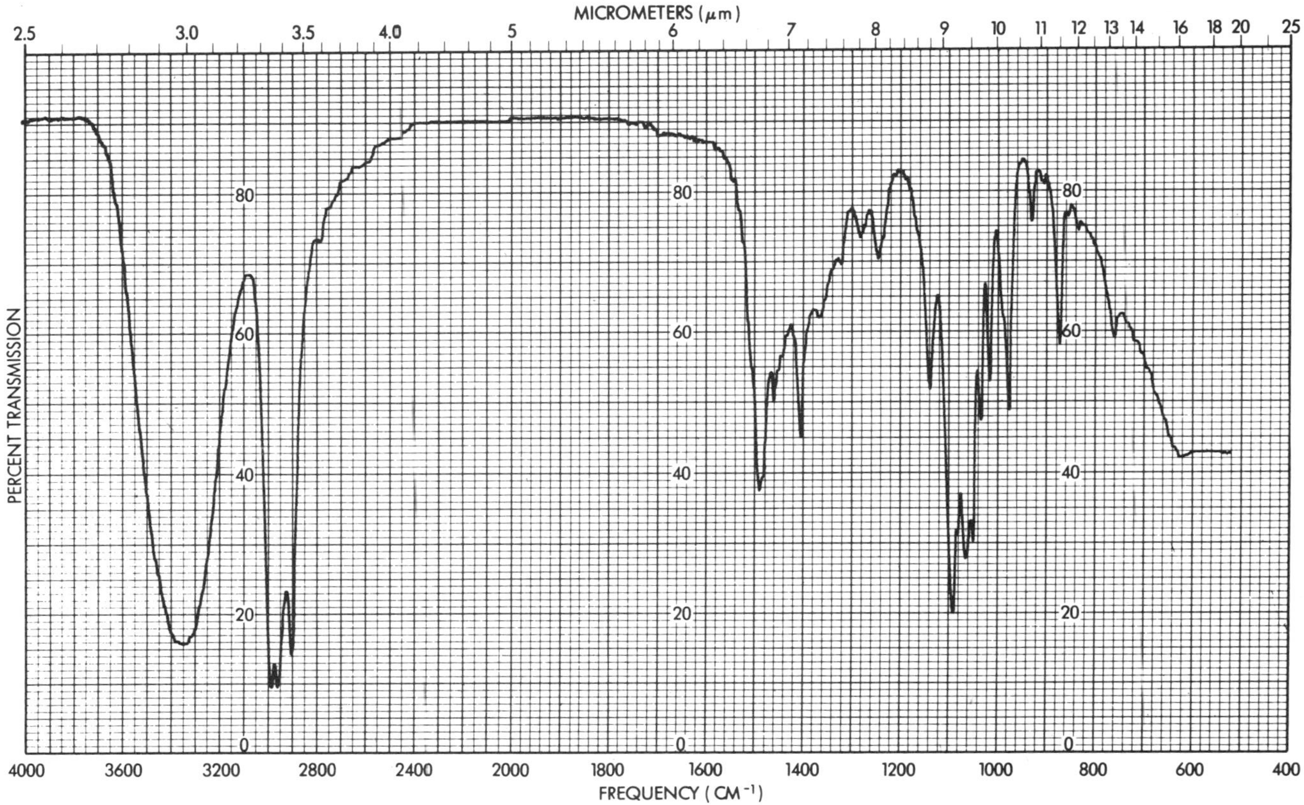

FIG. 6.96. Infrared spectrum of *n*-butanol.

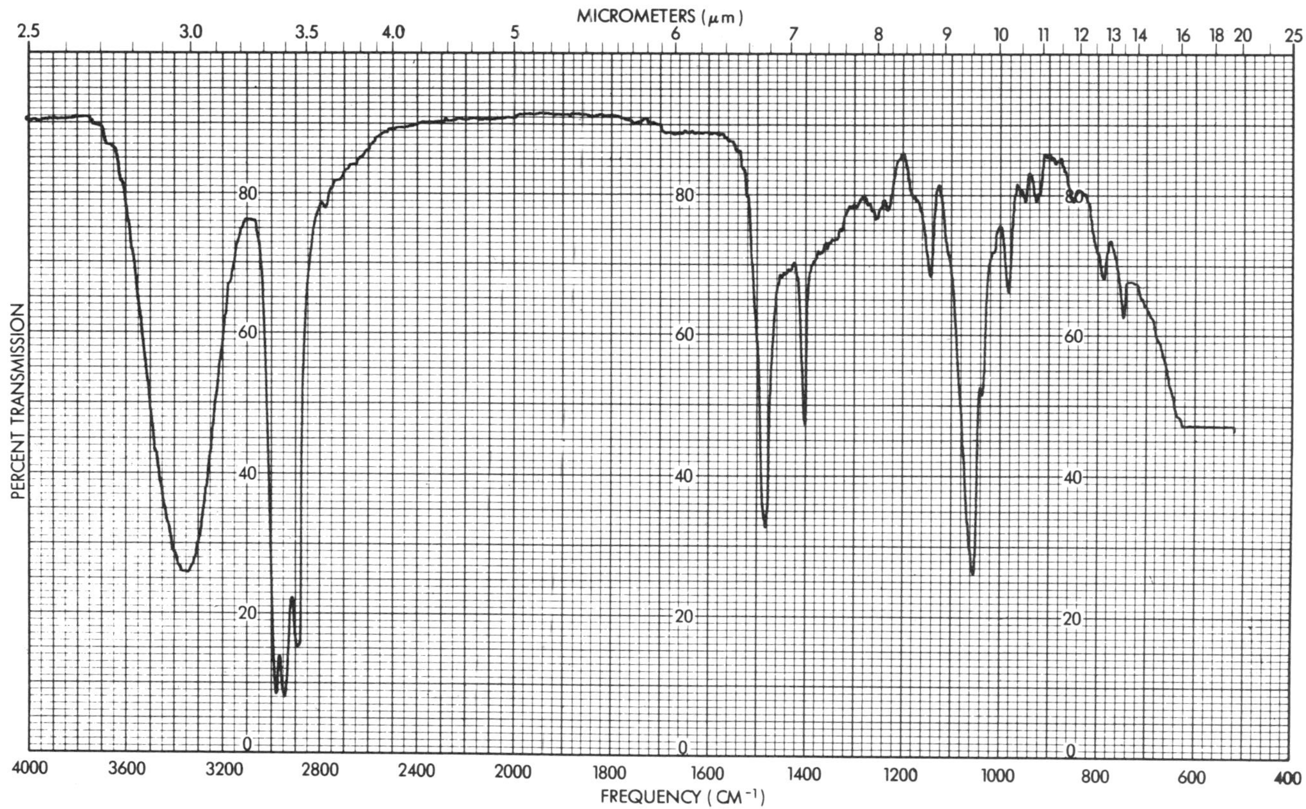

FIG. 6.97. Infrared spectrum of 2-ethyl hexanol.

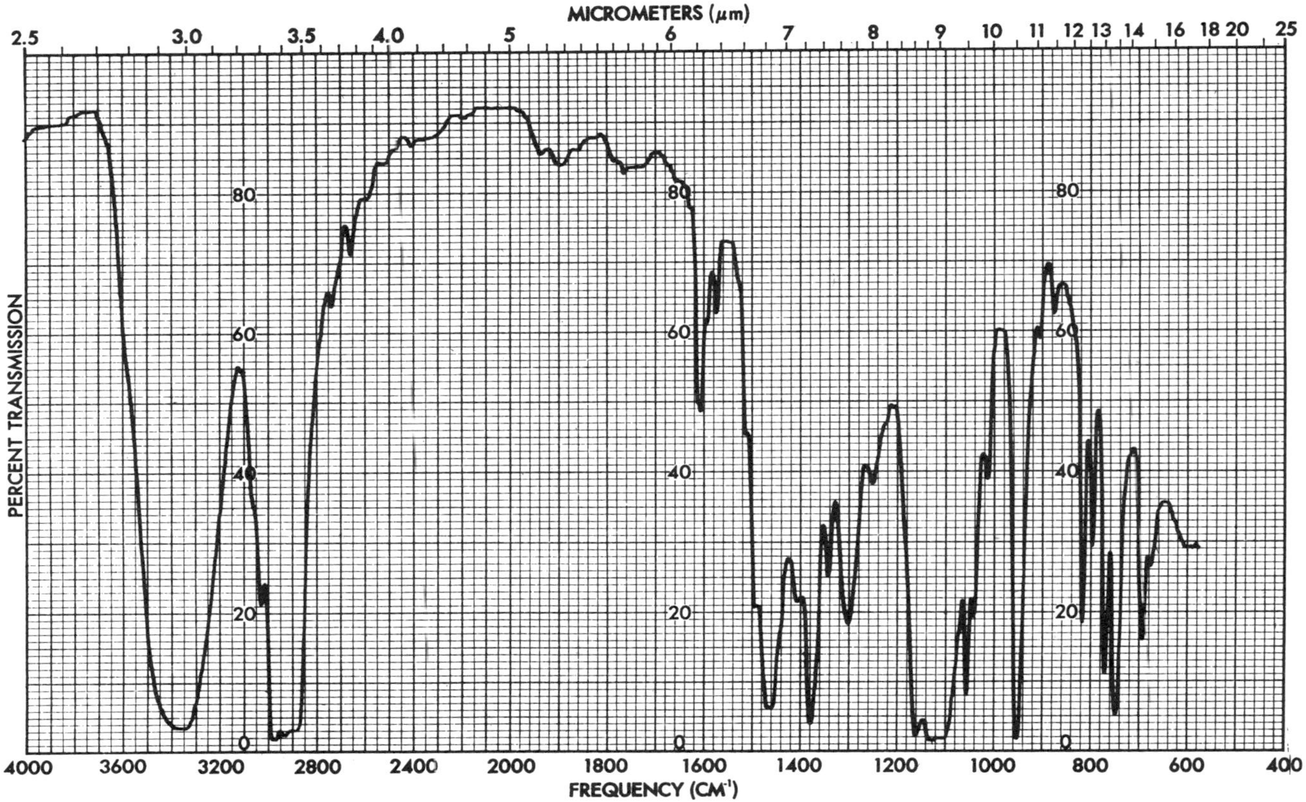

FIG. 6.98. Infrared spectrum of xylene–IPA–alkylaryl.

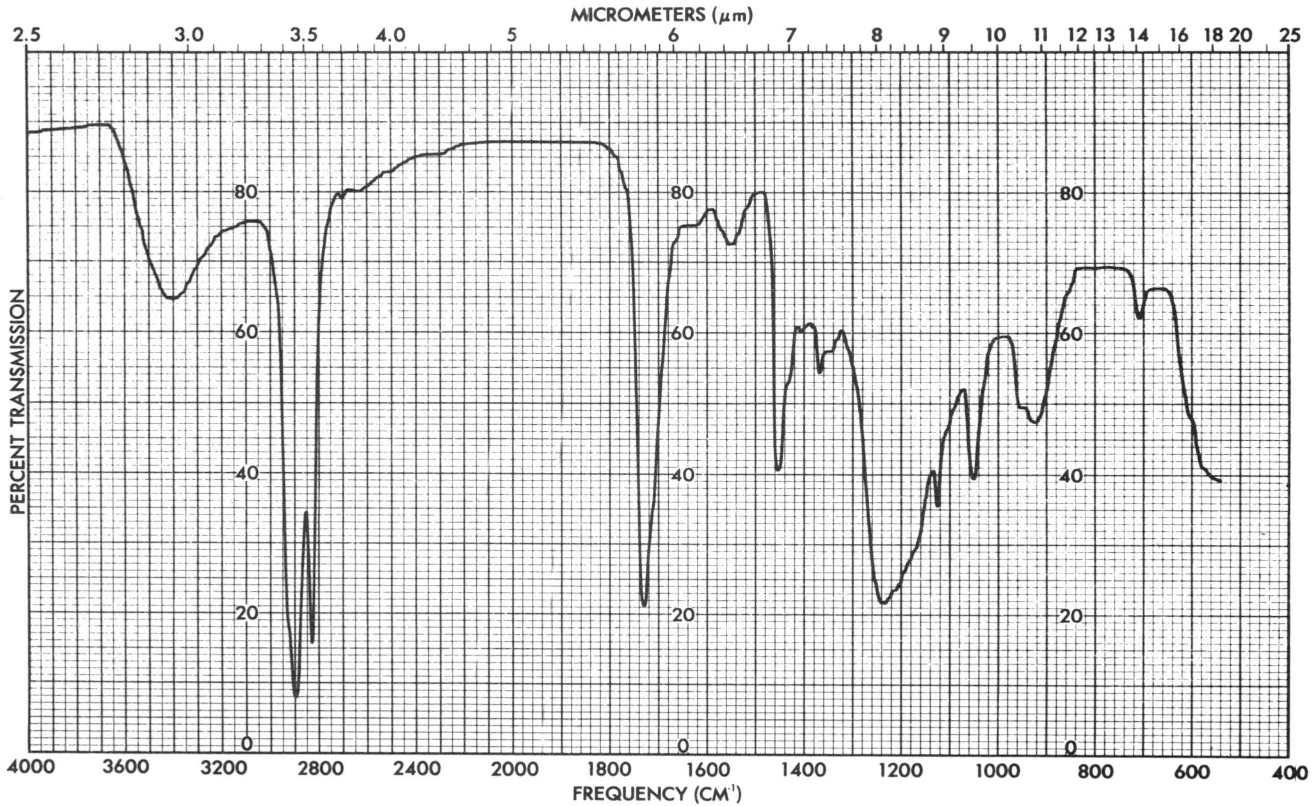

FIG. 6.99. Infrared spectrum of sulfonated propyl oleate–sulfonated castor oil.

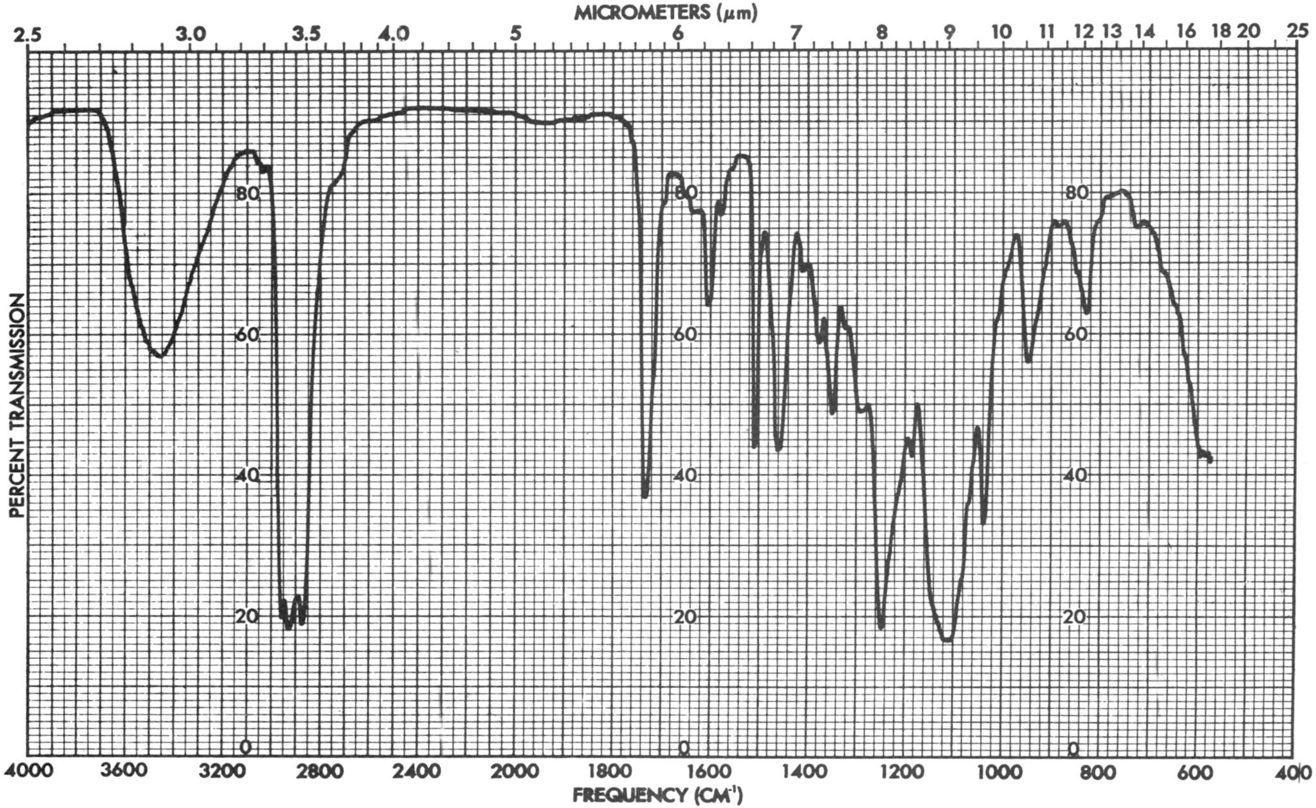

FIG. 6.100. Infrared spectrum of $(EO)_9$ nonyl phenol–sulfosuccinate.

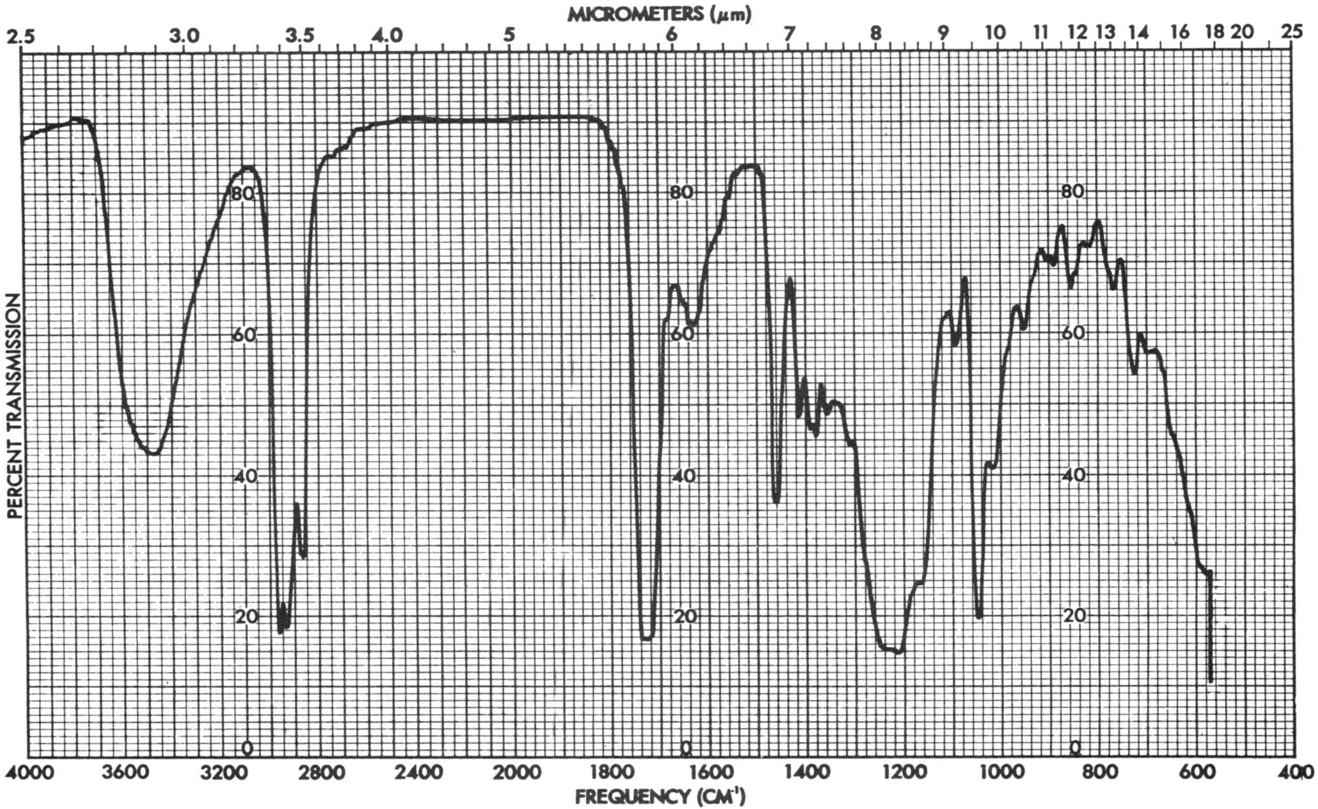

FIG. 6.101. Infrared spectrum of sulfosuccinate.

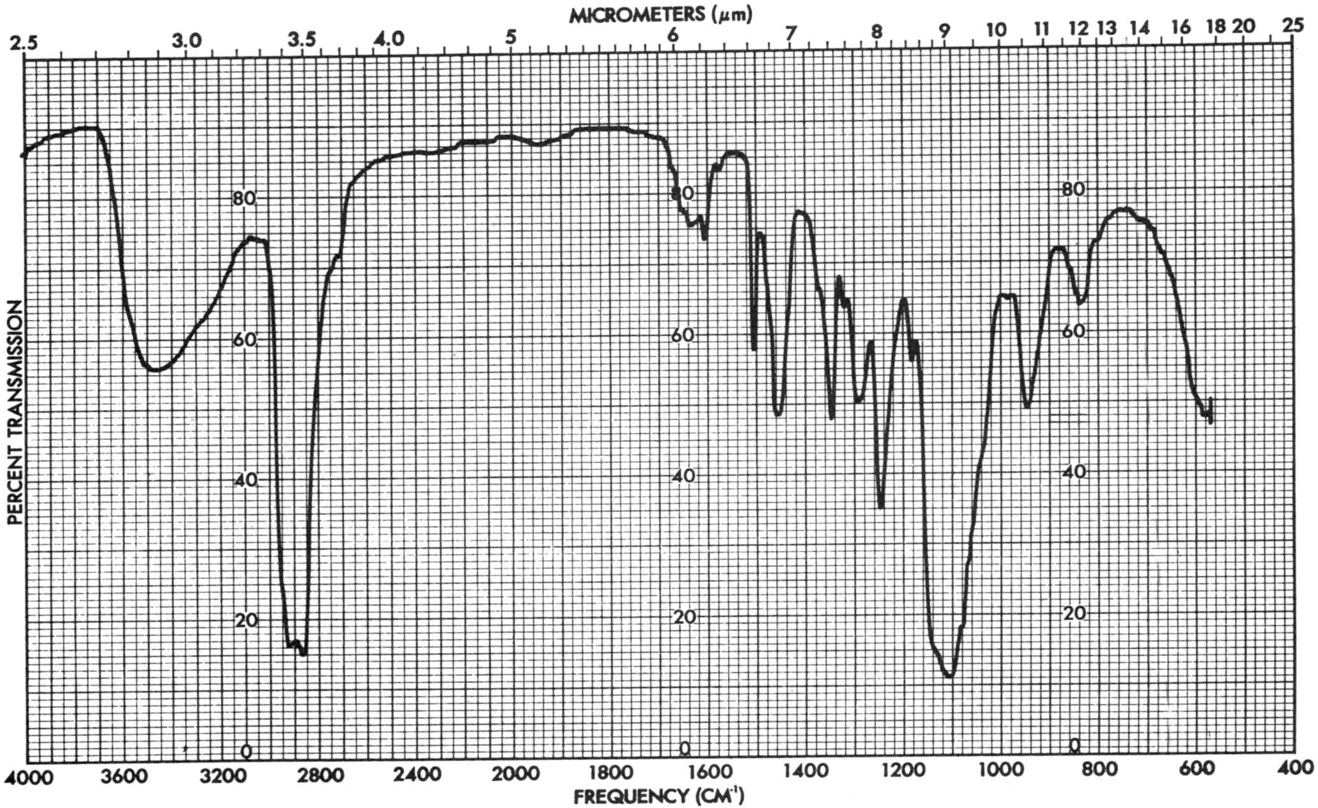

FIG. 6.102. Infrared spectrum of $(EO)_9$ nonyl phenol phosphate–oleyl alcohol.

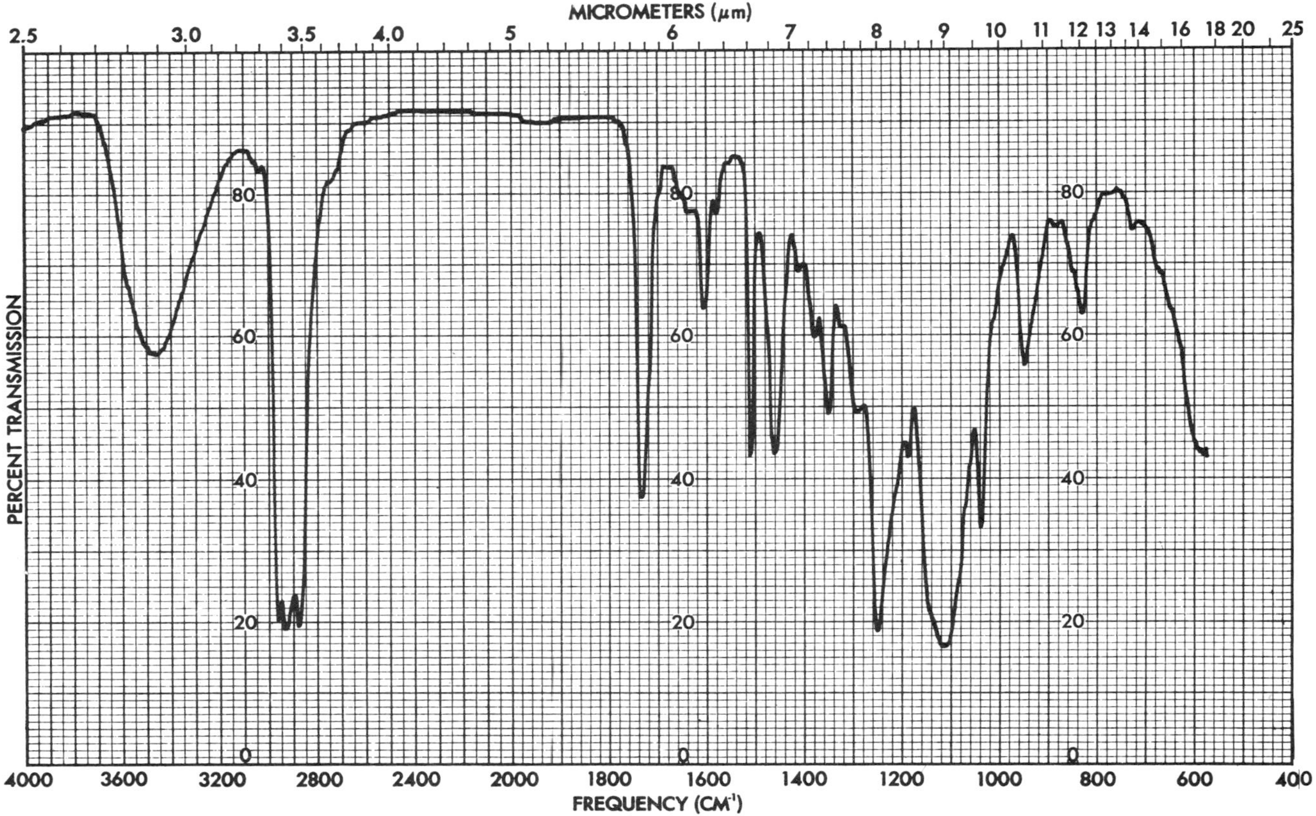

FIG. 6.103. Infrared spectrum of $(EO)_{30}$ nonyl phenol–sulfosuccinate.

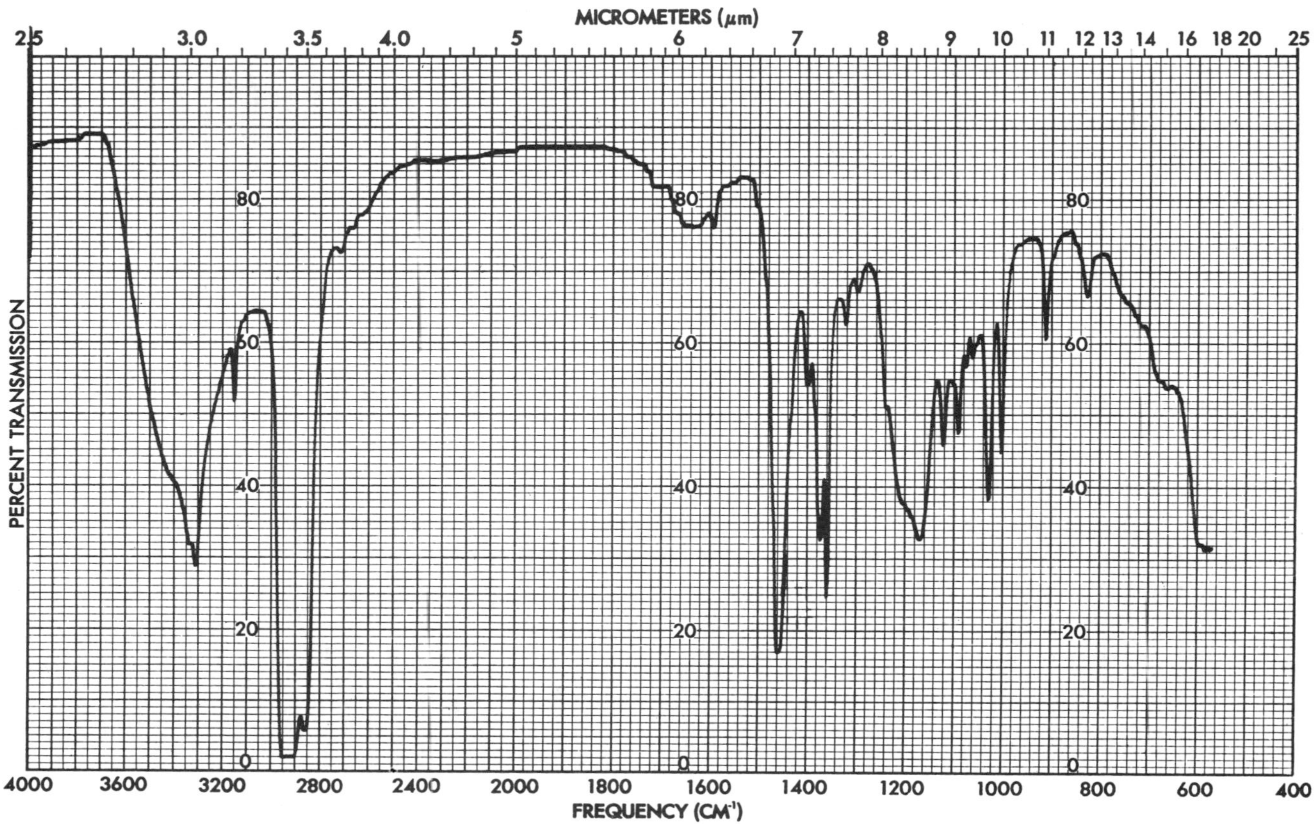

FIG. 6.104. Infrared spectrum of TEA–alkylaryl–pine oil.

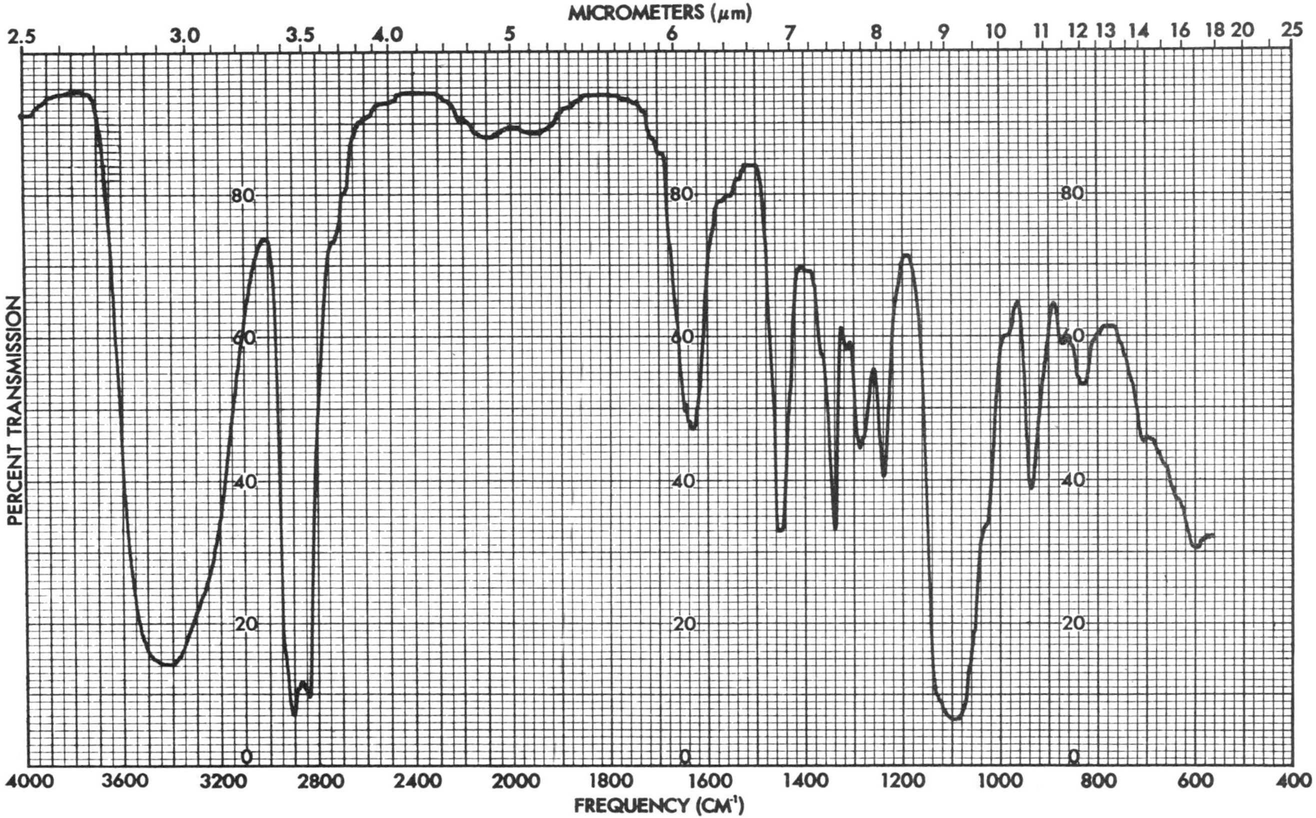

FIG. 6.105. Infrared spectrum of EO linear alcohol.

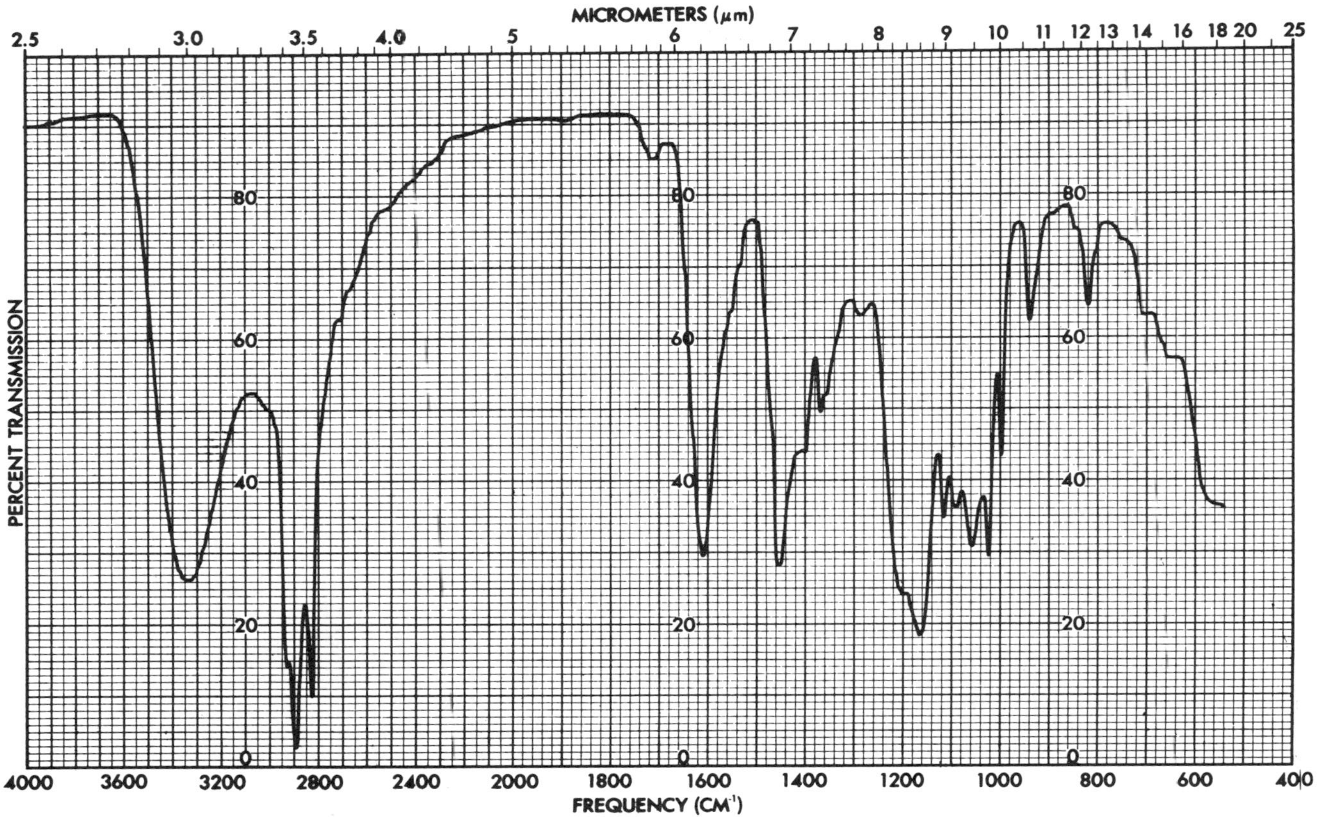

FIG. 6.106. Infrared spectrum of amine condensate + DDBSA.

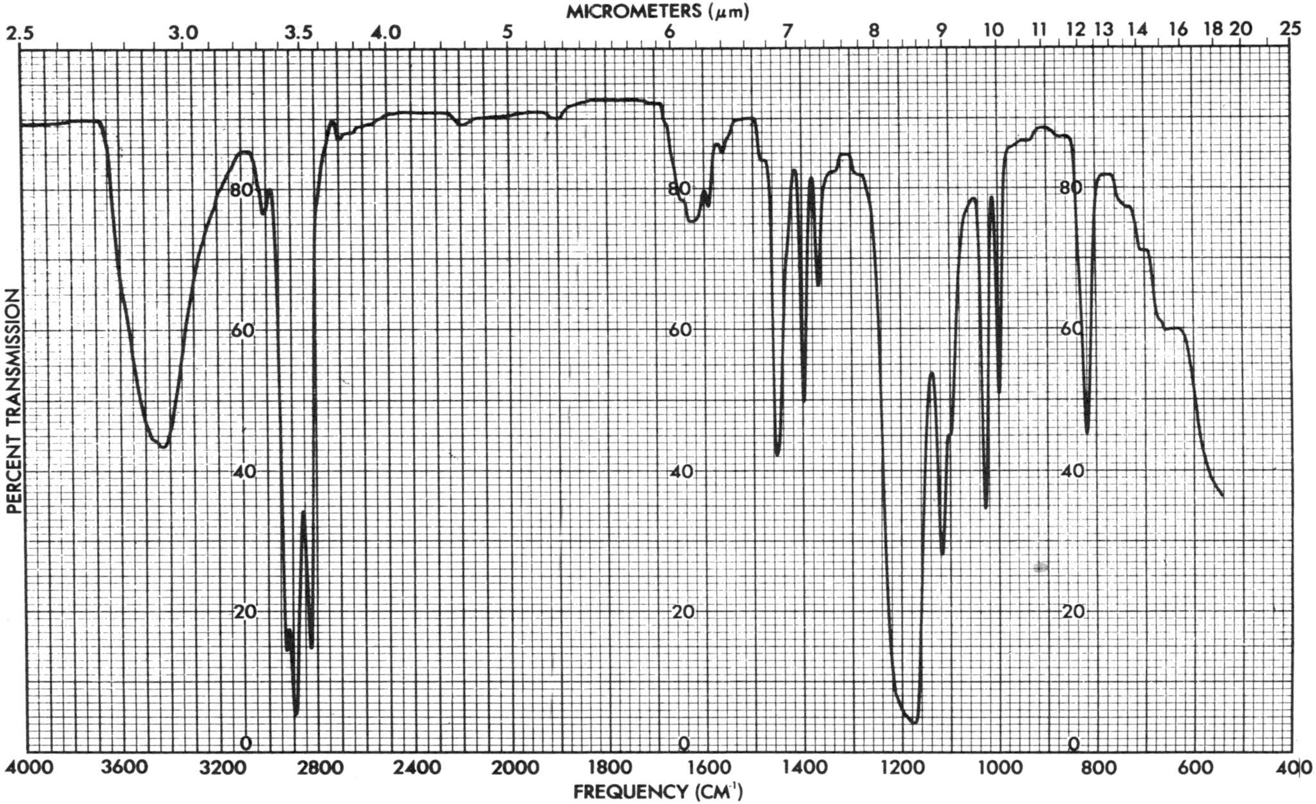

FIG. 6.107. Infrared spectrum of DDBSA–sodium salt.

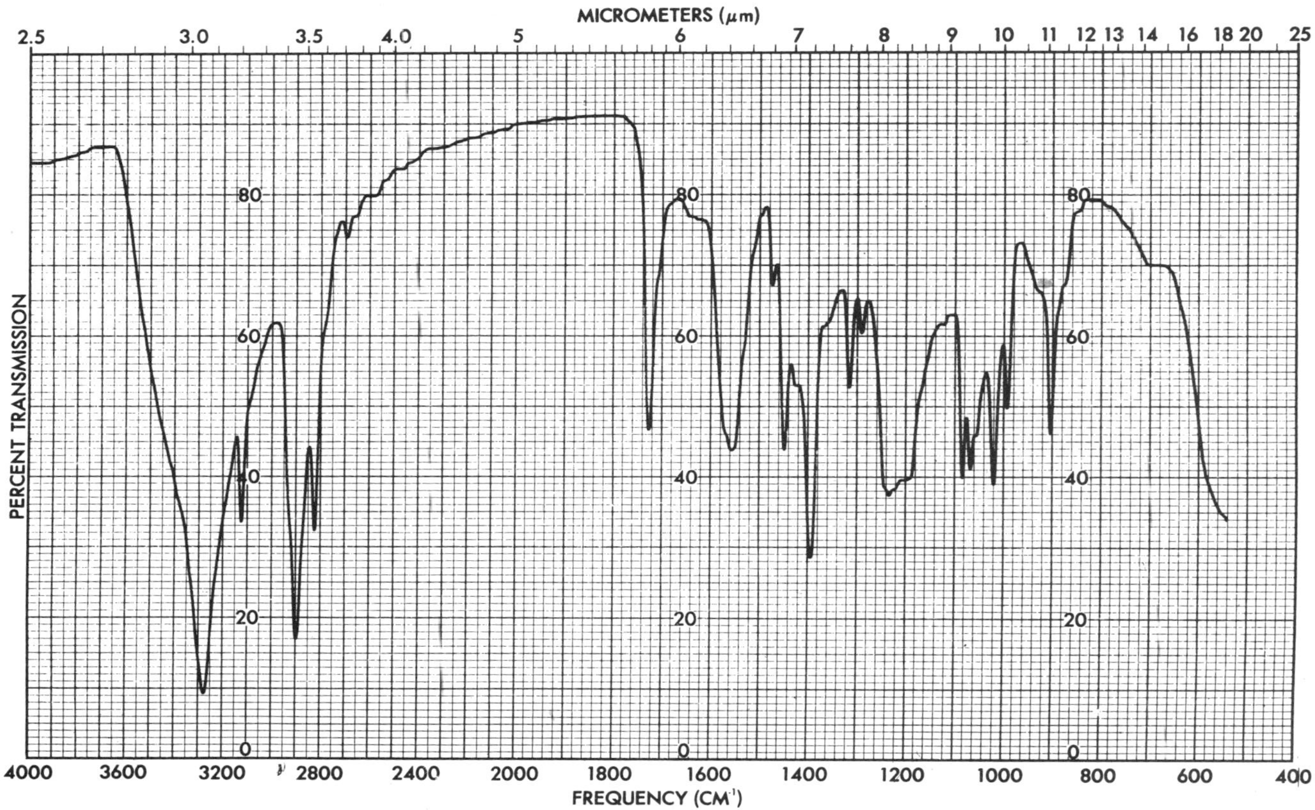

FIG. 6.108. Infrared spectrum of TEA salt–sulfonated butyl oleate.

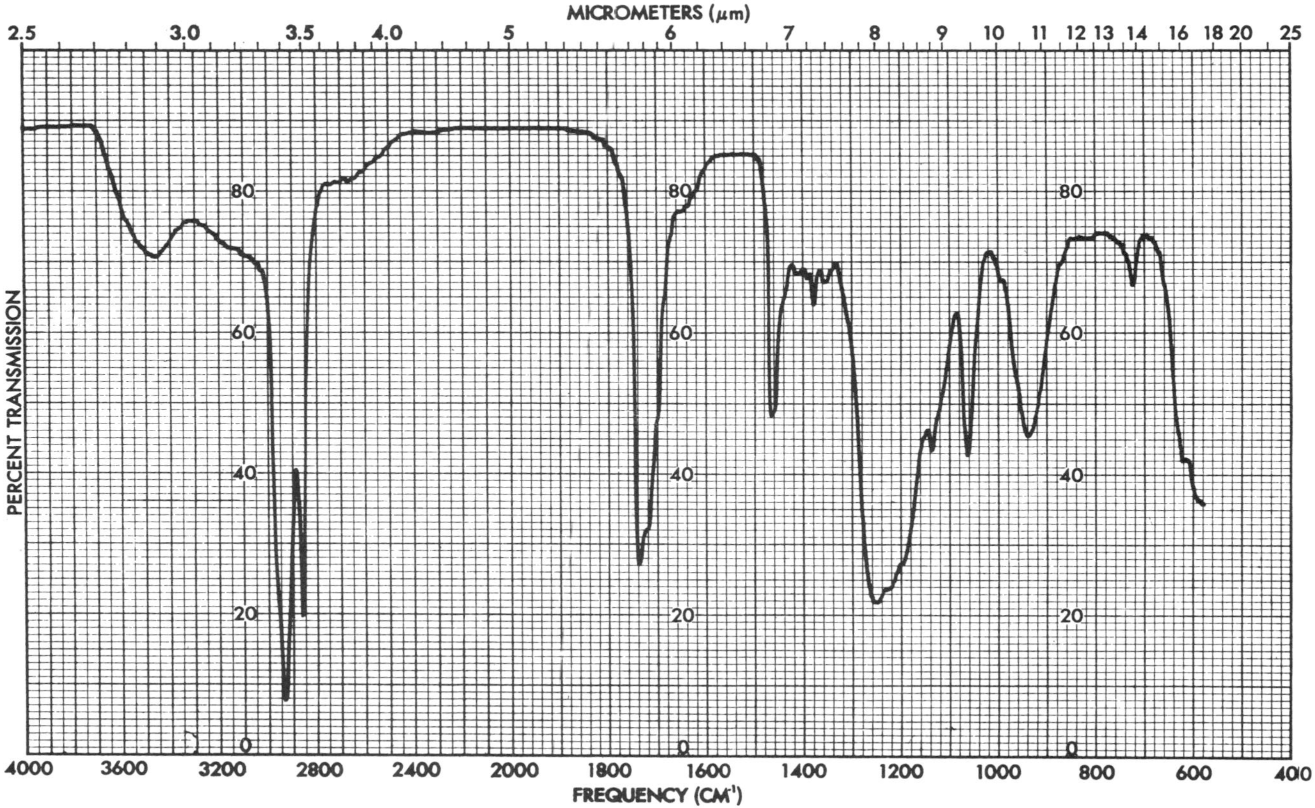

FIG. 6.109. Infrared spectrum of sulfonated butyl oleate–sulfosuccinate.

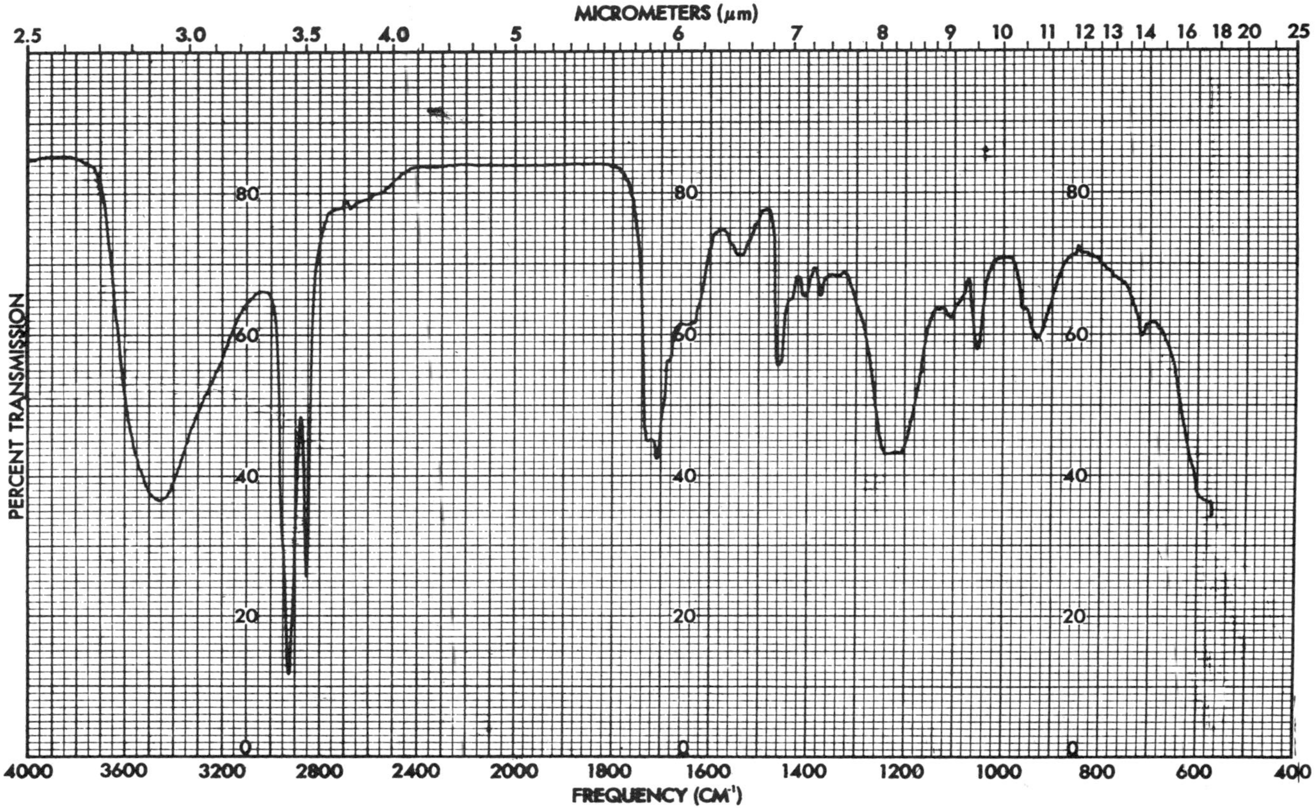

FIG. 6.110. Infrared spectrum of sulfonated red oil.

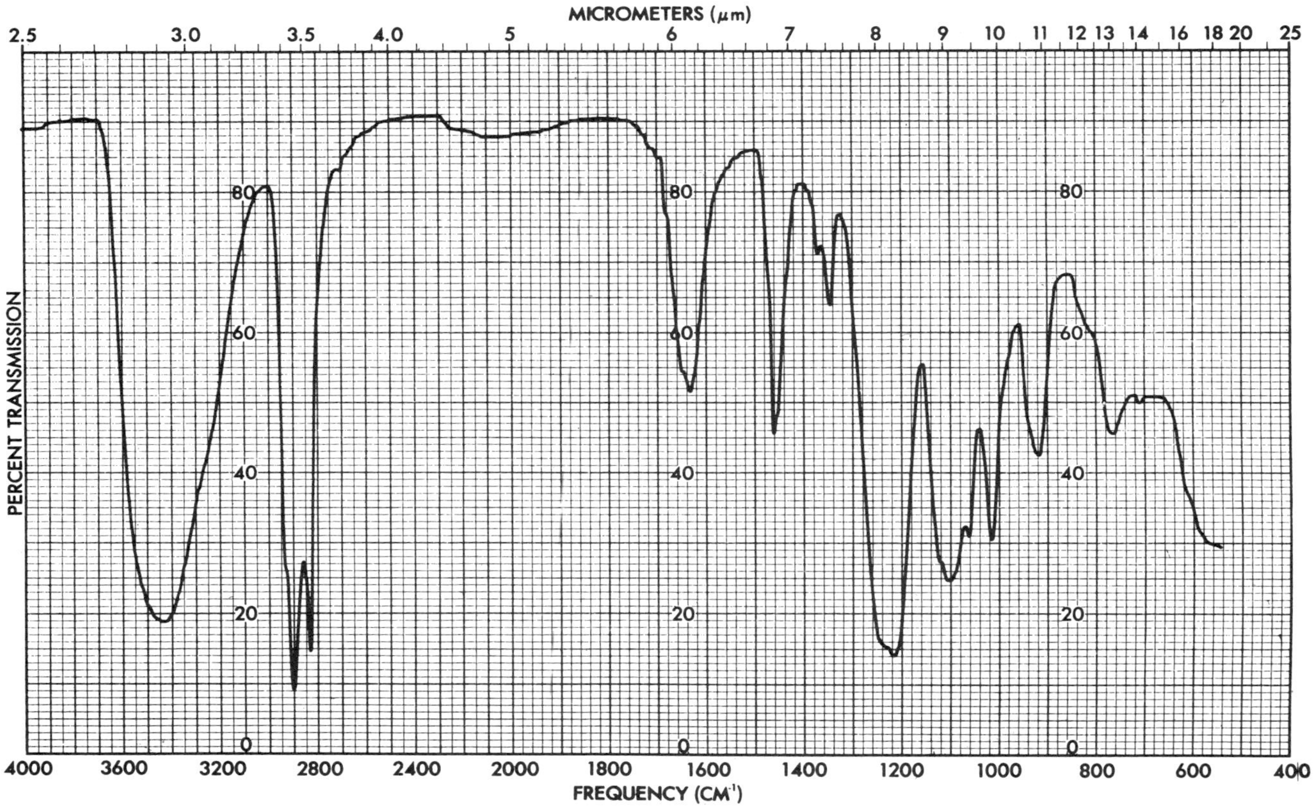

FIG. 6.111. Infrared spectrum of sodium tridecyl ether sulfate.

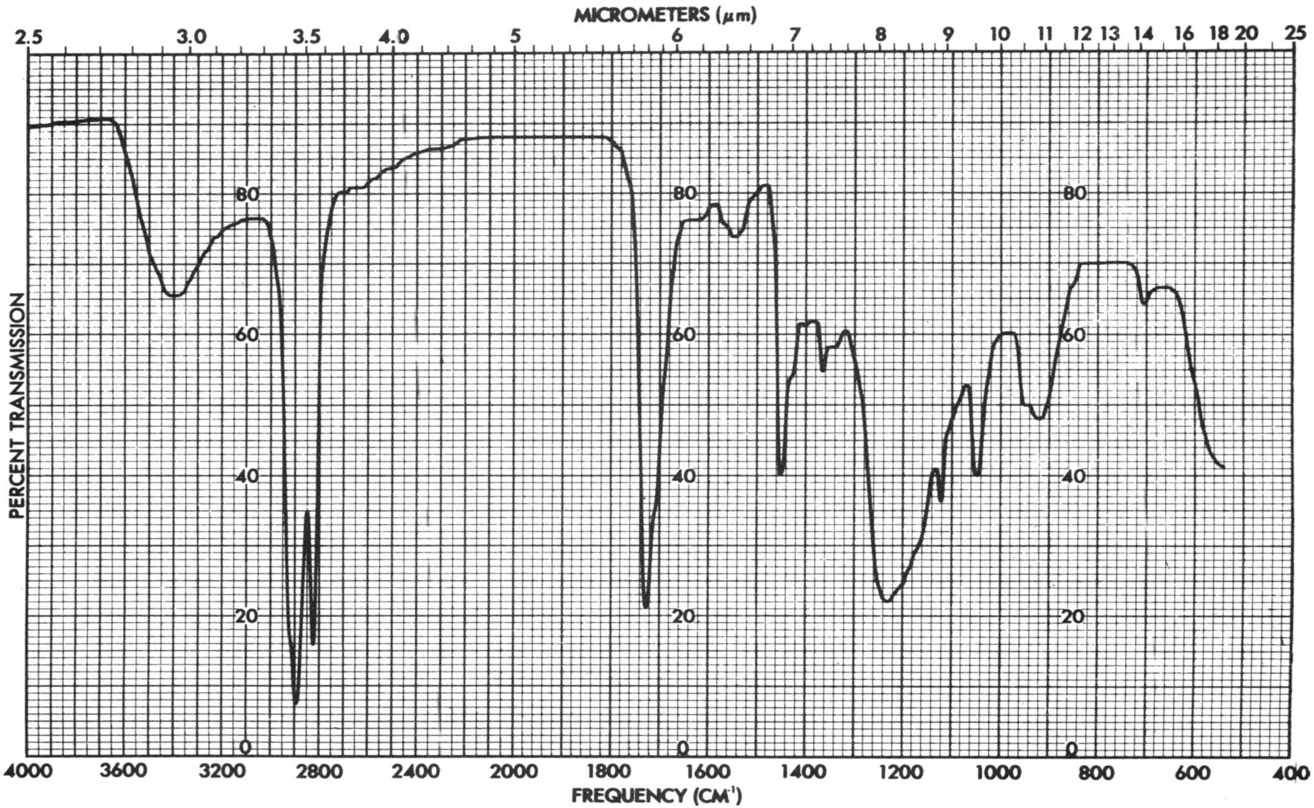

FIG. 6.112. Infrared spectrum of sulfonated glyceryl trioleate.

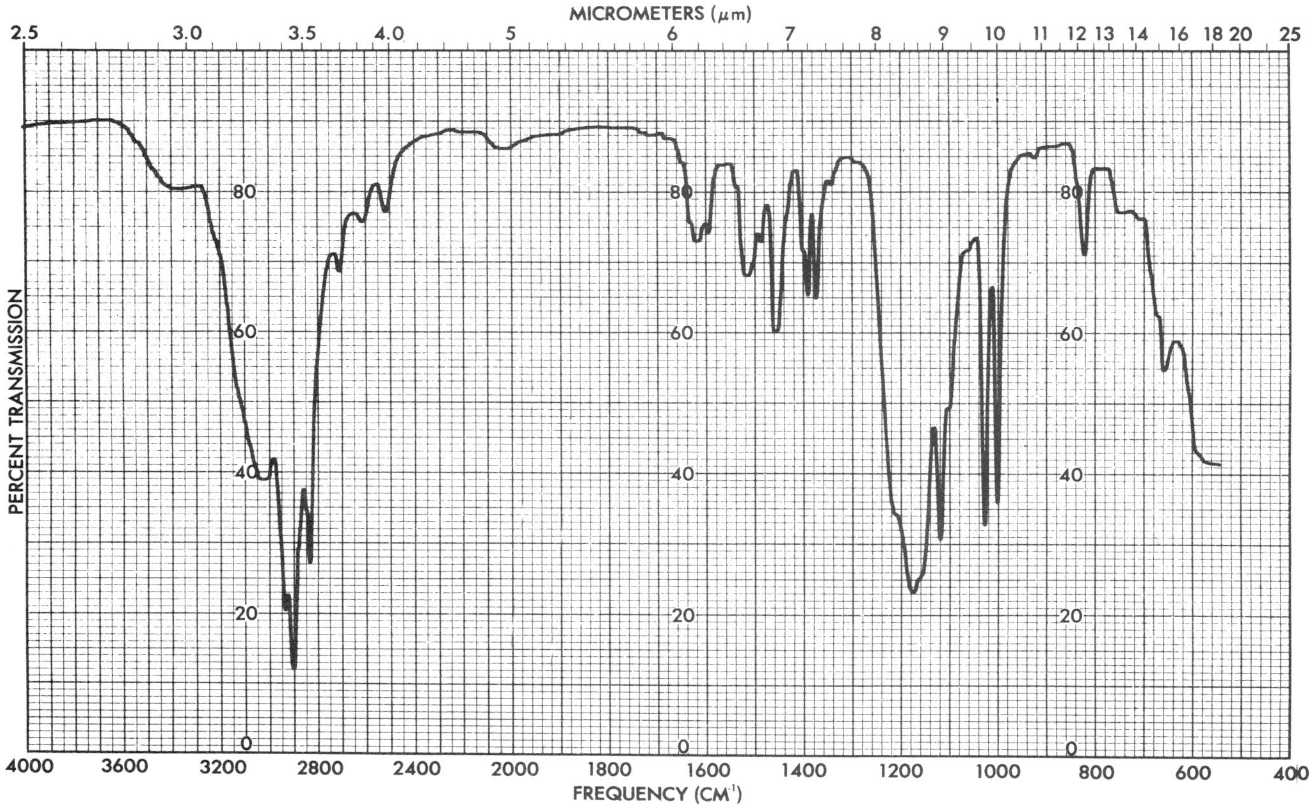

FIG. 6.113. Infrared spectrum of DDBSA–isopropyl amine salt.

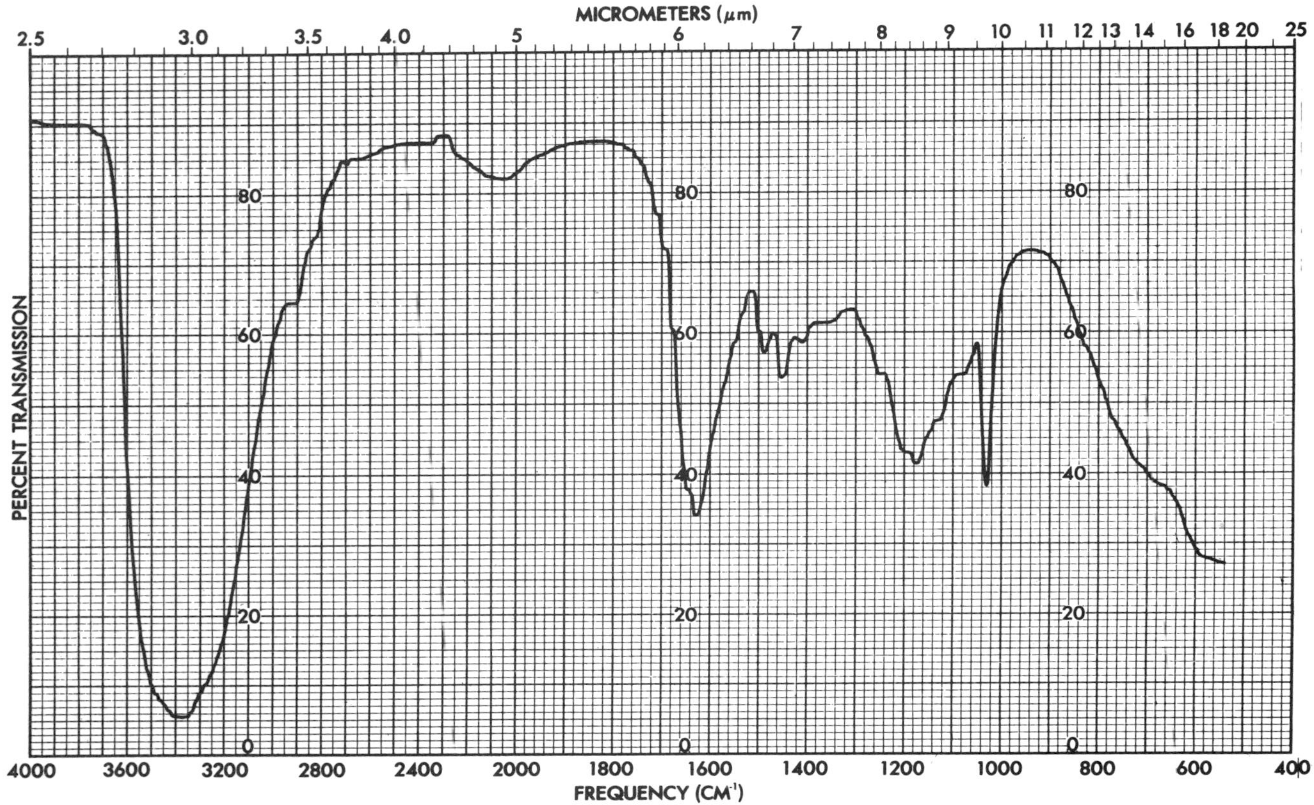

FIG. 6.114. Infrared spectrum of lignin sulfate.

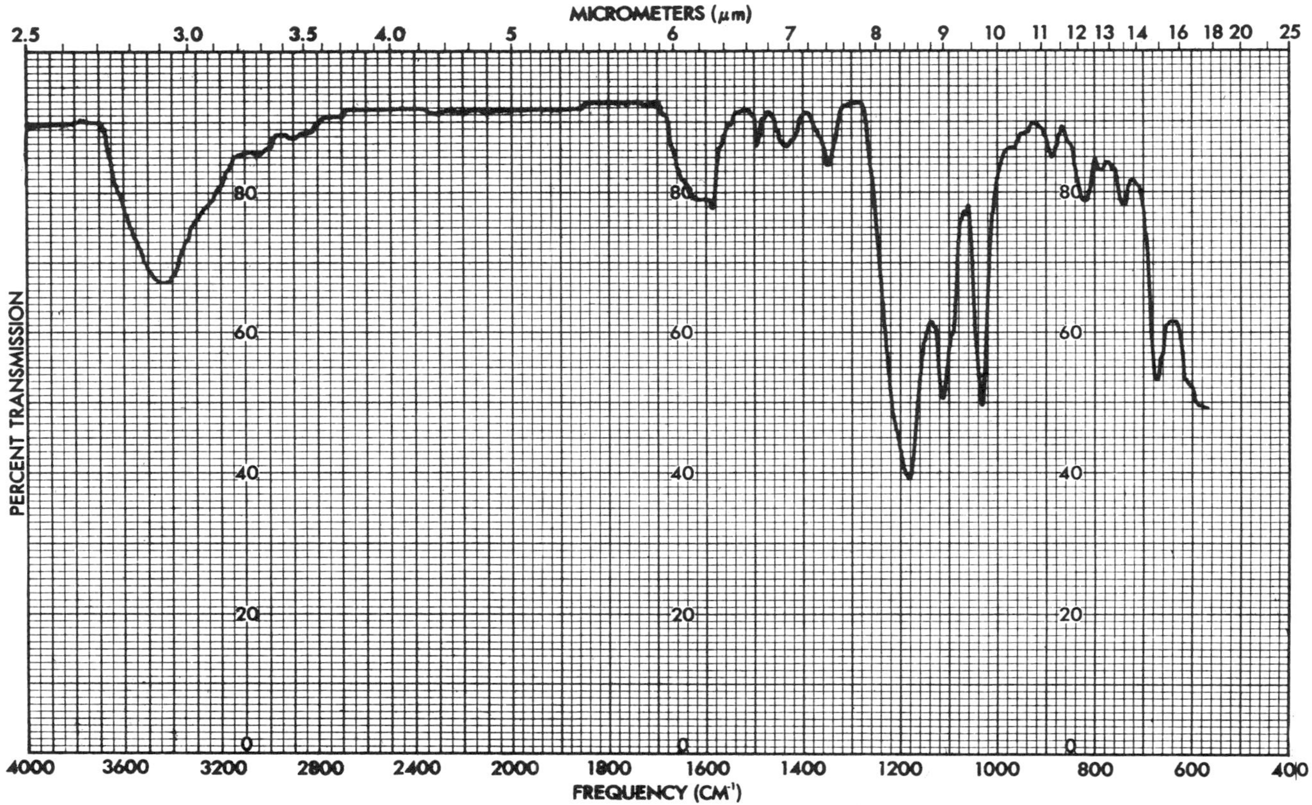

FIG. 6.115. Infrared spectrum of naphthalene formaldehyde condensate sulfate.

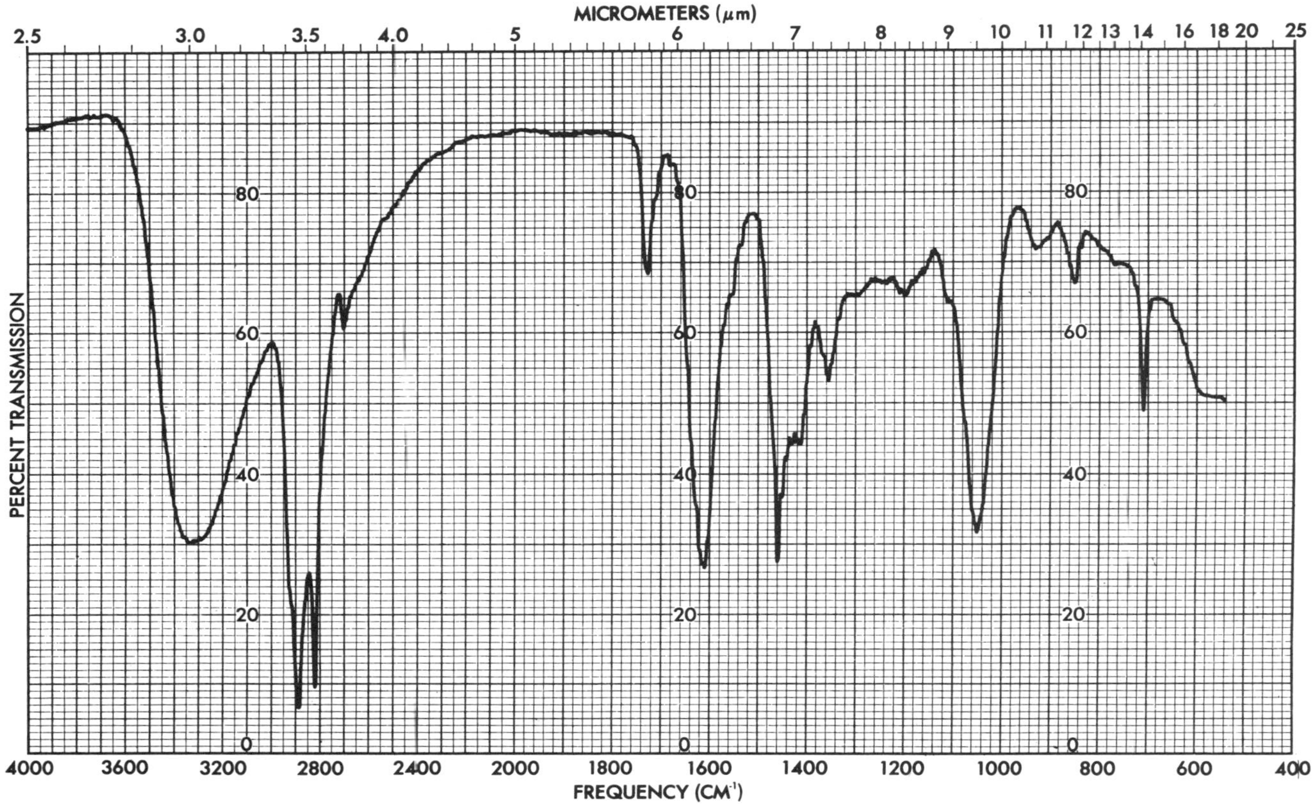

FIG. 6.116. Infrared spectrum of 1:2 amine condensate.

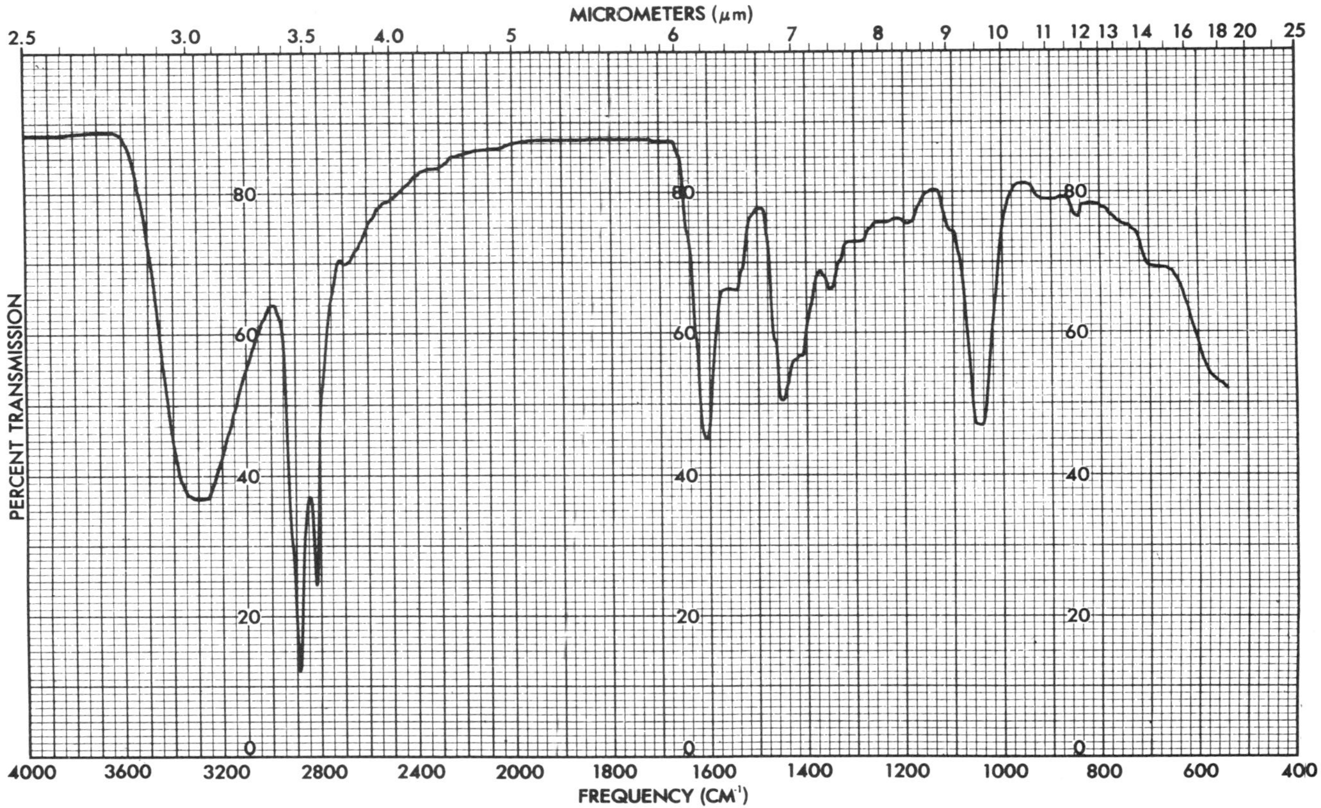

FIG. 6.117. Infrared spectrum of 1:1 amine condensate.

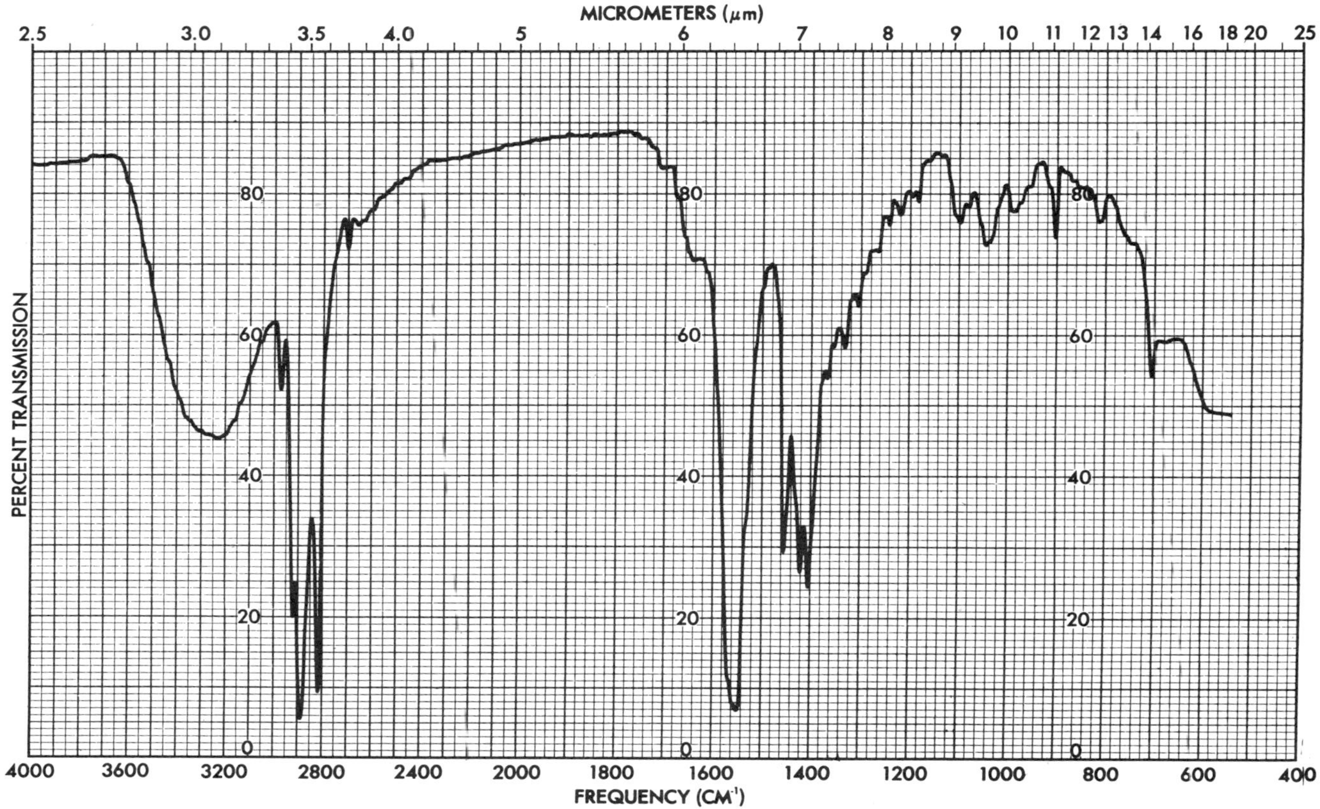

FIG. 6.118. Infrared spectrum of TEA–fatty acid soap.

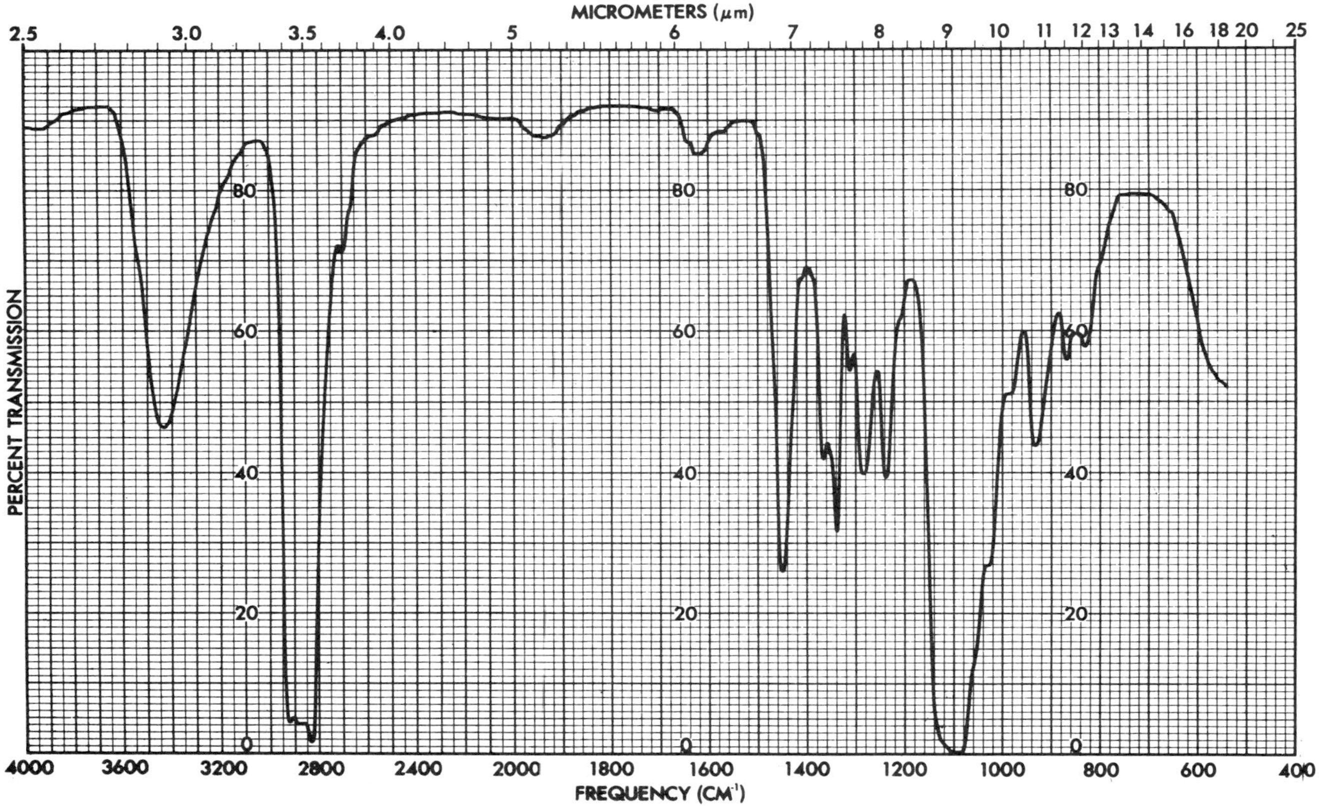

FIG. 6.119. Infrared spectrum of phosphated thioether.

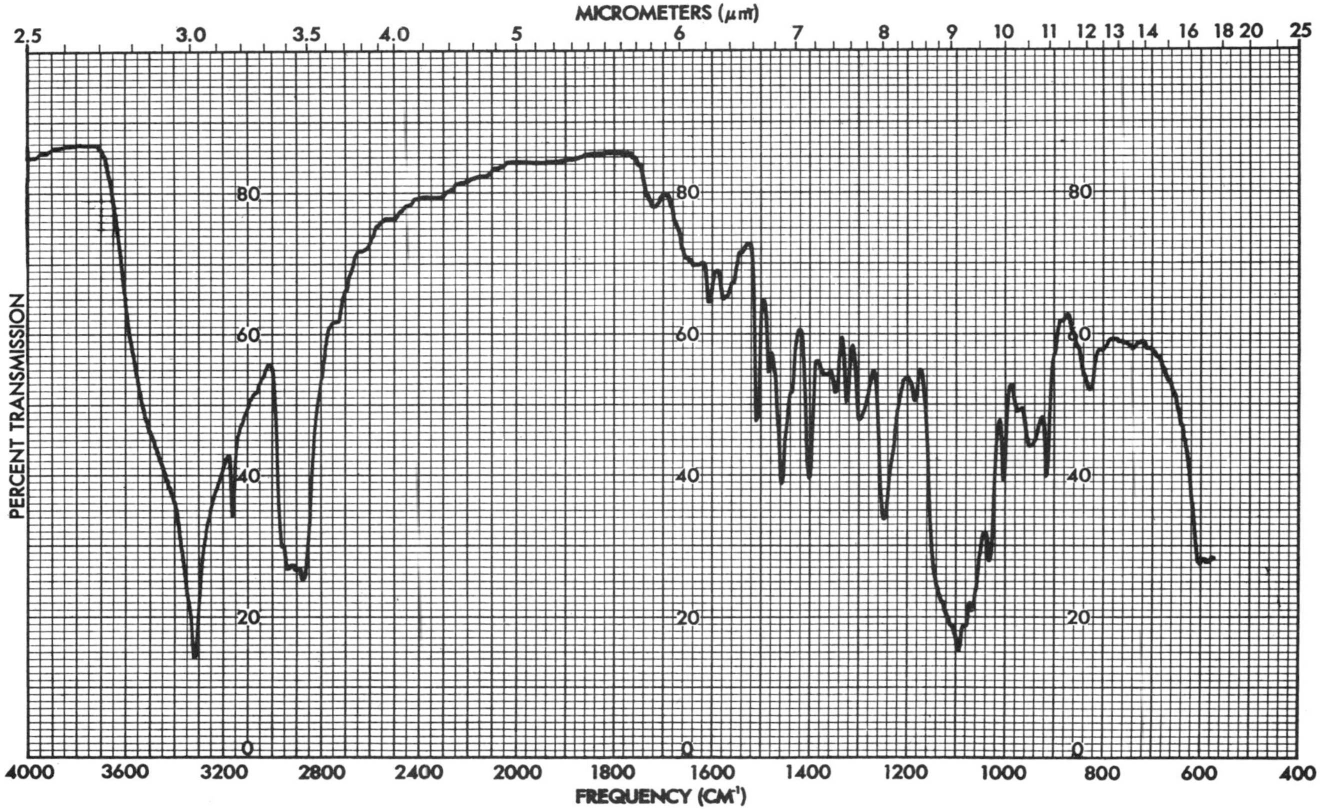

FIG. 6.120. Infrared spectrum of TEA salt–$(EO)_9$ nonyl phenol phosphate.

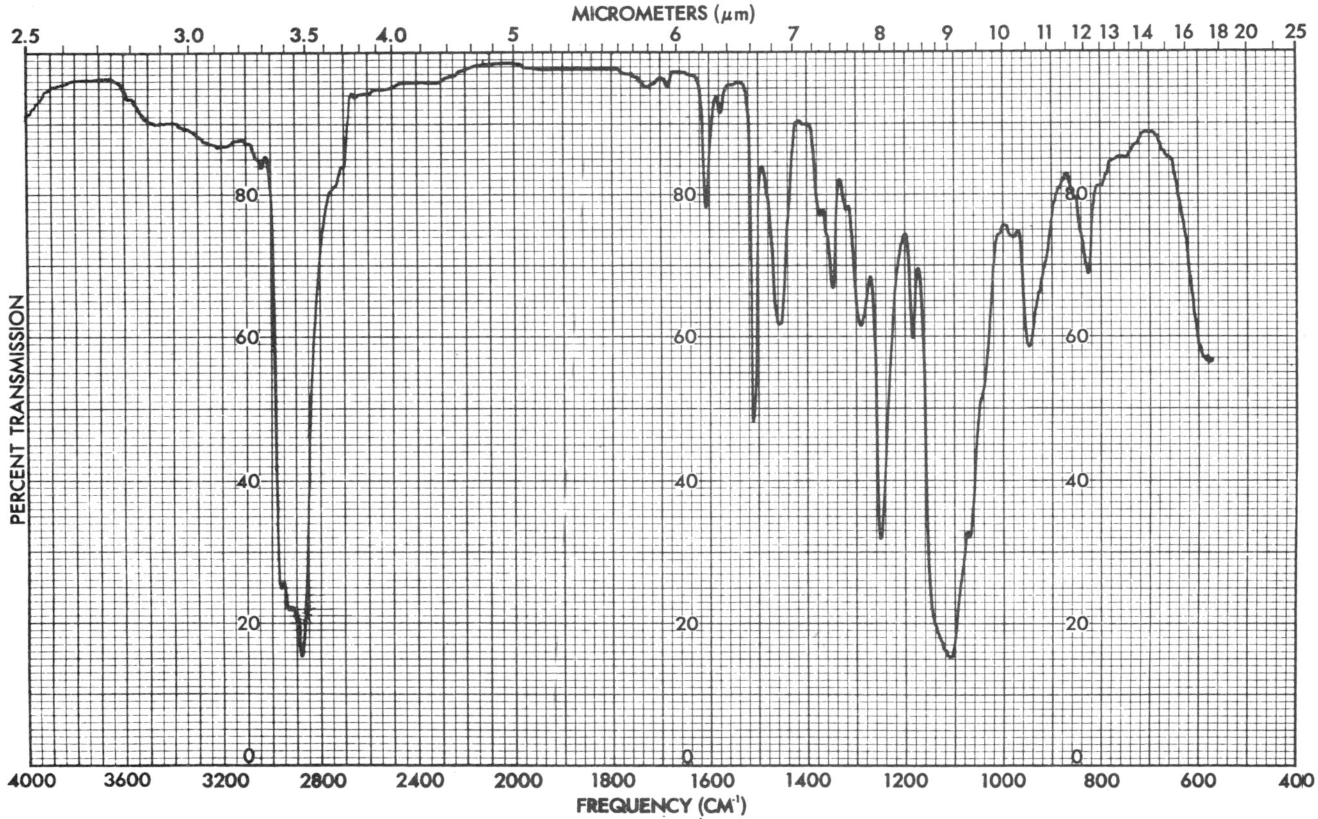

FIG. 6.121. Infrared spectrum of potassium salt–$(EO)_9$ nonyl phenol phosphate.

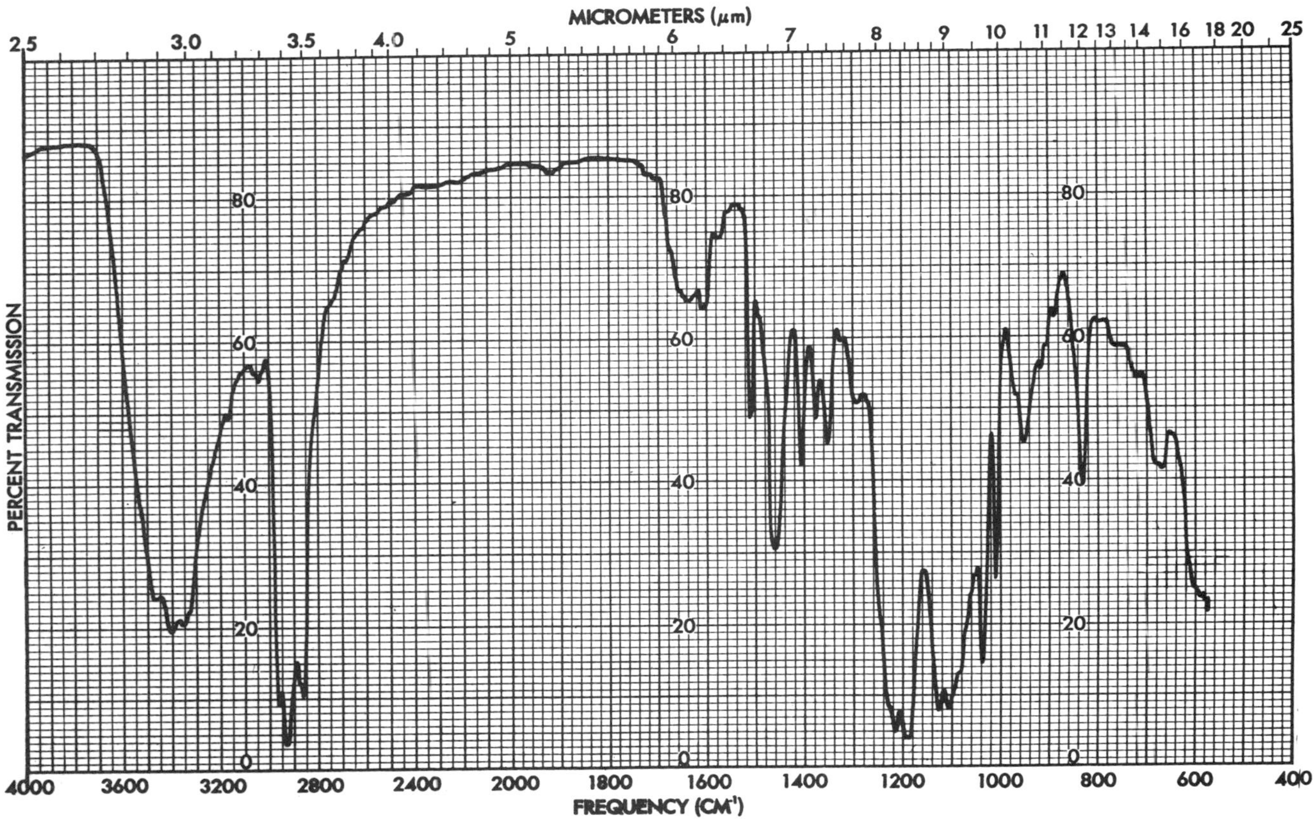

FIG. 6.122. Infrared spectrum of mixed TEA–potassium salt DDBSA.

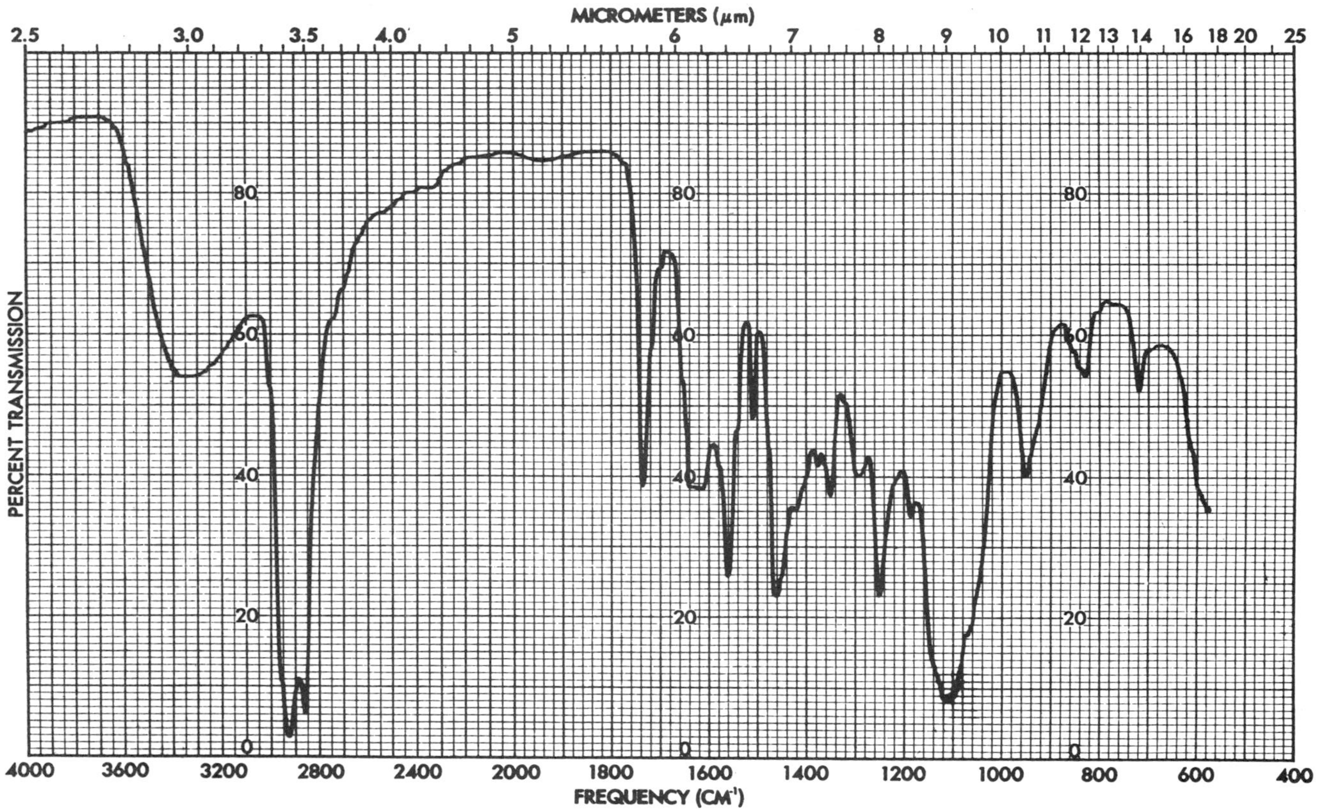

FIG. 6.123. Infrared spectrum of mixed sulfonated castor oil–(EO)$_9$ nonyl phenol–(EO)$_9$ nonyl phenol phosphate.

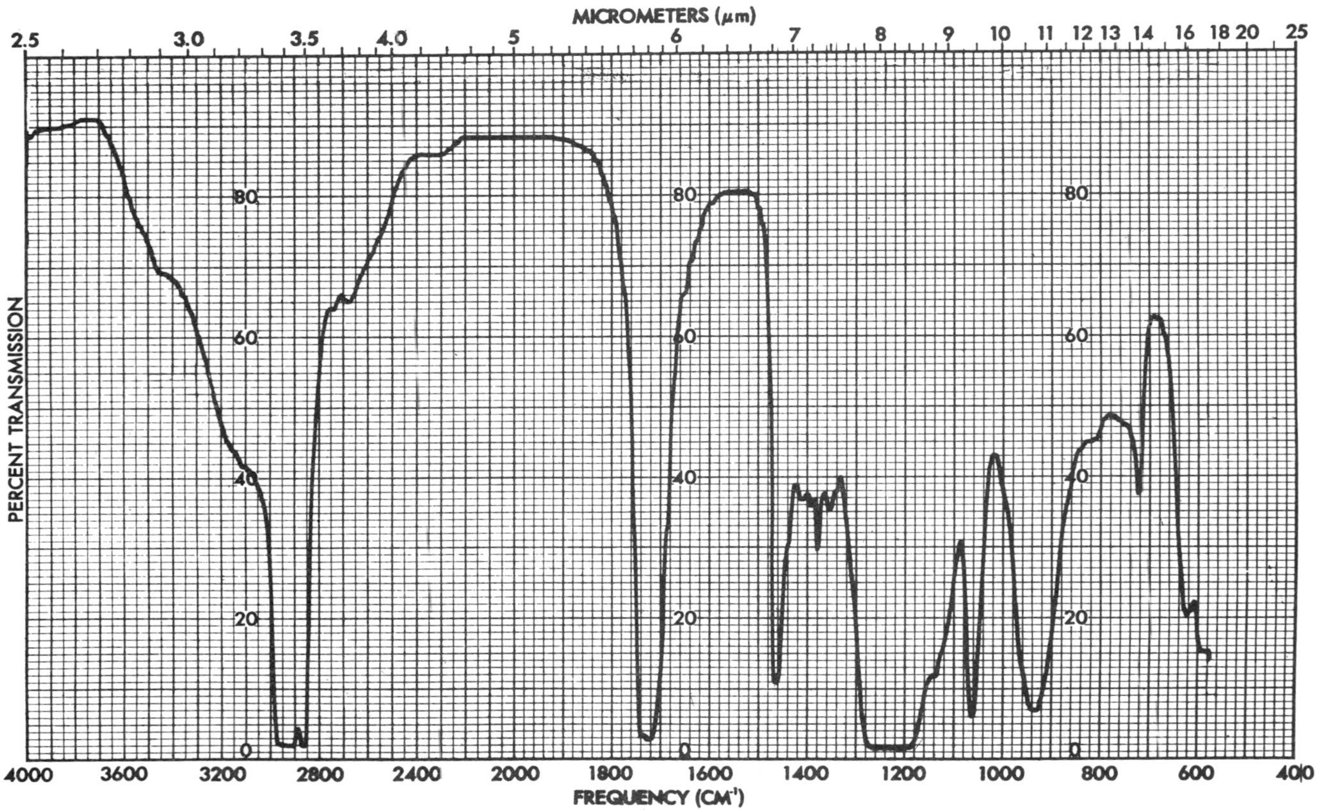

FIG. 6.124. Infrared spectrum of sulfonated red oil–sulfonated castor oil–sulfosuccinate.

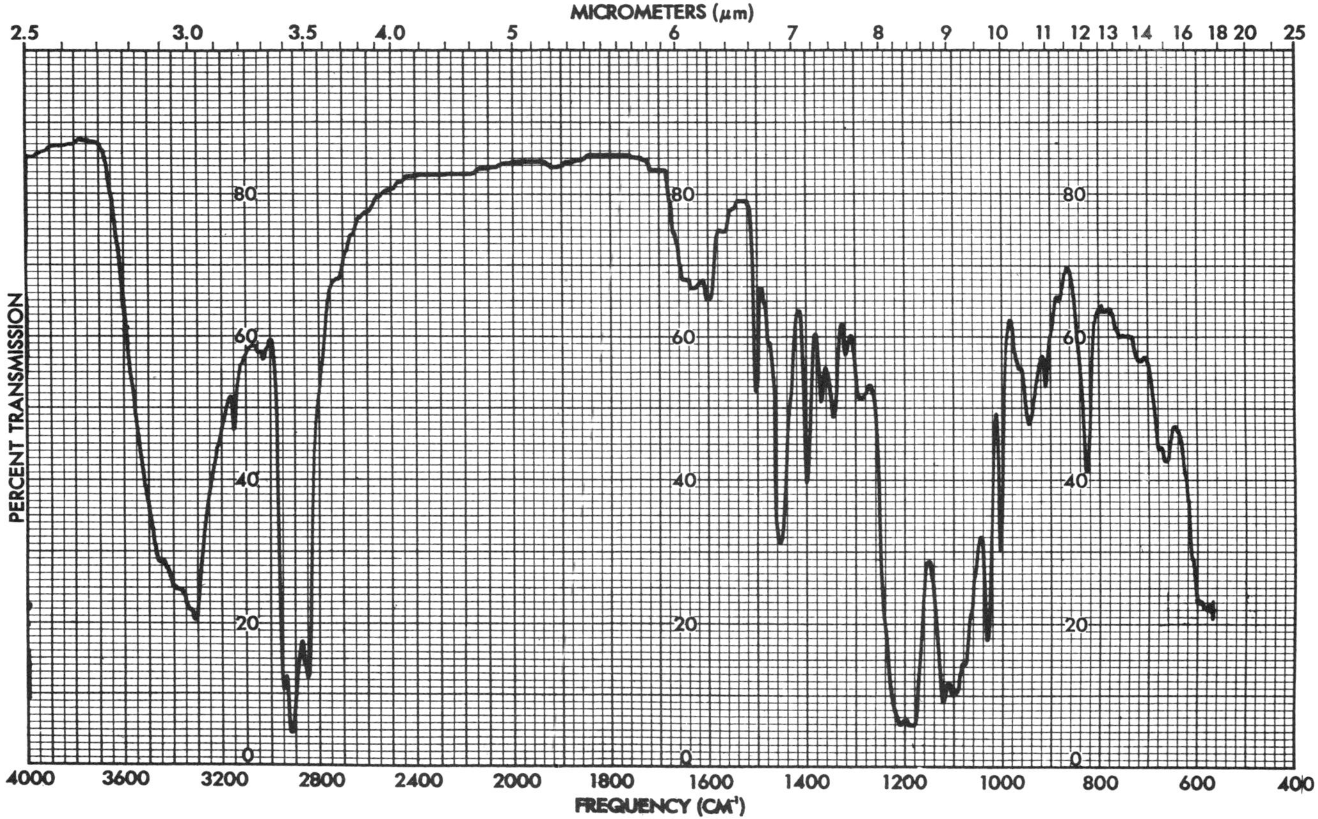

FIG. 6.125. Infrared spectrum of mixed alkanolamine salt DDBSA.

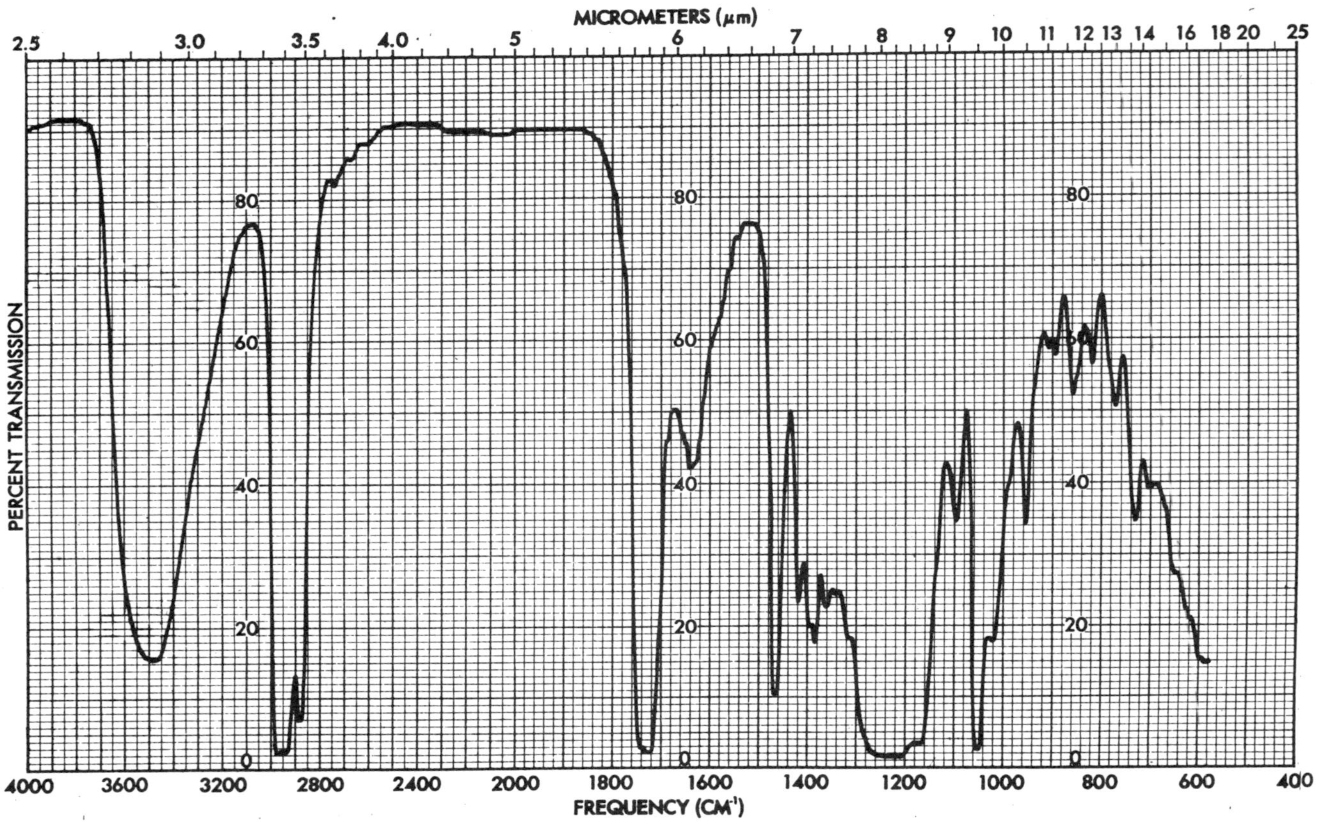

FIG. 6.126. Infrared spectrum of sulfonated butyl oleate–amine condensate–sulfosuccinate.

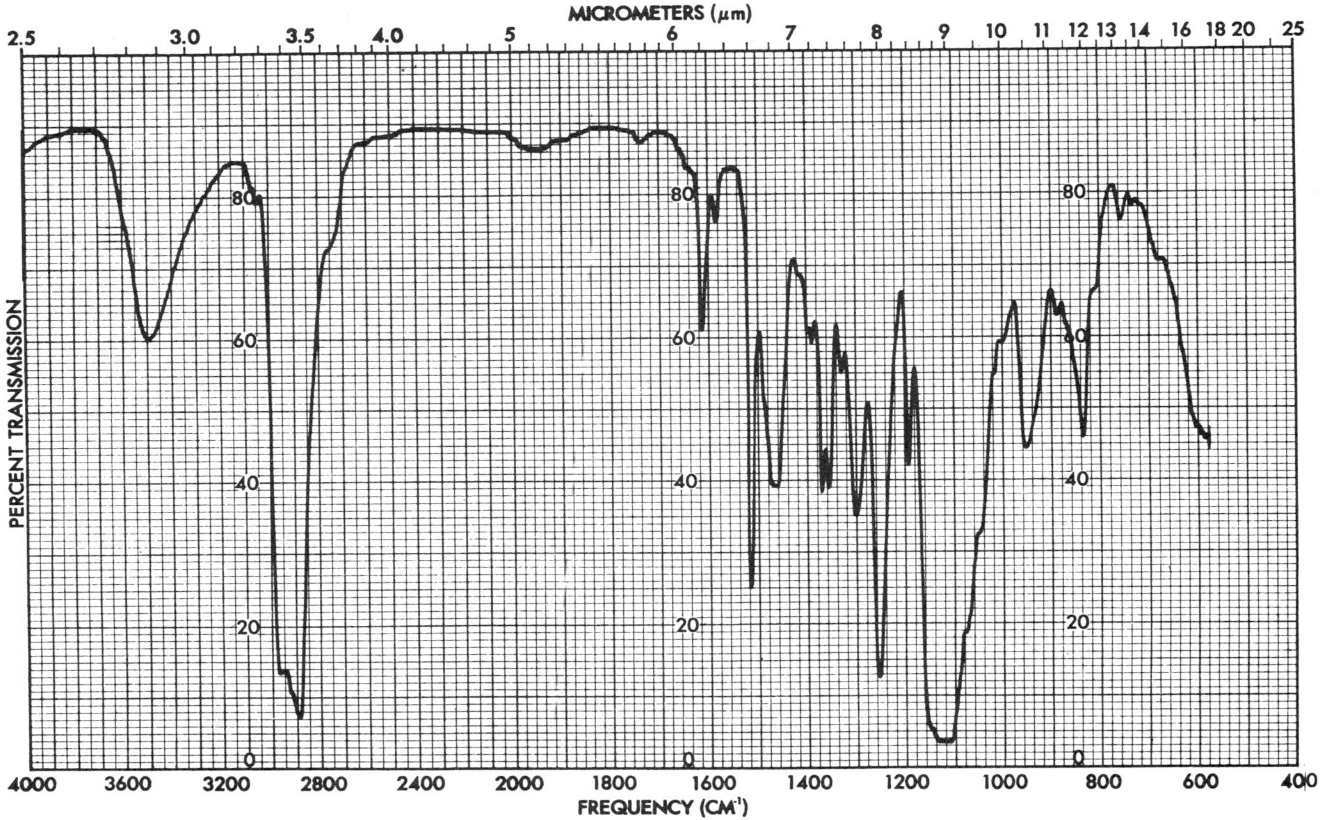

FIG. 6.127. Infrared spectrum of propylene dichloride–(EO)$_9$ nonyl phenol.

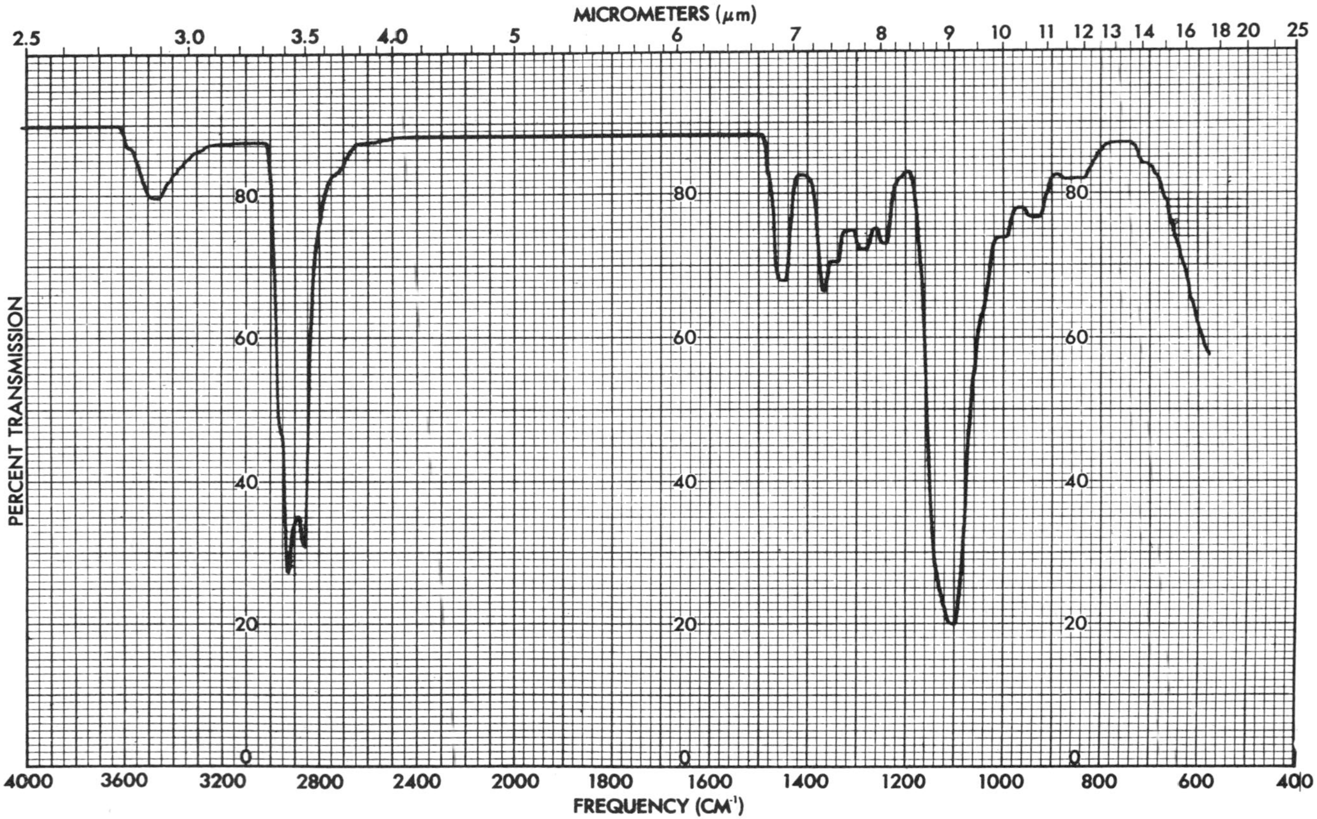

FIG. 6.128. Infrared spectrum of secondary alcohol ethoxylate.

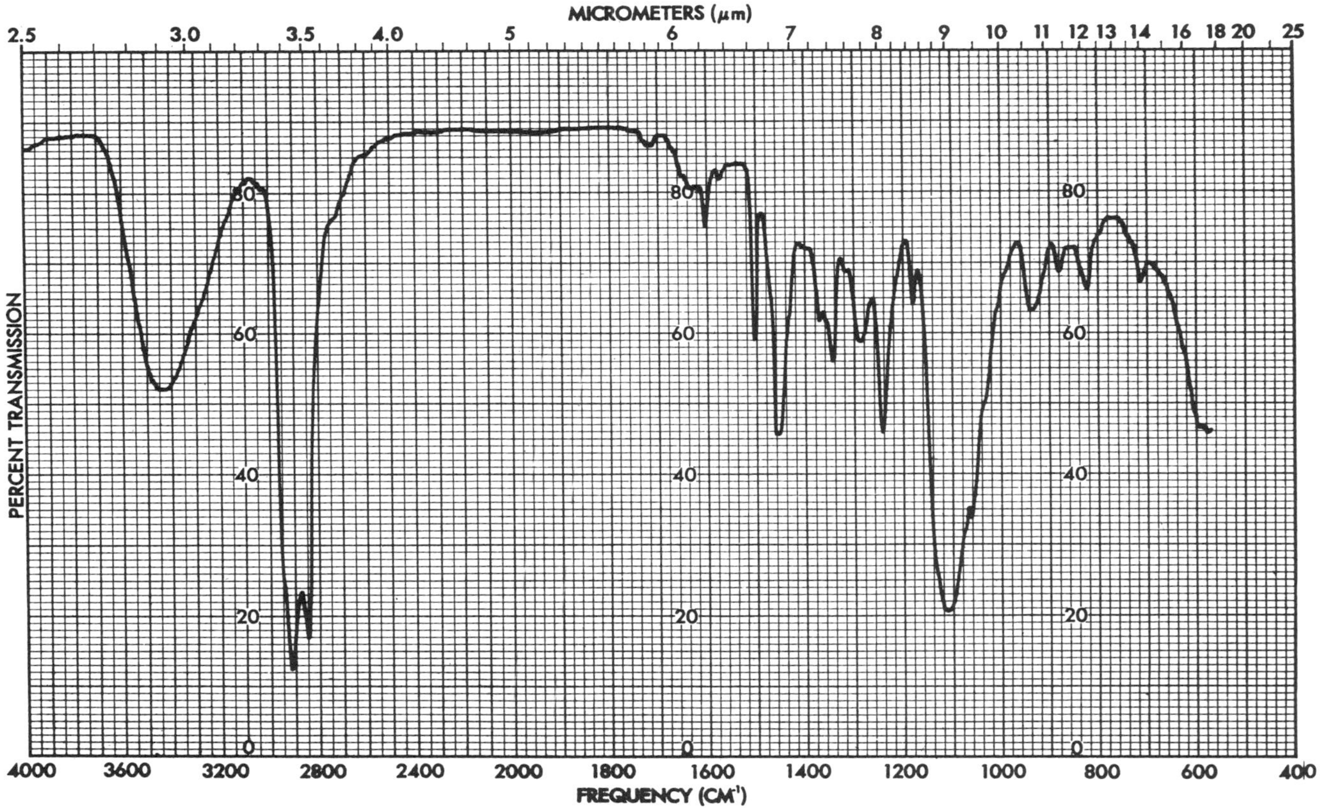

FIG. 6.129. Infrared spectrum of nonyl phenol–fatty alcohol EO–PO copolymer.

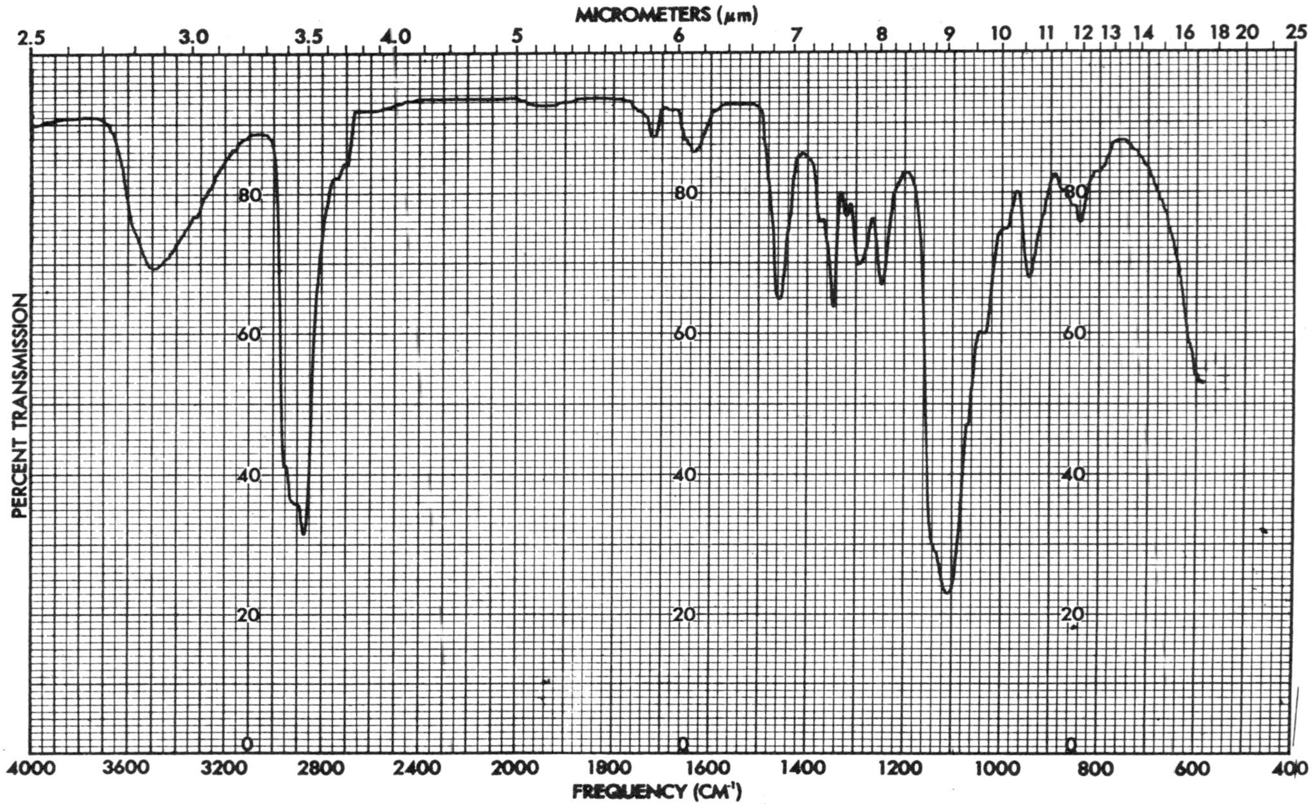

FIG. 6.130. Infrared spectrum of $(EO)_{10}$ tridecyl alcohol.

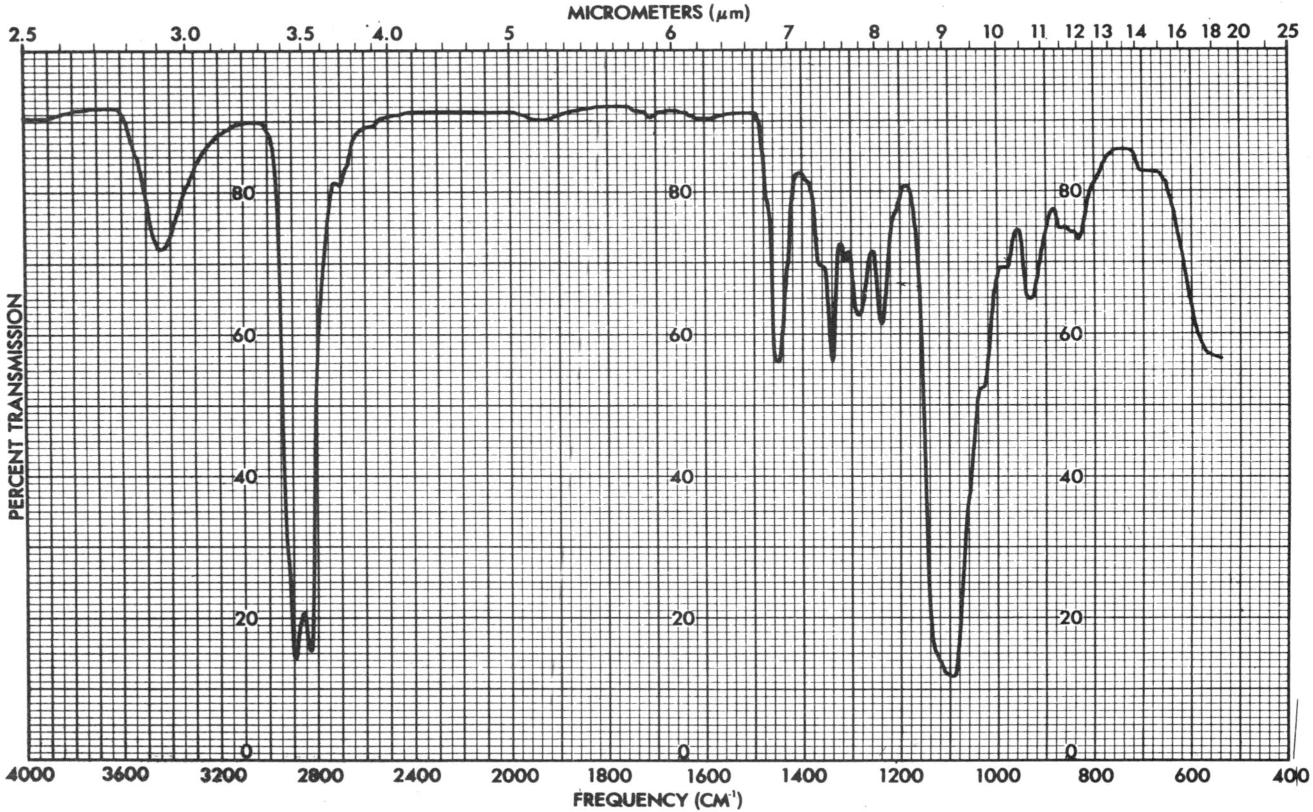

FIG. 6.131. Infrared spectrum of secondary alcohol $(EO)_9$.

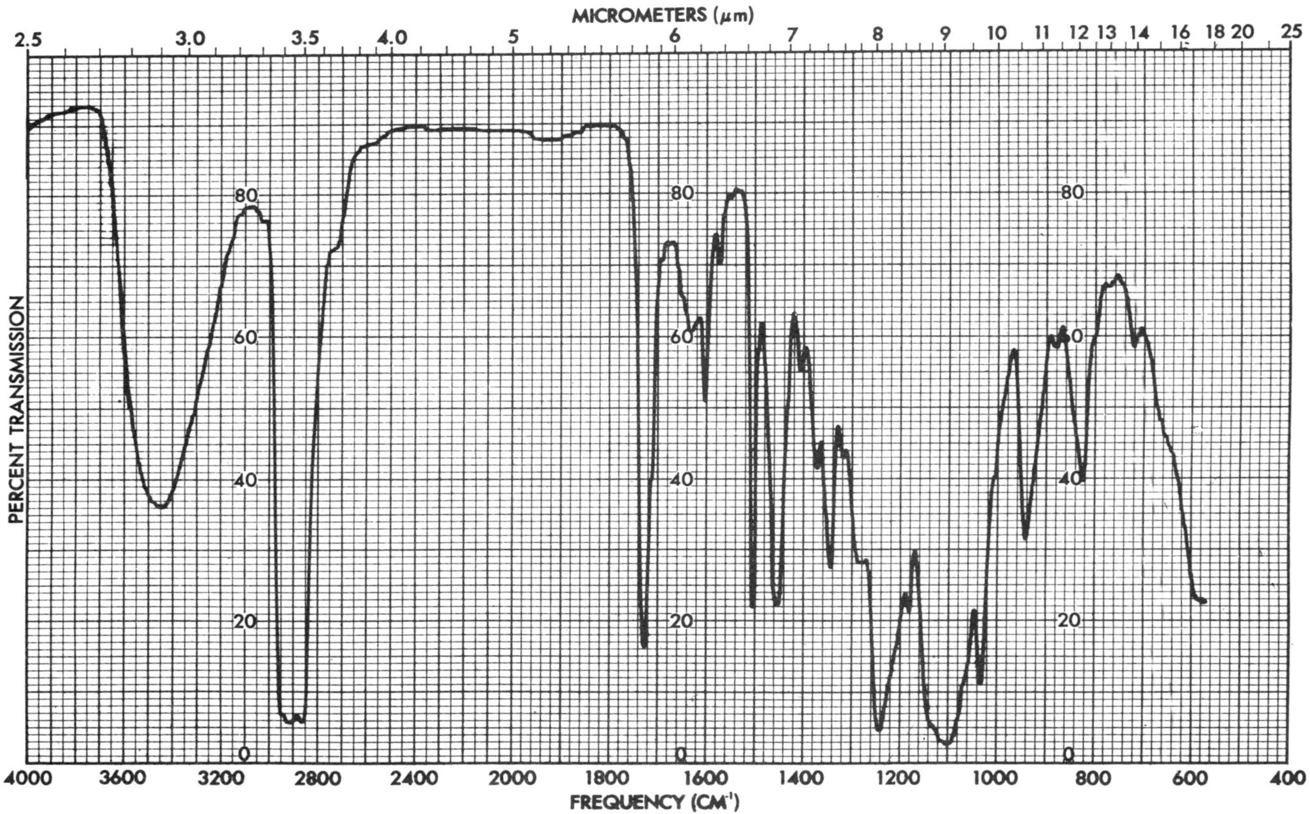

FIG. 6.132. Infrared spectrum of sulfosuccinate–$(EO)_9$ nonyl phenol–$(EO)_9$ nonyl phenol phosphate.

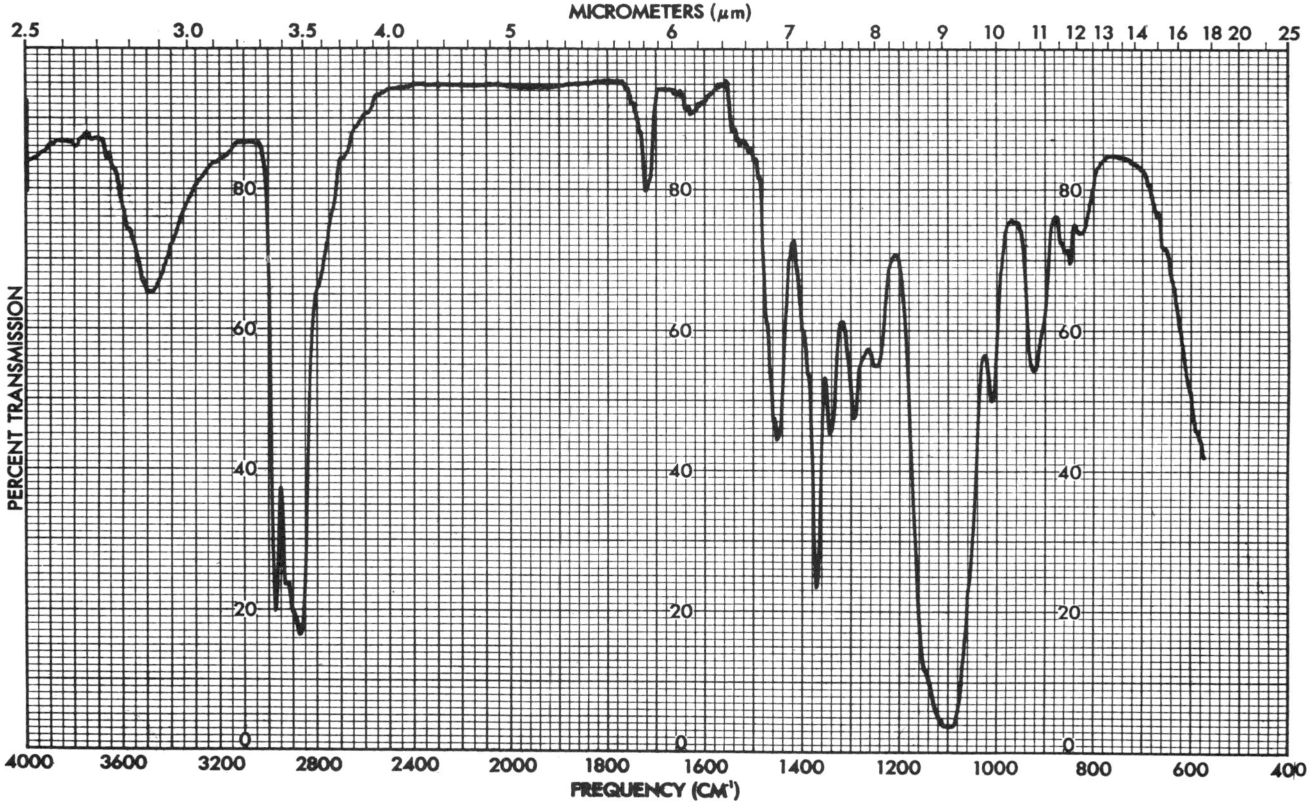

FIG. 6.133. Infrared spectrum of EO–PO block copolymer.

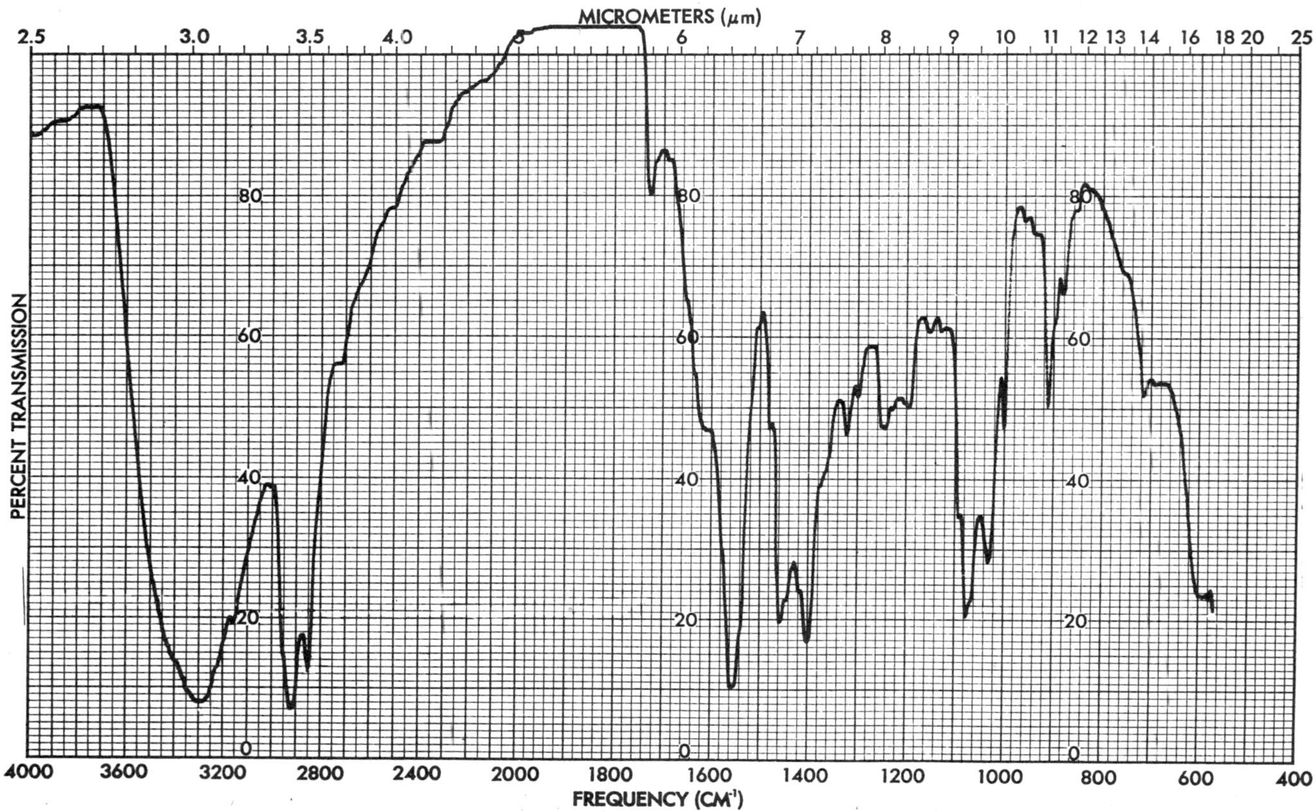

FIG. 6.134. Infrared spectrum of TEA salt–fatty acid alkyl aryl blend.

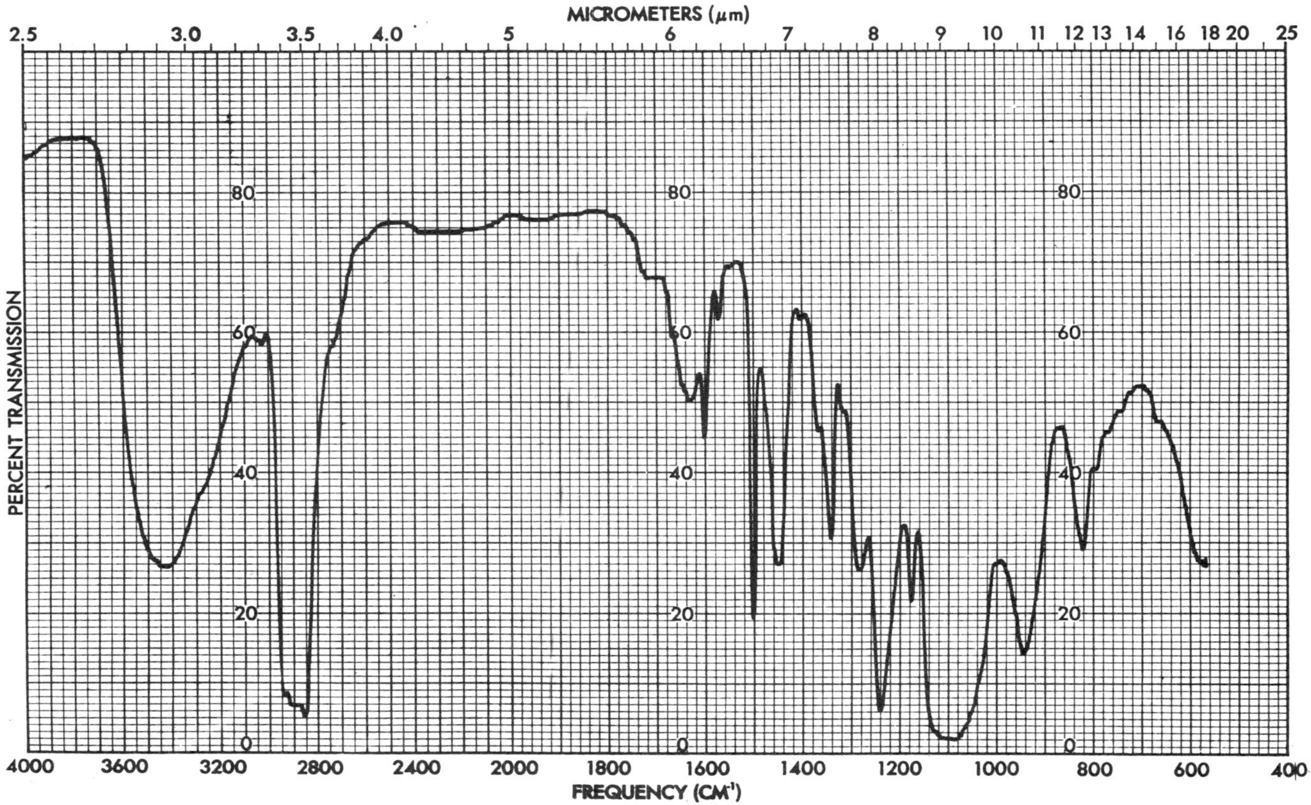

FIG. 6.135. Infrared spectrum of EO–PO nonyl phenol $(EO)_9$ phosphate.

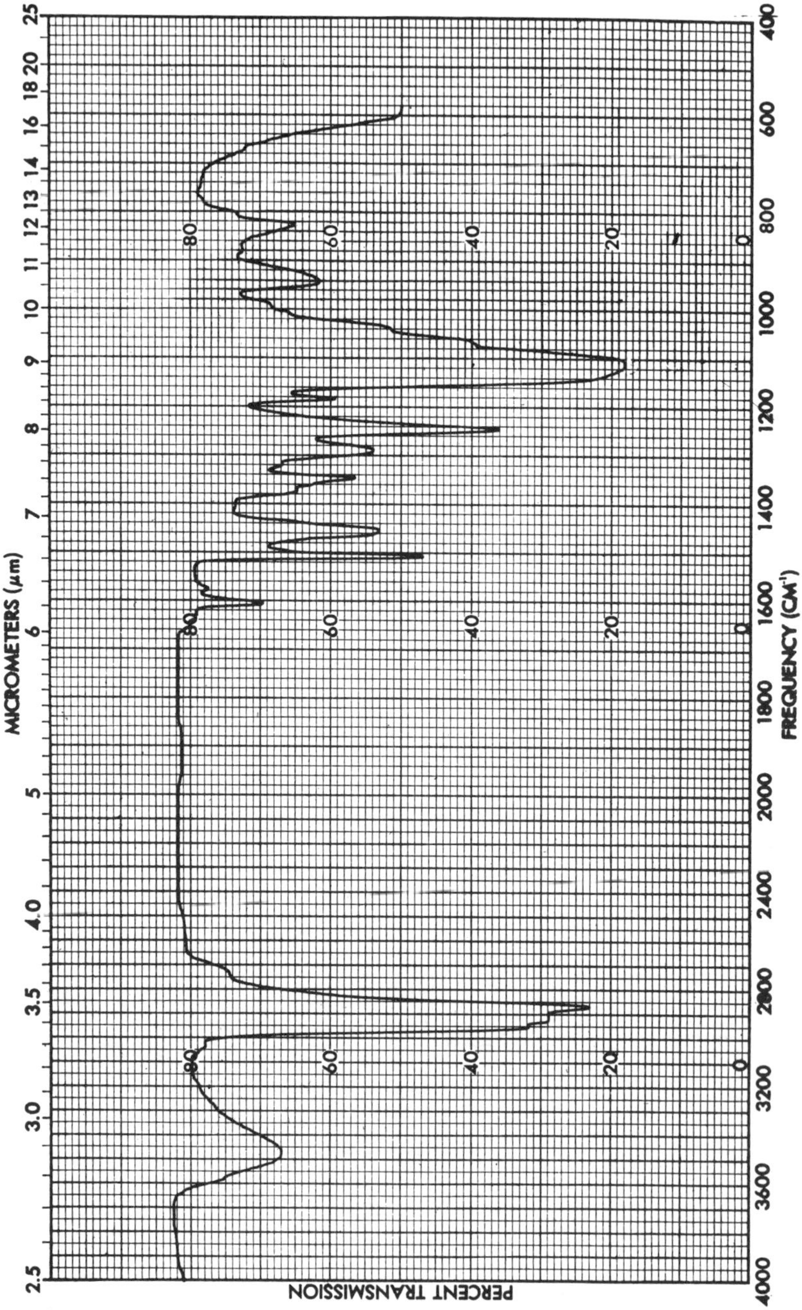

FIG. 6.136. Infrared spectrum of $(EO)_9$ nonyl phenol.

CHAPTER SEVEN

ALKALIS AND BUILDERS USED IN PREPARATION

ALKALIS COMMONLY USED IN PREPARATION

In addition to surfactants, an alkali is normally added to the preparation bath. The purpose of the alkali is to (1) saponify any residual oils or neutralize carboxylic acids remaining in the preparation bath and (2) solubilize sizing materials and cause dispersion of naturally occurring impurities in natural fibers. One must be careful in the choice of the alkali used, as free NaOH can have a deleterious effect on some fibers. For example, wool and silk are dissolved by alkali, whereas acetate and triacetate are converted back to their original cellulosic form. The following alkalis are commonly used in preparation: sodium hydroxide, potassium hydroxide, sodium metasilicate, sodium silicate, trisodium phosphate, sodium carbonate, tetrasodium pyrophosphate, sodium tripolyphosphate, sodium bicarbonate, borax, and ammonia.

Figure 7.1 shows the relationship of pH to concentration among some of these alkalis.

Table 7.1 compares some of the properties of various phosphates.

Table 7.2 gives weight percents of equivalent P_2O_5 and phosphorus in common sodium and potassium phosphates.

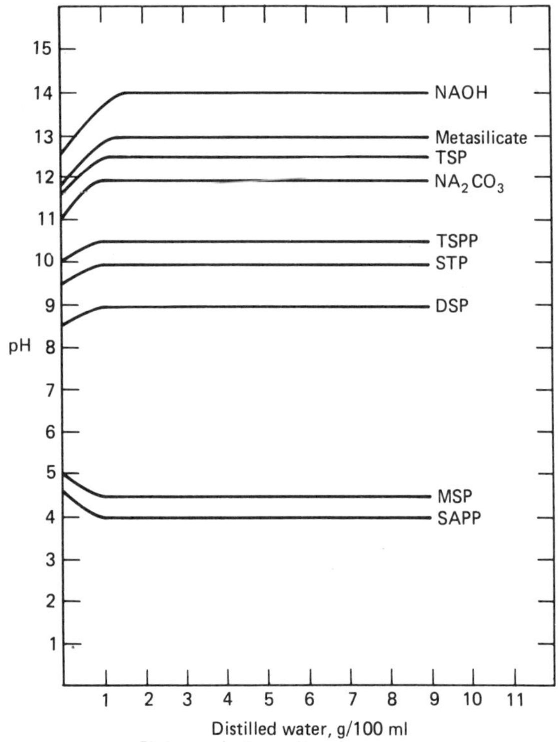

FIG. 7.1 Relationship between pH and concentration of alkalis. (NAOH = caustic soda; NA_2CO_3 = soda ash, TSP = trisodium phosphate, TSPP = tetrasodium pyrophosphate, STP = sodium tripolyphosphate, DSP = disodium phosphate, MSP = monosodium phosphate, SAPP = sodium acid pyrophosphate).

TABLE 7.1. Comparison of Phosphates

Phosphate	pH at 1%	Percentage as P_2O_5	Percentage as Na_2O
Monosodium phosphate	4.6	59	26
Disodium phosphate	9.0	50	43
Trisodium phosphate	12.0	19	27
Tetrasodium pyrophosphate	10.0	53	47
Sodium tripolyphosphate	9.9	57	42
Sodium acid pyrophosphate	4.2	64	—
Sodium hexametaphosphate	6.9	67	33

TABLE 7.2. Phosphates Conversion Factors[a]

Name	Chemical Formula	Molecular Weight	P_2O_5, %	P, %
Monosodium phosphate	NaH_2PO_4	119.98	59.2	25.8
Disodium phosphate	Na_2HPO_4	141.96	50.0	21.8
Trisodium phosphate	Na_3PO_4	163.94	43.3	18.9
Trisodium phosphate hemihydrate	$Na_3PO_4 \cdot \frac{1}{2}H_2O$	172.9	40.2	17.5
Trisodium phosphate crystals	$Na_3PO_4 \cdot \frac{1}{4}NaOH \cdot 12H_2O$	390.13	18.2	7.9
Sodium acid pyrophosphate	$Na_2H_2P_2O_7$	221.94	64.0	27.9
Tetrasodium pyrophosphate	$Na_4P_2O_7$	265.91	53.4	23.3
Tetrasodium pyrophosphate crystals	$Na_4P_2O_7 \cdot 10H_2O$	446.06	31.8	13.9
Sodium tripolyphosphate	$Na_5P_3O_{10}$	367.87	57.9	25.3
Sodium tripolyphosphate hexahydrate	$Na_5P_3O_{10} \cdot 6H_2O$	475.96	44.7	19.5
Sodaphos®	$(NaPO_3)_n = 6$	674	63.8	27.8
Hexaphos®	$(NaPO_3)_n = 13$	1387	67.5	29.5
Glass H®	$(NaPO_3)_n = 21$	2203	68.7	30.0
Monopotassium phosphate	KH_2PO_4	136.10	52.2	22.8
Dipotassium phosphate	K_2HPO_4	174.19	40.7	17.8
Tripotassium phosphate	K_3PO_4	212.29	33.4	14.6
Tetrapotassium pyrophosphate	$K_4P_2O_7$	330.35	43.0	18.8
Potassium tripolyphosphate	$K_5P_3O_{10}$	448.43	47.5	20.7

[a]Courtesy of FMC Corporation.
Sodaphos, Hexaphos, Glass H–FMC trademarks.

Figure 7.2 shows the pounds of phosphate per 100 gal to soften a given parts per million (ppm) value of calcium hardness in process water for various phosphates.

Figure 7.3 shows the synergistic effect when phosphates are combined with alkylaryl sulfonates and are used as detergents in processing water.

The three remaining classes of auxiliaries commonly used in preparation

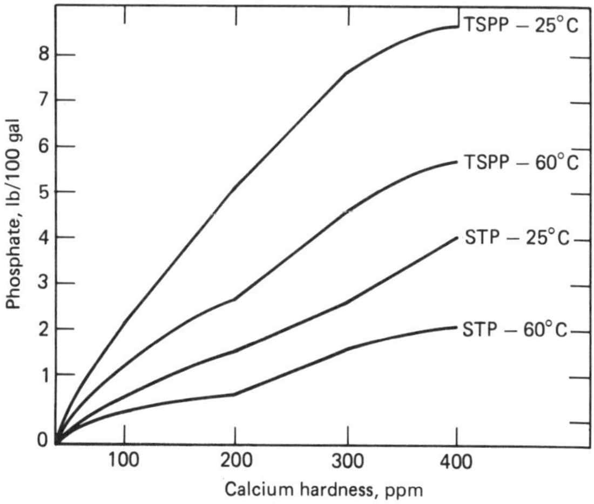

FIG. 7.2. Water Softening of Phosphates.

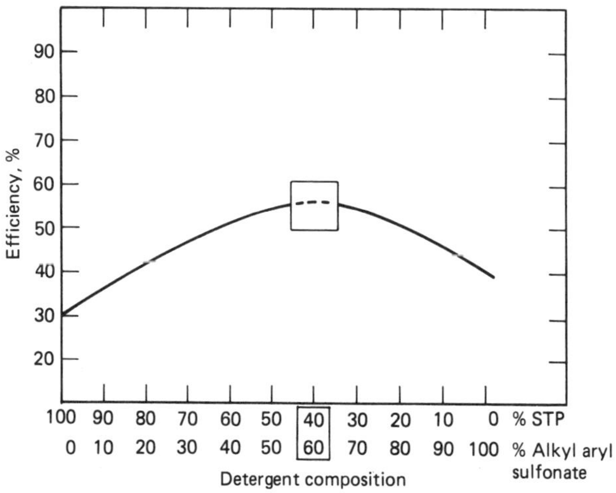

FIG. 7.3. Effect of sodium tripolyphosphate on detergency.

and that need discussion at this time are (1) antiredeposition agents, (2) defoamers, and (3) sequestering agents, or chelates.

ANTIREDEPOSITION AGENTS

Antiredeposition agents are compounds that function by adsorbing particulate matter and preventing the redepositing of this type of soil on the surface of the fiber. With the advent of synthetic fibers and synthetic detergents, a graying effect was noticed on both white and colored goods with prolonged laundering. It was found that the addition of carboxymethyl cellulose (CMC) and/or polyvinyl pyrrolidone (PVP) would minimize the graying condition. This was accomplished by an increase in the viscosity of the wash solution and by their action as scavengers for loose dye and particulate matter. Carboxymethyl cellulose is used primarily in surfactants, whereas PVP is used primarily with compositions used as one-bath scouring–dye auxiliaries. The generic composition for CMC and PVP are:

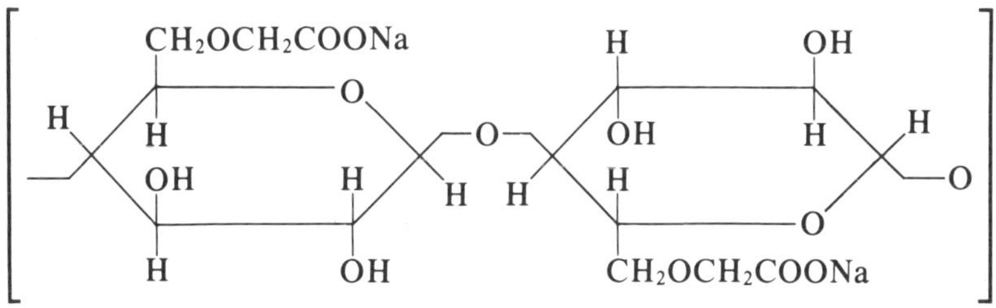

Generic composition of CMC

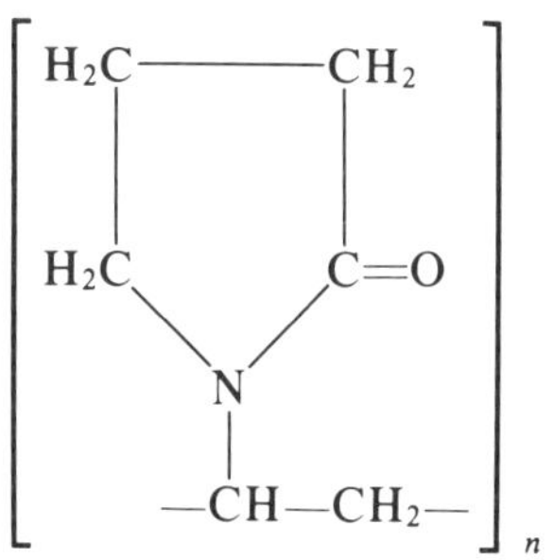

Generic composition of PVP

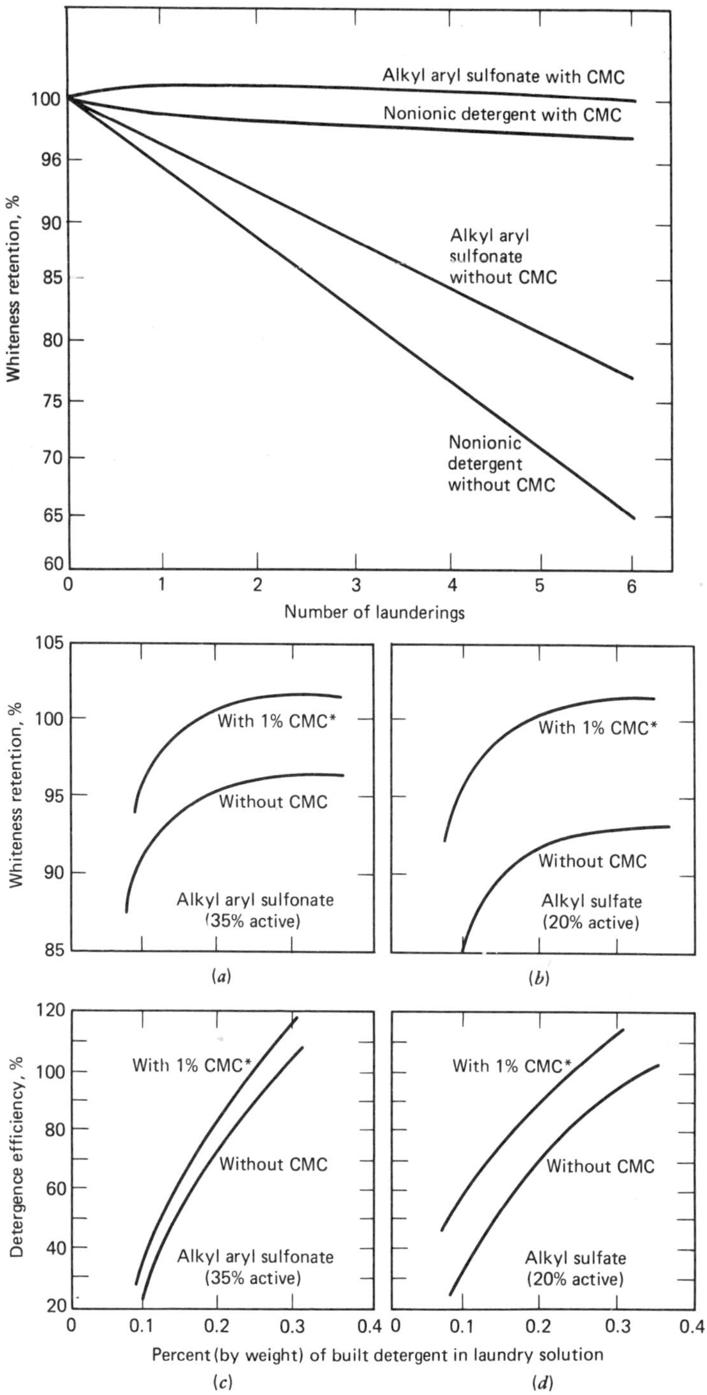

*On weight of built detergent.

FIG. 7.4. Effect of soil redeposition.

Figure 7.4 shows the effect on soil redeposition through various laundry cycles.[116]

DEFOAMERS

Defoamers are, as a rule, classified under one of two generic headings—silicone or nonsilicone. Silicone defoamers are, for the most part, emulsions of silicone oils and nonionic or cationic surfactants. The silicone oils themselves may be reacted with silica.[117] These compounds, if not properly prepared, can separate and be deposited on the fabric and be the cause of oil spotting during later processing. Therefore, products currently used in high-temperature dyeing contain segments that have been ethoxylated, propoxylated, or both. The following is the generic structure of the conventional polydimethyl siloxane.

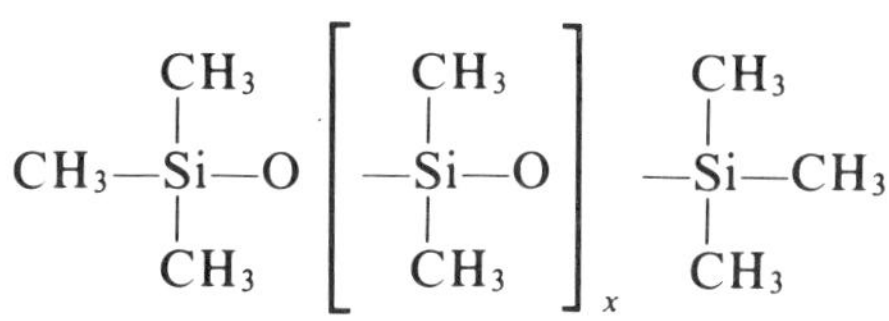

Polydimethyl siloxane[118]

The following structures show organofunctional groups attached to polydimethyl siloxanes.

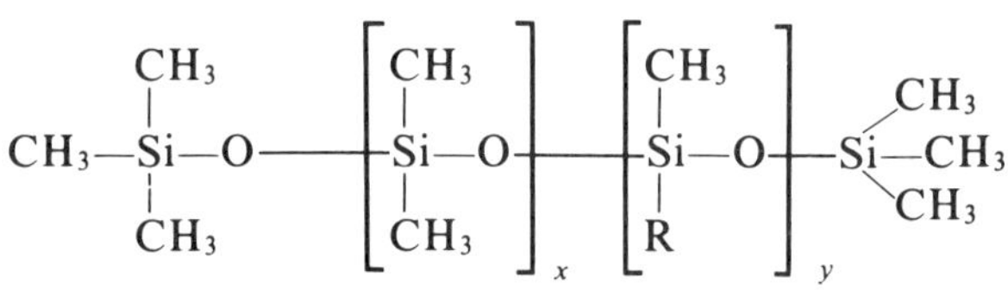

Functional group—polydimethyl siloxane[119]

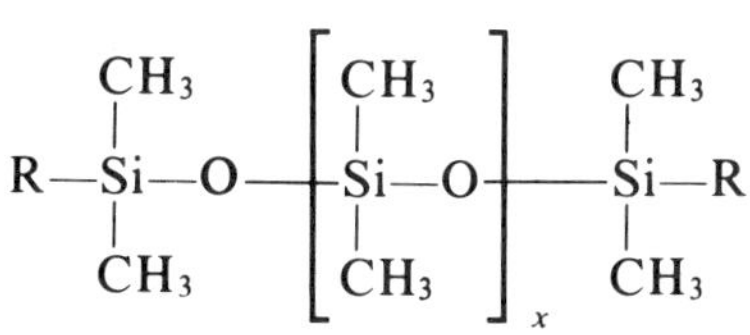

R = $(CH_2)\bar{y}$ functional group

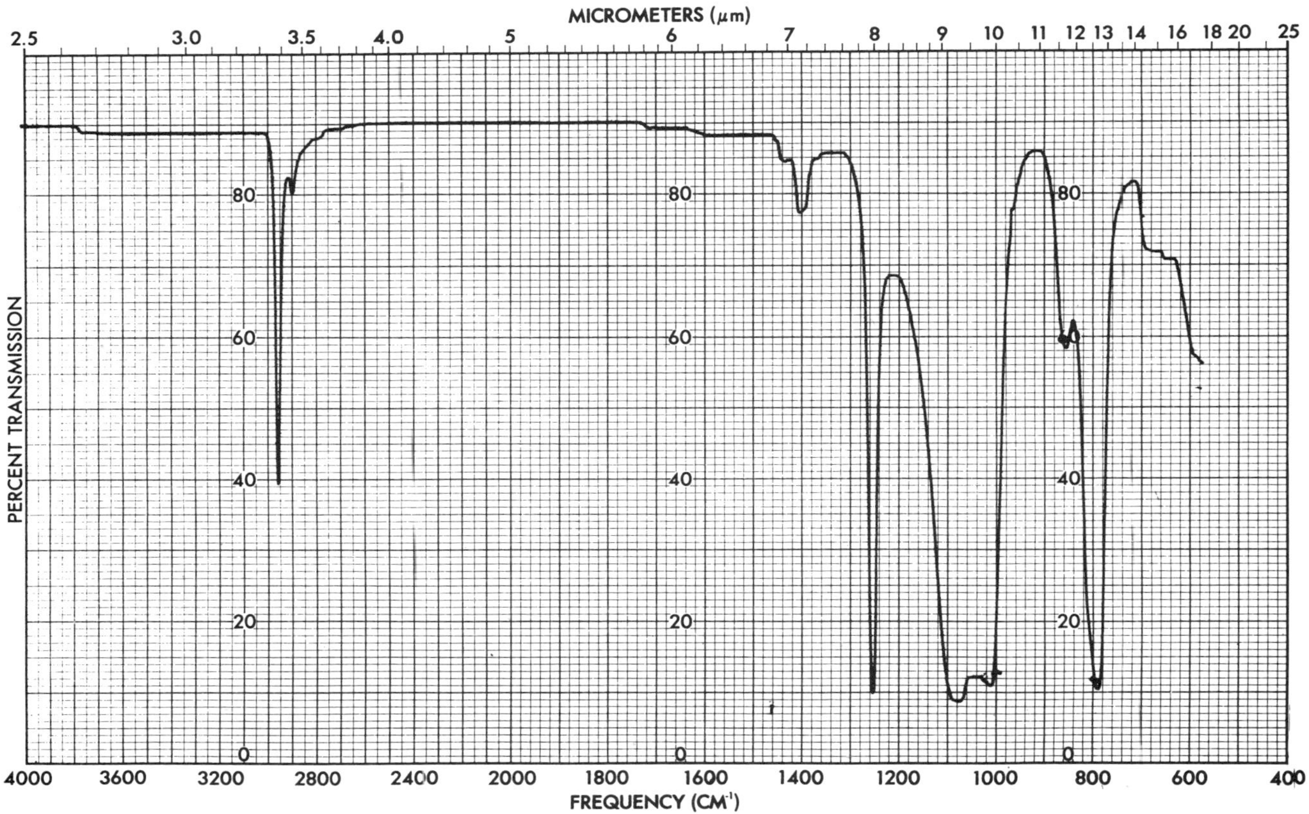

FIG. 7.5. Infrared spectrum of silicone–silica.

Figure 7.5 shows the infrared spectrum of a typical silicone compounded with a silica.

Although the terms "defoamer" and "antifoam" are used interchangeably, in reality they serve different functions. As a general rule, products are compounded from components so that the finished product will serve both functions. By definition, an *antifoam* prohibits the formation of foam, whereas a *defoamer* reduces foam that has already formed. Experience has shown the best antifoam agents to be silicones, waxes, and special esters, whereas aliphatic hydrocarbons and branched-chain alcohols are excellent defoamers. In today's specialty chemical market, organosilicones or compounds that are composed of both types of ingredient are commonly used. A defoamer–antifoam, to be effective, must be dispersible in the medium in which it operates because if it is solubilized, it will not function. Much of the difficulty encountered with defoamers and antifoams is that to be effective, the emulsion must be borderline and operate on the surface of the medium in which it operates but at the same time remain semidispersed on the surface of the liquid. Although it would appear that it would be a simple matter to manufacture effective defoamer–antifoam formulations, the heartache and grief experienced by many specialty chemical formulators belies that statement.

Unlike surfactants that can work synergistically, many defoamer–antifoam ingredients will produce just the opposite effect when combined. For this reason, it is usually more economical to add defoamer to a given operation as a separate entity than to combine it in other specialty products, such as scouring agents.

Factors that affect the efficiency of a defoamer are generally considered to be the pH of the system in use, the temperature at which foaming occurs, the method by which foam is generated, the percentage of surfactant available in the system, and the chemical makeup of the foamer requiring defoaming. An additional factor to be considered is whether silicone residues can be tolerated at later stages of processing. United States Patent 3,730,907 discusses the use of ethylene bisdistearamide, calcium DDBSA, and dimethyl polysiloxane containing silica with an aliphatic mineral oil being used as a nonsilicone type defoamer. United States Patent 3,677,963 discusses nonsilicone defoamers containing paraffin oil, castor oil, stearic diamide of ethylene diamine, vinyl acetate fumaric acid copolymer, and silicone oil for improved performance. (Although the patent literature shows these compounds to contain silicone, similar types of product may be manufactured without the use of the silicones.) United States Patent 3,652,453 shows the use of branched-chain alcohols in conjunction with bisstearamides emulsified with naphthalene sulfonic acids condensed with formaldehyde. Additional patent references[121] are listed at the end of this book.

A summary of components used in defoamers by class might be shown as outlined below:

Silicones	Emulsified oils
	EO–PO oils
	Filled oils
Nonsilicones	Oils: Castor, mineral, synthetic
	Waxes: Paraffin, castor
	Esters: Butyl stearate
	Amides: Ethylene bisstearamide
	Metallic stearate: Calcium, magnesium, aluminum
Organosilicones	Blends of silicones and nonsilicones

Nonsilicone and organosilicone infrared spectra vary according to the composition; Figs. 7.6 through 7.32 show infrared spectra of components and blends of components commonly used in the compounding of these types of defoamer.

A specialty type of defoamer compound is manufactured by Air Products and Chemicals, Inc., under the trade name of Surfȳnol®. These products are based on acetylenic glycols and various diluants. The structure of the base glycol is:

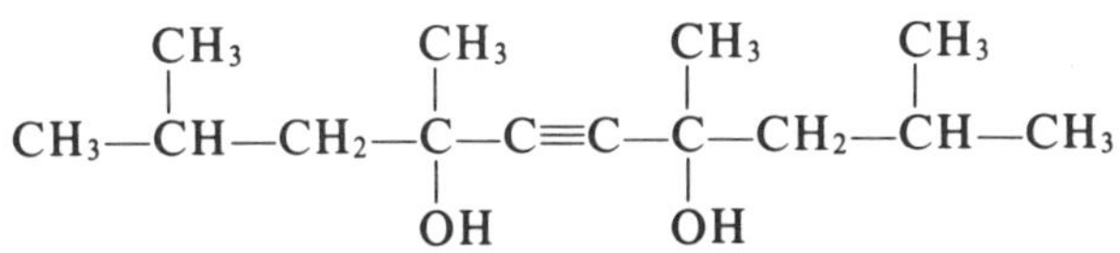

2,4,7,9-Tetramethyl-5-decyn-4,7-diol

SEQUESTERING AGENTS (CHELATES)

Sequestrants, or chelates, as they are sometimes known to the specialty trade, are compounds that promote an equilibrium reaction between a metal

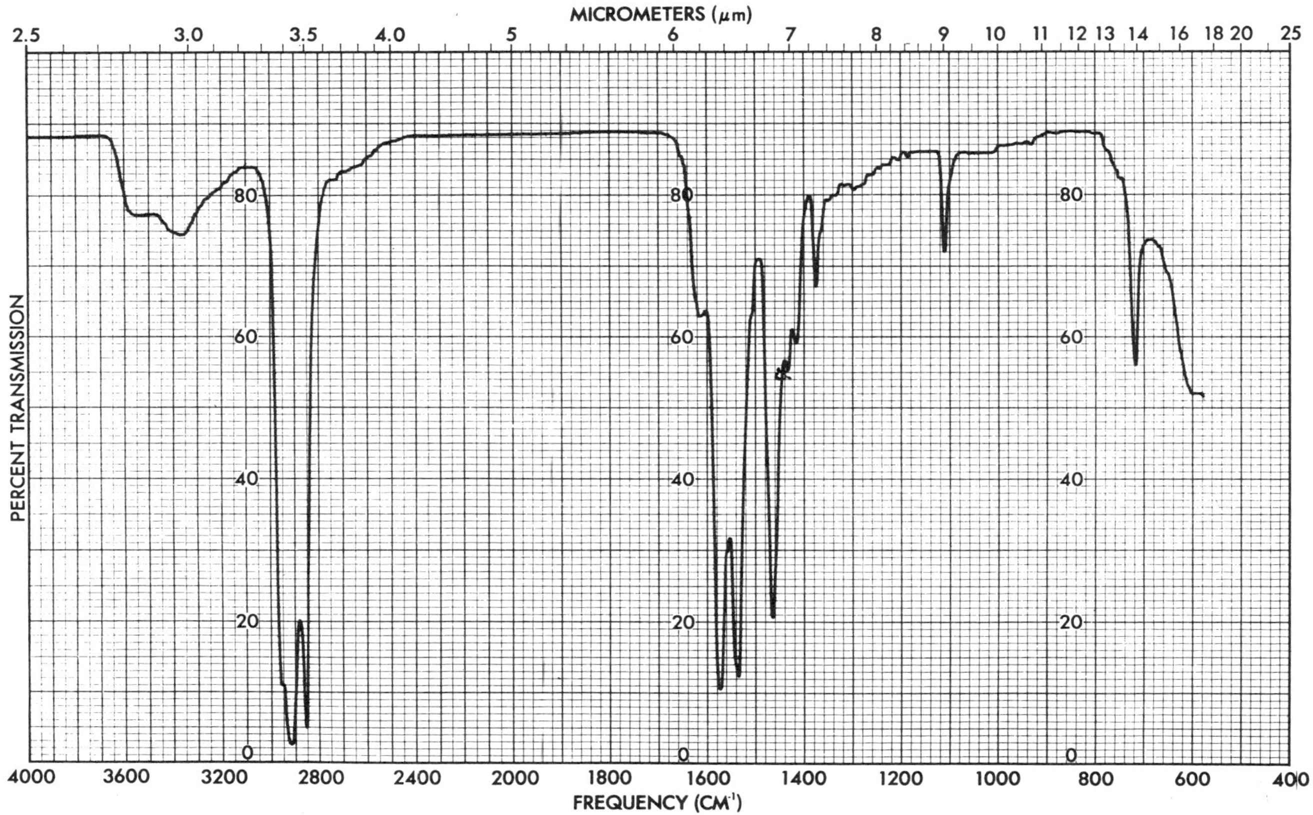

FIG. 7.6. Infrared spectrum of calcium stearate.

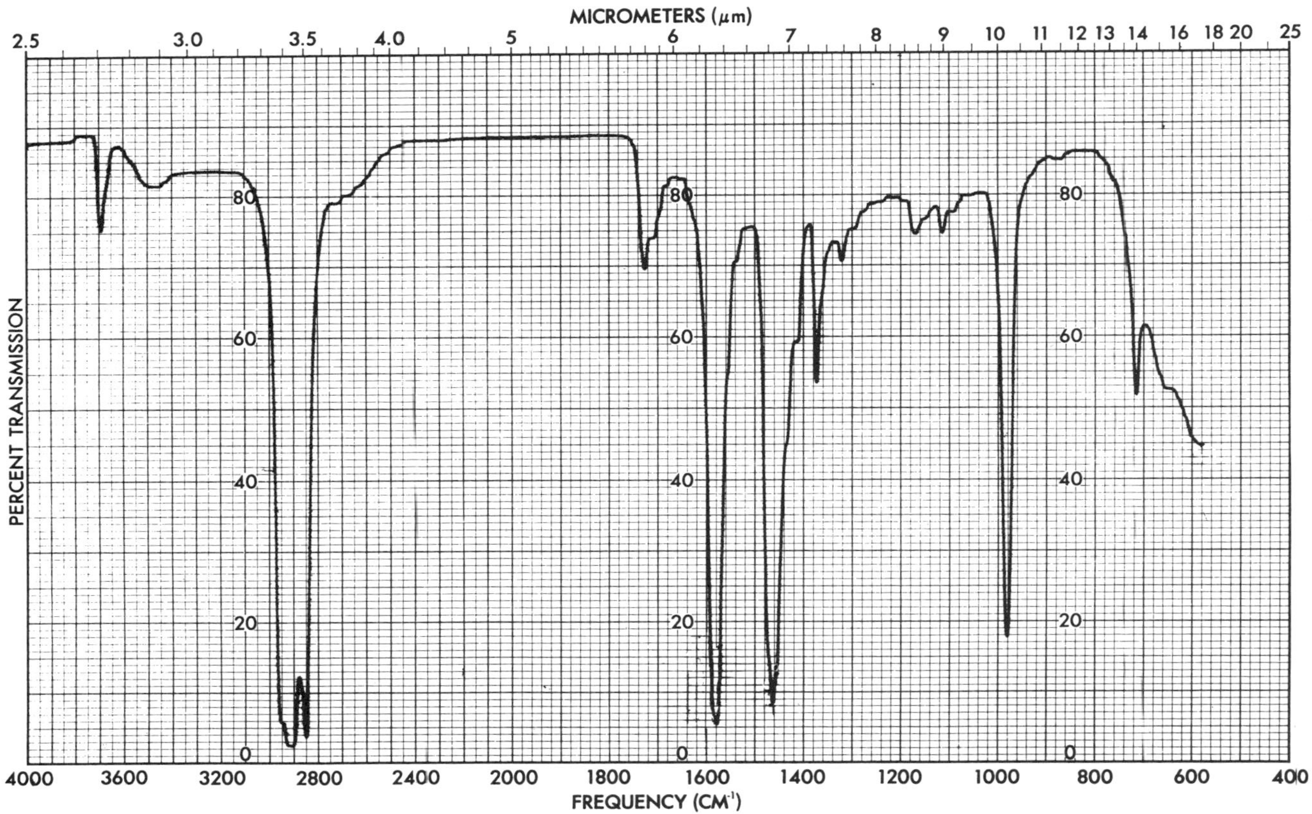

FIG. 7.7. Infrared spectrum of aluminum stearate.

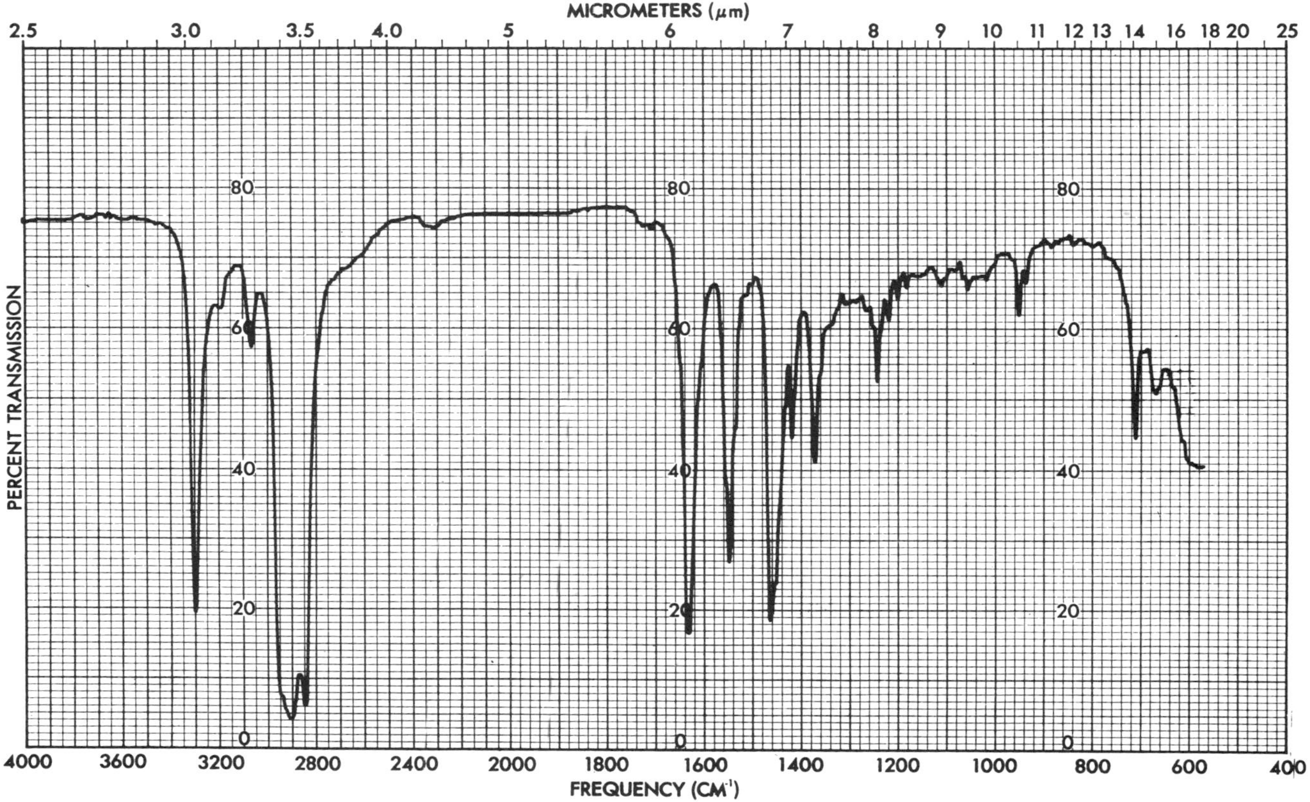

FIG. 7.8. Infrared spectrum of ethylene diamine bisstearamide.

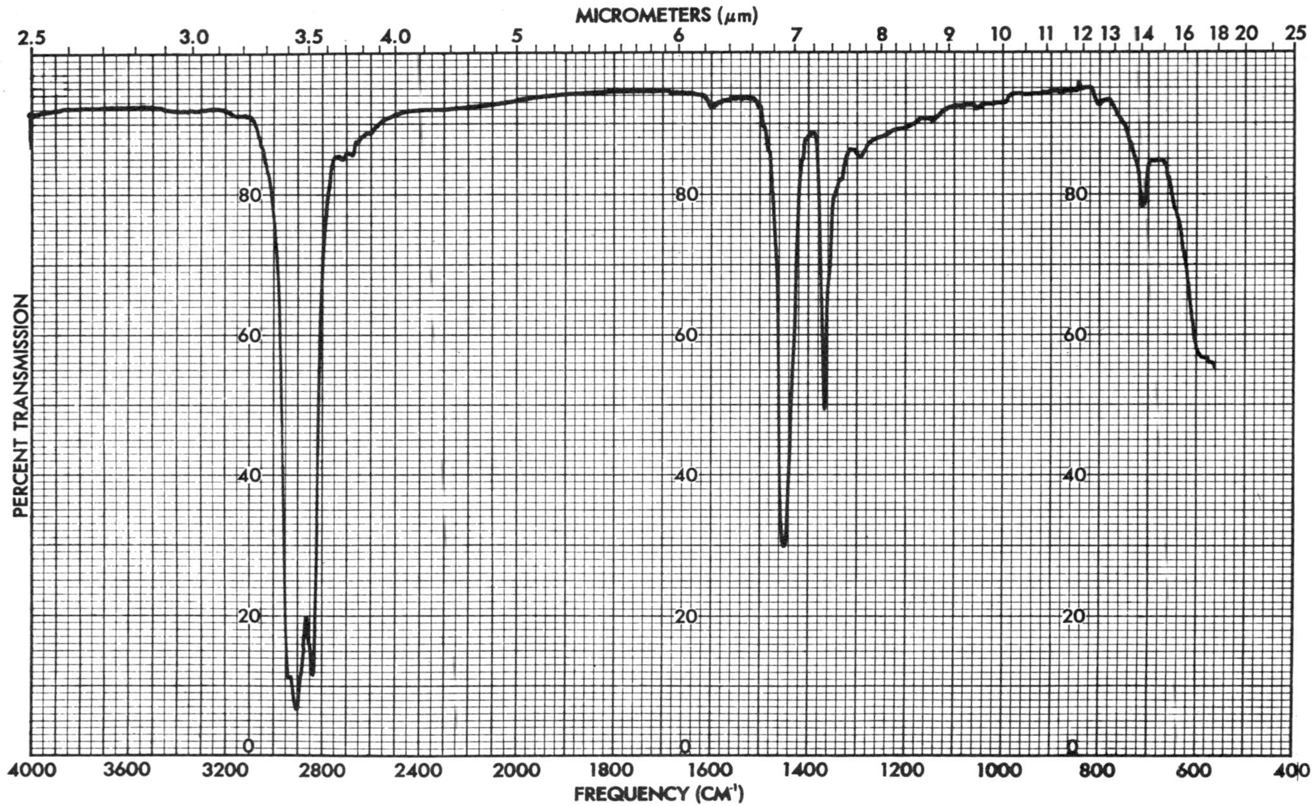

FIG. 7.9. Infrared spectrum of light hydrocarbon oil.

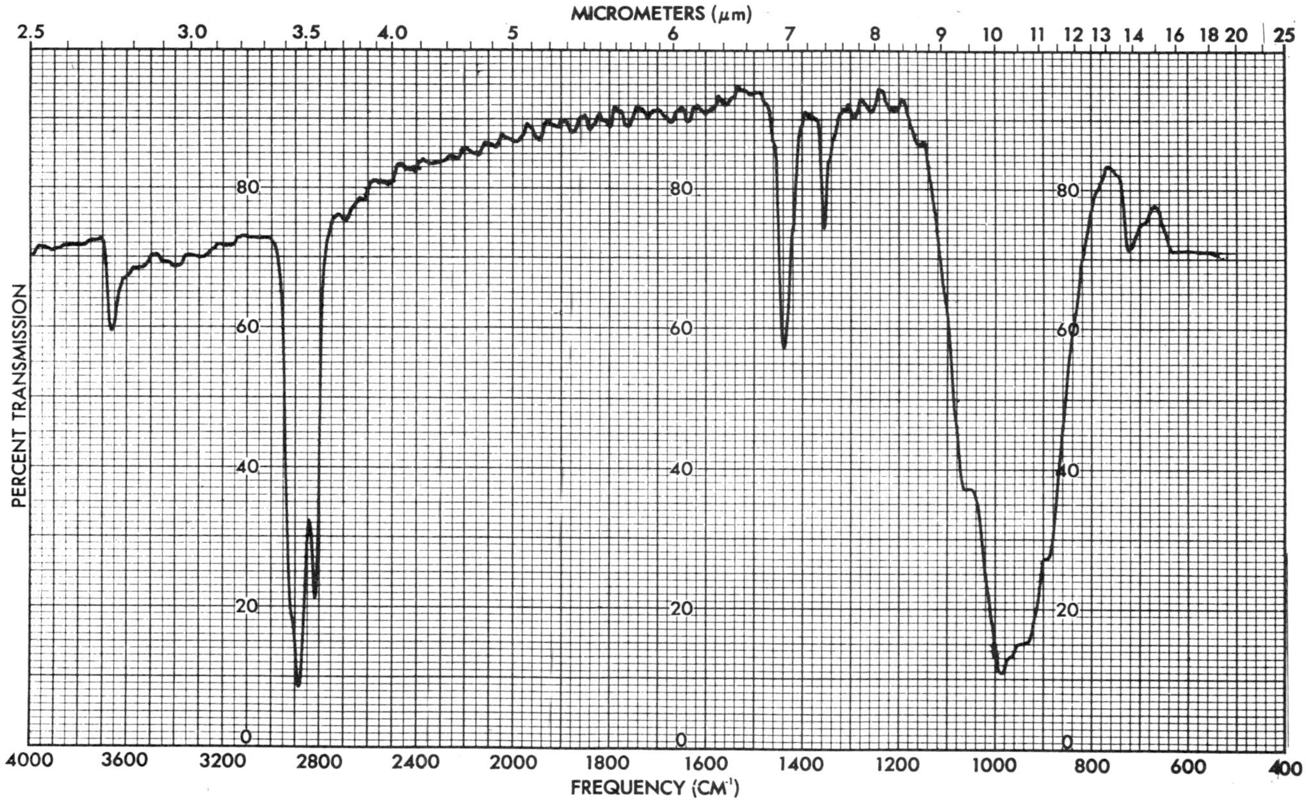

FIG. 7.10. Infrared spectrum of magnesium silicate.

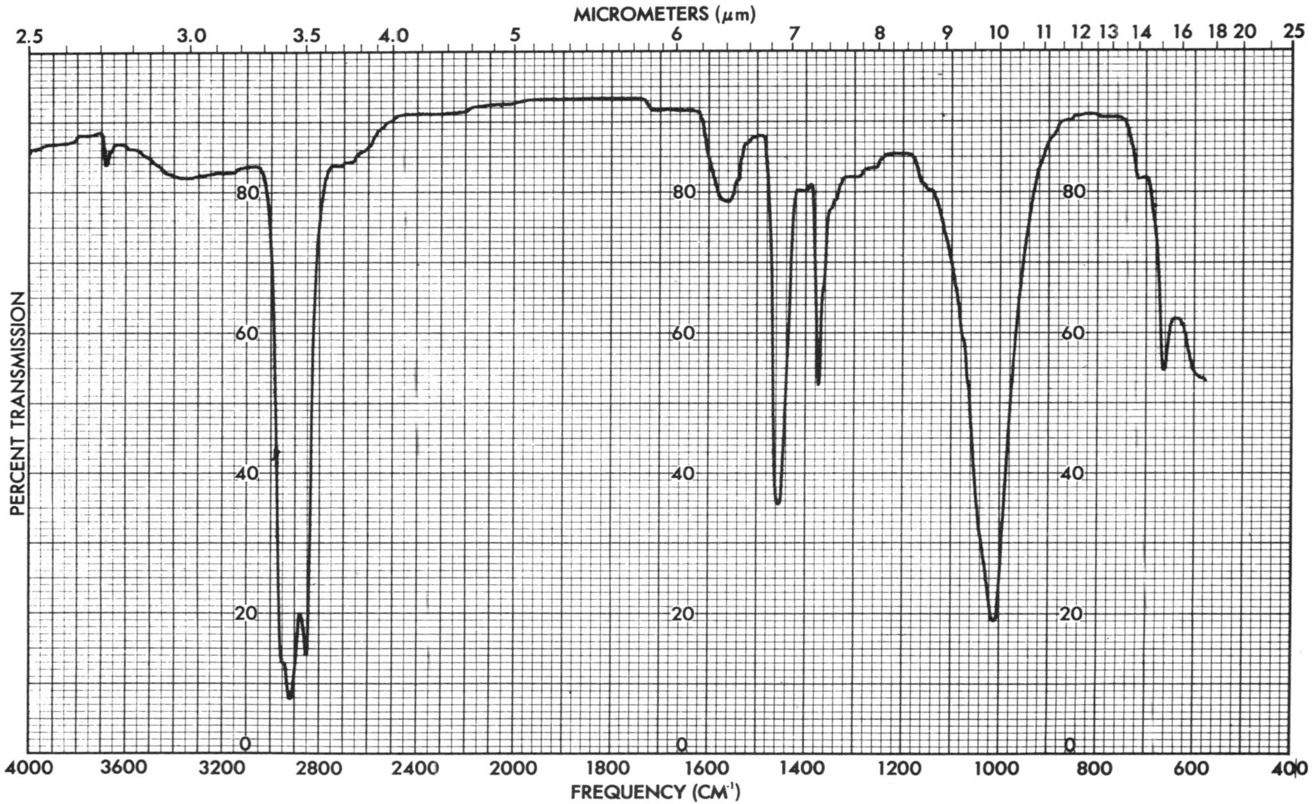

FIG. 7.11. Infrared spectrum of magnesium silicate plus light hydrocarbon oil.

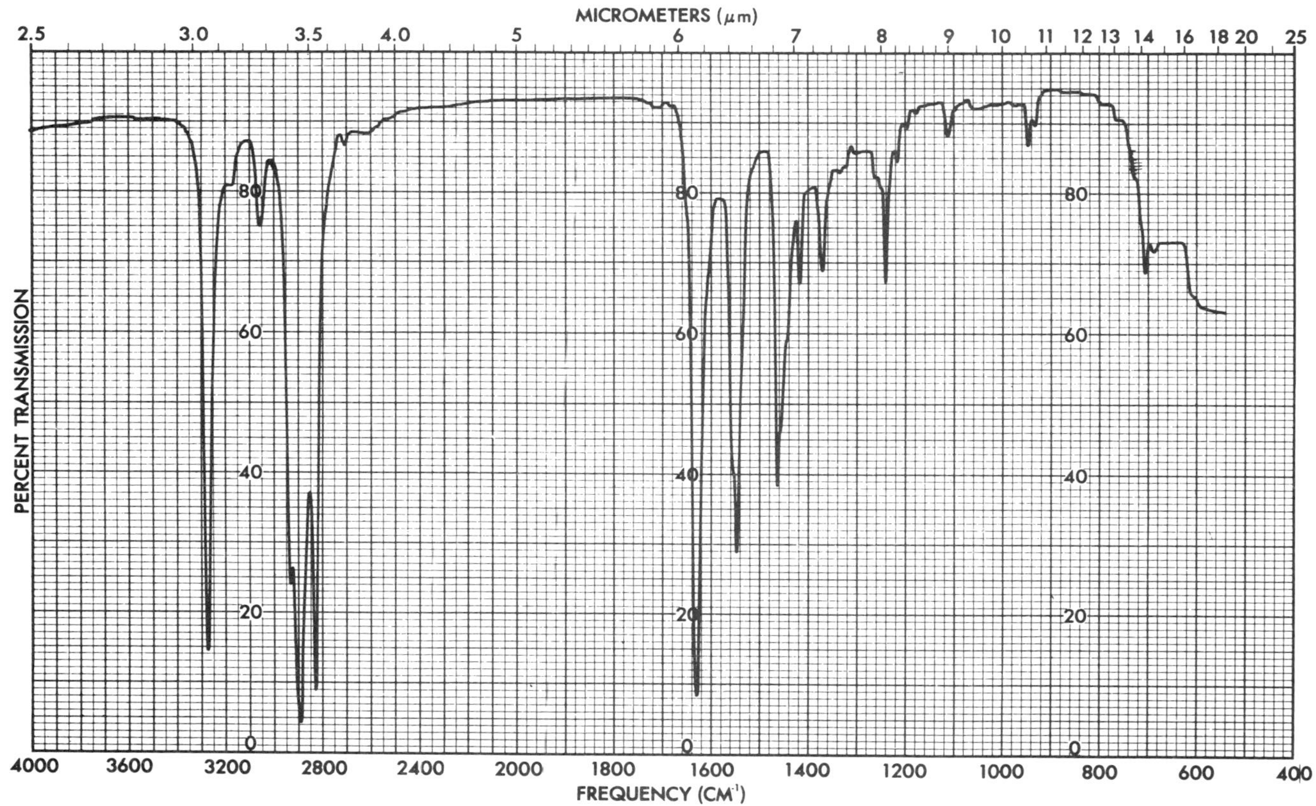

FIG. 7.12. Infrared spectrum of bisstearamide plus light hydrocarbon oil.

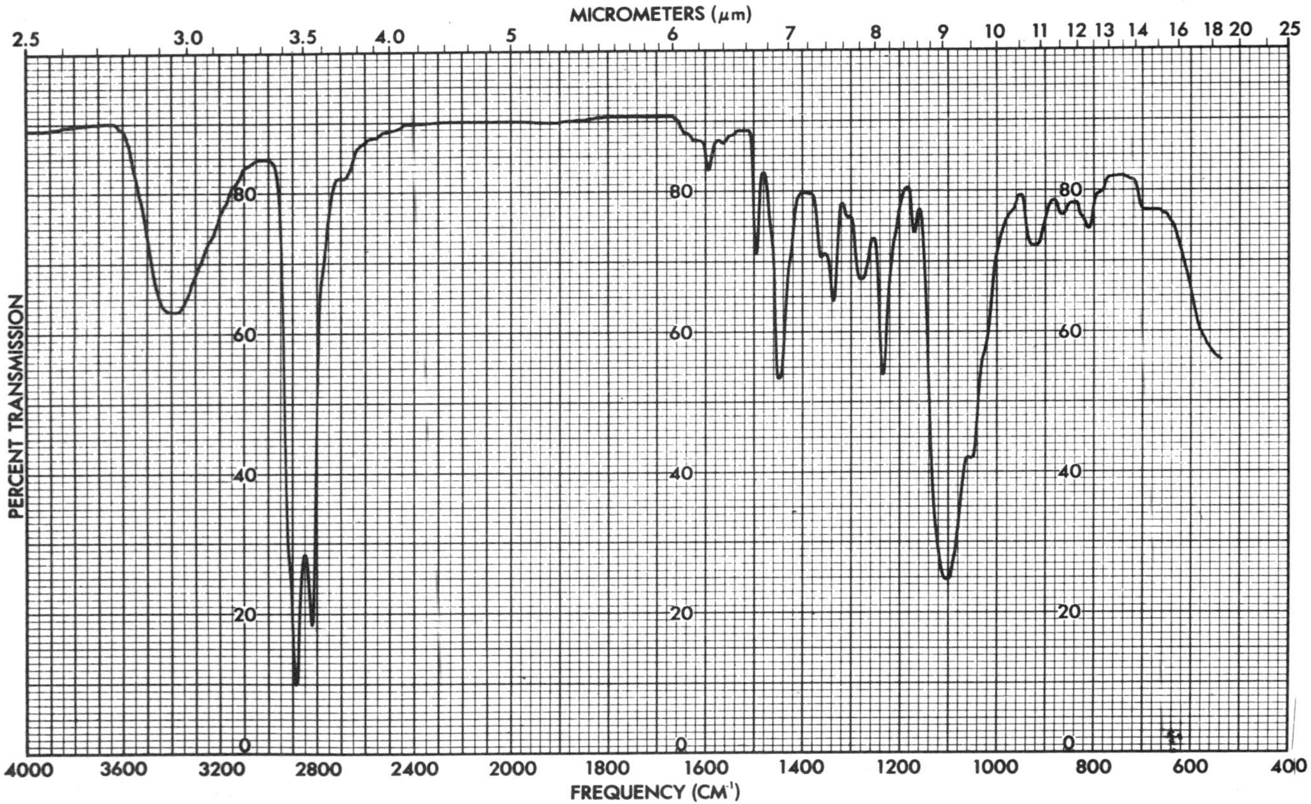

FIG. 7.13. Infrared spectrum of low-mole EO emulsifier.

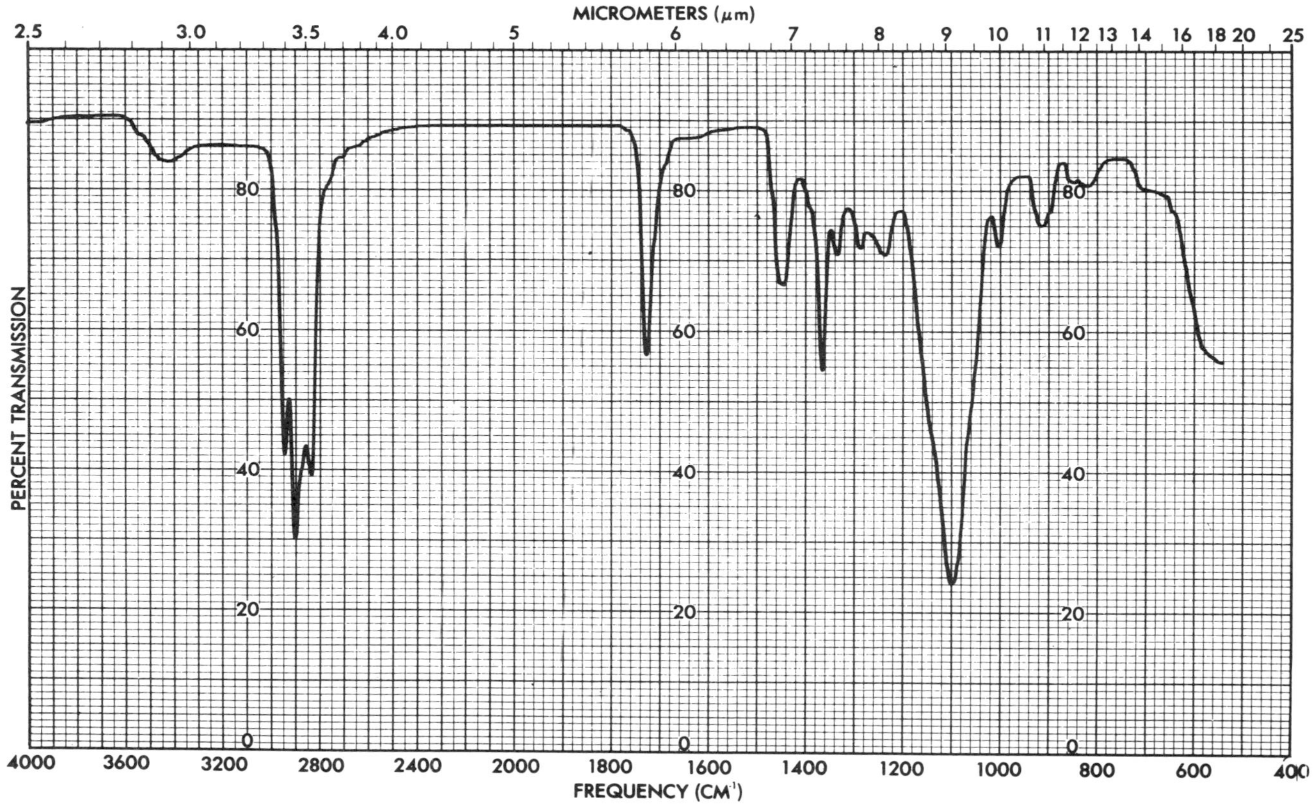

FIG. 7.14. Infrared spectrum of propylene oxide–fatty acid emulsifier.

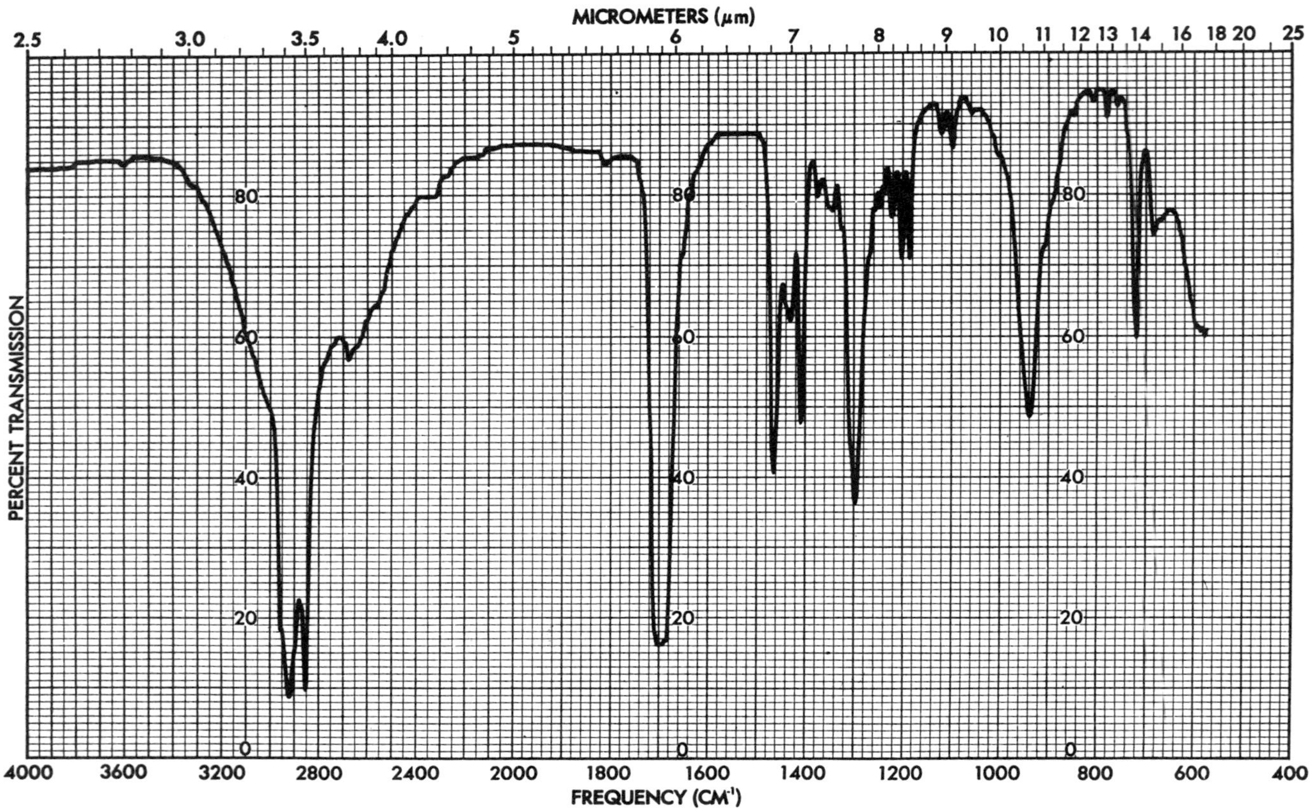

FIG. 7.15. Infrared spectrum of high-molecular-weight fatty acid.

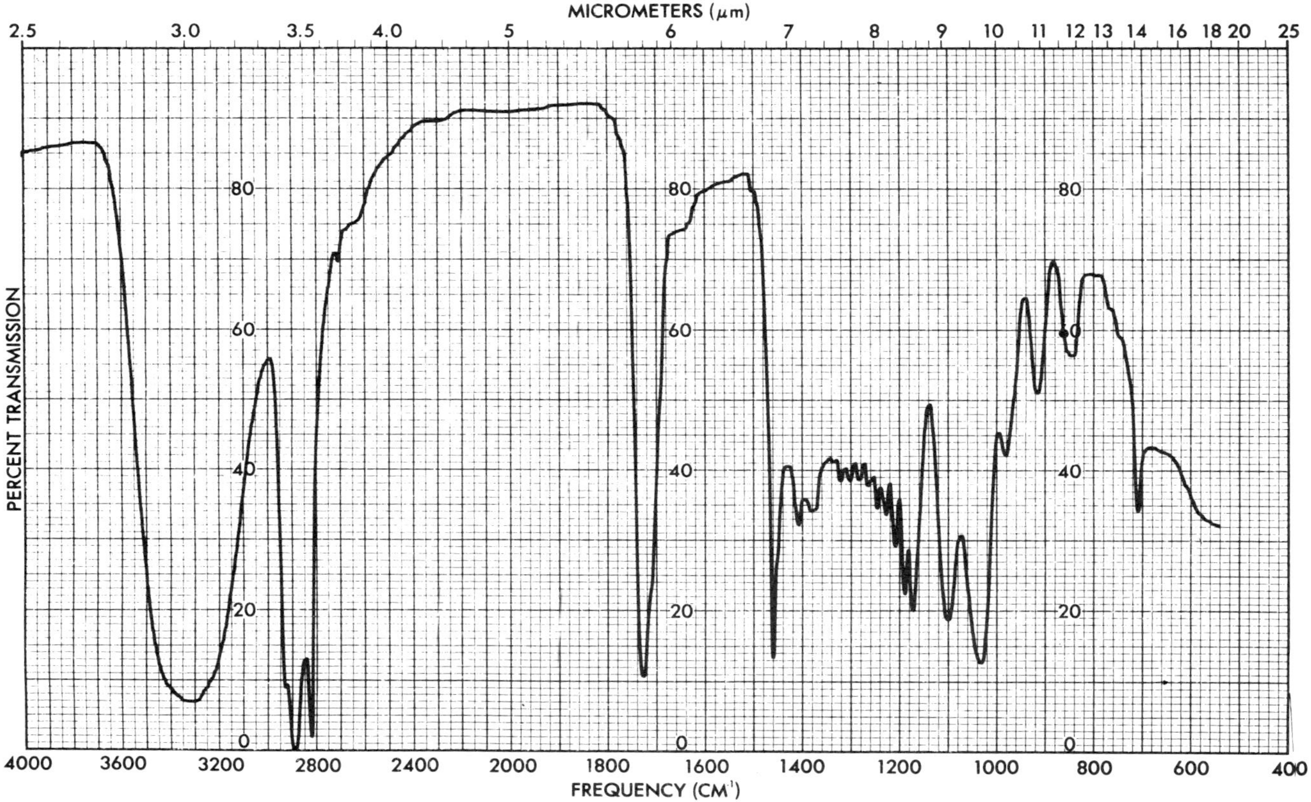

FIG. 7.16. Infrared spectrum of glyceryl monostearate.

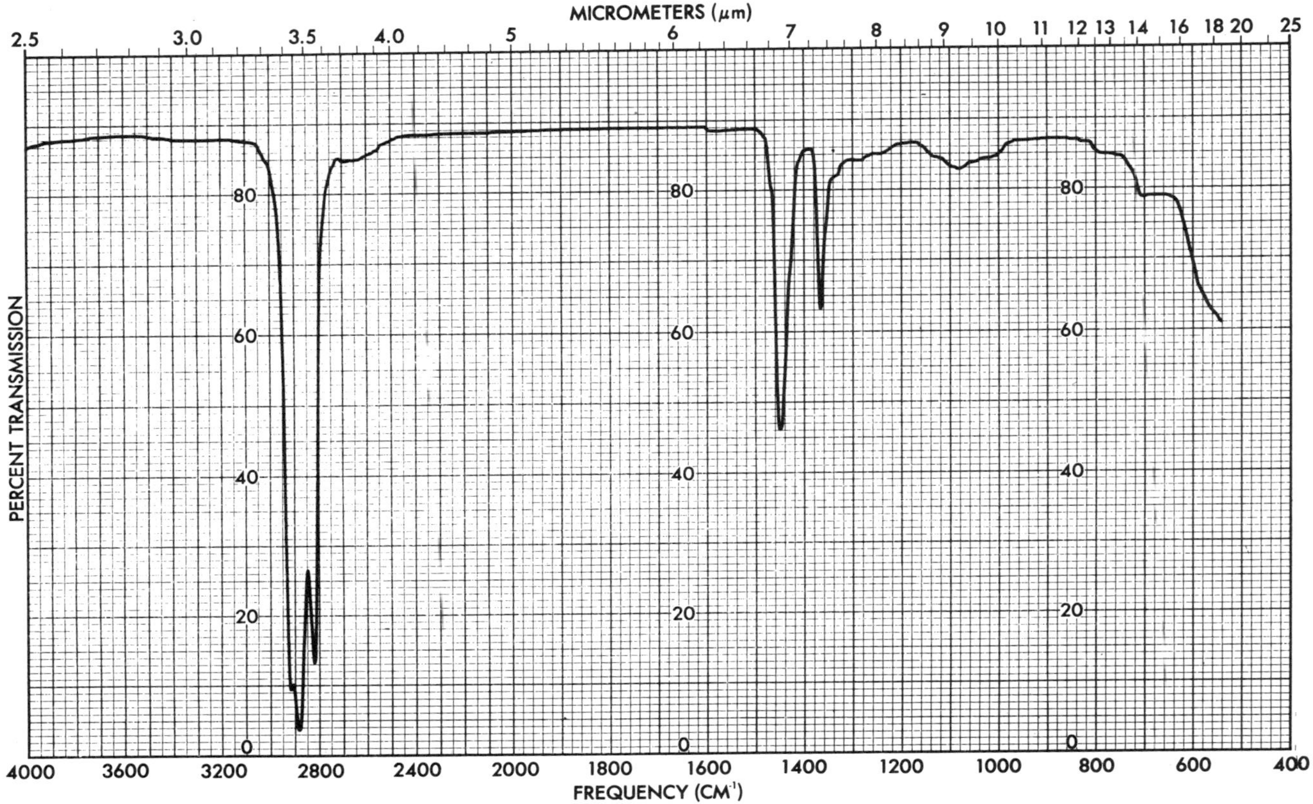

FIG. 7.17. Infrared spectrum of aliphatic hydrocarbon.

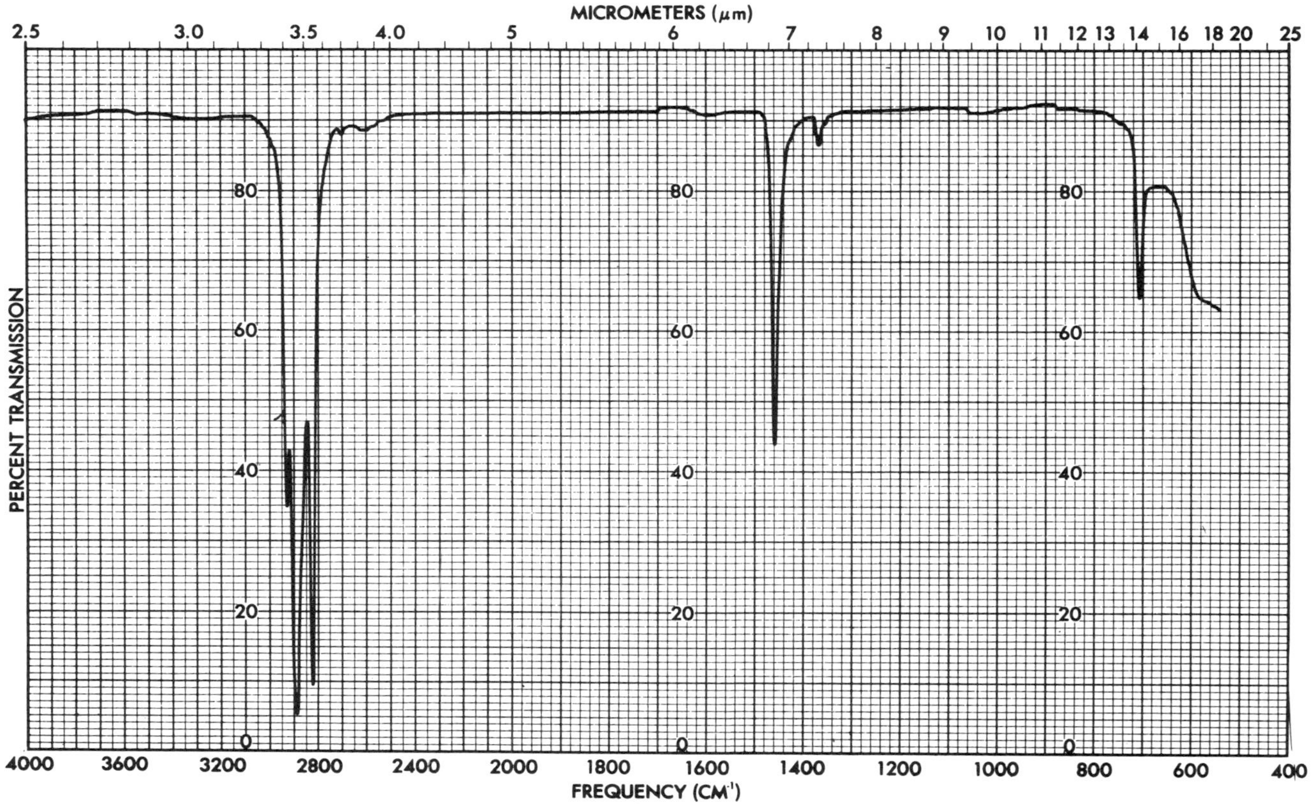

FIG. 7.18. Infrared spectrum of high-molecular-weight aliphatic hydrocarbon.

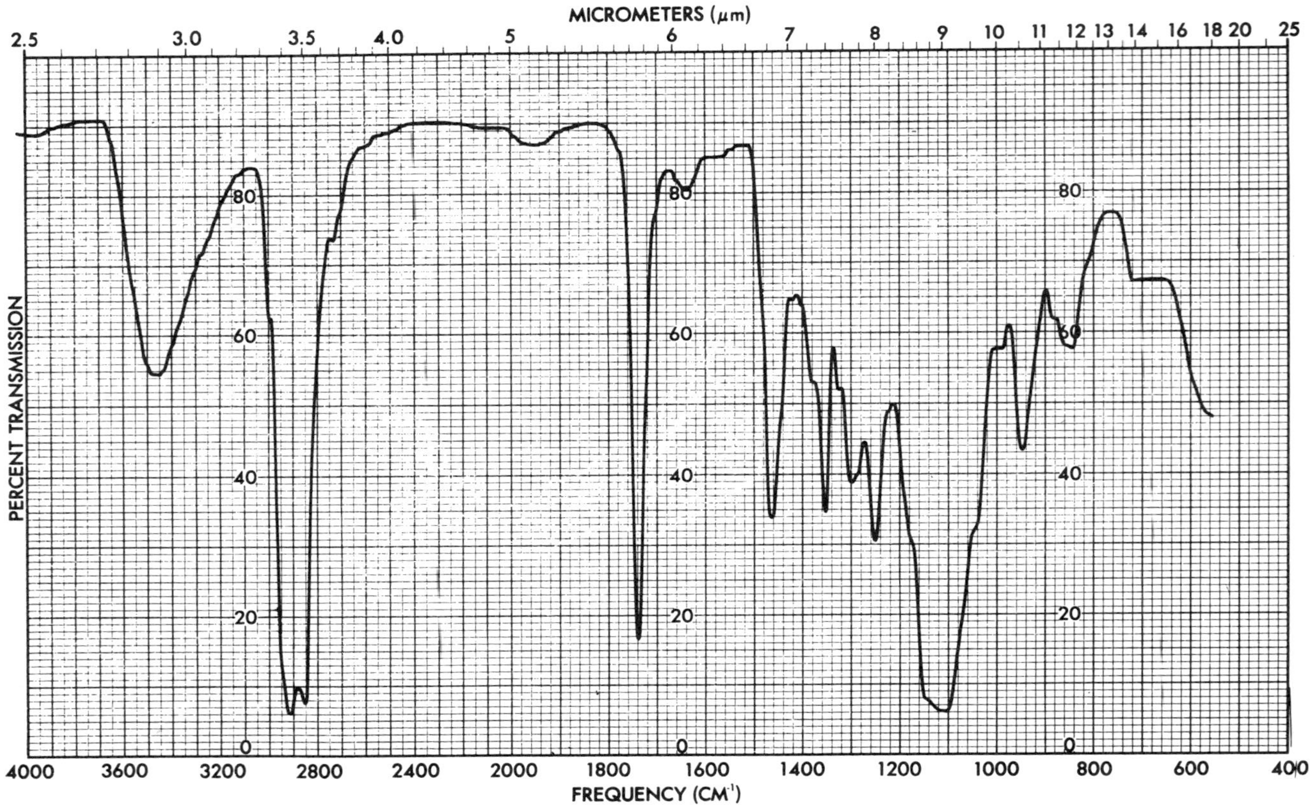

FIG. 7.19. Infrared spectrum of PEG-400 monotallate.

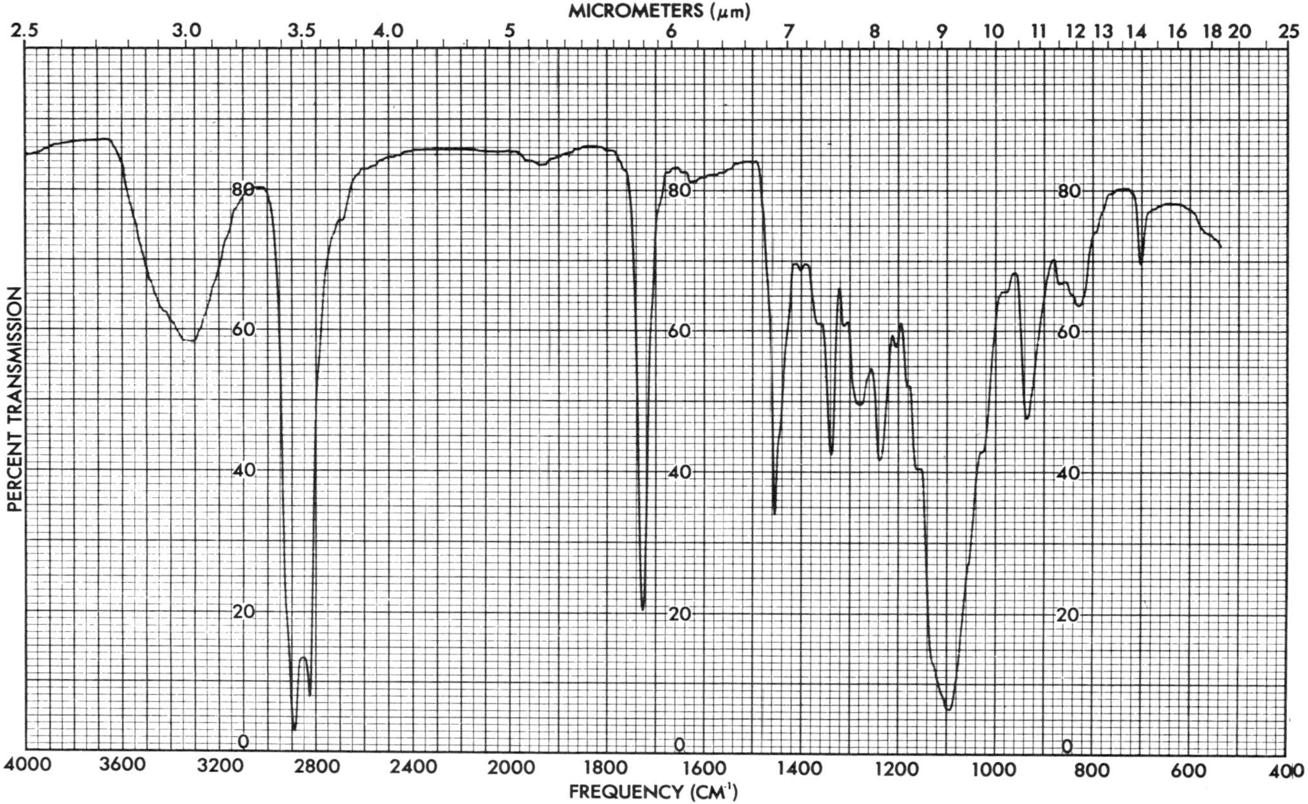

FIG. 7.20. Infrared spectrum of PEG-400 monostearate.

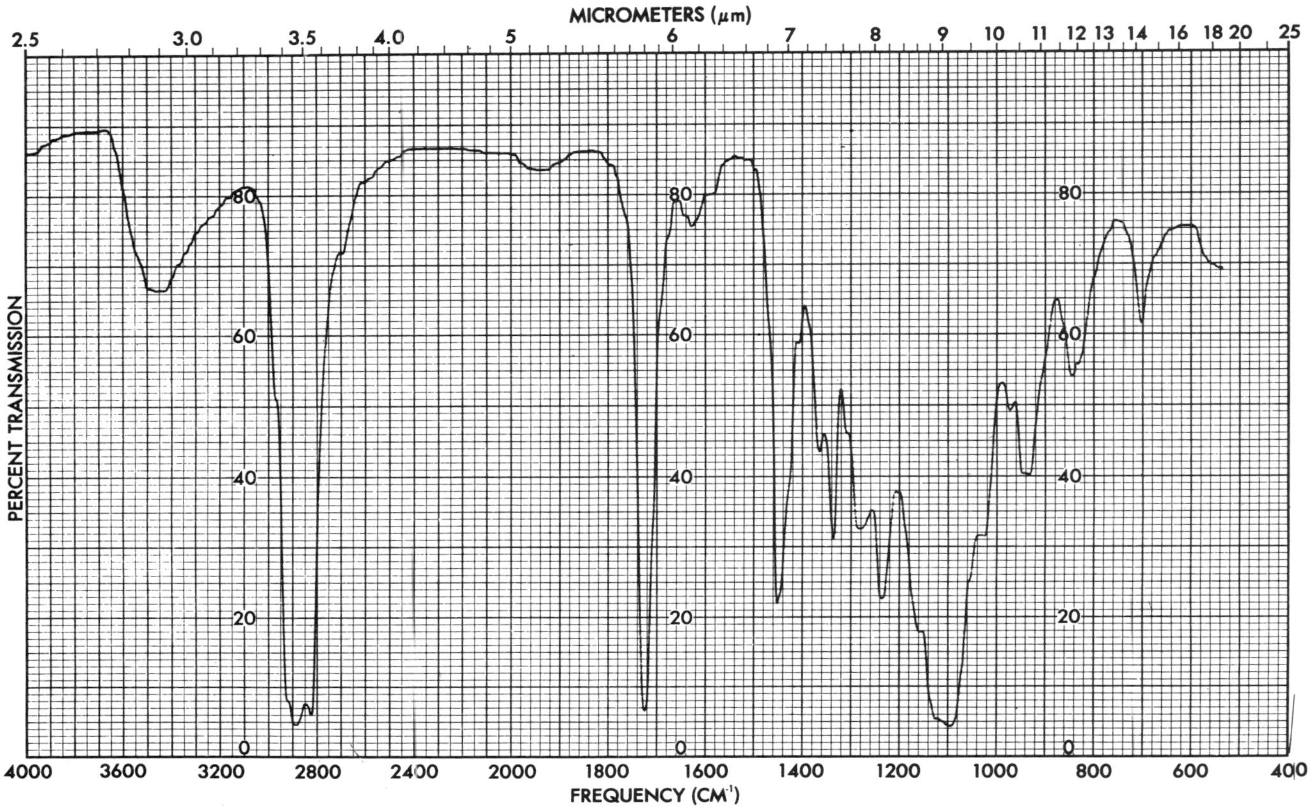

FIG. 7.21. Infrared spectrum of PEG-400 dioleate.

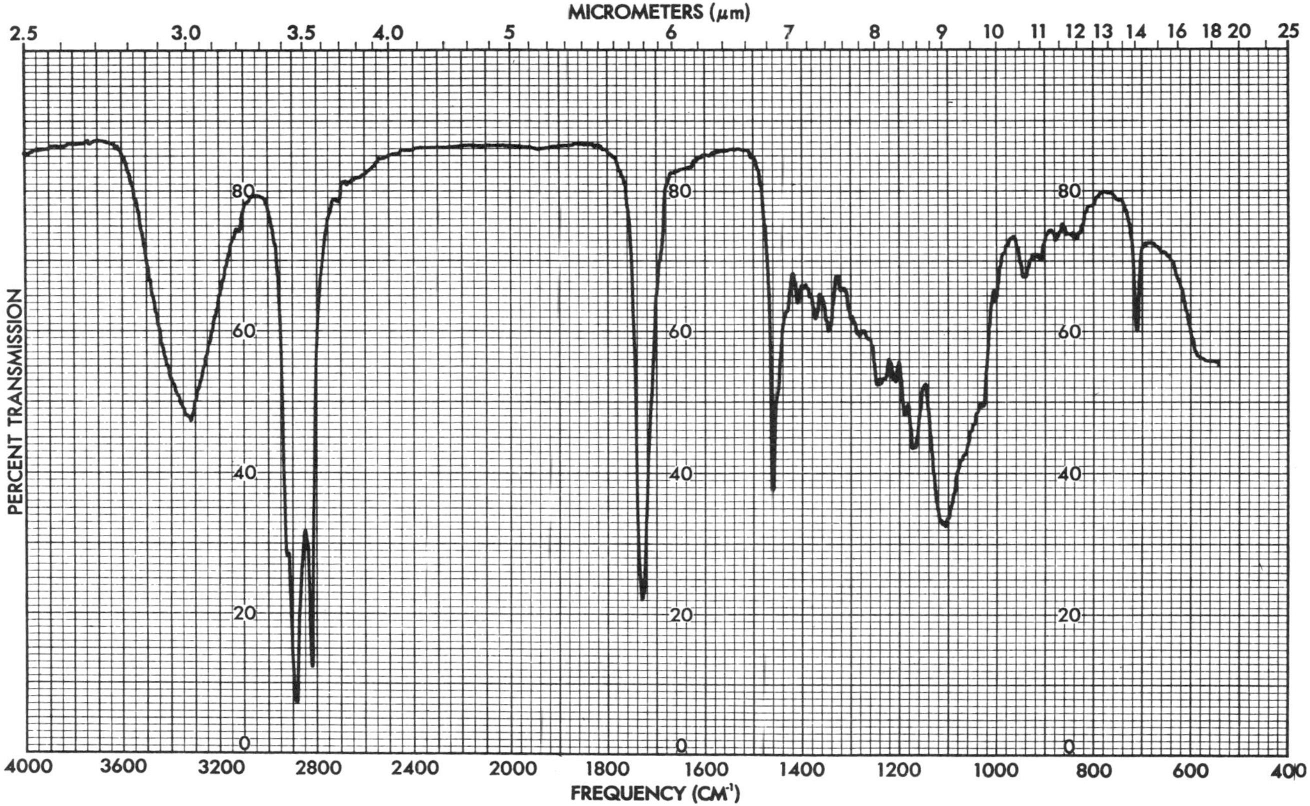

FIG. 7.22. Infrared spectrum of diglycol stearate.

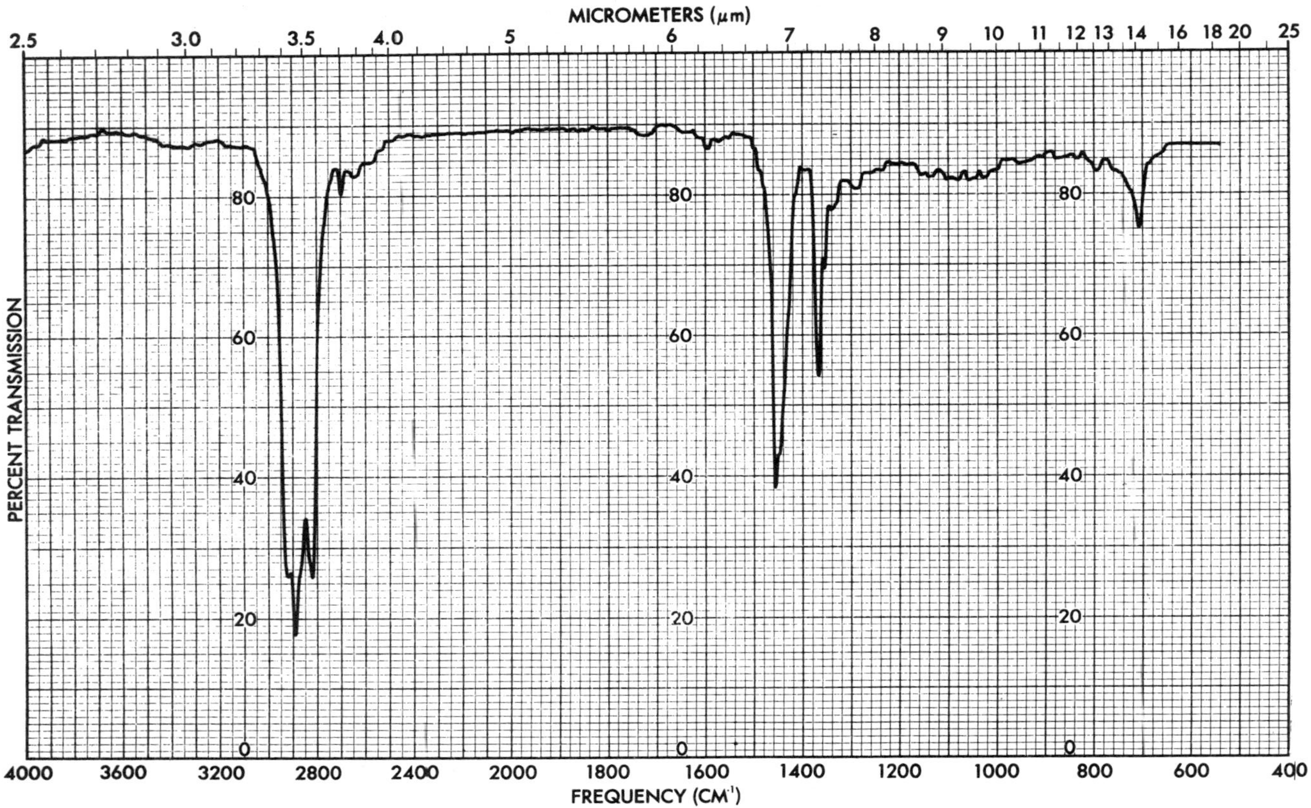

FIG. 7.23. Infrared spectrum of white mineral oil.

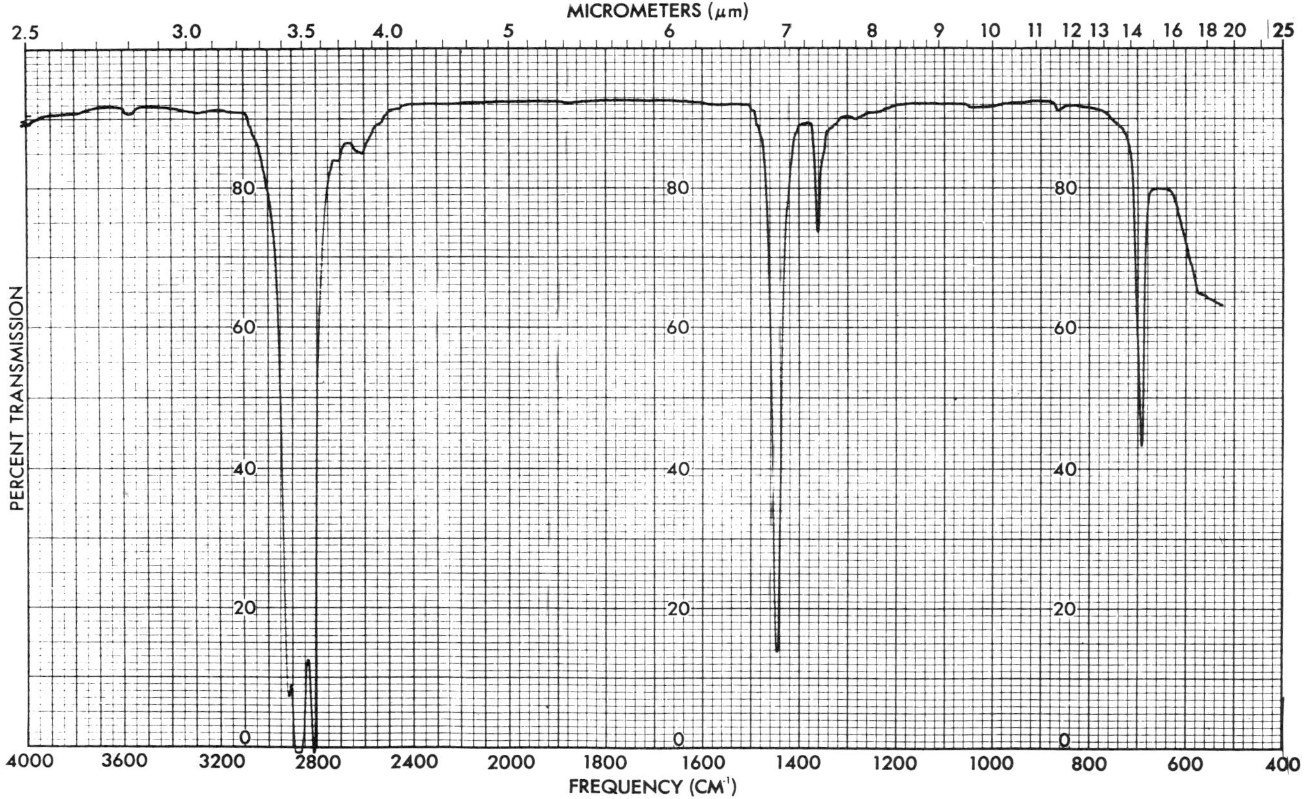

FIG. 7.24. Infrared spectrum of hydrocarbon wax.

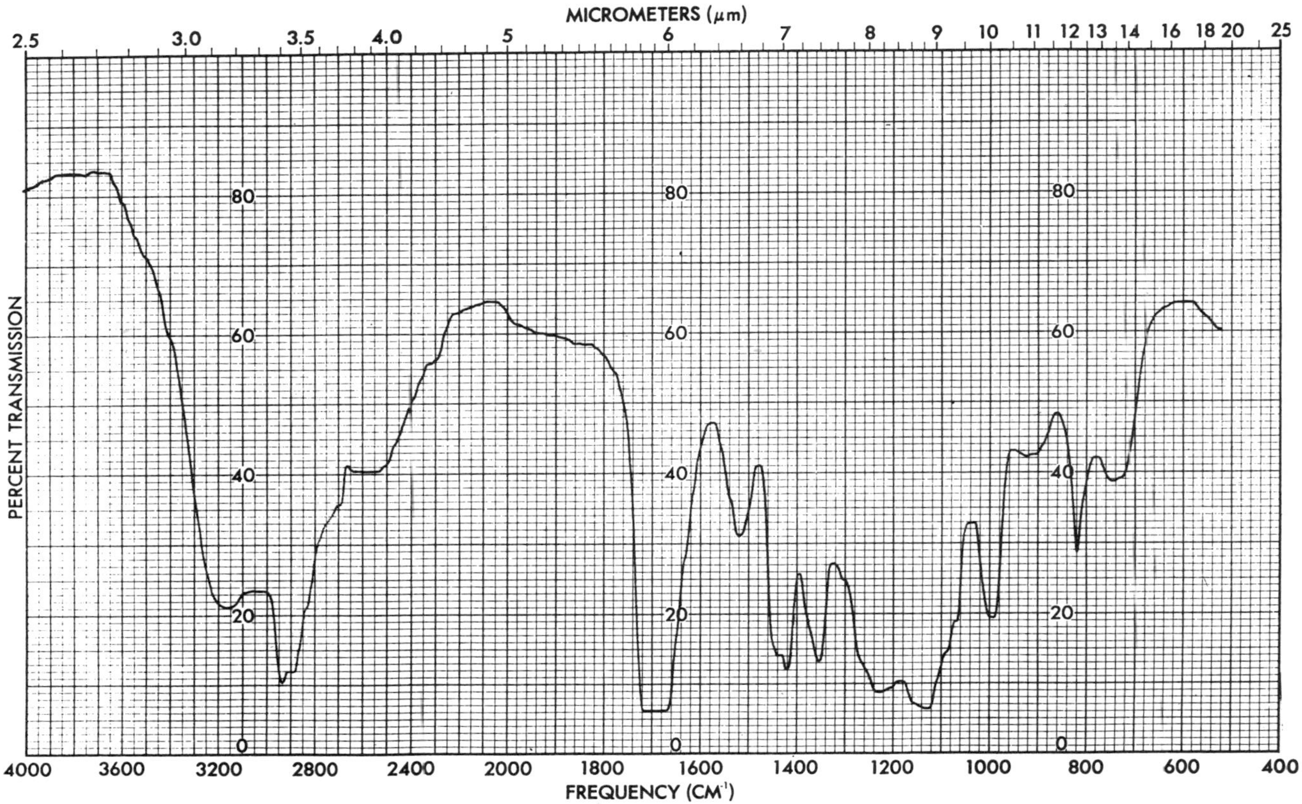

FIG. 7.25. Infrared spectrum of acrylic thickening agent.

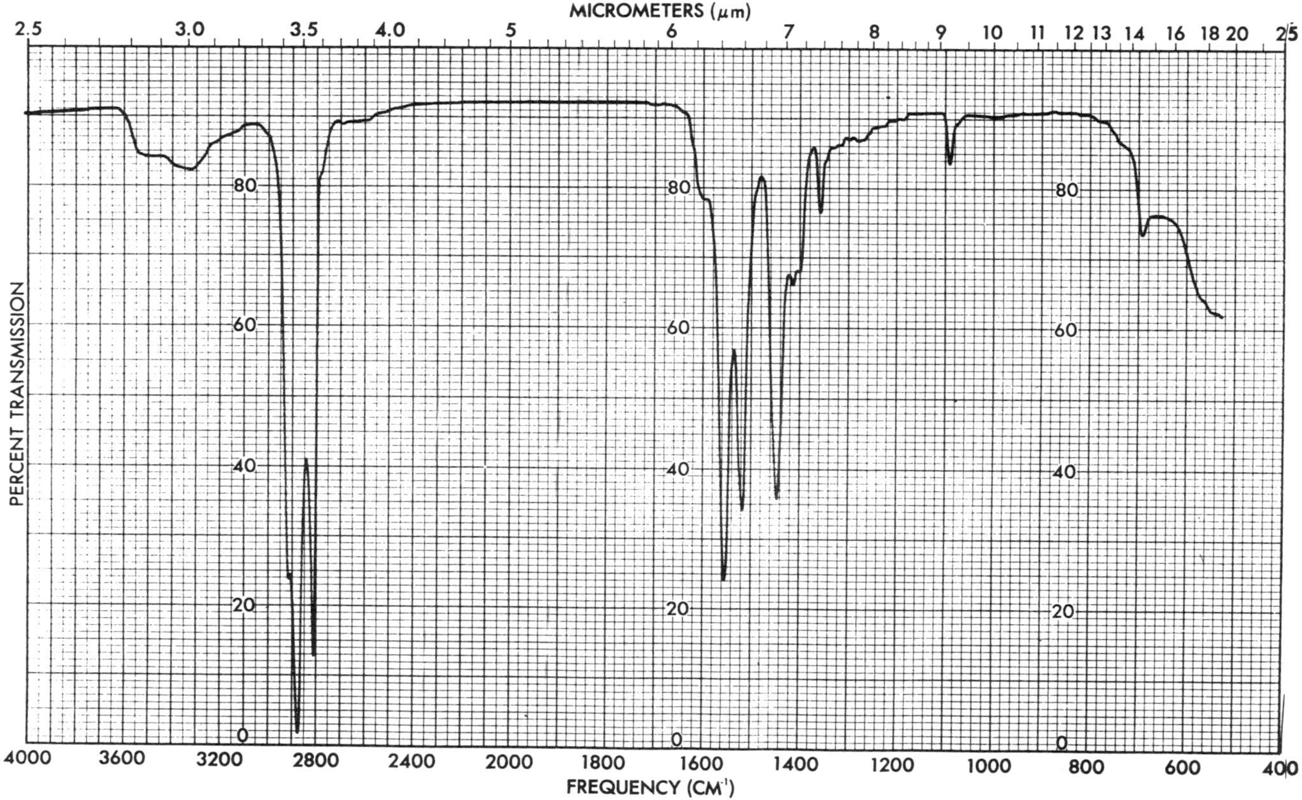

FIG. 7.26. Infrared spectrum of calcium stearate.

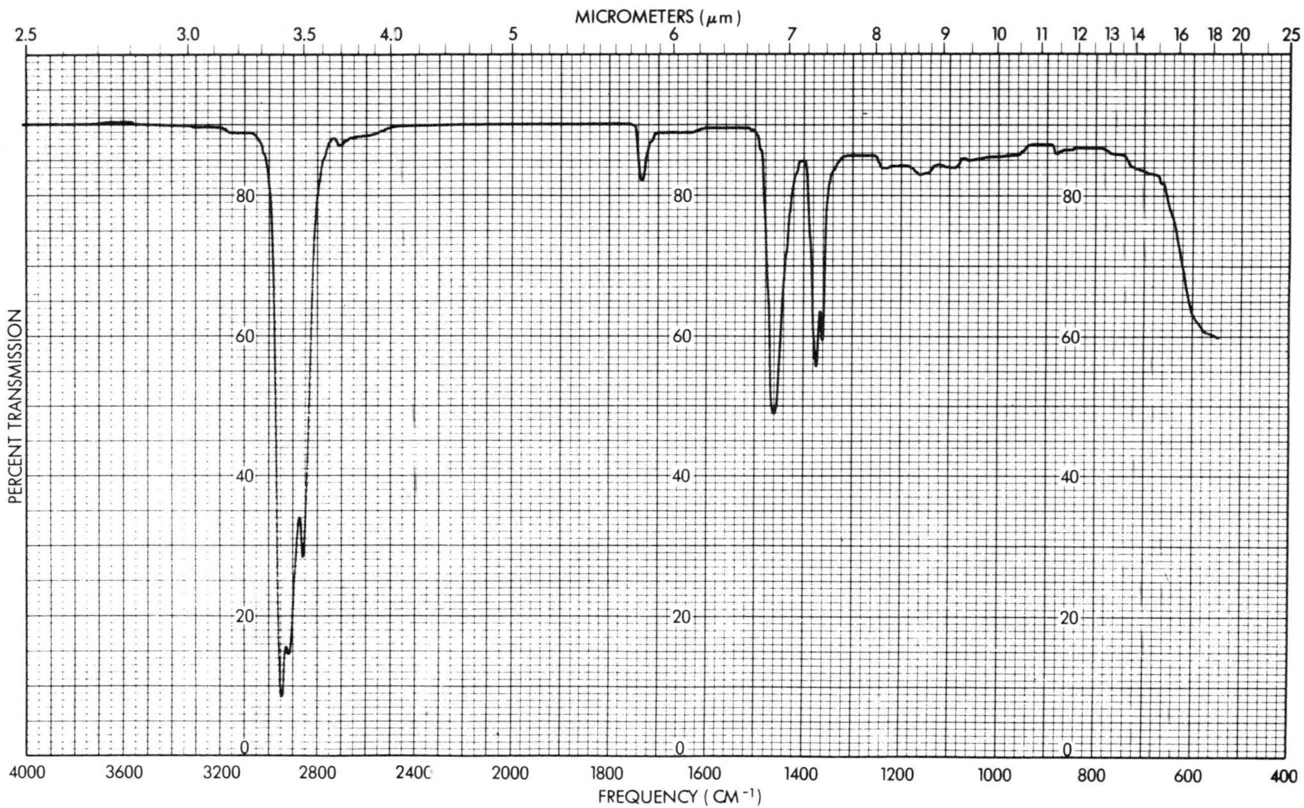

FIG. 7.27. Infrared spectrum of calcium stearate.

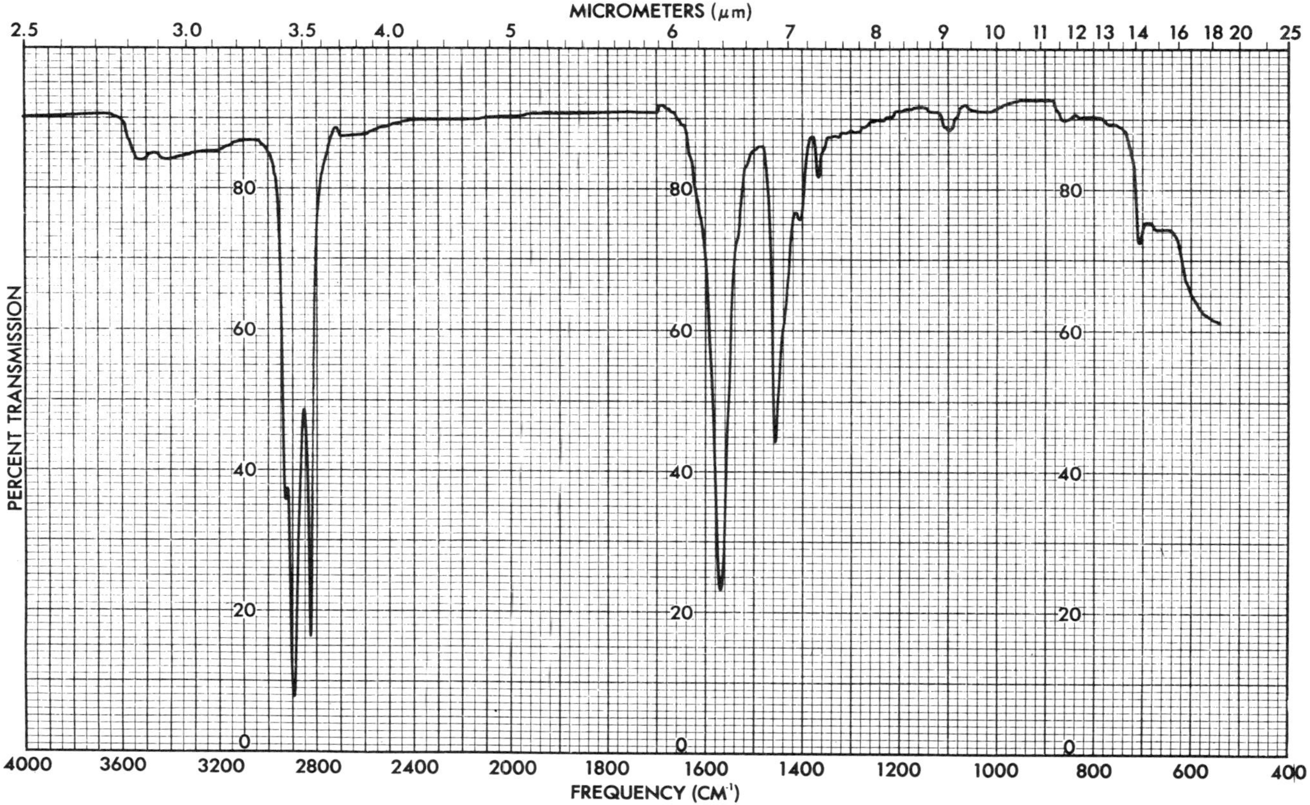

FIG. 7.28. Infrared spectrum of polyisobutylene.

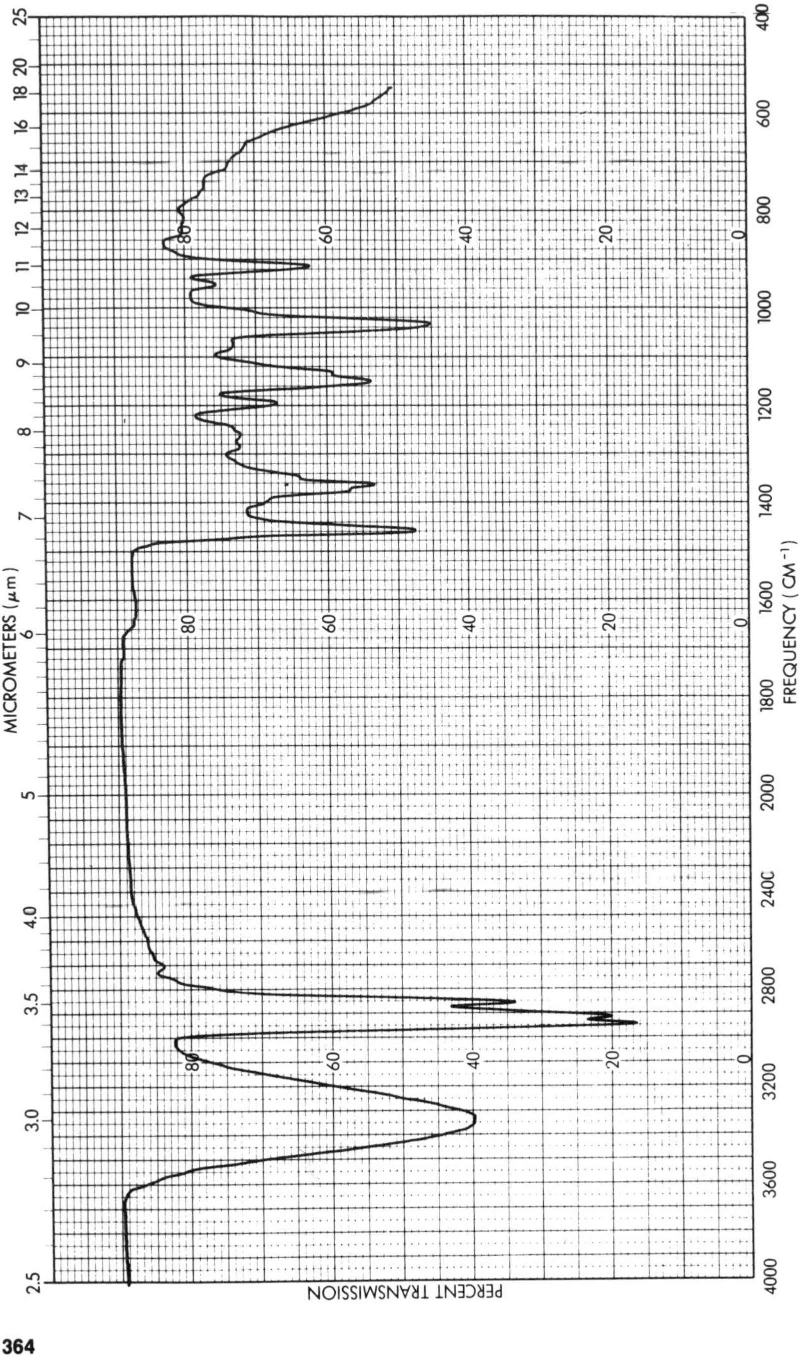

FIG. 7.29. Infrared spectrum of magnesium stearate.

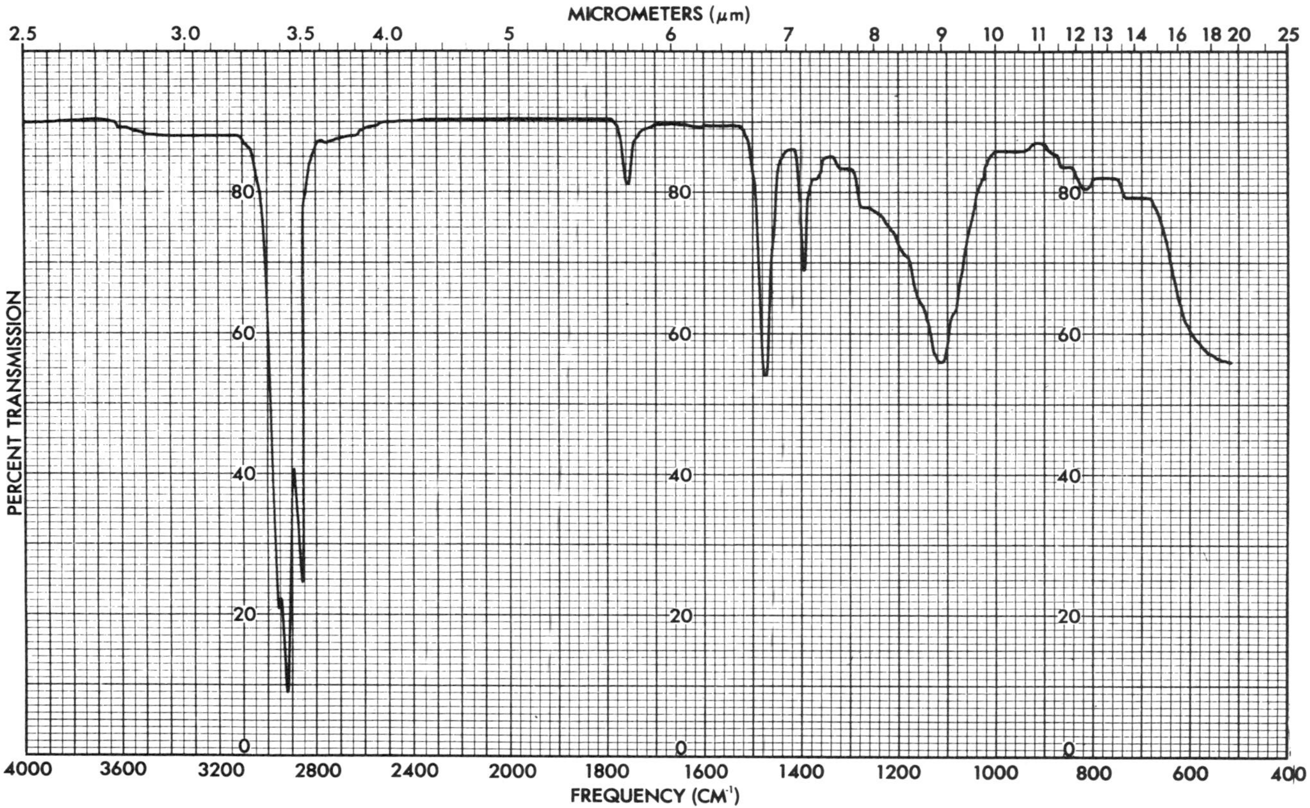

FIG. 7.30. Infrared spectrum of acetylenic derivative mineral oil hydrophobic silica.

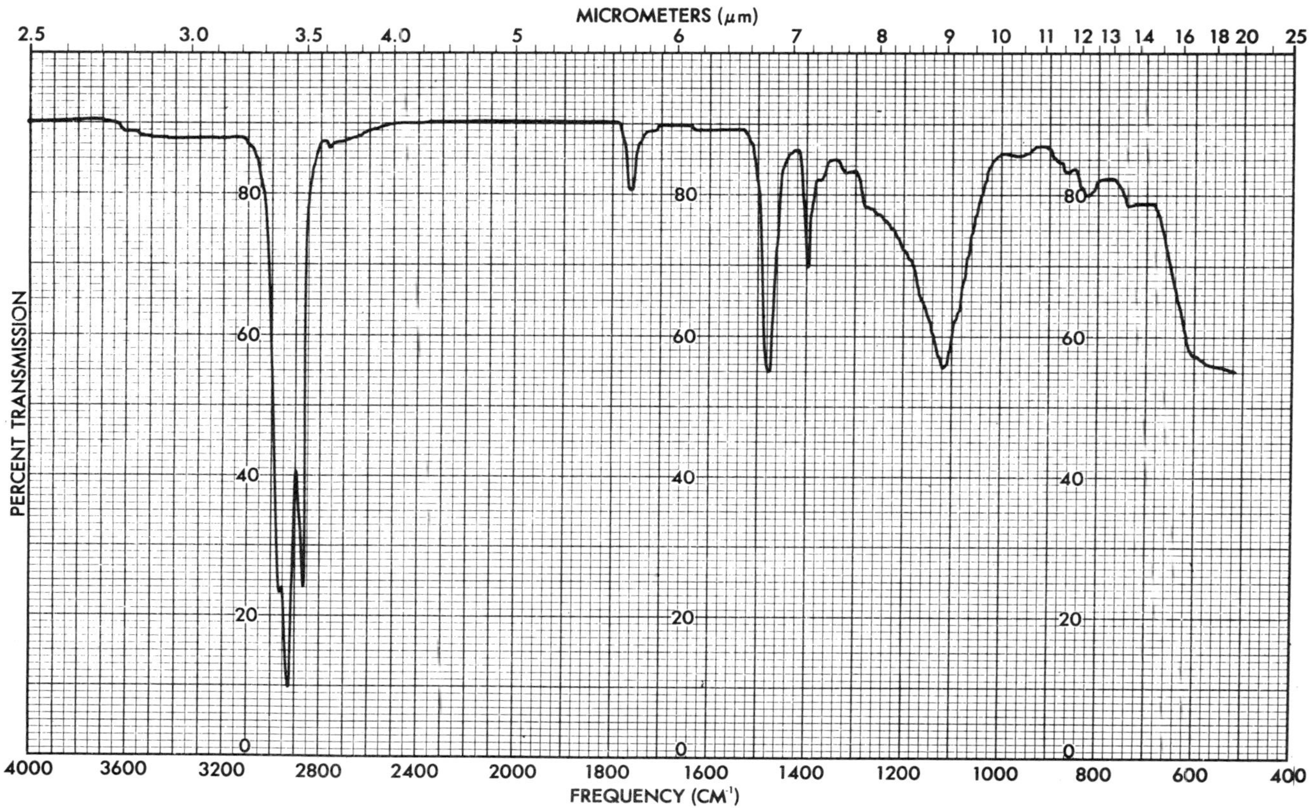

FIG. 7.31. Infrared spectrum of mineral oil silica low-mole (EO) emulsifier.

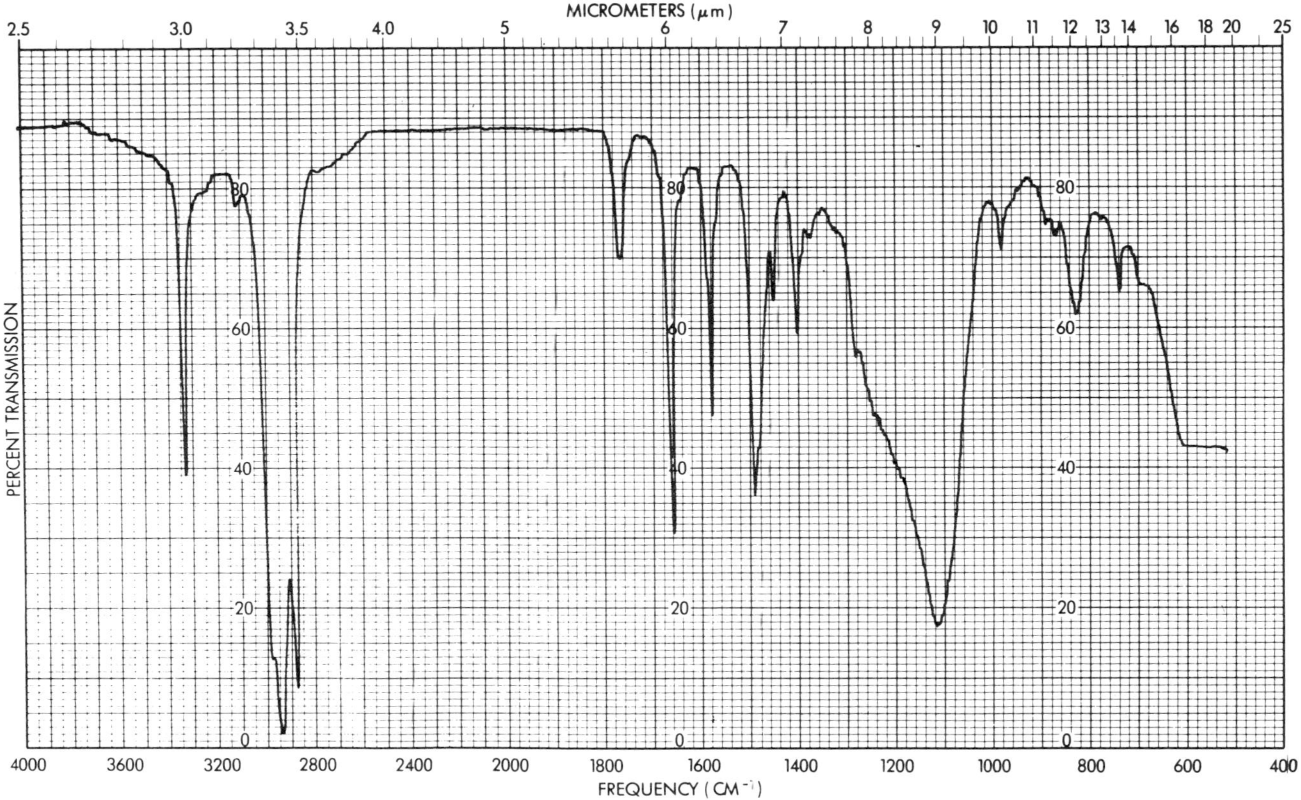

FIG. 7.32. Infrared spectrum of mineral oil silica bisstearamide low-mole (EO) emulsifier.

ion and a complexing agent. They are characterized by the ability to form more than one bond between the metal and the molecule of the complexing agent with the resulting formation of a ring structure incorporating a metal ion. Sequestrants control the metal ion by blocking the reactive site of the metal and preventing it from entering into its normal, and many times, undesirable reactions.

Those sequestrants that form the most stable bonds with metal ions will be the most effective in preventing metal ion activity.[122] A typical chelate structure is as follows:

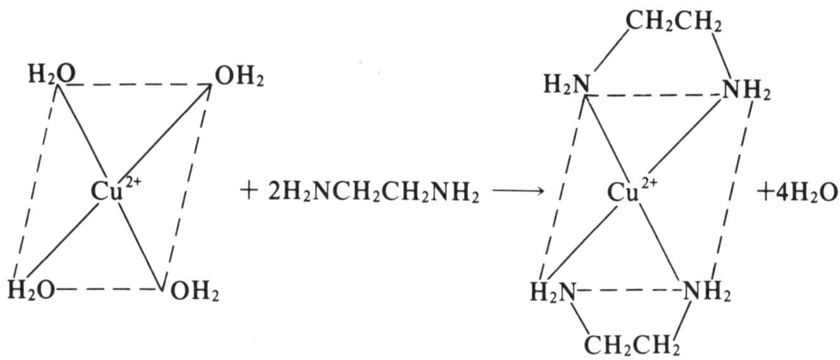

Courtesy of Dow Chemical Company

Chelates, or sequestrants, have become important textile auxiliaries and are vital to the preparation process because of their ability to react with and "tie-up" metal ions in a stable and soluble complex form. These metal ions are generally present in most mill water as the positively charged cation, particularly Fe^{3+} and Ca^{2+}. These metal ions in this valence form are readily available for reaction with any negatively charged anion such as hydroxide (OH^-) or carbonate (CO_3^{2}), which are insoluble compounds that can settle out or form discolorations on fabric.

Chelation agents or sequestrants are negatively charged and thus capable of forming strong ring structures with the metallic ion as previously shown. This ring structure alters the reactivity of the metallic ion by rendering it water soluble and changing the charge of the new compound to either neutral or negative.

There are many commercially available chelates, and the one used is usually chosen for its particular end-use characteristics and performance properties.

Some of the reasons for the use of chelates in soap and detergent formulations are the prevention of (1) film and scum formation, (2) precipitation of hard water soaps, (3) calcium and magnesium inhibition of foaming properties, (4) clogging in liquid dispensers, (5) haze and turbidity in liquid solutions, and (6) rancidity and oxidation that cause discoloration of formulations.

References are U.S. Patents 2,130,505; 2,240,957; 2,316,636; 2,407,645; 2,673,213; 3,022,343; 2,205,995; and 2,384,817.

Sequestering agents that find commercial usage in the specialty chemical market include ethylenediamine tetraacetic acid (EDTA), diethylenetriamine pentaacetic acid (DTPA), hydroxyethylene diaminetriacetic acid (HEDTA), nitrilotriacetic acid (NTA), organic acids (e.g., citric, gluconic), *N,N*-di(2-hydroxyethyl)glycine (DHEG), diaminopropanol tetraacetic acid (DPTA), and triethanolamine (TEA).

Examples of typical chelates and their structures are outlined below:

1. *Ethylenediamine Tetraacetic Acid and Salts* (*EDTA*). Thc most widely used chelating agent. This material is available in many various salt forms, as well as the free acid. It is a general-purpose chelating agent that chelates water hardness ions above pH 7 and most other common metal ions throughout the pH range. It is effective for controlling iron only below pH 9. Its structure is:[123]

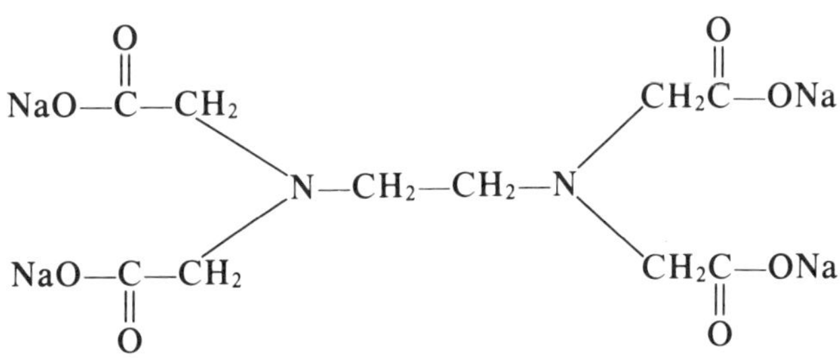

Courtesy of Dow Chemical Company

(See also Fig. 7.33.)

2. *Nitrilotriacetic Acid and Salts* (*NTA*). The most economical chelating agent, widely applied in many areas, especially in providing greater con-

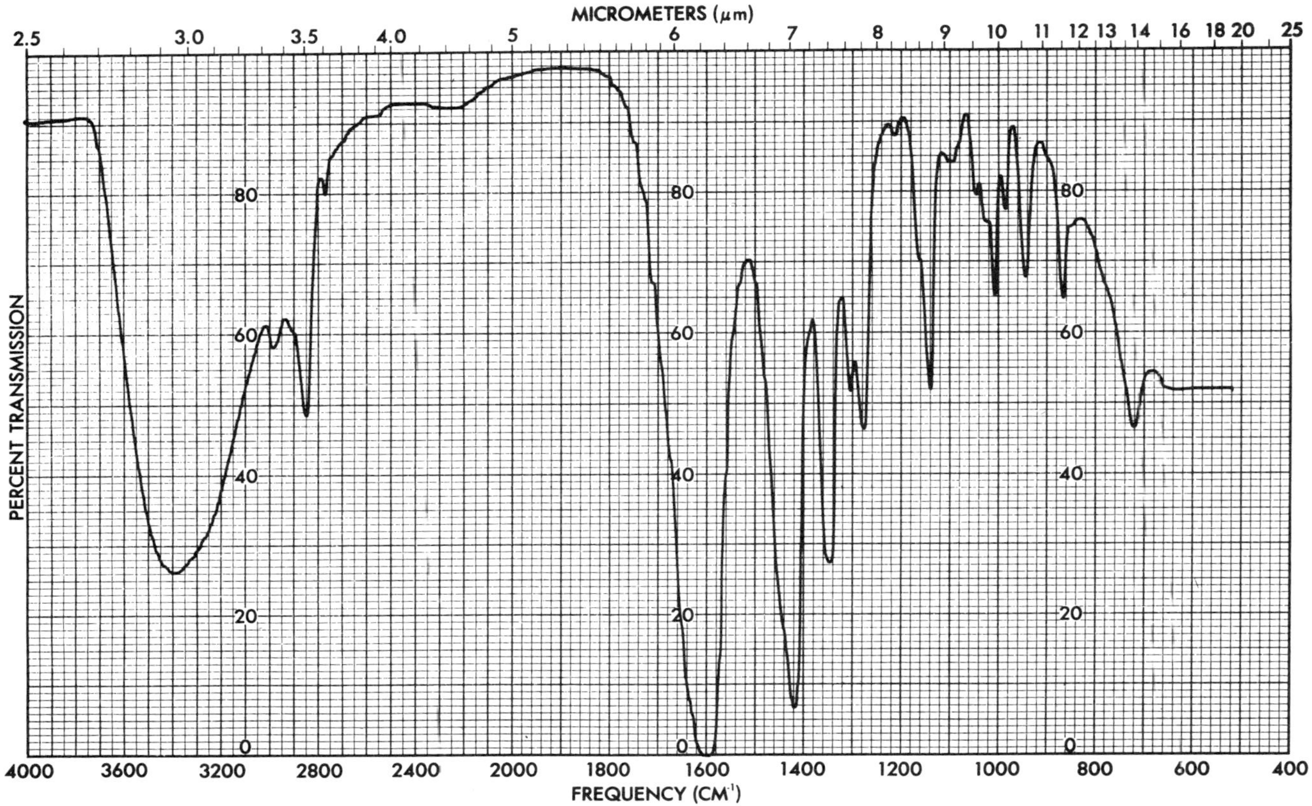

FIG. 7.33. Infrared spectrum of ethylenediamine tetrasodium acetate.

trol of water hardness than with the use of those obtained by polyphosphates. The structure of NTA is:

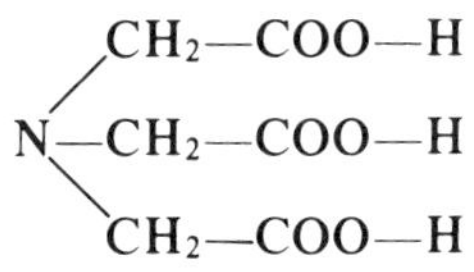

Courtesy of W. R. Grace

3. *N-(2-Hydroxyethyl)ethylenediamine Triacetic Acid (HEDTA).* A chelating agent closely related to EDTA and used principally in chelating iron in alkaline systems. It can prevent iron from precipitating up to pH 12. The structure of HEDTA is:

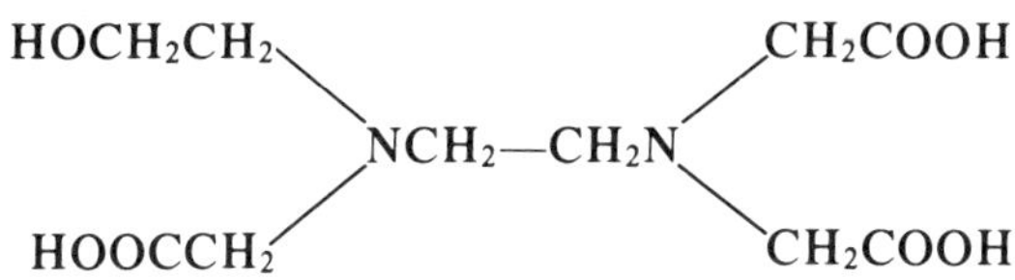

Courtesy of Dow Chemical Company

(See also Fig. 7.34.)

4. *Diethylenetriamine Pentaacetic Acid and Salts (DTPA).* A chelating agent that binds metal ions with a slightly higher affinity than does EDTA, often used where EDTA yields only marginal results. It is also used to stabilize paper and textile bleach baths because of its superior resistance to oxidation. The structure of DTPA is:

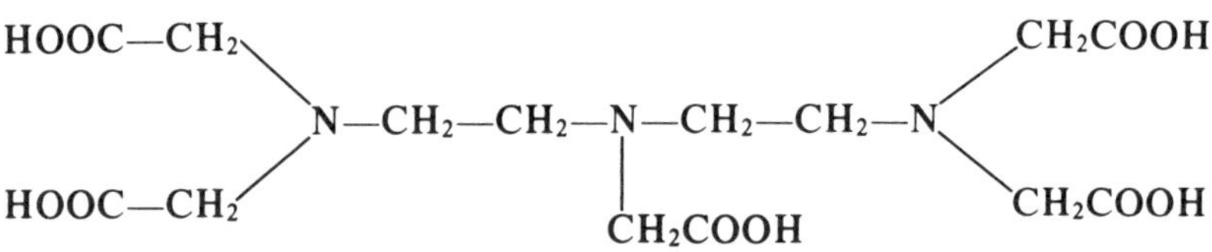

Courtesy of Dow Chemical Company

(See also Fig. 7.35.)

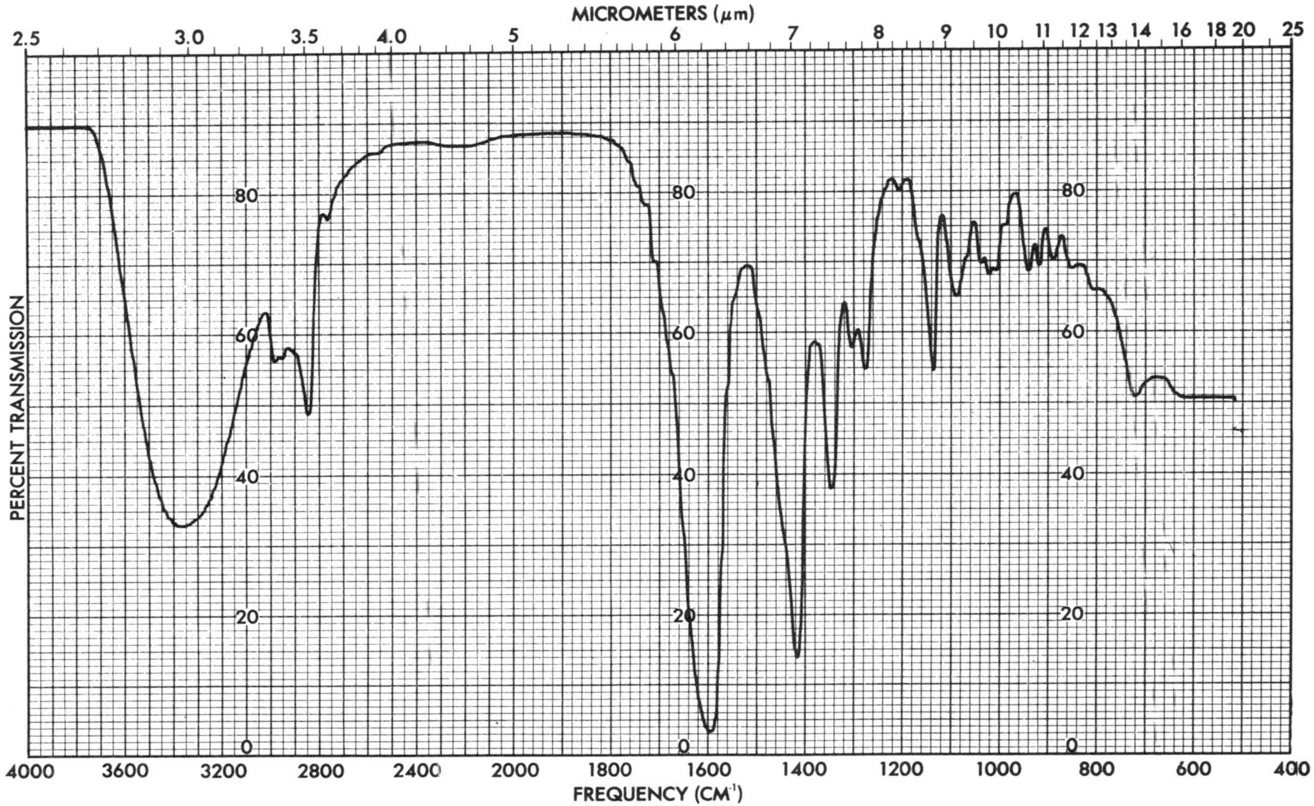

FIG. 7.34. Infrared spectrum of hydroxyethyl ethylenediamine triacetic acid.

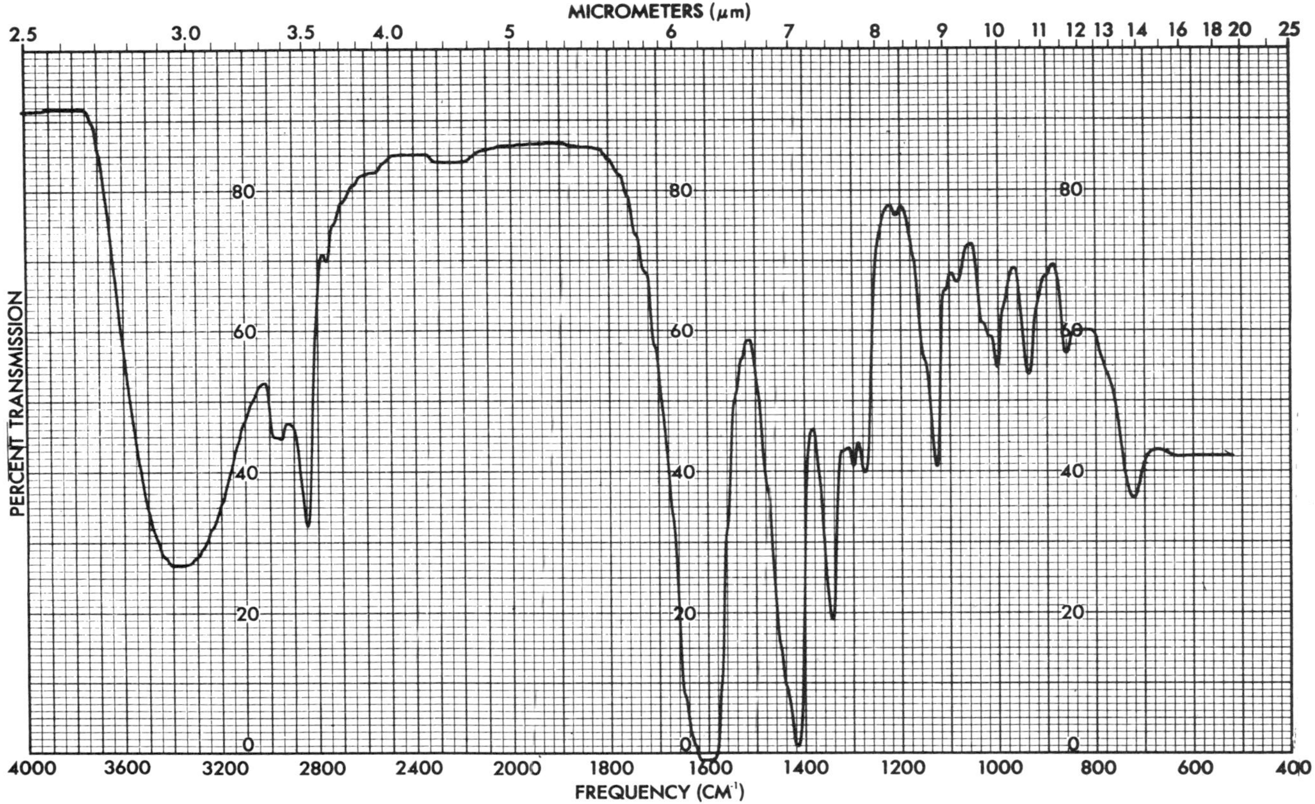

FIG. 7.35. Infrared spectrum of diethylenetriamine pentaacetic acid plus salts.

5. *N,N-Di(2-Hydroxyethyl) Glycine (DHEG).* Glycine type chelates that were originally marketed for the control of iron in alkaline pH ranges. They have been replaced largely by HEDTA or mixtures of HEDTA and triethanolamine. The structure of DHEG is:

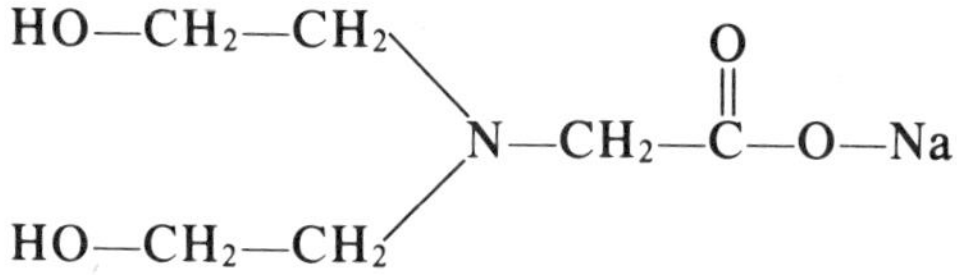

Courtesy of Dow Chemical Company

6. *Diaminopropanol Tetraacetic Acid (DPTA).* The structure of DPTA is:

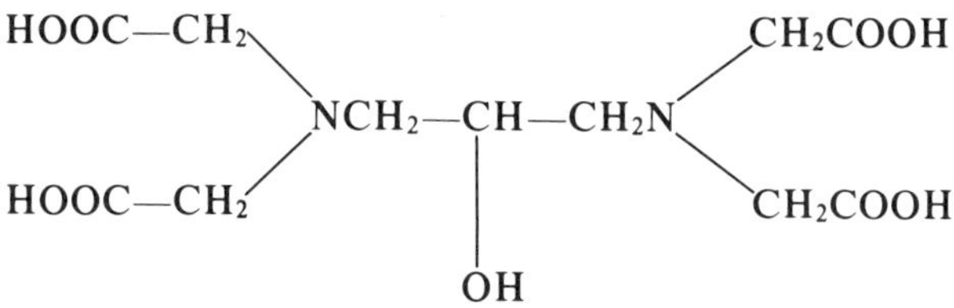

Courtesy of Dow Chemical Company

Structural formulas that represent metal chelates are represented as follows:

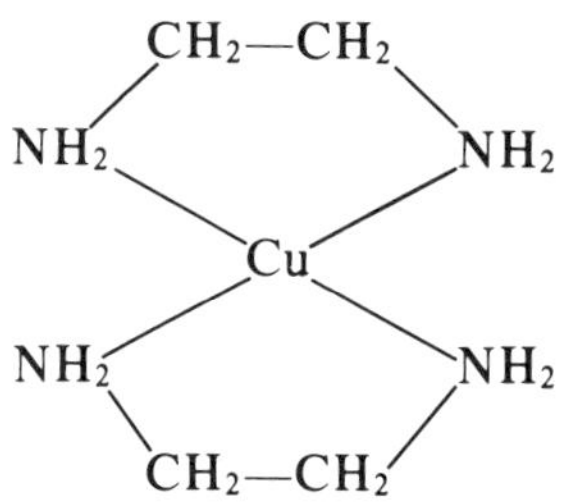

Ethylenediamine–copper complex

Courtesy of Société Protex

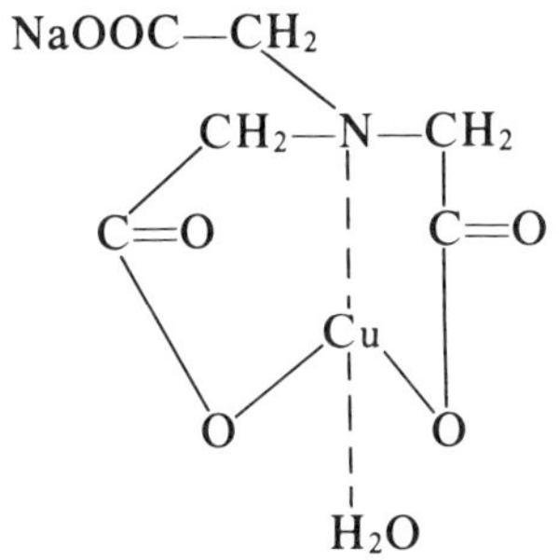

NTA-copper complex

Courtesy of Société Protex

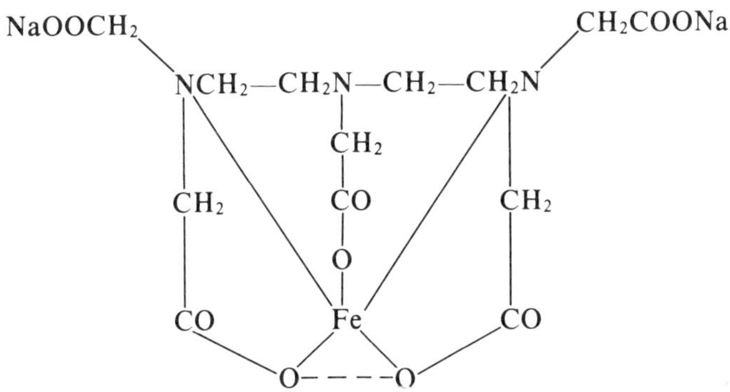

DTPA-iron complex

Courtesy of Société Protex

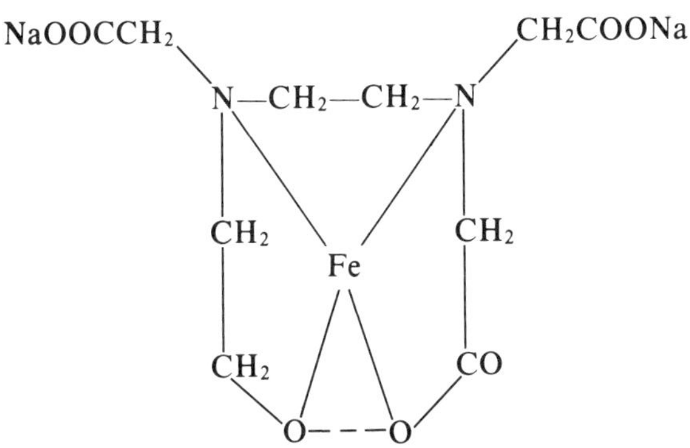

EDTA-iron complex

Courtesy of Société Protex

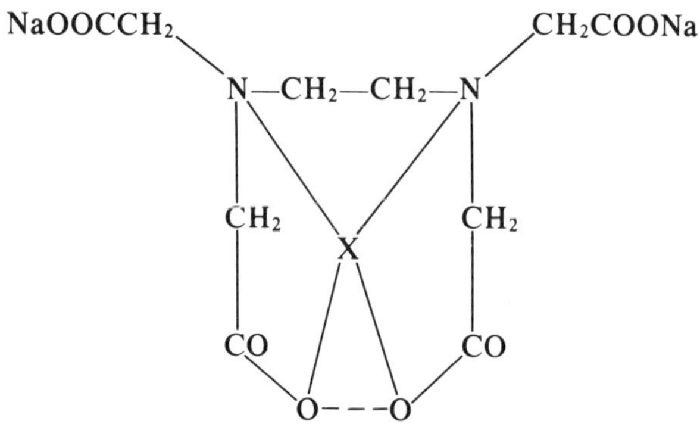

EDTA–metal complex

Courtesy of Société Protex

where x = Cu, Mn, Zn, or Mg.

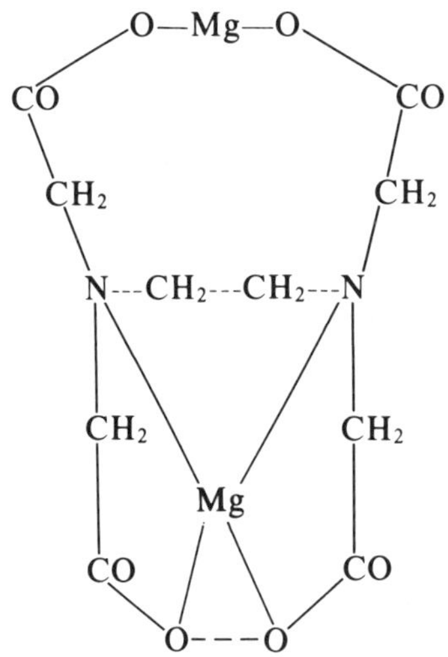

EDTA–dimagnesium complex

Courtesy of Société Protex

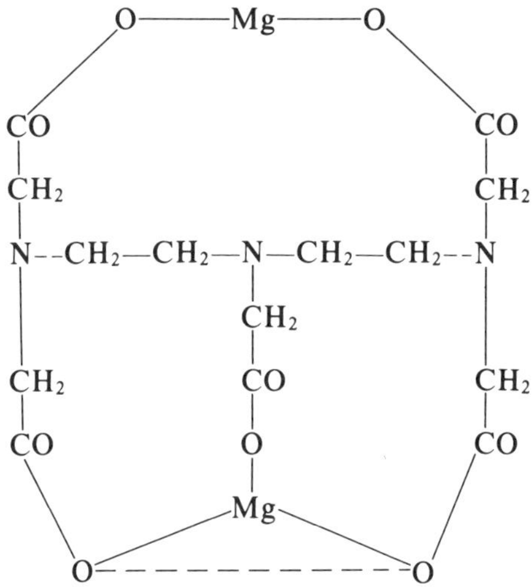

DTPA-dimagnesium complex

Courtesy of Société Protex

An important factor in choosing a specific chelate is to remember that in an aqueous medium there is a considerable tendency for reversal of the equilibrium reaction when one factor or the other is shifted. This may be either a change in the concentration of either chelate or metal and/or an upward or downward shift in pH. Tables 7.3 and 7.4 tabulate data that show the sequestering power in order of selection when influenced by pH.

At a given pH, different quantities of chelation agents are required to chelate a given amount of metal. Table 7.5 outlines the respective quantities of chelate required to inactivate an equivalent quantity of metal.

TABLE 7.3. EDTA[a]

pH 3.5	pH 3.5–5	pH 5–7.5	pH 7.5–9.5	pH 9.5–12	pH 12
Fe^{3+}	Fe^{3+}	Cu	Cu	Cu	Cu
Hg	Cu	Fe^{3+}	Pb	Ni	Fe^{2+}
Cu	Pb	Pb	Ni	Fe^{2+}	Ca
Pb	Ni	Ni	Zn	Ca	Ni
	Zn	Zn	Fe^{3+}	Pb	Mg
	Ca	Ca	Ca	Zn	
	Mg	Mg	Mg	Mg	

[a] Courtesy of Dow Chemical Company.

TABLE 7.4. NTA[a]

pH 3.5–5	pH 5–7.5	pH 7.5–9.5	pH 9.5–12
Cr	Ni	Ni	Co
Cu	Cu	Co	Ni
Ni	Co	Cu	Cu
Pb	Zn	Zn	Cd
Co	Cd	Cd	Zn
	Ca	Ca	Ca
	Fe	Fe	Fe
		Mg	Mg

[a]Courtesy of Dow Chemical Company.

TABLE 7.5. Amounts of Chelate Required to Chelate a Given Amount of Metal[a]

Metal(V)	Name	EDTA	DTPA	HEDTA	NTA
Al(III)	Aluminum	37.1	46.4	30.9	24.7
Ba(I)	Barium	7.3	9.1	6.1	4.9
Cd(II)	Cadmium	8.9	11.1	7.4	5.9
Ca(II)	Calcium	25.0	31.3	20.8	16.7
Co(II)	Cobalt	17.0	21.3	14.2	11.3
Cu(II)	Copper	15.8	19.8	13.2	10.5
Fe(III)	Iron (ferric)	17.9	22.4	14.9	11.9
Fe(II)	Iron (ferrous)	17.9	22.4	14.9	11.9
Ga(III)	Gallium	14.4	18.0	12.0	9.6
In(III)	Indium	8.7	10.9	7.3	5.8
Pb(II)	Lead	4.8	6.0	4.0	3.2
Mg(II)	Magnesium	41.1	51.4	34.3	27.4
Mn(II)	Manganese	18.2	22.8	15.2	12.1
Hg(II)	Mercury	5.0	6.3	4.2	3.3
Ni(II)	Nickel	17.0	21.3	14.2	11.3
Pd(II)	Palladium	9.4	11.8	7.8	6.3
Sc(III)	Scandium	22.2	27.8	18.5	14.8
Sr(II)	Strontium	11.4	14.3	9.5	7.6
Th(IV)	Thorium	4.3	5.4	3.6	2.7
Ti(III)	Titanium	20.9	26.1	17.4	13.9
TiO(II)	Titanium oxide	15.6	19.5	13.0	8.7
V(III)	Vanadium	19.6	24.5	16.3	13.1
V(II)	Vanadium	19.6	24.5	16.3	13.1
VO(III)	Vanadium oxide	14.9	18.6	12.4	9.9
Y(III)	Yttrium	11.3	14.1	9.4	7.5
Zn(II)	Zinc	15.3	19.1	12.8	10.2

[a]Courtesy of Dow Chemical Company.

CHAPTER EIGHT

THE PREPARATION PROCESS

Now that we have discussed the chemistry of preparation surfactants, auxiliaries, and alkalis, let us take a look at the way in which they are used together.

The first point at which the preparation process begins is desizing. In desizing, the starches, polymers, and natural gums that were applied during the sizing operation, and which are usually water insoluble, are converted to water-soluble compounds to ease their removal. This is accomplished by transforming them into either simple sugars or simple water-soluble polymers. As sizing compounds are applied primarily to warp yarns in weaving, as a rule, only oils and waxes are applied to knitting yarn, with the exception of warp knits. Desizing of cotton can be accomplished by physical, chemical, or a combination of physical and chemical mechanisms.

Cotton Desizing

Mechanisms of Desize	Common Desize Auxiliaries
Physical	Enzymes
Chemical	Sodium hydroxide
Physicaochemical	Sodium persulfate
	Sodium bromite
	Hydrogen peroxide
	Potassium perphosphate
	Wet-out surfactants and glauber salts

DESIZING

The desize procedure may be performed either by itself as a preliminary process or in conjunction with a scour operation that removes the oils and waxes coapplied with the polymers. In the normal desize operation, a dry fabric will be impregnated with enzymes, peroxygens, and/or alkalis. A typical enzyme formulation might be enzyme, detergent wet-out, or glauber salt, with or without solvent. The goods are padded, batched, and allied to lay per the recommendation of the enzyme manufacturer. They are then washed out to remove the water-soluble sizing material. However, they may be run continuously through a series of "J"-boxes or range-type operations for progressive desizing and scouring. They are then thoroughly washed to remove alkali and are generally given a mild acid sour to ensure that all alkali has been removed. They may then be taken directly to the dye house for dyeing, or they may be dried and stored for further preparation (Fig. 8.1).

Typical of equipment used in the desize and scour operation is the single-stage J-box shown in Fig. 8.2 and the multistage J-box arrangement shown in Fig. 8.3.

These newer-type J-boxes have made a more efficient, one-step desizing and scour possible and have facilitated the removal of synthetic sizes, such as

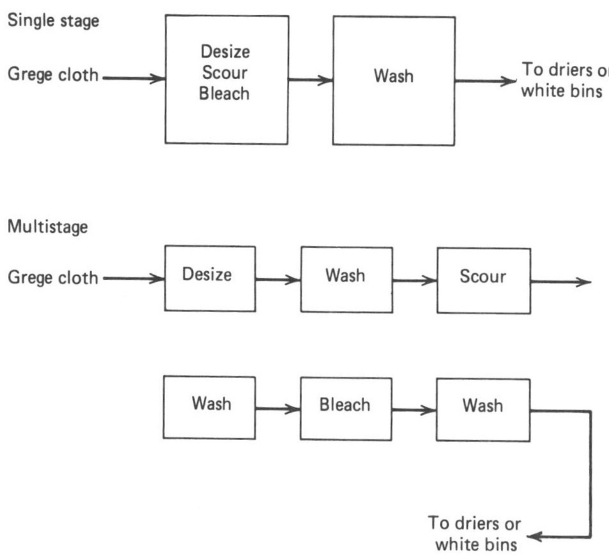

FIG. 8.1. Desize flowchart.

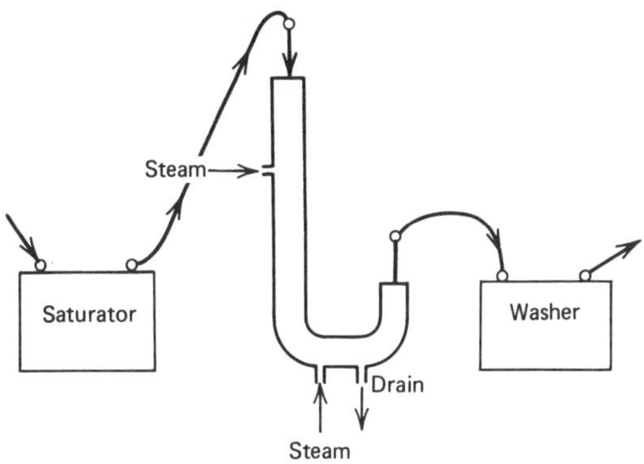

FIG. 8.2. Single-stage J box. Courtesy of FMC Corporation.

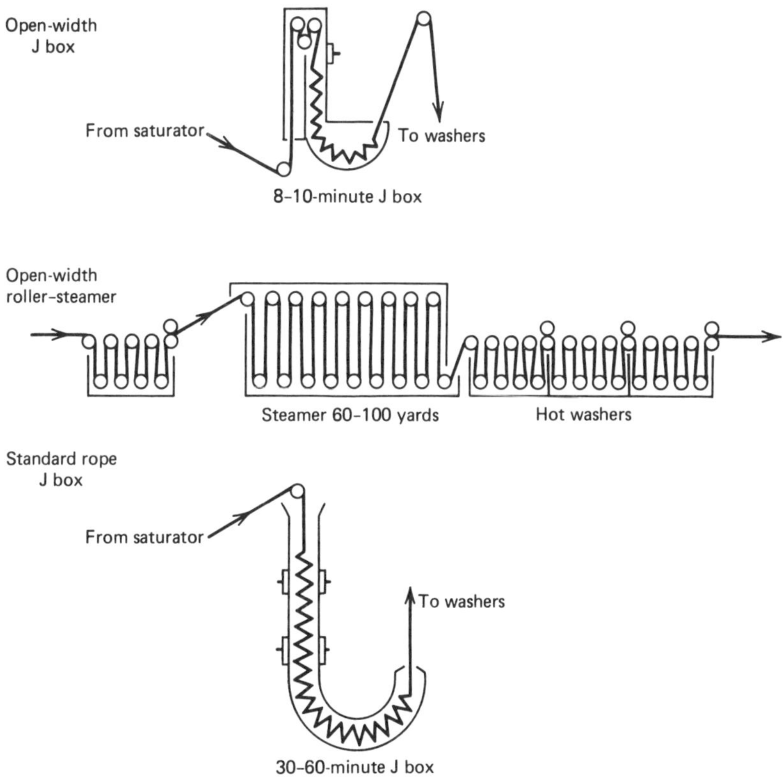

FIG. 8.3. Multistage J box. Courtesy of E. I. DuPont.

polyvinyl alcohol. The invention and use of the newer synthetic polymers, which we discussed in Chapter 1, have also provided an opportunity for the development of newer specialty chemicals. The use of these new compounds has been necessitated in part because of the problems created from the use of polymers that are not fully hydrolized. If fabrics that have been sized with nonhydrolyzed sizes are subjected to severe heat treatment, such as heat setting in the greige, the size can become set into the fabric and become almost impossible to remove.

Enzyme Desizing Procedure

The following is a typical enzyme desize procedure.

Per 1000-gal Mix

1. Saturate cloth with solution of enzyme at a concentration of 2.5–10 gal per 1000 gal of make-up bath, depending on bacterial content of enzyme solution.

2. The pH should be between 4.5 and 7.5, depending on which enzyme is used (see Table 8.1).

3. The temperature range should be 105–212°F, depending on the heat stability of enzyme used. As desizing is a function of time and temperature, a higher temperature will allow a shorter desize time (see Table 8.2).

4. The addition of calcium and/or sodium chloride will aid in the stability of the enzyme and increase efficiency.

5. Following saturation of cloth with enzyme, a digestion period will be required for the desizing solution to solubilize the starch components. The

TABLE 8.1. pH and Temperature Ranges of Different Amylases[a]

Enzyme	pH Range	Temperature Range	Effect of NaCl	Effect of Ca^{++}
Malt α-amylase	4.5–5.5	130–150°F	−	+
Malt β-amylase	4.5–5.5	105–120°F	−	−
Pancreatic amylase	6.8–7.5	105–130°F	+	+
Bacterial amylase	6.5–7.5	150–212°F	+	+

[a]Courtesy of Graham Fisher, G. B. Fermentation Industries.

TABLE 8.2. Relative Desizing Efficiency of Certain Desizing Enzymes at Various Temperatures[a]

Temperature, °F	110	120	130	140	150	160	170	180
Relative desizing efficiency, %	18.6	23.0	25.6	48.8	72.0	81.4	100	100

[a]Courtesy of Graham Fisher, G. B. Fermentation Industries.

length of time for digestion will vary with the concentration of enzyme used, the temperature of the desizing bath, and the types of goods being desized.

6. Following the digestion period, a hot-water wash, with or without a subsequent alkaline scour, is required for complete removal for the size that has been solubilized by the enzyme.

7. Particular care should be given to the surfactant used, as incompatibility may inactivate the enzyme. In some instances it is desirable to incorporate a solvent or a surfactant in with the enzyme to ensure thorough penetration and complete removal of wax from the warp size. Warps with high wax content, either from sizing or from naturally occurring conditions, will not wet out evenly and uneven desizing will take place. An uneven desize operation will lead to a more difficult bleaching step and in the case of goods not bleached, will produce blotchy dyeings.

8. Enzyme desizing can occur in a wet steam chamber at 200–212°F with a dwell time of 30 seconds. Washing should be carried out at 175°F or hotter. A second procedure would be to pad desizing enzyme, hold in bins (covering to prevent drying out) with a dwell time of 2–12 hours, and wash as in the preceding procedure. A third procedure would be the padding with enzyme, running continuously in a J box at 170–180°F with a dwell time of 10–20 minutes and then washing.

Figures 8.4 and 8.5 show the effect of pH and temperature on enzyme activity.

Persulfate Desize Procedure

It has been known for some time that persulfates could be used for the desizing of fabrics sized with starch and polyvinyl alcohol and have the added advantage of being effective in highly alkaline mediums. Therefore, they eliminate the need for a separate desizing operation.[124] Potassium perphosphate

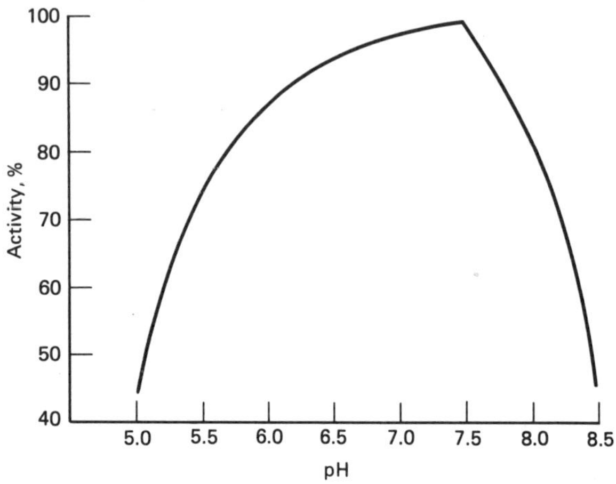

FIG. 8.4. Effect of pH on enzyme activity. Courtesy of Graham Fisher, G. B. Fermentation Ind.

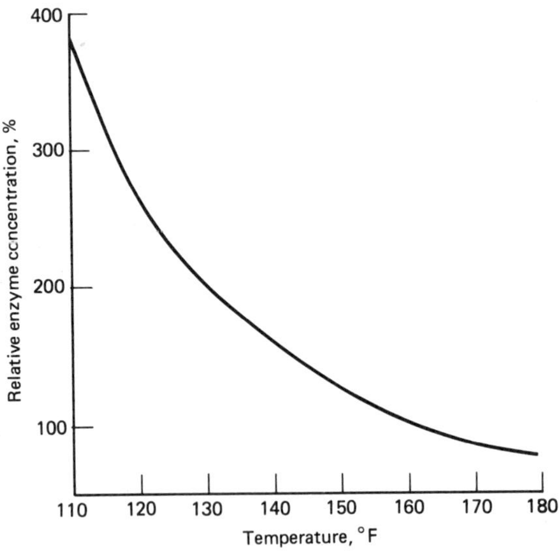

FIG. 8.5. Effect of temperature on enzyme activity. Courtesy of Graham Fisher, G. B. Fermentation Ind.

has also been reported to desize starch and polyvinyl alcohol when added to caustic solutions.[125] As potassium perphosphate requires a higher temperature for activation than does persulfate, it is used more frequently in the caustic saturator because of its stability at elevated temperatures. In addition to degrading all the commonly used sizing agents, perphosphates exhibit a synergistic effect on the caustic treatment of the fabric as evidenced by the lower solvent extractibles and the cleaner bottom.

Perphosphate Desize Scour Procedure[126]

The steps in this procedure are as follows:

1. Saturate the greige cloth (preferably pre-wet out) with a solution adjusted to provide a wet pickup of about 0.5% potassium perphosphate (OWM) in addition to the normal levels of caustic soda and wetting agent.
   a. The temperature of the saturating solution may be heated to as high as 160°F.
   b. A potassium perphosphate solution can be fed to the saturator separately or together with the caustic soda.
2. Heat thc saturated cloth is heated with steam to 190°F or higher for the required retention period. Steaming can be carried out in any of the conventional equipment, such as J-boxes or conveyor steamers.
3. Wash well with hot water (preferably above 160°F).
4. Proceed with bleaching operation.

Advantages

The following advantages are provided by the potassium perphosphate desize scour process:

1. Fabrics sized with PVA, starch, CMC, or any combination thereof can all be run with the same potassium perphosphate–caustic desize–scour procedure. This flexibility eliminates the necessity of checking each fabric lot to identify the size composition and changing process accordingly.

2. Potassium perphosphate is the *only* solid peroxygen compound that is both extremely water soluble and possesses a high active oxygen stability. Neutral or alkaline solutions retain almost 100% of their active oxygen content on standing for several months at room temperature or up to 6 hours near the boiling point.
3. The simultaneous desize scour treatment can result in:
   a. Fewer operations.
   b. Less production time.
   c. Less handling of the fabric.
   d. Less floor space required.
   e. Savings in equipment, steam, water, electricity, and maintenance.
4. The potassium perphosphate process will result in an equal or, in most cases, a superior *after-bleach* preparation as measured by absorbency, residual noncotton content (i.e., size, fats, oils, waxes), and reflectance. This can mean better dyeing, less rehandling of cloth, and reduced seconds.

Review of Areas to Watch During Desize Operation

1. *Select temperature* to prevent degradation of enzyme; the addition of common salt or glauber salt may help.
2. *Choose the surfactant* either nonionic, carboxymethylate, or thioether.
3. *Adjust time* to ensure complete removal of starch and/or polymer.
4. *Use solvent* if goods are heavily waxed, but be sure of compatibility with enzyme.
5. *Watch steam* to ensure even heating and even treatment of goods.
6. *Prevent contact* of goods with concentrated chemicals or alkalis.
7. *Use common sense* and adjust degree of desize to match size. Be sure that adequate starch conversion or polymer removal has occurred, or give additional preparation scour.

Surfactants Used in Desize Operations

Surfactants recommended as being typical of those used in desize operations include (1) carboxymethylates, (2) nonyl phenol (EO), (3) linear alcohol (EO), (4) secondary alcohol (EO), and (5) EO–PO copolymers.

SCOUR FORMULATIONS

Immediately following the desize procedure, or in the case of a combination one-bath scour and desize procedure, yarns and fabrics are washed with detergent and alkalis for the reasons discussed earlier in this chapter. Now that we have considered the generic chemistry of these compounds, let us look at several typical formulations as they might be encountered by the mills.

Nonsolvent Formulations

Intermediate	Percentage			
Water	83	81	57	60
IPA	—	—	2	5
Hexylene glycol	—	—	3	3
EDTA	2	2	3	2
Nonyl phenol (EO 10)	5	—	10	20
LDEA	5	7	10	5
DDBSA	5	10	15	5
Water	80	80	70	60
IPA	5	5	5	10
Soap	—	—	15	10
Nonyl phenol (EO 10)	6	5	—	—
LDEA	3	—	—	10
Sulfosuccinate	6	10	10	10
Water	55	55	55	75
Butyl Cellosolve®	10	10	10	5
Hexylene glycol	5	5	5	5
Phosphate ester NP-10	10	10	10	10
Nonyl phenol (EO 10)	10	—	5	—
LDEA	10	10	5	5
Coconut soap	—	10	10	—

Intermediate	Percentage			
EO alcohol	15	—	10	—
IPA	5	5	5	5
Water	60	60	60	60
Butyl Cellosolve®	5	5	5	—
LDEA	15	—	10	10
Thioether–EO	—	15	10	10
Blocked copolymer	—	15	—	15

Water	42
Flaked caustic	4
Hexylene glycol	2
Coconut oil soap	10
Sulfonated red oil	14
Tall oil (adjust to free-alkaline side	10
Potassium silicate	18

Water	30
Rosin soap	30
Sulfonated red oil	13
Nonyl phenol (EO 10)	15
Sodium hydroxide to free-alkaline side	
Potassium silicate	12

Nekal NS®	10
PVP K-45®	2
Red oil soap	10
Water	68

NP 10 phosphate	10
Flaked potash	2
PVP K-45®	3
Lomar PW®	5
Water	80

Solvent Formulations

Intermediate	Percentage
Orthodichlorobenzene	30
IPA, 10%	10
Sulfonated red oil	10
Sulfonated propyl oleate	10
Sulfonated castor oil	15
Water	25
Tea (to clear)	QS
Odorless mineral spirits	22
Orthodichlorobenzene	14
NP 10 phosphate	5
DDBSA	4
Butyl Cellosolve®	5
TEA	2
Water	50
K-SIL®	3
Odorless mineral spirits	80
Isopropyl amine–DDBSA	10
Tall oil (EO 20)	8
IPA	2
Deodorized kerosene	6
Odorless mineral spirits	6
IPA/DDBSA	24
IPA	4
Water	60
Pine oil	35.0
NP 10	6.0
Tall oil	3.0
IPA–DDBSA	6.0
Flaked potassium hydroxide	1.5

Intermediate	Percentage
EDTA	0.2
Water	48.3
Xylene	80.0
Thioether–EO	10.0
NP 10	10.0
1,2,4-Trimethylbenzene	70.0
TWIX NA-36®	20.0
Blocked copolymer	10.0

Note: Butyl Cellosolve, registered trademark of Union Carbide; Nekal NS, registered trademark of GAF; PVP K-45, registered trademark of GAF; Lomar PW, registered trademark of Diamond Shamrock; K-SIL, registered trademark of Philadelphia Quartz; TWIX NA-36, registered trademark of Whitestone Chemical.

MERCERIZING

There is one additional process to which cellulose or cellulose blends may be subjected to. This is an extremely alkaline immersion called *mercerizing.* It is sometimes performed prior to preparation, or it may be performed as the final step of preparation. Unfortunately, for the specialty chemical companies, few, if any, auxiliaries are used in this process. Prior to the advent of caustic recovery and reuse, surfactants were sometimes used as penetrants. However, this is rarely done except in the case of goods that have been heavily oiled or waxed. Chemicals that were commonly used were cresylic acid derivatives coupled with select solvents such as triethyl and tributyl phosphate. In the very early stages of synthetic surfactants, compounds known as *aspartates* were common; these are covered under U.S. Patents 2,438,092 and 2,252,401 to American Cyanamid.

At the present time, caustic mercerization is being replaced in some instances by subjugating the fabrics to liquid ammonia in the vapor phase. United States Patent 3,824,076 discusses a dyeing system, as does U.S. Patent 3,666,398. An additional patent reference on ammonia cellulose reactions is U.S. Patent 4,199,319.

CHAPTER NINE

THE BLEACHING PROCESS

Once the yarn or fabric has been made clean and uniform, preparation is considered to be complete. From here, yarn or fabric may go directly to bleach or dye. Unless goods are to be dyed in pastel shades, or are to be finished as white goods, Bleaching is not performed. In the case of pastel dyed goods, sometimes a process known as "half-bleaching" is performed. This, in reality, is merely a mild bleach.

HISTORY OF BLEACHING

The final step in preparing goods prior to dyeing is called "bleaching." Over the past few years, there have been many new machines designed and installed in the modern bleach house. Chemicals, however, have remained basically the same as in the beginning.

Archeological evidence would seem to indicate that chemical bleaching of cloth goes back to well before the time of Christ. Greeks and Romans soaked their cloth in alkaline seaweed–ash solutions and then soaked it in soured milk to remove calcium salts and neutralize the cloth. This scouring method remained predominant until the mid-eighteenth century. Cloth was then laid in the sun for days at a time to be bleached. Romans are also given credit for using the fumes of burning sulfur to bleach wool. However primitive bleaching may have seemed, it was a chemical process in which light served as an oxidation catalyst.[127]

By the late eighteenth century the demand for bleached cloth had caused some changes in the traditional methods. By then, lime and soda ash solutions were used in the scouring step, and dilute sulfuric acid replaced sour milk in the scouring step. Bleaching, however, still took months of time and acres of land to complete. In 1790 Bertholet developed the first commercial bleaching process using water solutions of chlorine. However, he found that this process weakened the cloth excessively. He later used caustic potash solutions and formed potassium hypochlorite. When used as a bleaching compound, potassium hypochlorite eliminated the need for long exposure to the sun. Tennant, in 1798, made further improvements by developing a solid bleaching powder that was the calcium salt called *calcium hypochlorite.*[128]

Chlorine-based materials, such as sodium hypochlorite, calcium hypochlorite, and sodium chlorite, predominated until the late 1920s, when hydrogen peroxide came into wider use with the cheaper electrolytic synthesis available. By 1940, peroxide was bleaching 65% of all cottons. Today, 90% of all cotton is bleached with hydrogen peroxide, and 85–90% of this is done continuously.[129] Bleaching is usually done only to natural fibers and/or blends of natural fibers. Fibers commonly bleached are cotton, flax, and wool. Occasionally, acrylics, polyamides, and other synthetics may be bleached as a special cleaning procedure as a result of improper size removal, off-specification fiber processing, long storage periods, or the contamination by mill oil and dirt prior to dyeing.

Chemical bleaching of natural fibers is aided by the addition of *optical brighteners.* Optical brighteners are compounds that have the ability to absorb light at one wavelength and reflect it in the blue–white band of the visible spectrum. They are discussed in some detail at the end of the section on bleaching.

STRUCTURE AND COMPOSITION OF COTTON

It is necessary for natural fibers such as cellulose to be bleached because they contain naturally occurring protein, pectins, waxes, and ash content that must be removed for satisfactory dyeing. As a natural fiber, cotton is influenced by climate and soil growing differences that affect the fiber length, luster, softness, and color of the fibers. Drought or frost can cause various degrees of yellowness, and tips of leaves or stalks coming in contact with the moist boll after opening will cause dark spots and discoloration. Discoloration can also come from dirt, dust, insects, or from harvesting or processing equipment in the form of oils and greases.

Chemical analysis of the cotton fiber has shown the following probable composition:

Turner, 1949[130]	Cellulose	82.7%	
	Pectin and hemicellulose	5.7 %	
	Wax	0.6%	
	Water	10.0%	
	Water insolubles	1.0%	
McCall and Jurgens, 1951[131]		Mature Cotton	Immature Cotton
	Cellulose	96.41%	92.44%
	Protein	1.00%	2.00%
	Wax	0.45%	1.14%
	Ash	0.79%	1.32%
	Undetermined	1.35%	3.10%
		American Cotton	Egyptian Cotton
	Cellulose	91.00%	90.80%
	Encrusted substance	0.53%	0.68%
	Wax	0.35%	0.42%
	Water	8.00%	7.85%
	Ash	0.12%	0.25%

When considering these various chemical analyses, one must note that differences exist not only in methods of classification, but also in results. Values obtained for cottons of different origin and maturity can often be different even when analyzed by the same method and the same system. The nature of the analytical method and the means by which it is carried out also play an important role. This becomes particularly clear when one observes the stated

TABLE 9.1. Composition of Cotton Wax[133]

Wax ester	22%
Phytosterols	12–14%
Polyterpenes	1–4%
Hydrocarbons	7–8%
Free-wax alcohols	42–46%

values for the wax content of cotton. Soxhlet extraction is the usual procedure for separation of the wax from the cotton, with percentage composition being obtained by gravimetric means after the evaporation of the solvent. Different values are sometimes encountered for the same source of cotton when different solvents are used. Therefore, any figures given should be carefully referenced. Solvents commonly used are ethanol, ether, benzene, benzene–methylene chloride, chloroform, carbon tetrachloride, and acetone.

Studies carried out by Hartmann would seem to indicate that the length of time an extraction is run has an influence on the results obtained.[132]

Cotton wax extracted by the above process has been analyzed by Zerbe, and the results of his analysis are reported in Table 9.1.

Although early researchers achieved considerable success in their attempts to separate the individual fractions of cotton wax, detection of the individual components that are present in extremely low concentrations was not possible until the invention of modern analytical techniques such as gas chromatography, mass spectrometry, and X-ray defraction. X-ray research by Hartmann confirmed that a waxy substance is present within the crystalline cellulose chain. Cuoxam was used to dissolve the cotton fiber, the wax portion was separated through the use of organic solvents, and greige cotton was found to contain approximately 2% wax. By continuing the extraction process, an additional 1% wax was discovered, even though normal extraction processes had suggested that this batch of cotton contained only 0.6–0.7% wax. As newer techniques are developed, additional findings concerning the composition of cotton constituents will come to light.

As government regulations and control of the textile industry become more commonplace, additional research into the isolation of individual components of cotton and cotton dust will be necessary if we are to determine the causitive factors of byssinosis, or "brown-lung disease."

BLEACHING OF COTTON AND COTTON BLENDS

Most cotton bleaching done today involves a two-stage process consisting of an alkaline pretreatment followed by a peroxide, chlorine, chlorite bleaching stage. Through the absorption of alkali, the noncrystalline part of the cotton fiber is swelled while the wax esters are saponified and certain parts of the insoluble calcium pectinate are converted into a soluble sodium pectinate by ion exchange. Degradation of naturally occuring pigments accounts for the brownish color that the bleaching liquor develops. Inasmuch as the saponified waxes produce soaps during degradation, they, too, can have an emulsifying effect on the impurities in the bleach liquor. This explains the foam encountered in a caustic saturator, even when no additional detergent has been added.

The change from 100% cotton fabrics to blends of synthetic fibers with cotton has resulted in the increased use of polyvinyl alcohol in warp sizing to improve the efficiency of weaving. In a commission bleachery, a typical fabric mix will probably contain sizings composed of starch, polyvinyl alcohol, CMC, or blends of these. Inadequate preparation and bleaching, accompanied by nonuniform removal of these sizings, will create havoc during the dyeing and finishing operation. Although enzymes are effective in solubilizing naturally occurring starch sizes, they are ineffective in the removal of polyvinyl alcohol or the other synthetic polymers we have discussed. Therefore, one of the major purposes of bleaching is the removal of these polymers and the formation of an absorbent, clean, uniform bottom for dyeing and finishing.

Table 9.2 shows in outline form some of the compounds and specialty chemicals commonly used in typical bleach formulations.

Advantages and Disadvantages of Major Bleaching Agents

Peroxide

Advantages	Nonyellowing; odorless; safe to handle; safe for fibers; nontoxic; can be used on wool
Disadvantages	Can cause white burns on skin; pinholes from iron; can support combustion on storage

TABLE 9.2. Chemicals Used in Typical Bleach Formulations

1. *Bleaching Agents*

a. *Oxidative*	b. *Reductive*
Peroxide (hydrogen, urea, metal salts)	Sulfur dioxide
Hypochlorite	Sulfurous acid
Chlorite	Hydrosulfite
Peracetic acid	Thiourea dioxide
Persulfate	Formaldehyde sulfoxylate
Perphosphate	

2. *Buffering Agents*

a. *Alkali*	b. *Acid*
Silicates	Oxylates
Phosphonates	Sodium nitrate
Metal salts	Metal salts
Sodium hydroxide	
Ammonium hydroxide	

3. *Surfactants*

Anionic	Amphoteric
Nonionic	Blends
Cationic	

4. *Defoamers*

Silicone
Organic
Organosilicone

5. *Antiredeposition Agents*

PVP
CMC
HEC

6. *Sequestrants* (*Chelates*)

EDTA
DTPA
HEDTA

7. *Optical Whiteners*

As recommended per fiber

Peracetic Acid

Advantages	Does not tender polyamides; extremely efficient
Disadvantages	Odor; corrosive; expensive

Sodium Chlorite

Advantages	Does not tender cellulose; extremely potent; relatively inexpensive; does not cause pinholes; can be run on acid side
Disadvantages	Odor; cannot be used on protein fibers; corrosive; produces toxic fumes if not properly used

Sodium Hypochlorite

Advantages	Inexpensive; efficient; easy to handle liquid
Disadvantages	Corrosive; harmful to skin; cannot be used on protein fibers; loses efficiency on storage; produces toxic fumes on acid side

Mechanism of the Bleaching Reaction

Sodium hypochlorite	$Cl_2 + 2NaOH + H_2O \rightarrow 2NaCl + (O) + H_2O$
Calcium hypochlorite	$2Cl_2 + 2Ca(OH_2) \rightarrow 2CaCl_2 + H_2O + (O)$
Hydrogen peroxide	$H_2O_2 + NaOH \rightarrow HO_2^- + H^+$
Peracetic acid	$CH_3COOH \rightarrow CH_3COOH + (O)$
Persulfate	$H_2O_2 + S_2O_8 \rightarrow O_2 + H_2O + OH^-$
Acid sodium chlorite	$H^+ + NaClO_2 \rightarrow HClO_2 + Na \quad (ClO_2)$
Alkaline sodium chlorite	$2NaClO_2 + NaOCl + H_2O \rightarrow 2ClO_2 + NaOH + NaCl$

Now that we have looked at the chemistry of the bleaching reaction and have discussed the basic advantages and disadvantages of the most common systems, let us study each one of them in closer detail. Included in this discussion are actual mill formulations by fiber type that either are being used or have been used commercially by a production bleach house. With the return of natural fiber and natural fiber blends, some of these formulations may offer a good beginning to a technician unfamiliar with the bleaching of natural fibers.

MAJOR BLEACHING SYSTEMS

Sodium Hypochlorite Bleaching

Sodium hypochlorite bleaching is one of the oldest known methods used by the textile industry, dating back to the mid 1700s. The bleaching mechanism consists of the following reaction:

$$2NaOH + Cl_2 + H_2O = NaOCl + NaCl + 2H_2O$$
$$NaOCl = NaCl + (O)$$

The mechanism of these reactions and the history of hypochlorite bleaching is discussed in great detail in the September 26, 1966, issue of the *American Dyestuff Reporter.* Those interested should refer back to this article.[134]

Advantages of Using Sodium Hypochlorite

1. The most obvious advantage is cost. Hypochlorite bleaching is extremely inexpensive as most bleacheries prepare their own.
2. Goods treated with NaOCl need a less severe alkaline prescour.
3. A pH of 9.5–11, which is necessary for hypochlorite stability, is an efficient pH for scouring and bleaching. At this pH, minimum fiber degradation occurs.
4. A bleaching temperature of 55–70°F with NaOCl shows minimum degradation of cotton fiber; this is an excellent advantage.
5. Since a rise of 14°F doubles the activity of the NaOCl, a drop to a lower strength bleach and a higher operating temperature can be utilized, resulting in not only a cost savings, but a shorter bleaching operation.
6. The acid scour, which is used for the removal of metal contaminants such as calcium or iron, can also act as an antichlor agent.
7. It is stated that fabrics pretreated with hypochlorite appear to be more easily bleached by alkaline peroxide than those not pretreated. (The theory is that the products formed by the reaction of the hypochlorite and

nitrogenous proteins are more easily oxidized than untreated materials; i.e., a cheaper peroxide bleaching system can be devised.) Since peroxide is also an effective antichlor, this has created interest in one-shot scour and bleach formulations by use of both hypochlorite and peroxide.

8. Bleach users have found that the pink color obtained when fibers or elastic material have an aniline content can be overcome by a NaOCl bleach followed by peroxide.

This brings us to a discussion on why hypochlorite can be effectively used as an auxiliary in the preparation of cotton. Although it is generally accepted that an improvement in whiteness is one of the basic requirements in bleaching, it is by no means the sole objective. For example, extremely high whiteness is not worthwhile if it is obtained at the expense of degrading the material being bleached. Decreasing absorbency, increasing ash content, or creating the need for a highly purified wash system could make a bleaching agent unacceptable. Sodium hypochlorite creates none of the above problems.

Evaluation of Effective Bleaching

1. *Absorbency.* No adequate method has been found for a realistic evaluation of absorbency.
2. *Fluidity.* Tested by ASTM Standard, 1952, Part 5, pp. 187–194.
3. *Whiteness.* Hunter Color Difference Meter.
4. *Analysis of Impurities*
   a. Infrared or gas analysis of solvent extraction residues, the total of which should be 2–3% or less.
   b. An ash content of 0.20–0.25%.

Three types of sodium hypochlorite bleaching are commonly used: continuous chemic bleach; activated bleach; and straight chemic bleach.

A. *Continuous Chemic Bleach.* Essential fundamentals of the process include:

1. Purification of the cloth to a high degree prior to bleaching.
2. Scouring after NaOH and before initial NaOCl treatment.

3. Application of NaOCl at optimum available chlorine content and pH value in two or three successive steps.
4. Control of conditions as carefully as in peroxide bleach.

Procedure

The method now includes the following steps:

1. Singeing.
2. Desizing with either NaOH or enzymes.
3. Washing with water at room temperatures.
4. Scouring with 0.5% of H_2SO_4 at room temperature for 10–15 minutes (omitted in the case of cotton of high natural purity).
5. Extracting with 4% NaOH either by heating with steam in a J-box continuously for 0.5–3 hours or by kiering batchwise.
6. Washing with water first at 130°F and then at room temperature.
7. Scouring with 0.5–1.0% H_2SO_4 or HCl at room temperature.
8. Washing with water at an even temperature; for example, maximum summer temperature.
9. Bleaching by saturation with NaOCl at 2.5–5.0 gm/l available chlorine at pH 9.5 and at an even temperature (e.g., maximum summer temperature) and holding for 15–30 minutes. Specific bleaching limits should be established by trial to meet bleachery requirements and then carefully controlled, preferably by instrumentation.
10. Repeating operation 1 (above) once or twice if required, depending on the cloth quality desired.
11. Washing with water at room temperature.
12. Antichlor with SO_2.
13. Washing with water at room temperature.

Continuous Hot Chemic Bleaching

Wet out cloth with 0.75 gm/l of nonionic and 0.25 gm/l of 15% chemic. Run at 180°F for 5–7 seconds. (Cold chemic may be used in a saturator if the cloth is then run through a steam tube that heats it up to 180°F). Chemic with a pH of 10.5 or above is quite stable at high temperatures. If the time of treatment of scoured cotton with a chemic liquor or given composition is re-

TABLE 9.3. Effect of Continuous Chemic Bleach Concentration and Temperature on Fluidity and Brightness

Grams/Liter	Temperature, °F	Temperature, J-Box	Brightness	Fluidity
1.0	100	100	80.2	1.92
2.0	100	100	80.9	2.57
3.0	100	100	81.4	3.23
1.0	110	110	81.7	1.86
2.0	110	110	82.7	2.69
3.0	110	110	82.5	3.53
1.0	120	120	81.9	2.66
2.0	120	120	83.1	4.07
3.0	120	120	83.5	5.14
1.0	130	130	84.5	3.25
2.0	130	130	84.1	5.50
3.0	130	130	85.2	6.51
1.0	160	160	84.9	5.15
2.0	160	160	84.9	8.16
3.0	160	160	86.0	5.76
Unbleached fabric			66.0	1.43

duced by a factor of 2.7 for each 10°C rise in temperature, the consumption of oxygen and the extent of chemical modification of the cellulose will be very nearly constant. The rate of bleaching with chemic at pH of 11 increases by a factor of 2.3 for each 10°C rise in temperature. Rate of cloth oxidation increases 2.7 for each 10°C rise in temperature. Table 9.3 shows the effect of continuous chemic bleach concentration and temperature on fluidity and brightness.

B. *Activated Bleach.* The cloth is singed, desized, and purified with 4% caustic in the normal manner. Good cloth preparation is essential. Following caustic purification and washing and before hydrogen peroxide bleaching, the cloth is activated or prebleached with chemic.

Procedure

In a J-box, saturate material for 1 minute at room temperature with chemic at 1–4 g/liter so that 0.1% (OWG) is retained on the cloth. Follow with a 10–30-minute J-boxing at room temperature. Without washing, it is bleached in the hydrogen peroxide solution by use of reduced amounts of hydrogen peroxide and additives. (Caustic is cut down as a result of the caustic brought over in the chemic.) The hydrogen peroxide acts as an antichlor on the residual chemic. The resultant oxygen released is used in bleaching to no hydrogen peroxide is wasted:

$$NaOCl + H_2O_2 \rightarrow NaCl + H_2O + (O)$$

The hydrogen peroxide consumption is reduced because the chemic accomplishes some of the bleaching. (If soda ash is used to make the chemic, deposition of silica in the cloth will occur, increasing the washing requirements for removal of residual alkali.)

Procedure (In a Kier)

After washing and boiling in the kier, the cloth is run through the chemic saturator and J-box, and then into another kier that contains the hydrogen peroxide bleaching solution. When bleaching takes place in one kier, after a light scour, the cloth is saturated with chemic and recirculated for 15–30 minutes. Hydrogen peroxide is then introduced. A stainless-steel kier is preferred in the above case.

Advantages of Activated Bleaching

1. Reduction in hydrogen peroxide usage and bleach liquor additives.
2. Increased whiteness and whiteness retention by two to five times.
3. Increased absorbency, decreased fiber degradation, and lower ash and metal content.

C. *Straight Chemic Bleach.* Production formulations as practiced by various segments of the textile industry (bleaching procedures) are given below.

1. *Spandex (Vyrene)®/Cotton Bleach*

a. *Prescour:*	Nonionic	3.00%
Scour cold and drop. Rinse cold.		
b. *Bleach:*	Chemic	12–15.00%
Run cold for 30 minutes. Drop. Rinse cold.		

c.	*Antichlor:*	Peroxide, 35%	8.00%
		Sodium silicate	4.00%

Raise temperature to 180°F and run for 20 minutes. Drop. Cold rinse.

d.	*Optical:*	Tinopal 4BMA	1.25%

e. *Finishing:* Extract—tumble partially dry. Board at 15# psi (220°F).

2. *Scouring and Bleaching of Spandex (Lycra)®—Hosiery*

a. *Prescour* (make up bath at 100–120°F)

Pine oil scour	5–8.00%
Built detergent	1–2.00%
Soda ash	1–2.00%
TSPP	1–2.00%
DTPA—sequestrant	1.00%

Raise temperature to 170°F and run for 15 minutes. Drop. No rinse.

b.	*Bleach:*	Chemic (12–15% available chlorine)	12.00%
		Soda ash	0.50%
		TSPP	0.50%

Run cold for 10 minutes. Raise temperature slowly to 115°F. Run for 15 minutes. Drop bleach bath and give rinse for 15 minutes.

c.	*Antichlor:*	Peroxide, 35%	6–8.00%
		TSPP	2.00%
		Built detergent	0.25%
		DTPA—sequestrant	0.25%
		Blancophor® SVT	0.25% (Add at 120°F)

Bring slowly to 170°F for 20 minutes. Give warm rinse and soften.

3. *Bleach on Cotton Socks*

a.	*Scour:*	Pine oil gel	5.00%
		Built detergent	2.00%
		DTPA	1.00%

Heat at 160°F for 15 minutes.

b. *Bleach:* Chemic 20.00%

Run cold for 10 minutes. Heat to 130°F for 15 minutes. Then raise to 140°F for 5 minutes. Drop. No rinse.

Hydrogen peroxide	2.00%
DTPA	0.50%

Heat at 160°F for 15 minutes. Drop and rinse. Add optical and softener at 110–120°F.

4. *Activated Bleaching Formula for Smith Drum Machine (100# Hosiery Unit—280 gal of water)*

a. Lightly scour goods. Dump scouring solution. Without washing, add:

Chemic, 15%	3 gal

Run cold for 10 minutes. Raise temperature to 120°F and run for 10 minutes. Drain and fill with water.

b. *Add:*

Peroxide, 35%	1 gal
Optical dye	

Heat to 200°F and run for 10 minutes. Two hot rinses and finish.

5. *Borax Bleaching Formula for Smith Drum Machine (Cotton Bleaching)*

a. Take 100 lb dry cotton and add:

Water	220 gal
Borax	1 lb
Nonionic	$\frac{1}{2}$ lb
Chemic, 15%	33 lb

b. Raise temperature to 130°F and maintain for 10 minutes. Don't dump, but add:

Sodium silicate 43°Be	6 lb
Peroxide, 35%	$\frac{1}{2}$ lb
Optical dye	10 oz

Raise the temperature to near boil and maintain for 30 minutes. Drain and rinse at 180°F for 10 minutes and then rinse with softener at 110–120°F.

6. *Tub Bleach (Hose or Piece Goods)—Cotton*

a. Fill tub with water to dye level and add:

Pine oil gel	2.00%
Soda ash	1.00%

Load tub and run for 5 minutes or until wet. Add:

Chlorine	20.00%

Run cold for 5 minutes. Bring to 120°F and run for 15 minutes. Bring to 140°F in 30 minutes. Drop. No rinse.

b. Follow with: Peroxide 2.00%

Run for 10 minutes cold and then add:

Blancophor® SVT or Tinopal® 3BSA	0.50%

Bring to 180–190°F and run for 20 minutes. Drop. Add softener. Run at 120°F for 20 minutes.

7. *One-Bath Bleach for Bedspreads (All Percentages are OWG)*:

Pine oil gel	2.00%
Built detergent	1.00%
Chemic	30.00%

Start cold and raise temperature to 120°F for 20 minutes. Raise temperature to 180°F and add:

Peroxide, 35%	2.50%

Run for 15 minutes at 180°F and add:

Optical	0.50%

Run for 10 minutes. Drop. Wash.

8. *Activated Bleach on Cotton in Package Form (500–550 lb of Goods)*

a. In bath, add:

Solvent scour	7.5 lb
Chemic	20.0 lb

Run at 130°F for 30 minutes. Drop. Refill. No rinse.

b. Set second bath and add:

Silicate	15.0 lb
Peroxide, 50%	2.0 gal
DTPA	2.0 lb
Caustic, flake	5.0 lb

Heat to 195°F and run for 15 minutes. Add:

Tinopal® 3BSA	3.75 lb

Run for additional $1\frac{1}{4}$ hours at 195°F. Drop to 180°F and rinse for 10 minutes. Drop to 160°F and rinse for 10 minutes. Drop and rinse 5 minutes in and 5 minutes out. Then add:

Acetic acid	1.0 pint

Run at 120–130°F for 20 minutes.

9. *Bleaching of Burlap*

Scour lightly in neutral soap and nonionic. Rinse. Saturate with 5 g/l of chemic and, without squeezing, let lie for 20 minutes at room temperature. Rinse and then immerse in a solution containing:

Peroxide, 35%	6.00%
Sodium silicate 42°Be	5.00%

Boil for 2 hours. Rinse and scour with acetic acid or scour with potassium permanganate and neutralized with sodium bisulfite.

10. *Bleaching of Supima Cotton (All Percentages are Dry Weight of Goods)*

a.	In first bath:	Nonionic	0.50%
		Caustic, flake	8.00%

Boil for 1 hour. Drain and rinse.

b.	Add:	Chemic	2.00 g/l

Bring to 140°F in 10 minutes. Hold for 15 minutes.

c.	Add:	Caustic, flake	0.60%
		Sodium silicate 42°Be	0.50%
		TSPP	0.50%
		Peroxide, 35%	5.00%
		Optical dye	

Run for 5 minutes before adding peroxide and optical dye. Run at 180°F for 1 hour. Drain and rinse. (Regular procedure uses 1.00% caustic and 3.00% peroxide).

11. *Activated Bleach for Cotton Tape:*

a. *Scour:*

Caustic, dry	3.00%	9.0 lb
Alkyl–aryl amine salt	0.50%	1.5 lb

After filling the machine with water and chemicals, the tape is loaded into the tub. The temperature is then raised to 200°F and held for 45 minutes. Dump, then rinse for 15 minutes at 160°F.

b. *Bleach:*

Caustic, dry	0.50%	1.5 lb
Alkyl–aryl amine salt	0.50%	1.5 lb
Sodium silicate 42°Be	2.00%	6.0 lb
TSPP	2.00%	6.0 lb
Chemic, 15%	15.00%	45.0 lb
Peroxide, 35%	5.0%	15.0 lb

After refilling machine with cold water, add the above chemicals except for the hydrogen peroxide. Raise temperature to 150°F in 10 minutes and hold for 20 minutes. Then slowly add the peroxide and raise temperature to 200°F for 45 minutes. Dump and refill.

c. *Finish:*

Acetic acid, 56%	0.50%	1.5 lb

Rinse goods twice at 160°F for 10 minutes each. Then refill the machine and add the acetic acid and rinse for 10 minutes at 140°F. *Note:* If an optical brightener is to be used, add 0.25% to the bleach after the peroxide has been added and 0.25% to the third rinse. (Approximately 3 hours of running time.)

12. *Activated Bleach for Package Machine (All Percentages are OWG)*

a. *Scour:*

Wetting agent	1.00%
TSPP	1.00%
Chemic	10.00%

Fill machine with cold water and add chemicals while machine is running, over a 5-minute period. Raise temperature to 150°F in 20 minutes and hold for 15 minutes.

b. *Bleach:*

Peroxide, 35%	4.00%
Caustic, flakc	0.40%
TSPP	1.50%
Wetting agent	0.25%
Optical brightener	0.35%
Chelating agent	0.10% (if needed)

Fill machine with cold water and add above chemicals, except for optical. Add the optical after 5 minutes, one-half to each machine cycle. Raise temperature to 190°F and hold for 30 minutes. Dump.

c. *Finish:*

Sodium bisulfite	0.20%
Optical brightener	0.10%

Rinse the yarn for 10 minutes at 160°F. Dump, refill, and add the sodium bisulfite. Run for 10 minutes at 140°F. Dump. Refill and add the brightener. Run at 120°F for 20 minutes. Dump and unload.

13. *Activated Bleach—Jig (All Percentages are OWG)*

a. *Chlorine Bath:*

Solvent scour	3.00%
TSPP	1.00%
Chemic	15.00%

Run two ends cold and two ends at 140°F. Drop bath, two ends overflowing wash.

b. *Peroxide Bath:*

Peroxide, 35%	4.00%
Caustic, flake	0.50%
TSPP	1.50%
Solvent detergent	3.00%
Optical brightener	(if needed)
DTPA	0.10%

Start cold. Run two ends at 120°F, two ends at 140°F, and two ends at 180°F. Bleach until white. Overflow and neutralize as in regular procedure.

Disadvantages of Sodium Hypochlorite

Does not produce completely satisfactory whites.

Produces slight damage to cellulosic fibers.

Greater damage for synthetic fibers.

Corrosion-resistant equipment is necessary.

Produces unpleasant odors in working environment.

Stabilization is difficult to achieve where pH varies.

Produces harsh hand on fabric.

Harmful to skin in concentrated form.

Cannot be used on natural animal fibers.

Manufacture of Small Quantities of Sodium Hypochlorite Bleach[64]

Equipment Needed (Courtesy of Westvaco, Inc.)

1. Ceramic mix tank and/or fiber glass, approximately 3 ft in diameter by 3 ft deep, equipped with a cooling coil of eight or more turns of 1-in. pipe. (Titanium, although expensive, is probably cheaper in the long run because of the heat transfer and the relative nonreactivity of the metal.)

2. A suitable mixer with a direct drive, having a 3-in. anchor-type propeller.
3. Rubber-lined steel tank, 3 ft in diameter by 3 ft deep, for settling of the crude bleach.
4. One 50-gal-capacity ceramic filter, with small vacuum pump if filtration is desired.
5. One 6-ft by 12-ft by 2-ft steel tank for making up caustic concentration, and pump for transfer.
6. Rack for holding chlorine cylinder in such position that liquid chlorine may be withdrawn.
7. One 12-gal carboy to be used as trap between the vacuum pump and vacuum filter.
8. Chlorine cylinder.

Chemicals Required (for 120 gal of 10% Bleach)

Chlorine	103 lb
Caustic soda, 50%	20 gal
Water	95 gal

Equation for Reaction:

$$Cl_2 + 2NaOH \rightarrow NaOCl + NaCl + H_2O$$

Procedure

1. Prepare ahead of time the total caustic soda and water solution, and allow to cool to 85°F or less in the caustic holding tank. (As reaction of chlorine and caustic will generate heat, the cooler the caustic is prior to reaction, the faster the chlorine may be added later.)

2. Charge chlorine at a rate such that temperature remains below 85°F. Cooling water at 40–50°F is run through coils to absorb heat as generated. As much cooling is required per batch as heat generated is equivalent to that required to melt 200 lb of ice.

3. Continue reaction until caustic solution reaches neutral point. (Caustic is consumed as bleach is manufactured and would eventually be totally taken up and decomposed unless bleach solution remains on the alkaline side. Most bleach solutions contain a 1% excess of caustic as a buffer.)

4. When batch has reached neutral point, an additional 10 gal of caustic solution is added from the caustic make tank.

5. Test for strength of bleach. Take 1 ml of bleach, dilute to 100 ml, and add 10 ml of 10% potassium iodide and 10 ml of $4N H_2SO_4$. Titrate the liberated iodine to a pale straw color with 0.1 N $Na_2S_2O_3$. Add starch solution and complete titration to colorless: 1 ml 0.1 N $Na_2S_2O_3$ = 0.00355 g available Cl_2. If the batch is too strong, dilute with water.

6. Test for alkalinity of bleach. Take 5 ml of bleach, dilute to 100 ml, and add 25 ml of 10-volume hydrogen peroxide. Shake to expel liberated oxygen and add 6 drops of phenolphthalein solution. Titrate with 11 N HCl. Run a blank on hydrogen peroxide solution with 0.1 N NaOH solution and add blank to titration: 1 ml 0.1 N HCl = 0.004 g NaOH.

7. Add 0.25 lb of sugar to the bleach. This prevents the formation of any ions in the solution that would color the bleach brown. Also, add 1.5 lb of filtercel. The batch is then dropped to the settling tank.

8. After settling, the solution is dropped into the ceramic vacuum filter and filtered through Jeybecco No. S3600, a canvas twill manufactured by John Boyle Company. After filtering, the bleach is completed and ready for use.

Hydrogen Peroxide Bleaching

Because of the ease of operation, the minimal effect on synthetic fibers, and the positive effect on the removal of synthetic sizes, hydrogen peroxide has become, in most mills, if not the primary bleaching system, definitely part of the bleaching operation. It has generally replaced the chemic, or sodium hypochlorite, bleach as the primary source of bleaching oxygen. Some mills even use an activated bleach that combines both sodium hypochlorite and hydrogen peroxide in their bleach system. In this case, the hydrogen peroxide acts not only as an antichlor, but also as an activator for the residual chlorine bleaching agent.

The mechanism by which hydrogen peroxide is reported to function is as follows:

$$(H_2O_2 + NaOH \rightleftharpoons NaOOH + H_2O)$$

$$(NaOOH \rightleftharpoons Na^+HO_2^-)$$

TABLE 9.4. Size Removal by Peroxide

		Residual Size, %		
Type of Size	Percentage on Greige	After Desize	After Scour	After Bleach
Polyvinyl alcohol	3–7	0.0–0.5	0.0–0.2	0.0–0.2
Starch	10–14	0.8–1.5	0.6–1.2	0.2–0.8
CMC	3–7	0.0–0.2	0.0–0.2	0.0–0.2

Although preparation for dyeing, printing, or finishing is the prime reason for using peroxide as a bleaching agent, its effect on the removal of synthetic sizes is extremely beneficial. The efficiency of peroxide in size removal is outlined in Table 9.4.[136]

Hydrogen peroxide is a weak acid that is thermodynamically unstable and readily decomposes into water and molecular oxygen. It would appear that a peroxide bleaching bath is stabilized by compounds that either absorb or eliminate the formation of free radicals.[137]

It is impossible to give strict guidelines for the optimum bleach bath because of the nature and quality of the goods to be bleached. The amount of bleaching required, and the available equipment in which bleaching is to take place. However, the following general variables are considered important: (1) time; (2) temperature; (3) concentration of peroxide; (4) alkalinity (which activates peroxide); and (5) stabilization.

Advantages of Hydrogen Peroxide

1. Economical bleaching system.
2. White effect good and permanent.
3. Minimum damage to cellulosic, protein, or synthetic fibers.
4. No danger of equipment corrosion.
5. No unpleasant odors.
6. No limitation as to processing techniques.
7. Elimination of need to segregate greige goods according to type of size.
8. Minimum accumulation of size residue on processing rolls.
9. Instantaneous absorbency achievable.

10. Lower overall cost.
11. No need for organic solvents in the desize solution.
12. Improvement of mote removal of 100% cotton fabrics.
13. Allowance for some fabrics to be dyed dark and medium shades without bleaching.
14. Prevention of lay marks on some scoured only goods.

Disadvantages of Hydrogen Peroxide

1. Limitation in white obtained on acrylic fibers.
2. Stabilizer system important for optimum results.
3. Deleterious effect on skin when used in a concentrated form.
4. A catalytic effect exhibited on synthetic fibers in the presence of iron, which causes pinholing and fiber weakness.

Procedure for Use of Hydrogen Peroxide as a Desize Prior to Bleaching

	Percentage	lb/100 gal
Caustic soda, 50%	4.0	33.6
Hydrogen peroxide, 50%	0.4	3.4
Surfactant	0.2	1.7

1. Fill the tank about half with water.
2. Start stirring and heating to 140°F (60°C).
3. Add caustic soda.
4. Add hydrogen peroxide.
5. Fill to level with water.
6. Add surfactant with minimal stirring; stop stirring.
7. Test the concentration.
8. Hold the solution temperature at 140°F (60°C) in the saturator.

The equipment used in hydrogen peroxide bleaching is shown in Fig. 9.1.

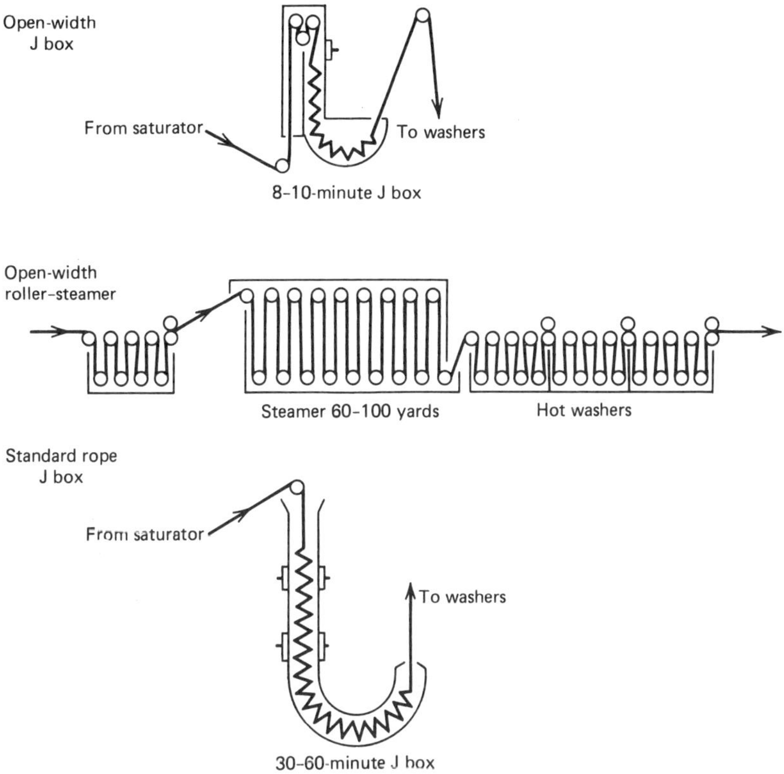

FIG. 9.1 J-box as used in hydrogen peroxide bleaching. Courtesy of E. I. DuPont.

Typical of some of those procedures that either are or have been used commercially by production bleaching departments are as follows:

A. *Triacetate-Cotton—Bleach Range* (Goods to be Dyed)

	lb/100-gal Mix
Hydrogen peroxide	59.0
Sodium silicate 42°Be	24.0
TSPP	8.0
DTPA—sequestrant	3.0
Pine oil detergent	2.0
Water to 100 gal	

B. *Triacetate-Cotton—Bleach White for Pad Jig Process*

Hydrogen peroxide	30.0
Sodium silicate 42°Be	24.0
Soda ash	40.0
Nonionic detergent	2.0
Water to 100 gal	

(Runs 5500 yards)

C. *Triacetate-Cotton—Enzyme Pad Jig Bleach*

1. Pad through: 2 dippers enzyme
   ½ pint gel detergent made up to 100 gal
2. Set jig at 120°F and add: 3 dippers hydrogen peroxide
   1 dipper sodium silicate 42°Be
   1 pint DTPA sequestrant
   ½ pint nonionic detergent
3. For dark shades, run 4 ends at 180–190°F.
4. For light to medium shades run 6 ends at 180–190°F.
5. Drop bath and add ½-pint detergent to new bath.
6. Run two ends at the boil. Give one overflow wash.
7. If goods have nonfugitive tint, add ¾ dipper hydro, ½ dipper sodium tripolyphosphate to detergent–boil step, and increase overflow wash to two ends.
8. Shell out through acetic acid.

D. *Triacetate-Cotton and Polyamide-Cotton—Cold Pad Bleach*

		Per 100-gal Mix
Triacetate	Sodium silicate 42°Be	4 dippers
	Hydrogen peroxide, 35%	6 dippers
	Soda ash	40 lb
	Nonionic detergent	2 oz
Polyamide	Caustic, 50% liquid	5 dippers
	Hydrogen peroxide, 35%	8 dippers
	Sodium silicate 42°Be	5 dippers
	Enzyme	½ pint
	Nonionic detergent	1 pint

1. Slop pad and let lay for 8 hours. (Cover to prevent premature drying of fabric ends.)
2. When ready to boil out, run down on jig or range through regular boil-off.
3. If nonfugitive tint is encountered, add hydrogen and sodium tripolyphosphate to boil off.

E. *Solvent Scour Bleach* (400-gal Bath at 110°F)

Hydrogen peroxide, 35%	125.0 lb
Caustic, 50% liquid	75.0 lb
Sodium silicate 42°Be	40.0 lb
Varsol (mineral spirits)	256.0 lb
Emulsifier	35.0 lb
DTPA—sequestrant	3.0 lb

F. *Acetate Bleach*

Hydrogen peroxide, 35%	1.00%
Sodium tripolyphosphate	1.50%
DTPA—sequestrant	0.30%
Magnesium sulfate	0.10%
Borax	0.50%

1. For acetate, pH should be 10.0–10.5.
2. For triacetate, pH should be 10.5–11.0.
3. For white goods, use trisodium phosphate in place of silicate or total alkali of above formulation.

G. *Wool Bleach*

Hydrogen peroxide, 35%	2.00% OWG
Sodium silicate to a pH of 9.0 (use ammonia for softer hand)	
TSPP or STP	4 lb/100 gal

Run at 100–120°F for 1–6 hours.

H. *Polyamide-Cotton—Range Bleach (200 gal)*

Hydrogen peroxide, 35%	40.00 lb
Sodium silicate 42°Be	24.00 lb
Caustic, 50% liquid	16.00 lb
TSPP	4.00 lb
DTPA—sequestrant	3.00 lb
Pine oil gel	2.00 lb

I. *Polyester-Cotton*

Hydrogen peroxide, 35%	40.00 lb
Caustic, 50% liquid	35.00 lb
Sodium silicate 42°Be	30.00 lb
Nonionic detergent	2.00 lb

The effect of alkali on hydrogen peroxide is represented by the following reaction:

$$2H_2O_2 \xrightarrow[\text{(strong alkali)}]{} 2H_2O + O_2\uparrow$$

$$H_2O_2 \xrightarrow[\text{pH 10.5 (good)}]{} HO_2^-H^+$$

Peracetic Acid Bleaching

Peracetic acid (CH_3COOOH), an acid bleaching agent, is used primarily for polyamide fibers and the stripping of acid and premetallized colors from these fibers. The mechanism of reaction is somewhat analogous to that of hydrogen peroxide. In fact, one of the processes for the manufacture of peracetic acid is the reaction of acetic anhydride and hydrogen peroxide. When used as a bleaching agent or stripping agent, the following reaction occurs (U.S. Patent 2,377,038):

$$CH_3COOOH \rightarrow CH_3COOH + (O)$$

The use of peracetic acid for the bleaching of cotton has been investigated in the literature but has never reached commercial success in the way that sodium hypochlorite, sodium chlorite, or hydrogen peroxide has.[138]

Advantages and Disadvantages of Peracetic Acid Bleaching

Peracetic acid has no major advantages or disadvantages except that it is extremely useful in hard-to-bleach polyamides, but it is expensive and is used only as a last resort. It is corrosive to the skin and has an obnoxious odor. Storage is difficult and requires care.

Typical Peracetic Acid Running Procedures

A. *Polyamide*

1. Prescour goods in normal manner and give usual rinse.
2. Set bath cold in the following order:

0.75% (OWG) Calgon®	Predissolved
0.75% (OWG) Sodium bicarbonate	Predissolved
2.00% (OWG) Flake caustic	Predissolved
5.00% (OWG) Peracetic acid	Prediluted (use separate container; do not mix concentrated acid with concentrated alkali)

3. Run cold for 10–15 minutes and heat slowly to 170°F.
4. Hold at 170°F for 20 minutes.
5. Cool; drop; rinse.
6. If desired, after-scour with 1.00% bisulfite at 140°F.
7. Rinse, and finish as normal.
8. For nylon whites, an optical whitener such as Tinopal® CF may be used with 0.50% acetic acid at 160–190°F.

B. *Acrylic*

1. Prescour goods in normal manner and give usual rinse.
2. Set bath cold in the following order:

1.00% (OWG) Sodium hydroxide, 50%	Predissolved

1.00% (OWG) Sodium bicarbonate	Predissolved
1.00% (OWG) Calgon®	Predissolved
5.00% (OWG) Peracetic acid	Prediluted (use separate container; do not mix concentrated acid with concentrated alkali)

3. Run cold for 10–15 minutes and heat slowly to 170°F.
4. Hold at 170°F for 20 minutes.
5. Cool; drop; rinse.
6. If desired, after-scour with 1.00% bisulfite at 140°F.
7. Rinse, and finish as normal.
8. For acrylic whites, an optical whitener such as Calco® White RW may be used.

Sodium Chlorite Bleaching

Sodium chlorite ($NaClO_2$) was discovered in the middle of the nineteenth century as one of the components of stored bleaching powder. It was first prepared synthetically in 1939 by the Mathieson Alkali Works, now Olin-Mathieson Chemical Corporation.

Sodium chlorite utilizes the unique capability to bleach cotton and synthetic fibers to high whiteness without adversely affecting the physical properties of the original fiber. Sodium chlorite, however, cannot be used with protein fibers such as silk or wool. Sodium chlorite is also an extremely effective means to strip dye from both natural and synthetic fibers. Misuse or misapplication of sodium chlorite in the bleaching operation can cause excessive chlorine dioxide gas to be released. This should be kept in mind and adequate ventilation should be provided to prevent over exposure of personnel in the event of an accidental release of concentrated chlorine dioxide. The TLV of this compound is 0.1 ppm and shows the necessity for adequate care and ventilation during usage.[139]

The principal method of activating the bleaching power of sodium chlorite is with acid. The actual bleaching agent is chlorine dioxide (ClO_2). Reactions that occur during bleaching are:

$$H^+ + NaClO_2 \rightarrow HClO_2 + Na^+$$
$$5HClO_2 \rightarrow 4ClO_2 + HCl + 2H_2O$$

The conditions for acid sodium chlorite bleaching may vary as follows: sodium chlorite concentration 0.2–7.5 g/liter; pH, 2.5–4.0; temperature, 172–212°F. Under these conditions, chlorine dioxide, a gas, is evolved from sodium chlorite solutions. Escape of ClO_2 from the system represents a loss of bleaching power and may present an odor problem. It is desirable, therefore, to minimize loss of ClO_2. Some metals, such as iron and nickel, catalyze the release of chlorine dioxide and appreciably affect the rate, even if present in only a few tenths of a part per million.[140]

The rate of chlorine dioxide formation at any pH and temperature is proportional to the concentration of sodium chlorite in the solution. In the absence of metals and cloth, the rate approximately doubles for each 0.4-pH drop, in the 3.0–5.0 range, at 185°F. At 158°F the rate doubles for each 0.55-pH drop. On the basis of the foregoing, it is possible to make certain recommendations for batch-bleaching operations to aid in reducing the release of chlorine dioxide to the atmosphere: (1) use only the quantity of sodium chlorite necessary for the actual bleaching; (2) prepare sodium chlorite solutions at ambient or slightly higher temperature—do not increase the temperature until the goods are circulating; and (3) add the acid in steps, if possible. This is especially important with pH values below 3.0.[141]

Since most bleaching is conducted in metallic equipment, corrosion inhibitors, particularly sodium nitrate, can be used. Inhibitors protect the goods by minimizing formation of corrosion products in the bath and increase the useful life of the equipment.[142]

Cotton goods that have been prepared in kiers are advantageously bleached by an alkaline sodium chlorite process when particularly high quality bleached goods are desired. In this process sodium chlorite is activated by hypochlorite, probably according to the following reaction:[143]

$$2NaClO_2 + NaOCl + H_2O \rightarrow 2ClO_2 + NaOH + NaCl$$

The presence of chlorine dioxide protects the goods from the degrading action of hypochlorite.[144]

A third method of activating sodium chlorite may be of interest, although it has not yet found an application in the textile industry. This involves the reaction of sodium chlorite and chlorine. It goes to completion very rapidly:[145]

$$2NaClO_2 + Cl_2 \rightarrow 2ClO_2 + 2NaCl$$

Sodium Chlorite Bleaching—Acid Process[146]

Research and development have made possible the use of sodium chlorite in conventional 316 or 317 SS equipment as well as fiberglass and various other plastics. In addition to cotton, a wide variety of synthetic fabrics and blends are being bleached commercially in conventional rope ranges. Polyester, polyamide, acrylic, and rayon fabrics are bleached to high brightness with no pinholing and with improved physical properties over other bleaching processes.

Sodium Chlorite Bleaching—Alkaline Process[147]

Since this process operates under alkaline conditions, regular chemic bleach equipment is suitable. The goods pass through a saturator or chemic pit to pick up bleach solution having the proper available-chlorine concentration and pH value. As a general rule, the bleach solution contains 1–3 g/liter of available-chlorine mixture of sodium chlorite and sodium hypochlorite and a buffer mixture of bicarbonate of soda and soda ash. An available-chlorine pick up of 0.3%, based on the dry-cloth weight, is typical. After leaving the saturator, the goods remain in J-boxes or bins for the time necessary for bleaching, usually 1–2 hours. At the end of the bleaching cycle, the goods are washed, antichlored, and washed again before passing on to subsequent finishing operations.[148]

Adequate pretreatment is important to make the fabric absorbent and enable the bleaching chemicals to reach the inner fibers. The fabric must also be uniformly low in caustic soda content, since bleaching depends on the correct solution of pH. Variations in pH result in nonuniform bleaching.[149]

The hypochlorite:sodium chlorite ratio and buffer ratio are the major factors in preparing the bleach solution. In actual practice, the hypochlorite:sodium chlorite ratio can vary between 3:2 and 3:1, on an available-chlorine basis, without adversely affecting the result. The larger proportion of sodium chlorite provided by the 3:2 ratio is added protection against degradation of the fiber. On clean, well-prepared, highly absorbent goods, the 3:2 ratio produces bleached cloth of excellent quality. Goods that have not received adequate pretreatment require the 3:1 ratio. The rate of bleaching depends on pH of the bleach solution and increases at lower pH values. The acceptable pH for alkaline sodium chlorite bleaching varies from 8.8 to 9.7. The proper buffer ratio controls pH within a fairly narrow range, usually 0.2

of a pH unit. A 1:1 ratio of bicarbonate of soda to soda ash provides a pH range of approximately 9.4–9.7. Increasing the bicarbonate lowers the pH. The composition of the liquid carried into the pit by the goods also affects the buffer ratio. It is important to note that the pH of the bleaching bath should not be allowed to fall below 8.8. At lower pH values, chlorine dioxide is evolved, resulting in decomposition of the bleach bath and loss of bleaching power, as previously mentioned.[150]

Sodium Chlorite Bleaching—Semicontinuous and Batch[151]

Many fabrics, principally synthetics and blends, are processed in semicontinuous or batch equipment. Sodium chlorite bleaches these fabrics to high whiteness while maintaining their original physical properties, regardless of the equipment used. In all cases of bleaching fabrics containing synthetics, the recommendations of the fiber manufacturer should be followed.

Most noncontinuous operations are run in dyebecks, jigs, Burlington machines, and yarn package machines. Concentration of sodium chlorite and pH of the solution depend on the fiber to be bleached and the condition of the cloth. It is necessary to calculate sodium chlorite requirements on the basis of the dry-cloth weight, since liquor:cloth ratio is a factor in determining sodium chlorite concentration.

Advantages of Sodium Chlorite Bleaching[152]

1. Silicate free bleaching system.
2. Abrasive bruise marks on the fabric, which are caused by silicate deposits, are completely avoided.
3. Softer hand.
4. Smooth operation.
5. Versatility.
6. Improved physical properties of fabrics.
7. Pinholing eliminated.

Sodium Chlorite Bleaching Formulas

Typical sodium chlorite bleaching formulas are given in Table 9.5.[153]

TABLE 9.5. Sodium Chlorite Bleaching Formulas

Material	Sodium Chlorite	Acid	pH	Additives, Sodium Nitrate	g/l Detergent
Rayon	0.5	Nitric	3.5–4.0	0.5	0.15
Acetate	1.0	Nitric	3.5–4.0	1.0	0.50
Polyamides	1.0	Acetic	3.0–3.5	1.0	0.20
Acrylics	1.5	Nitric	3.5–4.0	1.5	0.50

Typical Applications for Sodium Chlorite Bleaching[154]

A. *Polyester, Nylon, or Acrylic Fibers (100%)*

1. Liquor ratio 1:10–1:15

Sodium chlorite, 50%	2.00%
Acid stable surfactant	0.50 g/liter
Buffered acid activator, pH 4	1.2–1.50%

The bleaching liquor is heated to 200–205°F and the process is continued at this temperature for 1–2 hours.

2. Liquor ratio 1:15–1:30

Sodium chlorite, 50%	3–5.00%
Acid-stable surfactant	0.50 g/liter
Buffered acid activator, pH 4	1.5–3.00%

Heat up to 200–205°F; continue for 1–2 hours at this temperature.

B. *Polyester–Cotton Fibers*

1. Liquor ratio 1:10 yarn, package machine

Sodium chlorite, 50%	3.00%
Acid-stable surfactant	0.50 g/liter
Buffered acid activator, pH 4–4.5	2.00%

Heat up to 230°F; continue for 1 hour.

2. Liquor ratio 1:15–1:30

Sodium chlorite, 50%	4–5.00%
Acid-stable surfactant	0.50 g/liter
Buffered acid activator, pH 4	2–3.00%

Heat up to 200–205°F; continue for 1.5 hours at this temperature.

C. *Cotton Fibers (100%)*

1. Liquor ratio 1:10 yarn, package machine

Sodium chlorite, 50%	4.00%
Acid-stable surfactant	0.50 g/liter
Buffered acid activator, pH 4	2.50%

Heat up to 205°F; continue for 1–2 hours at this temperature.

2. Liquor ratio 1:20 (i.e., beck)

Sodium chlorite, 50%	5.00%
Acid-stable surfactant	0.50 g/liter
Buffered acid activator, pH 4	3–4.00%

Heat up to 190°F; continue for 2 hours at this temperature.

Miscellaneous Peroxygen Bleaching Compounds[155]

In addition to those compounds that we have already discussed, there are several additional peroxygen derivatives that occasionally find use in the textile industry. These are potassium peroxymonosulfate (monopersulfate), potassium perphosphate, and urea peroxide. These compounds are rarely used by themselves, but are used primarily as bleach boosters. They are all powders and are usually compounded where a strong, odorless oxidizing agent is indicated. Along with sodium perborate, they may or may not be compounded with a surfactant.

The mechanism of potassium peroxydisulfate–peroxide is represented by the following formulas:[156]

$$
\begin{aligned}
S_2O_8^{2-} &\rightarrow 2SO_4^{-}\cdot \\
SO_4^{-} + H_2O &\rightarrow H^{+} + SO_4^{2-} + OH^{\cdot} \\
OH^{\cdot} + H_2O_2 &\rightarrow H_2O + HO_2^{\cdot} \\
HO_2^{\cdot} + S_2O_8^{2-} &\rightarrow O_2 + HSO_4^{-} + SO_4^{-}\cdot \\
HO_2^{\cdot} + H_2O_2 &\rightarrow O_2 + H_2O + OH^{\cdot}
\end{aligned}
$$

to give the net reaction

$$S_2O_8^{2-} + H_2O_2 \rightarrow O_2 + 2HSO_4^{-}$$

In base, above pH = 5.0, the additional steps are written:

$$
\begin{aligned}
HO_2^{\cdot} &\rightleftharpoons O_2^{-} + H^{+} \\
O_2^{-} + S_2O_8 &\rightarrow O_2 + SO_4^{2-} + SO_4^{-}\cdot \\
O_2^{-} + H_2O_2 &\rightarrow O_2 + OH^{-} + OH^{\cdot}
\end{aligned}
$$

Use of Peroxygen Chemicals to Overcome Polyamide Yellowing[157]

In the commercial finishing of nylon fabrics, the heat setting operation usually causes yellowing that is difficult to remove. The use of a monopersulfate/hydrogen peroxide mixture in the ratio of 3–5% monopersulfate to 0.5% hydrogen peroxide bleach bath started at 100°F and run for 10 minutes with temperature being raised over 15 minutes to 180°F will usually remove all traces of yellowing.

Urea peroxide bleaching formulations are as follows:

A. *Pad Jig:*	Urea peroxide	8.00 lb
	Caustic, flake	4.00 lb
	Anionic gel detergent	1.00 lb
	Tinopal®* RBNAD, 0.25%	1.00 lb

1. Run one end at 130°F.
2. Run five ends at 170–180°F.
3. Drop; give two overflow washes.

* Registered trademark of CIBA-Geigy.

B. *Jig* (65-gal mix):	Urea peroxide	1.00%
	Caustic, flake (sets pH at 9.0)	2.00 lb

1. Set bath at 160°F and add:

Nonionic detergent	2.00 oz

2. Run two ends at the boil.
3. Overflow wash; shell through acetic acid.

A typical persulfate bleach procedure would be: (1) singe, (2) saturate with 0.3% persulfate and caustic and 0.1% surfactant, (3) steam, (4) J-box, and (5) wash.

CHEMICALS USED AS BUFFERS, ACTIVATORS, AND STABILIZERS

1. Alkali:
   a. Silicates.
   b. Phosphonates.
   c. Metal salts.
   d. Metallic hydroxides.
   e. Amine hydroxides.
   f. Amino carboxylates.
2. Acid:
   a. Oxylates.
   b. Sodium nitrate.
   c. Metal salts.
   d. Aldehyde/ketone–bisulfite.
3. Miscellaneous:
   a. Hydrazine alkali phosphonate.
   b. Hydrazine phosphonate.
   c. Picoline derivatives.
   d. Cyanamide derivatives.
   e. Metal salts of amino polycarboxylic acids.
   f. Sodium metaborate.

g. Aluminum phthalocyanines.
h. Sulfonyl oxime.
i. Acylphosphonate.

Stabilizers for peroxide normally function by controlling the formation of free radicals. In effect, stabilization may be effected by elimination of heavy metal ions that are responsible for radical formation. This appears to be an anticatalytic reaction. Sequestering agents for heavy metals should theoretically make the best stabilizers. However, when used by themselves, this has not proved to be correct.

Organophosphonates offer a new technology that, when combined with or used as a replacement for the older silicate buffers, offers new possibilities. In determination of which stabilizer is most suitable for use with peroxide, determining factors should be degree of whiteness to be achieved, determination of the active oxygen content as a function of time and temperature, the length of time bleach mix is to be run, and the equipment in which the bleaching operation is to be carried out.

Metallic silicates offer good stabilizing effects because they are oxidation resistant, are thermally stable, offer a buffering effect, are colloidal in nature, and are not greatly affected by the presence of ionic iron.

Amino polycarboxylic acids used in the form of their magnesium or calcium salts have been investigated for use as peroxide stabilizers and have the advantage over conventional amino polycarboxylic acids in that they are not reduced to amino oxides at high temperature and thus do not lose their sequestering power.

One of the major drawbacks to the use of metallic silicates by themselves is the deposit on bleaching machinery of insoluble silicates. The use of organic stabilizers, such as those that we have discussed, indicates that a combination of phosphonate and/or metallic amino polycarboxylic acid salts will minimize this deposition.

The environmental aspect of inorganic stabilizers is also of concern, and research has indicated that phosphonates would be acceptable to environmentalists.

In acid bleaching, oxylic acid and sodium nitrate are buffers that not only control pH, but also minimize chlorine degradation of metals. This is covered in more detail under the section on chlorite bleaching.

United States Patent 4,164,394 to FMC reveals a process for the removal of soil and/or stains from fabric using a peroxygen bleach that contains an acylphosphonate as the peroxygen activator. References are made to U.S. Patents 2,284,477; 3,532,634; 3,298,775; and 2,955,905 and various other American and foreign patents that disclose compounds which are representative of prior art activators for peroxygen bleaches.

United States Patent 4,164,395 to FMC discloses a process of removing soil and stains from fabric, utilizing a peroxygen bleach bath, which contains a sulfonyl oxime as the peroxygen activator. Additional references, both American and foreign, for classes of compounds representative of peroxygen activators are included.

United States Patent 4,166,718, to CIBA Geigy, discloses a process for bleaching textiles by treating the material in an aqueous bath containing water-soluble aluminum phthalocyanine, under irradiation with light and the presence of oxygen and other techniques. Water-soluble aluminum phthalocyanine-containing detergent compositions are also described.

United States Patent 4,185,960, to Interox, discloses a superoxidized solid sodium perborate that is activated by sodium metaborate.

United States Patent 4,199,320, to Chimiques Protex, discloses a cellulose fiber that has been whitened by boiling in an aqueous alkaline solution that contains hydrazine phosphonate or a hydrazine alkali phosphonate.

United States Patent 4,141,685, to Chimiques Protex, discloses the use of the reaction product of alkali bisulfite aldehyde–ketone as a stabilizer for sodium chlorite bleaching solutions.

SURFACTANTS USED IN BLEACHING PROCESS

The surfactants used in bleaching are a very important class of specialty chemicals because, as in preparation, they help to ensure the success of the process.

Surfactants used in bleaching fall into the same five basic categories that were used in preparation and frequently are, in fact, the same identical compounds. Surfactants may be anionic, nonionic, cationic, amphoteric, or blends of all four classes. As a rule, cationics are not used during the bleach cycle, but may be applied at the end of the bleach cycle as an antistat and/or softener. This is particularly true of knitted goods that are scoured and bleached in the same machine in which they are to be dyed. Acrylics also occasionally have a softener applied prior to dyeing, and the softener will act as a retarder for the dye cycle.

Surfactants are the basic components of wetting agents, solvent and nonsolvent scouring agents, and emulsifiers for the oils, waxes, and lubricants that are encountered during the bleaching process.

The most common bleaching surfactants in current use today appear to be anionics of the sulfate, phosphate, carboxymethylate, and methyltaurine types. Also in common usage, and possibly even surpassing the anionics in some cases, are the nonionics composed of ethoxylated primary alcohols, ethoxylated secondary alcohols, mercaptan thioethers, and ethoxylated fatty acids. All of these have been discussed in detail in the section of surfactants at the beginning of this chapter.

OPTICAL WHITENERS OR FLUORESCENT BRIGHTENING AGENTS*

The whiteness that we regard as standard today would have been impossible without the discovery and use of fluorescent brightening agents commonly referred to as optical whiteners. This is true because there are physical limits to the increase in reflectance that can be achieved with normal bleaching. Additional whitening was accomplished in early times by the use of a blue dye, or tint, which was so faint that it could not be discerned by the naked eye but was strong enough to mask any yellowing remaining on the white goods. Optical whiteners function by absorbing ultraviolet light and reflecting it in the blue-white spectrum, thereby exerting a synergistic, or intensifying, effect on whiteness.

The earliest reference to the application of these compounds to textiles was in 1929, when Krais reported in the *Melliands Textilber.* Volume 10, p. 468, that *aesculin,* a glucoside of aesculetin found in horse chestnuts, improved the whiteness of cotton when applied to the fiber.[158]

Because of the lack of affinity of cellulosic fibers for aesculin, interest in the subject was dropped for approximately 10 years. In 1939, Meyer was issued British Patent 522,672 for a compound named *β-methyl umbelliferone,* which possessed a structure similar to aesculetin, but that did possess a greater affinity for cellulosic fibers.[159] Additional structures that give an optical whitening effect (i.e., fluorescent whitening agents) are:[160]

*All material in this section is reproduced by permission of the Publishers, Charles Griffen & Company, Ltd., London and High Wycombe, from Trotman, *Textile Scouring and Bleaching* (1968).

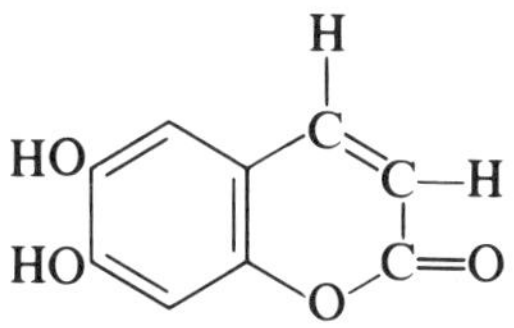

Aesculetin

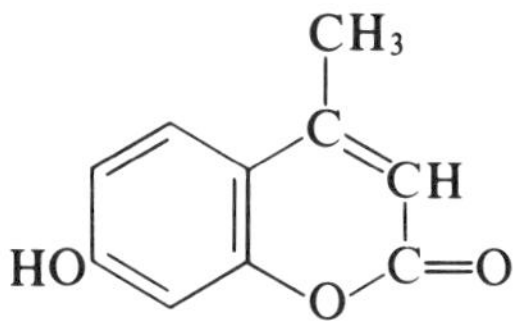

β-Methyl umbelliferone

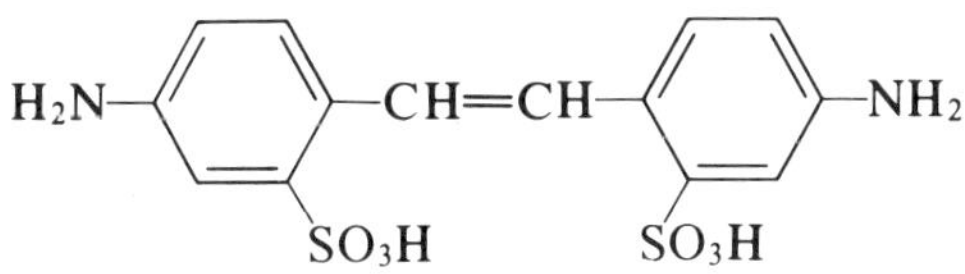

4:4′-Diaminostilbene-2:2′-disulfonic acid

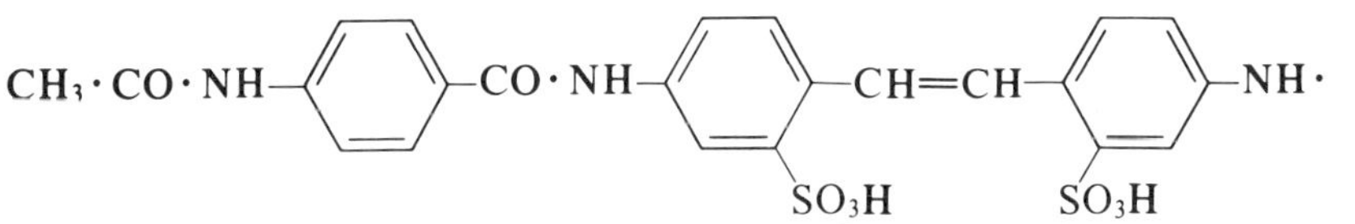

British Patent 584,484

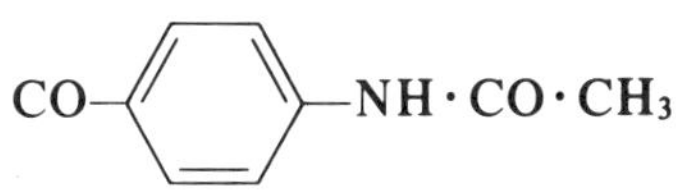

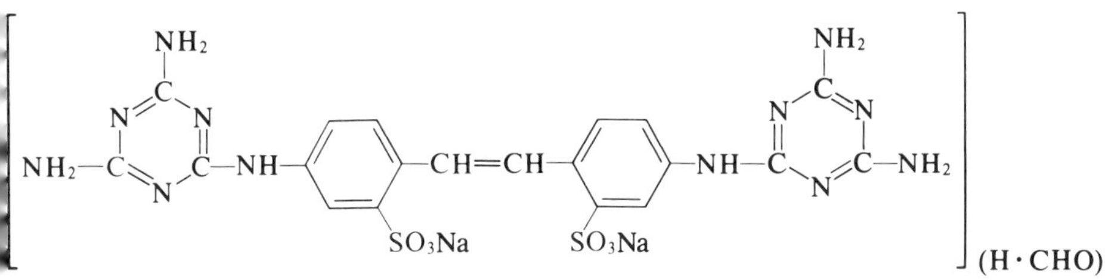

4:4′-Diaminostilbene-2:2′-disulfonic acid, reacted with cyanuric chloride, ammonia, and formaldehyde

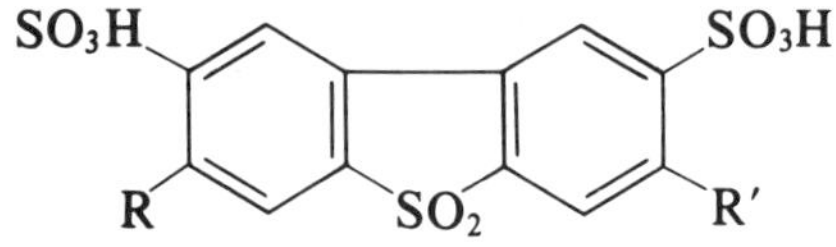

2:8-Disulfonic acid of benzidene sulfone of American Cyanamid

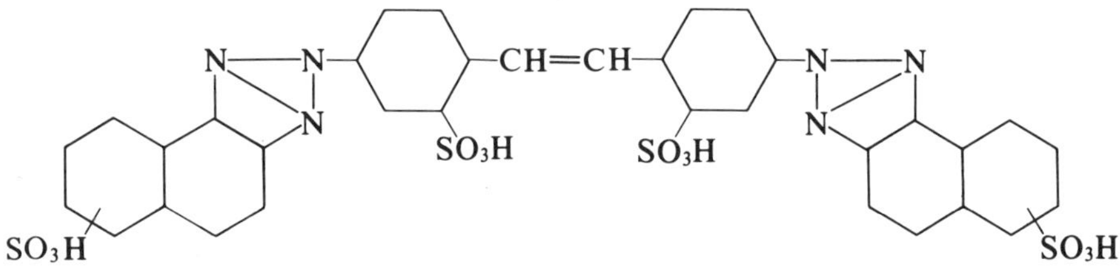

British Patent 683,895 to Bayer

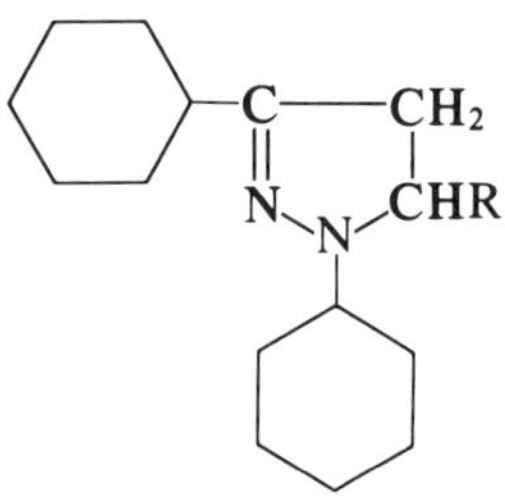

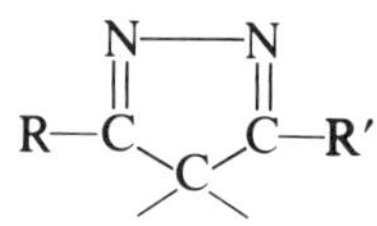

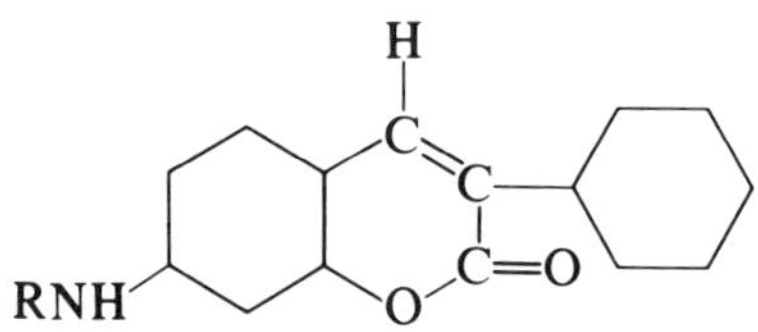

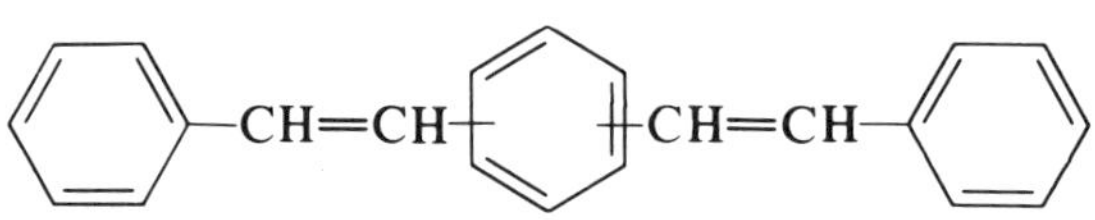

Optical whiteners for nylon

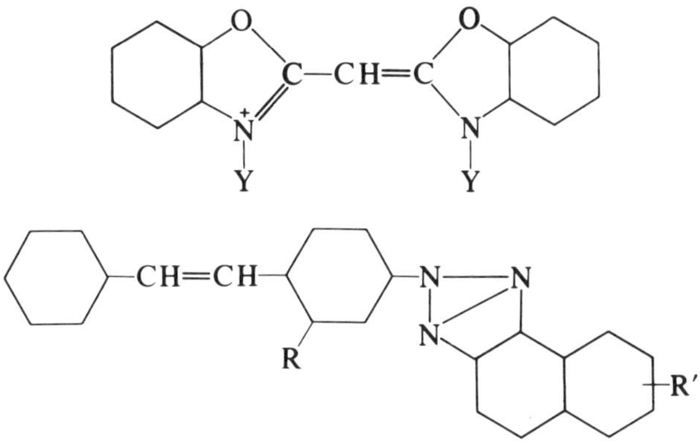

Optical whiteners for polyacrylonitriles

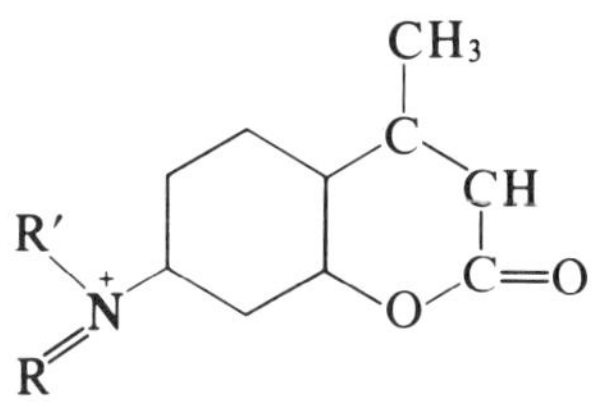

Optical whiteners for secondary cellulose and triacetate

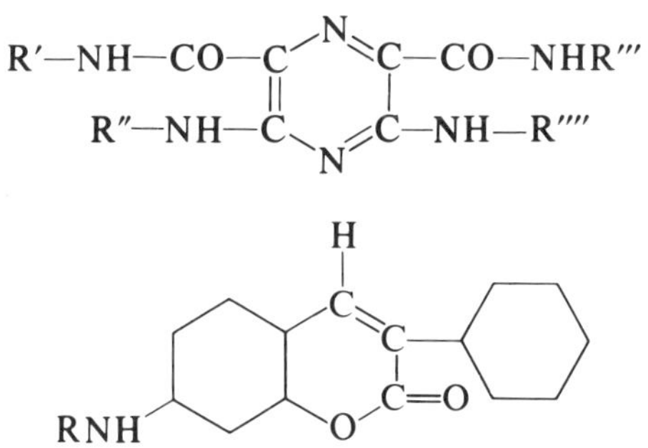

Optical whiteners for wools

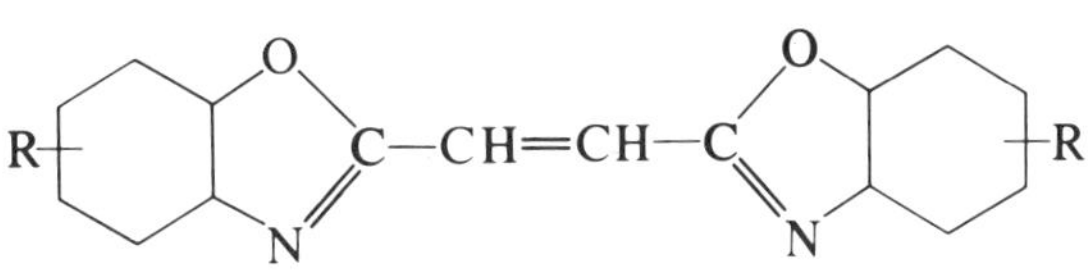

Optical whiteners for polyester

Factors that affect the effect of fluorescent brightening agents are temperature, time, salt sensitivity, and migration properties. Additional properties to be considered in selecting an optical are fastness properties required and compatibility with other processing agents.

More detail on the composition, use, and performance properties of optical brightening agents can be found in U.S. Patent 4,169,702; Japanese Patents 6,723,124; 7,110,871; 7,341,707; and 7,401,036; and British Patents 1,028,923; 1,031,750; 1,045,977; 1,052,179; 1,108,865; 1,116,619; and 1,256,660.

PROBLEM AREAS

No discussion of preparation, scouring, and bleaching would be complete without a look at those areas that have potential problems for bleachers. Several factors that need to be considered are outlined as follows:

1. *Alkali*—proper balance.
2. *Temperature* in excess of the level that is actually needed.
3. *Side effect* of rewetting surfactant versus absorbency.
4. *Solvent* spotting.
5. *Metal* scale buildup.
6. *Uneven* or incomplete size removal.
7. *Metal* catalysis on synthetics.
8. *Odor* in working area—OSHA.
9. *Toxicity*—TOSCA.
10. *Disposal* and/or chemical treatment resource recovery.
11. *In continuous* operations, many yards damaged before error is discovered.
12. *Burrs* and picks ruining delicate fabrics.
13. *Improper choice* of sequestrant, minimizing effectiveness.
14. *Improper choice* of defoamer, causing spotting.
15. *Nonfugitive* tints.

REFERENCES AND NOTES

1. Moncrieff, R. W., *Man-Made Fibres,* Wiley, New York, 1957. *"Collected Papers of Wallace H. Carothers on Polymerisation,* Interscience, New York, 1940; Trotman, E. R., *Textile Scouring and Bleaching,* Charles Griffin, London, 1968.
2. U.S. Patent 2,515,834, du Pont, and U.S. Patent 2,515,889, du Pont.
3. British Patent Numbers 582,518; 582,520; 582,522; 583,014. U.S. Patents to du Pont 2,130,523; 2,130,947; 2,130,948; and 2,163,636.
4. British Patents to du Pont 579,887 and 603,873; British Patents to ICI 583,939 and 584,548.
5. U.S. Patents to Chemstrand 2,749,325 and 2,744,086.
6. British Patents to du Pont 689,801 and 686,438.
7. U.S. Patent to Union Carbide 2,161,766.
8. British Patent 578,079 and U.S. Patent 2,578,899 to du Pont.
9. U.S. Patents 3,528,947 to Eastman Kodak; 3,018,272 to du Pont, and 3,164,566, 3,164,567.
10. Koenig, H. Steve, "Topical Lubricants and Their Use as Spin Finishes," Henkel Corporation, Charlotte, N.C., 1980.
11. Crossfield, U.S. Patent 4,105,569 to Goulston Company, Ltd.
12. Koenig.
13. Products of ICI–Atlas.
14. Formulation Courtesy of Whitestone Chemical, Spartanburg, S.C.
15. Rutenberg, M. W., "Starch and Its Modifications," in *Handbook of Water-Soluble Gums and Resins,* Robert L. Davidson, Ed., McGraw-Hill, New York, 1980, pp. 22.64–68.
16. Ibid.

17. Ibid.

18. Patterson, E. F., "Search for a Magic Ingredient," *America's Textiles Reporter/Bulletin,* November 1973, p. 41.

19. Rutenberg, pp. 22.64–68.

20. Guiseley, Kenneth B., Norman F. Stanley, Phillip A. Whitehouse, "Carrageenan," in *Handbook of Water-Soluble Gums and Resins,* Robert L. Davidson, Ed., McGraw-Hill, New York, 1980, p. 5.5.

21. Meer, William, "Gum Arabic," in *Handbook of Water-Soluble Gums and Resins,* Robert L. Davidson, Ed., McGraw-Hill, New York, 1980, pp. 8.15–16.

22. Meer, William, "Gum Karaya,"in *Handbook of Water-Soluble Gums and Resins,* Robert L. Davidson, Ed., McGraw-Hill, New York, 1980, p. 10.2.

23. Ibid.

24. Stauffer, Kenneth R., "Gum Trag," in *Handbook of Water-Soluble Gums and Resins,* Robert L. Davidson, Ed., McGraw-Hill, New York, 1980, p. 11.3.

25. Ibid., p. 11.2.

26. Seaman, James K., "Locust Bean Gum," in *Handbook of Water-Soluble Gums and Resins,* Robert L. Davidson, Ed., McGraw-Hill, New York, 1980, p. 14.1.

27. Ibid., p. 14.8.

28. Ibid., pp. 14.9, 14.10.

29. Gerard, Thomas, "Tamarind Gum," in *Handbook of Water-Soluble Gums and Resins,* Robert L. Davidson, Ed., McGraw-Hill, New York, 1980, p. 23.2.

30. Ibid., p. 23.1.

31. Ibid., p. 23.3.

32. Ibid., p. 23.2.

33. Ibid.

34. Cottrell, Ian W., Kenneth S. Kang, and Peter Kovacs, "Xanthan Gum," in *Handbook of Water-Soluble Gums and Resins,* Robert L. Davidson, Ed., McGraw-Hill, New York, 1980, p. 24.2.

35. Ibid.

36. Ibid., p. 24.20.

37. Ibid., p. 24.8.

38. Blumenstein, Carl R., "The Chemistry of Polymeric Materials Used in Sizing," Texchem Consulting Labs, Greenville, S.C.

39. Modi, T. W., "Polyvinyl Alcohol," in *Handbook of Water-Soluble Gums and Resins,* Robert L. Davidson, Ed., McGraw-Hill, New York, 1980, p. 20.3.

40. Ibid.

41. Ibid., p. 20.15.

42. Ibid., p. 20.18.

43. Ibid., p. 20.19.

44. Ibid., p. 20.24.

45. Ibid.

46. Stelzer, Glen I. and E. D. Klug, "Carboxymethylcellulose," in *Handbook of Water-Soluble Gums and Resins,* Robert L. Davidson, Ed., McGraw-Hill, New York, 1980, p. 4.1.

47. Hercules, Incorporated, Technical Service Bulletin No. VC-506, "CMC Warp Size."

48. Ibid.

49–52. Ibid.

53. EPA-660/2-75-014, June 1975.

54. Union Carbide, Technical Service Bulletin No. F-40525H, "Cellosize Hydroxyethyl Cellulose," October 1978.

55. Greminger, G. K., Jr. and K. L. Krumel, "Alkyl and Hydroxyalkylalkylcellulose," in *Handbook of Water-Soluble Gums and Resins,* Robert L. Davidson, Ed., McGraw-Hill, New York, 1980, p. 3.5.

56. Powell, George M., III (Consultant for Union Carbide), "Polyethylene Glycol," in *Handbook of Water-Soluble Gums and Resins,* Robert L. Davidson, Ed., McGraw-Hill, New York, 1980, p. 18.6.

57. Blumenstein.

58. Ibid.

59. Ibid.

60. Formulation Courtesy of Intex Products, Inc.

61. Ibid.

62. Hercules, Incorporated, "CMC Warp Size."

63. Ibid.

64. Ibid.

65. Ibid.

66. Formulations Courtesy of Intex Products, Inc.

67. Ibid.

68. West Point Foundry and Machine Company, Management Technical Bulletin, "High Pressure Sizing," Vol. 2, No. 6.

69. Ibid.

70. Schwartz, A. M. and Perry, J. W., *Surface Active Agents,* Vol. 1, p. 48, Interscience, New York, 1963.

71. Ibid., pp. 8, 17.28, 48.

72. "Sulfonating–Sulfating Agents: Their Characteristics and Uses," Technical Bulletin A-50244, p. 2, E. I. du Pont de Nemours & Co., Inc., Wilmington, Delaware, August 1966.

73. U.S. Patent 2,163,133 to Unichem; U.S. Patent 2,275,413 to American Hyalsal Corp.; U.S. Patents 2,203,641 and 2,203,642 to National Oil Products Co.

74. Schwartz and Perry, pp. 55, 57.

75. U.S. Patent 1,918,373 to Bertsch; U.S. Patents 2,355,503; 2,167,931; 2,372,786; U.S. Patent 1,981,792 to John Orelup; U.S. Patent 1,932,180 to I. G. Farben; British Patent 523,466 to ICI; Canadian Patent 427,726 to Bertsch; Schwartz, pp. 53, 73, 74.

76. Schwartz and Perry.

77. Trotman, E. R., *Textile Scouring and Bleaching,* pp. 66–67, Charles Griffin, London, 1968.

78. Huddleston, R. L. and Dr. A. M. Nielsen, "Las Biodegradation: The Fate of the Aromatic Ring," *Soap/Cosmetics/Chemical Specialties,* March 1979.

79. Schwartz and Perry, pp. 118–119.

80. Courtesy of Jefferson Chemical, Technical Service Bulletin No. 10, "A Discussion of Alkyl Phenol–Ethylene Oxide Adduct Sulfation Procedures.
81. Schwartz and Perry, p. 79.
82. Courtesy of FMC Corporation, Technical Bulletin, "FMC Superphosphoric Acid/FMC Polyphosphoric Acid."
83. Ibid.
84. Ibid.
85. Ibid.
86. Registered Trademark of Union Carbide.
87. U.S. Patent 3,235,627.
88. U.S. Patents 3,004,056 and 3,004,057.
89. Courtesy of Sandoz Colors and Chemicals, Product Bulletin 7-525/78, "Sandopan Carboxylated Surfactants."
90. Courtesy of GAF Corporation, Product Bulletin, "Igepon Anionic Surfactants," No. 2303-027.
91. Kastens, M. L. and J. J. Ayo, *Ind. Eng. Chem.*, 42,1626 (1950).
92. Shapiro, S. H., in *Fatty Acids and Their Industrial Applications,* E. Scott Pattison, Ed., Marcel Dekker, New York, 1968, pp. 102–103.
93. Schwartz and Perry, pp. 102–103.
94. Registered Trademark of GAF Corporation.
95. Courtesy of GAF Corporation.
96. Longman, G. F., *The Analysis of Detergents and Detergent Products,* Wiley, New York, 1975, pp. 11–12.
97. Longman, p. 12.
98. Trotman, p. 50.
99. Ibid., p. 53.
100. Ibid., p. 55.
101. Nettles, John E., "Infrared Spectroscopy for Identifying Surfactants," *Text. Chem. Colorist,* September 24, 1969, Vol. 1, No. 20.
102. Courtesy of Emery Industries, Technical Bulletin No. 201-C, "Fatty Acids and Liquid Soaps," March 1961.
103. U.S. Patents 2,285,333; 2,327,132; and 2,383,289 to Hercules, Incorporated.
104. Cook, E. Fullerton, quoted in "Soft Soap Manufacture" by J. M. Vallance, *Soap and Sanitary Chemicals,* November 1949.
105. U.S. Patent 2,174,761 to Schuette and Wittwer of I. G. Farben.
106. "Standard Draves Wetting Test," AATCC Method No. 17-1974.
107. Courtesy of Union Carbide, Product Bulletin No. F-46879, "Tergitol Surfactants for the Textile Industry," March 1978.
108. Courtesy of Dow Chemical Company, *The Alkanolamines Handbook,* 1980.
109. DiMaria, Tony, "Weight Reduction of Polyester Fabrics with Sodium Hydroxide," paper presented at Palmetto Sectional AATCC meeting.
110. Schwartz and Perry, p. 219; Pattison, E. Scott, Ed., *Fatty Acids and Their Industrial Applications,* Marcel Dekker, New York, 1968, pp. 106, 179. Also U.S. Patents 2,468,012; 2,619,467; and 2,816,920.

111. U.S. Patent 2,273,068.

112. Schwartz and Perry; Pattison and Scott; Jungermann, Eric, Ed., *Cationic Surfactants*, Vol. 4, Marcel Dekker, New York, 1970.

113. U.S. Patent 2,199,397 to du Pont; Swiss Patent 179,662 to CIBA Geigy; German Patent 692,417 to I. G. Farben.

114. *The Alkanolamines Handbook*, pp. 78–80; Pattison, pp. 248–249.

115. "Tergitol Surfactants for the Textile Industry," p. 6.

116. Davidson, Robert L., Ed., *Handbook of Water-Soluble Gums and Resins*, McGraw–Hill, New York, 1980, pp. 4.17.

117. Noll, Walter, *Chemistry and Technology of Silicones*, Academic Press, New York, 1968.

118. Dow Corning, "Formulators Forum," Vol. II, No. 1, Bulletin 22-758-79.

119. Ibid.

120. Air Products and Chemical, Inc., "Surfynol® 104," Technical Service Bulletin 120–211.

121. U.S. Patents 2,854,417; 3,180,786; 2,230,556; 2,366,727, and 3,180,836.

122. Courtesy of Societe Protex, Technical Bulletins, "Oligoelements, Chelates, Masquolates," and "Masquol."

123. Courtesy of Union Carbide, Product Booklet, "Ethyleneamines from Union Carbide," p. 19.

124. Courtesy of E. I. du Pont, Product Bulletin No. A-69742, "Oxone Monopersulfate Compound," April 1970.

125. Sitver, Leonard A., Bernard K. Easton, and Robert E. Yelin, FMC Corporation, "Effect of Potassium Perphosphate on Peroxide Bleaching."

126. Courtesy of FMC Corporation, Technical Data Bulletin, "Desize–Scouring of Textile Fabrics with Potassium Perphosphate (KPP)."

127. Cousineau, David T., "New Approaches and Developments for Bleaching Cellulose and its Blends (Organic Stabilizers," American Hoechst Corporation.

128. Nettles, John E., "Sodium Hypochlorite Bleaching: an Aid to Textile Processing," *American Dyestuff Reporter*, p. 31, September 26, 1966.

129. Cousineau.

130. Ibid.

131. Ibid.

132. Hartmenn, O., Dissertation Thesis, Stuttgard, 1927.

133. Cousineau.

134. Nettles, "Sodium Hypochlorite Bleaching."

135. Courtesy of Westvaco Chemical, Technical Data Bulletin No. C-240, "Making Small Quantities of Sodium Hypochlorite Bleach."

136. Courtesy of E. I. du Pont, Technical Data Bulletin E-12989, "Albone De Peroxygen Compound for Desizing and Scouring," January 1977.

137. Steinmiller, W. G. and D. M. Cates, "Bleaching Cellulose: Part II. Effect of Impurities," and *Text. Chem. Colorists*, Vol. 8, No. 1, p. 14 (1976).

138. U.S. Patent 3,528,115, "Bleaching Fabrics with Peracetic," to E. I. du Pont.

139–152. Courtesy of Olin Chemicals, Product Data Bulletin, "Textone: A Sodium Chlorite Product for Textile Bleaching and Processing."

153. Courtesy of American Hoechst.

154. Ibid.

155. FMC Corp., "Desize-Scouring of Textile Fabrics with Potassium Perphosphate (KPP)," and du Pont, "Oxone Monopersulfate Compound."

156. FMC Corp., Sitver, Easton, and Yelin.

157. du Pont, "Oxone Monopersulfate Compound."

158. Trotman, p. 146.

159. Ibid, p. 147.

160. Ibid. pp. 147–152.

161. Formulation courtesy of Emery Industries, Inc.

162. Formulation courtesy of Arizona Chemical Company Technical Service Bulletin No. 309, "Sulfation of Acintol® Tall Oil Fatty Acids," which states that while permission is given to use this formulation, recommendations and instructions are made without warranty or inducement to violate any patent owned by Arizona Chemical Company or others.

FURTHER READING

"Adhesive Maker Adds Defoaming Agent to Eliminate Customer Problems," *Adhesives Age,* **18**(10) (1975).

Badische Aniline and Soda Fabrik, British Patent 661,146 (1951).

Baird, G. S. and Griffiths, A. L., "The Advantages of CMC as a Warp Size," *Mod. Text.,* **46**(6), 34, 59–60, 65 (1965).

Balch, R. H., "Slashing of Acetate and Triacetate," *Text. Bull.,* **89**(2) 74–76 (1963).

Barrow, G. M., *The Structure of Molecules,* W. A. Benjamin, New York, 1963.

Baum, Burton M., et al., *Bleaching Agents,* FMC Corporation.

Bauman, R. P., *Absorption Spectroscopy,* Wiley, London, 1962.

Baylar, *Chemistry of the Coordination Compounds,* Reinhold, New York, 1956.

Beck, G., *Helv. Chim. Acta,* **29,** 357–360 (1946).

Beck, G. and Gasser, A., *Anal. Chim. Acta,* **3,** 51–52 (1949).

Beck, G., *Mikrochem.,* **33,** 344–347 (1948).

Beckman, W., "Recent Developments in Dyeing Textured Polyester," *Text. Chemist Colorist,* 350–355 (October 7, 1970).

Beio, D. V. and McCarthy, J., "Raw Materials Primer: The Fatty Alkanolamides," *Cosmetic Technol.* (March 1980).

Bellamy, L. J., *Infrared Spectra of Complex Molecules,* Wiley, New York, 1954.

Berger, Paul D., and Gast, John A., "Analysis and Measurement of Foaming Phenomena in Water-Based Coating Systems," *J. Coatings Technol.* **48**(621) 55–59 (1976).

Bikerman, J. J., "Surface Chemistry: Theory and Applications," in *Foams,* Academic Press, New York, 1958, Chapter 12.

Bjerrum, J., "Stability Constants—Organic Ligands," Chemical Society, London, 1957.

Blumenstein, C. R., "Sizing Synthetics and Blends," *Text. Ind.,* **127**(8) 58–68 (1963); "Sizing with PVA," *Text. Ind.,* **130**(7) 63–9, 97 (1966); "Size and Fiber/Yarn Relationships," *Text. Ind.,* **133**(6–8) (1969).

Bowden, F. P. and Tabor, D., *The Friction and Lubrication of Solids,* Oxford University Press, London, 1954 (Part I) and 1964 (Part II).

Boulanger, F., *Chim. Anal.,* **35,** 253–254 (1953).

Boys, C. V., *Soap Bubbles and the Forces Which Mould Them, Volume for Promoting Christian Knowledge,* reprinted by Doubleday Anchor Books, New York, 1959.

Cairns, T., et al., *Spectroscopic Problems in Organic Chemistry,* Heyden, London, 1964.

Calhoun, L. M., "Sizing Filament Yarns," *Text. Ind.,* **128**(9) 136–138, 153–154 (1964).

Catalogue of the Monsanto Chemical Company, *Sodium Phosphates for Industry.*

Chaberek and Martell, *Organic Sequestering Agents,* Wiley, New York.

"CMC Kayoed Stream Pollution," *Text. Ind.,* **124,** 161–162 (1960).

"CMC Warp Size—Abates Stream Pollution, *Hercules Chemist,* **26,** 14–17 (1956).

Colthup, N. B., Daley, L. H., and Wiberley, S. E., *Introduction to Infrared and Raman Spectroscopy,* Academic Press, New York, 1964.

Cook, J. G. *Handbook of Textile Fibers,* 3rd ed., Merrow, Walford, England, 1963.

Coward, C. M., "Chemical Properties of Cotton Cellulose," in Kyle Ward, Jr., Ed., *Chemistry and Chemical Technology of Cotton,* 1955, p. 15.

Czerwin, E. P., "Warp Size Trends Favor PVA," *Mod. Tex.,* **47**(12), 29–30, 32, 34 (1966).

Davidson, A. and Milwidsky, B. M., *Synthetic Detergents,* 4th ed., C. R. C. Press, Cleveland, 1968, pp. 131–136.

Dayvault, J. A., "Effects of Carrier Concentration of Barre' Coverage in Knits."

"Defoamers and Slimicides," Part III, Multiple-Client Survey by Hull and Company, Bronxville, N.Y., 1976.

"Defoamers for Increased Plant Productivity and Profitability," Silicone Products Department, Dow Corning Corporation, Midland, Mich., 1969.

"Defoamers for Effluent Chemical and Industrial Processing," Silicone Products Department, Dow Corning Corporation, Midland, Mich., 1969.

DeMytt and Reed, U.S. Patent 2,506,492 (1950).

"Deriphats—Amphoteric Surfactants," Technical Bulletin 15-A, General Mills Corp., Kankakee, Ill.

Dreyfus, H., Dickey, W. A., and Sawter, P. F. C. (to Celanese Corp. of America), *Warp Sizing of Yarns with Aqueous Solutions of Sodium Carboxymethylcellulose,* U.S. Patent 1,950,664 (1924).

Dyer, John R., *Applications of Absorption Spectroscopy of Organic Compounds,* Prentice Hall, Englewood Cliffs, N.J., 1965.

E. I. du Pont, "Hydrogen Peroxide Bleaching Procedures."

Elliott, J. H. (to Hercules Powder Company), *Textile Sizing with Zirconium Salts of Water-Soluble Carboxymethylcellulose,* U.S. Patent 2,650,887 (1953).

Ender, *Fette und Seifen,* **45,** 144–146 (1938).

Ernst, R. (to Textilana Corp.), U.S. Patent 3,024,260 (1962).

Ferdinand School, A. G., *Size Solutions and Size Compositions Containing Salts of Cellulose Ether Carboxylic Acids as Their Essential Constituents,* British Patent 496,351 (1938).

FMC Corporation and E. I. du Pont, Persulfate and Perphosphate Desize Procedures.

Fischer, K. A. and Hecht, O. F., U.S. Patent 2,610,952 (1952).

Fitch, F. T. and Russel, D. S., *Anal. Chem.,* **23,** 1469 (1951).

Fitch, F. T. and Russel, D. D., *Can. J. Chem.*, **29**, 363 (1951).

Foil, Miller A., "Application of Infrared and Ultraviolet Spectra to Organic Chemistry," Vol. 3, in *Organic Chemistry*, H. Gilman, Ed., New York, 1953, Chapter 2.

Freeman, Stanley K., Ed., *Interpretive Spectroscopy*, Reinhold, New York, 1965.

Gard, A. J., *J. Am. Leather Chemists Assoc.*, **50**, 255 (1955).

Gardner, R. T., Jr., "CMC as Aerosol Fabric Size," *Soap Chem. Spec.*, **38**, 98–101, 122 (1962).

Golding, B., Polymers and Resins—Their Chemistry and Chemical Engineering, Van Nostrand, New York, 1959.

Goling, G. F., and MacGregor, J. H., Principles and Practice of Sizing Warp Yarns Made of Synthetic Fiber, *J. Text. Inst.*, **54**, 24–42 (1963).

Gosselin and Coghlan, *Archiv. Biochem. and Biophys.*, **45**, 301 (1953).

Hahl, H. and Kropp, V., U.S. Patent 1,517,003 (1925).

Hamby, D. S., Ed., *The American Cotton Handbook*, 3rd ed., Wiley-Interscience, New York, 1965.

Hansen, W. W. and Tabor, D., *Text. Res. J.*, **27**, 300 (1957).

Harrick, N. J., *Internal Reflectance Spectroscopy*, Interscience, New York, 1967.

Hart, R., "An Outline on Warp Sizing of Spun Rayon," *Rayon Text. Mon.*, **27**, 192–195 (1946).

Heath, O. and Clark, J., *Nature*, **177**, 1118–1121 (1956).

Hemphill, J. E., "Dyeing 100% Textured Polyester Fabrics with Disperse Dyes," *Am. Dyestuff Reporter*, 28–33 (February 24, 1969).

Henderson, L. F., U.S. Patent 371,623 (1945).

Herzberg, G., *Molecular Spectra and Structure*, Vol. II, *Infrared and Raman Spectra of Polyatomic Molecules*, Van Nostrand, New York, 1945.

High Point Chemical, High Point, N.C., *Sodium Hypochlorite and Peracetic Acid Bleaching Technology*.

Hummel, Dieter, *Identification and Analysis of Surface-Active Agents*, Interscience, New York, 1962.

I. G. Farben, British Patent 496,252 (1938).

"Infrared Correlation Chart," Barnes Engineering Co., Stamford, Conn.

Jones, R. N., and Sandorfy, C., "The Application of Infrared and Raman Spectroscopy to the Elucidation of Molecular Structure," in *Chemical Applications of Spectroscopy*, A. Weissberger, Ed., Vol. 9 of *Technique of Organic Chemistry*, Interscience, New York, 1956.

Jungerman, E. and Tabor, D., *Nonionic Surfactants*, Marcel Dekker, New York, 1967, pp. 209–231.

Kalle and Company, French Patent, 845,587 (1939).

Kalle and Company, Italian Patent 368,055 (1939).

Kelly, Lloyd, "How to Select the Right Defoamer," *Modern Paint Coatings*, **65**(6), 33–36 (1975).

Kendall, D. N., Ed., *Applied Infrared Spectroscopy*, Reinhold, New York, 1966.

Kitazawa, F., "Sorption of Carboxymethylcellulose (CMC) on Synthetic and Rayon Stable Yarns," *Bull. Nagoya Inst. Technol.* (*Anniversary Issue*), **4**, 155–158 (1952).

Kowalski, Xavier, "Sequestering Agents in Bleaching and Scouring," Clemson University, Clemson, S.C., December 1977.

Kritchevsky, W., U.S. Patent 2,088,212 (1937).

Kritchevsky, W., U.S. Patent 2,096,749 (1937).

Kroll, H. and Lennon, W. J., "Proceedings of the Scientific Section of the Toilet Goods Association," No. 25 (1956). This article also appeared in *Drug and Cosmetic Industry*, **79,** 186–187, 272–277 (1956).

Kroll, H., Symposium on Advances in Chelate Chemistry, Brooklyn, N.Y., 1965.

Lehman, R. A. and Sproull, R. C., U.S. Patent 2,348,984 (1944).

Lever Bros. and Unilever Ltd., British Patent 574,504 (1946).

Lissant, K. J., *Emulsions and Emulsion Technology,* Marcel Dekker, New York, 1974 (Parts I and II).

Lohr, J. W. (to Andrew Jergens Company), U.S. Patent 3,040,075 (1962).

Mangum, G. D., Jr., "A Review of Slashing Methods for Spuns and Filaments," *Mod. Text.* **44**(12) 52–53 (1963).

Loriers, J. and Carminati, D., *Compt. Rend.,* **237,** 1328–1330 (1953).

McCall, E. R. and Jurgens, J. F., *Text. Res. J.,* **21,** 19 (1951).

McGuire, E. Patrick, Adhesives Raw Materials Handbook, Part 7, Defoamers, Padric Publishing Co., Bernardsville, N.J. 1964.

Meade, E. M. (to Lankro Chemicals Ltd.), U.S. Patent 2,464,094 (1949).

Mellon Institute Report of S. M. Livingston and C. A. Johnson, "Analysis of High Amide Fatty Alkylolamides," December 2, 1956.

Meyer, K. and Brune, W., *Z. Wiss. Phot.,* **46,** 174 (1951).

Miller, A., German Patent, 720,630 (1952).

Miller, M. M. and Mallery, M. F., *Anal. Chem.,* **30**(11), 1884 (1958).

Mittal, K. L., *Miscellization, Solubilization and Microemulsions,* Vol. 1, Plenum Press, New York, 1977.

Monk, *J. Chem. Soc.,* **423,** 2693 (1949), 1314 (1952).

Muhr, Swiss Patent 276,557 (1951).

Munz, F., U.S. Patent 2,308,021 (1943).

Nakamoto, K., *Infrared Spectra of Inorganic and Co-ordination Compounds,* Wiley, New York, 1963.

Nakanisi, K., *Infrared Absorption Spectroscopy,* Holden Day, San Francisco, 1962.

Neugebauer, W., U.S. Patent 2,238,862 (1941).

Numajiri et al., "Chelatometric Titration of Metals with Nitrilo Triacetic Acid," *Nipponkagaku Zasshi* (1960).

O'Farrell, T. F., "Foams and Foam Inhibition," Drew Chemical Company, Parsippany, N.J., 1973.

Olin–Mathieson and American Hoechst, "Sodium Chlorite Bleaching Technology: Non-silicate Stabilizers."

Park, K., Seefried, C. G., and Bryant, G. M., *Text. Res. J.,* **44,** 692 (1974).

Partridge, Hicks, and Smith, *J. Am. Chem. Soc.,* **63,** 454 (1941).

Pattison, E. Scott, Ed., *Fatty Acids and Their Industrial Applications,* Marcel Dekker, New York, 1968.

Perkin–Elmer Handbook for Infrared Spectroscopy, the Perkin–Elmer Corp., Norwalk, Conn.

Petit, A., French Patent 964,397 (1952).

Pfeiffer and Offermann, W., *Ber.,* 75B, 1–12 (1942).

Ploetz, E., German Patent 800,401 (1950).

Potts, W. J., *Chemical Infrared Spectroscopy,* Wiley, New York, 1963.

Pratt, H. T., "Dyeing and Finishing of Double Knits of Textured Set Polyester," *Can. Text. J.,* 79 (March 1970); "Makes Barre' Passe," *Text. Ind.,* 244–246 (October 1968).

Prince, L. M., *Microemulsions,* Academic Press, New York, 1977.

"Preventing Soil Redeposition with CMC," *Soap Chem. Spec.,* **35,** 135 (1959).

"Proceedings of the 18th American Institute," *Infrared Handbook,* Fisk University, Nashville, Tenn., 1967.

Ramsbottom, T., *Warp Sizing Mechanisms,* Columbine Press, Manchester, 1964.

Rao, C. N. R., *Chemical Applications of Infrared Spectroscopy,* Academic Press, New York, 1963.

Ross, Sydney, "The Inhibition of Foaming," Bulletin No. 63, Rensselaer Polytechnic Institute, Troy, N.Y., May 1950.

Sabia, A. J., "Improved Foam Control," Union Carbide Silicones, N.Y., 1967.

Schaeffer, *Textil-Rundschau,* 489–498 (1951).

Schick, M. J., *Text. Res. J.,* **43,** 198 (1973).

Schildknecht, C. E., *Vinyl and Related Polymers,* Wiley-Interscience, New York, 1952.

Schlatter, C. and Demes, J. H., *Text. Res. J.,* **32,** 87 (1962).

Schmidlin, H. V., *SVF Fachorgan Textilberedlung,* **8,** 122–126, 183–188, 1953.

Smith, J. B., Ed., *The Technology of Warp Sizing,* Columbine Press, Manchester, 1964.

Schulte, K. E. and Schillinger, Z. Lebenam, *Untersuch & Forsch* **94,** 166–182 (1951).

Schurman, J. V. (to Colgate–Palmolive Company), U.S. Patent 2,863,888 (December 9, 1958). See also Monick, J. A., *J. Am. Oil, Chem. Soc.,* **39,** 213–215 (1962).

Schwartz, A. M. and Perry, J. W., *Surface Active Agents,* Vols. I and III, Interscience, New York, 1963.

Schwarzenbach, G., *Complexometri-Titrations,* Interscience, New York, 1967.

Schwarzenbach, G. and Ackermann, *Helv. Chim. Acta* (1947).

Seydel, P., "Cotton Warp Sizing IV," *Cotton* (*Atlanta*), **110**(12), 82–84, 149 (1946).

Seymour, R. B. and Schroeder, G. M. (to Henry H. Frede Co.), *Sodium Carboxymethylcellulose in Textiles,* U.S. Patent 2,486,803 (1949).

"Silicone Additives for Protective Coatings," Silicone Products Group, General Electric Company, 1975.

Société Protex, Paris, and Dow Chemical Company, Structures of Chelation Mechanism.

Sommer, H. and Linke, F., "New Possibilities of Sizing with Synthetic Resins," *Teintex,* **27,** 283–293 (1962).

Sparr, A. E., "Foam and Its Control at Activated Sludge Plants," paper presented at 1957 meeting of N.Y. State Association of Water and Sewage Plant Operators.

Sponsel, "Cellulose Ether Sizing and Finishing Agents," *Melliand Textilber.,* **19,** 738–739 (1938).

Steinmiller, W. G. and Cates, D. M., "Bleaching Cellulose: Part II, Effect of Impurities," *Text. Chemists Colorists,* **8**(1), 14 (1976).

Stewart, J., *Far-Infrared Spectroscopy,* Reinhold, New York, 1965.

Swinehart, J. S., *Infrared Absorption of Infrared and Ultraviolet Spectra,* Appleton-Century-Crofts, New York, 1968.

Syzmanski, Herman S., *Interpreted Infrared Spectra,* Vols. I–III, Plenum Press, New York, 1964.

Tesoro, G. C. (to Onyx Oil and Chemical Co.), U.S. Patent 2,844,609 (July 22, 1958).

Texchem Consulting Laboratories, Greenville, S.C., "Polymer Structure of Synthetic Sizes."

Thomas, E. B. and Oxley, H. F. (to Celanese Corp. of America), *Sizes and Coating Compositions,* U.S. Patent 2,135,128 (1938).

Trosken, O., and Pense, W., U.S. Patent 2,266,448 (1942).

Trost, H. B. and Bush, H. B., "Sizing Up Warp Sizes," *Text. Ind.,* **134,** 127–134 (1970).

Trowbridge, J. R., Falk, R. A., and Krems, I. J., *J. Org. Chem.,* **20,** 990–995 (1955).

Turner, S. J., *Text. Ind.,* **40,** 973 (1949).

Ulrich, H., Saurwein, K., Goldacker, P., and Maiser, G. L., U.S. Patent 2,168,181 (1939).

Valko, E. I. and Tesoro, G. C., *Encyclopedia of Polymer Science and Technology,* Vol. 2, Wiley-Interscience, New York, 1965, pp. 204–229.

Van Wazer, *J. Am. Chem. Soc.,* **77,** 287 (1955).

Vickery, R. C., *J. Chem. Soc.,* 4357–4363 (1952).

Welcher, Frank, *Analytical Uses of EDTA,* Van Nostrand, New York, 1957.

Whetsel, Kermit, "Infrared Spectroscopy," *Chem. Eng. News,* 83, (February 5, 1968).

Whistler, R. L. and Bemiller, J. N., Eds., *Industrial Gums,* Academic Press, New York, 1959.

Whistler, R. L. and Paschall, E. F., Eds., *Starch: Chemistry and Technology,* Vol. 2, Academic Press, New York, 1967.

White, Robert G., *Handbook of Industrial Infrared Analysis,* Plenum Press, New York, 1964.

Widely, *Seifen-Seider* Ztg., **72,** 75–77 (1946).

Wiegerink, J. F., *Ironing Aid and Textile Refinishing Composition,* U.S. Patent 2,645,584 (1953).

Williams, H. L. and Mitchell, J. M., U.S. Patent 2,631,142 (1953).

Willingham, Lewis, "Effect of Carrier on Barre'."

Wurz, E. and Wurz, O., "Applications of Carboxymethylcellulose in the Textile Industry," *Text. Rundsch.,* **8,** 557–560 (1953).

Zussman, H. W., "Factors Involved in Sequestering 2,4D," *Agric. Chem.* (1949).

PATENTS

PATENTS RELATING TO FIBERS, YARN LUBRICANTS, AND SIZING COMPOUNDS

U.S. Patents

4,192,649	4,043,923	3,896,050	3,591,500
4,178,253	4,038,197	3,894,992	3,575,859
4,169,062	4,019,990	3,859,122	3,575,856
4,169,061	3,993,571	3,850,819	3,563,892
4,167,497	3,977,979	3,836,589	3,549,530
4,166,872	3,977,978	3,829,291	3,528,947
4,137,209	3,969,353	3,781,202	3,518,184
4,137,182	3,963,628	3,785,973	3,505,220
4,137,181	3,959,229	3,770,701	3,503,880
4,134,841	3,954,631	3,755,321	3,493,425
4,134,839	3,951,825	3,704,225	3,470,095
4,105,569	3,925,588	3,701,784	3,468,920
4,092,105	3,919,097	8,689,501	3,441,953
4,063,024	3,914,241	3,676,199	3,423,235
4,051,299	3,907,689	3,663,236	3,421,935

U.S. Patents

3,418,236
3,418,160
3,921,828
3,389,085
3,387,996
3,357,844
3,341,452
3,338,830
3,321,398
3,313,728
3,265,516
3,248,258
3,224,633
3,223,624
3,223,623
3,214,376
3,170,876
3,164,567
3,164,566
3,140,198
3,113,956
3,113,369
3,102,130
3,058,954
3,041,360
3,033,787
3,030,410
3,018,272
2,988,507
2,967,151
2,964,470
2,943,110
2,923,682
2,910,374
2,908,645
2,898,296
2,880,174
2,861,043
2,805,963
2,805,173
2,803,567
2,767,681
2,744,094
2,743,193
2,719,126
2,714,897
2,714,842
2,694,021
2,690,426
2,665,443
2,662,008
2,647,104
2,620,304
2,618,605
2,618,603
2,614,013
2,564,768
2,540,726
2,533,100
2,465,319
2,457,139
2,244,192
2,425,755
2,400,720
2,385,110
2,329,089
2,304,369
2,304,113
2,300,074
2,300,044
2,260,543
2,242,139
2,188,166
2,176,423
2,167,931
2,028,091
1,970,578
1,945,264

British Patents

1,115,149
1,110,389
1,055,344

Polish Patents

93,271
74,426
70,481

Belgian Patents

860,430
866,078
631,196

German Patents

528,734

Japanese Patents

51-60796
46-098159
96,234
78,189
78,111
54,547

French Patents

1,363,976

PATENTS RELATING TO THE BLEACHING PROCESS

U.S. Patents

4,315,092	4,321,052	4,005,182	3,308,067
4,300,897	4,294,575	3,979,313	3,723,341
4,312,634	4,323,465	3,979,312	3,617,206
4,313,804			

PATENTS RELATING TO SURFACTANTS, SCOURING AGENTS, AND GENERAL PREPARATION COMPOSITIONS

U.S. Patents

4,199,320	4,145,183	4,113,644	4,086,256
4,199,319	4,141,685	4,113,630	4,083,813
4,195,972	4,139,357	4,113,438	4,082,682
4,185,960	4,135,878	4,100,358	4,080,391
4,179,268	4,134,850	4,094,778	4,080,372
4,177,391	4,133,637	4,092,348	4,079,015
4,176,453	4,129,511	4,092,261	4,077,897
4,175,922	4,126,586	4,090,973	4,076,653
4,171,267	4,126,574	4,088,680	4,076,647
4,169,702	4,126,572	4,088,611	4,075,118
4,168,143	4,124,734	4,088,598	4,075,117
4,166,718	4,123,383	4,088,595	4,075,116
4,166,717	4,120,812	4,088,594	4,072,730
4,164,395	4,120,653	4,088,591	4,072,622
4,164,394	4,118,333	4,088,593	4,072,621
4,161,449	4,117,251	4,087,369	4,071,523
4,156,944	4,115,308	4,087,368	4,071,476
4,148,603	4,115,292	4,086,279	4,071,377

U.S. Patents

4,070,410
4,070,298
4,066,394
4,066,393
4,066,063
4,062,814
4,062,647
4,061,603
4,058,489
4,057,506
4,056,355
4,055,605
4,054,541
4,053,431
4,051,046
4,049,586
4,049,585
4,049,558
4,048,433
4,048,244
4,046,717
4,043,930
4,041,135
4,040,988
4,040,972
4,040,781
4,029,562
4,039,464
4,038,218
4,034,058
4,033,894
4,032,465
4,031,024
4,031,022
4,029,608
4,028,283
4,028,262
4,025,427
4,024,200
4,022,839
4,022,704
4,021,460
4,021,376
4,021,360
4,020,100
4,020,015
4,019,999
4,019,998
4,018,841
4,018,720
4,018,696
4,014,808
4,014,806
4,013,729
4,013,574
4,013,573
4,011,264
4,011,169
4,009,114
4,009,110
4,009,076
4,007,124
4,005,030
4,005,025
4,005,024
4,001,132
4,000,127
4,000,094
4,000,083
4,000,080
3,997,576
3,997,575
3,996,399
3,996,149
3,994,966
3,993,605
3,991,208
3,990,517
3,990,515
3,989,711
3,989,636
3,989,635
8,986,987
3,985,669
3,985,642
3,984,480
3,983,171
3,983,079
3,983,078
3,982,999
3,980,588
3,979,470
3,979,314
3,979,286
3,977,471
3,976,675
3,975,634
3,975,294
3,974,208
3,973,574
3,972,851
3,971,815
3,971,728
3,971,727
3,970,698
3,969,417
3,969,375
3,969,282
3,969,261
3,969,258
3,968,177
3,968,058
3,966,630
3,966,398
3,965,024
3,963,710
3,963,688
3,963,649
3,962,419
3,962,132
3,960,787
3,960,781
3,959,165
3,957,863
3,957,695
3,957,680
3,957,160
3,956,401

U.S. Patents

3,956,198	3,933,672	3,918,217	3,899,472
3,955,920	3,932,531	3,917,601	3,898,448
3,954,886	3,932,495	3,916,003	3,898,436
3,954,882	3,932,295	3,915,879	3,898,187
3,954,846	3,931,350	3,915,878	3,898,186
3,954,845	3,931,337	3,914,407	3,898,020
3,954,643	3,931,273	3,914,272	3,897,362
3,953,522	3,931,272	3,914,271	3,897,356
3,953,382	3,931,271	3,912,649	3,897,348
3,952,080	3,931,057	3,909,649	3,896,086
3,951,879	3,931,031	3,909,460	3,896,057
3,951,594	3,928,323	3,909,437	3,896,056
3,950,448	3,928,251	3,907,909	3,896,040
3,950,439	3,928,249	3,907,856	3,894,960
3,950,417	3,927,089	3,907,745	3,894,093
3,950,331	3,927,061	3,907,702	3,893,955
3,950,260	3,926,441	3,906,048	3,893,954
3,948,790	3,926,375	3,905,909	3,893,940
3,944,440	3,925,297	3,904,687	3,893,930
3,944,051	3,925,262	3,904,685	3,893,809
3,943,160	3,925,241	3,904,661	3,892,681
3,943,059	3,923,856	3,904,543	3,892,680
3,941,817	3,922,272	3,903,410	3,892,676
3,941,705	3,922,271	3,903,191	3,891,688
3,940,433	3,920,731	3,903,012	3,891,563
3,939,217	3,920,706	3,903,008	3,890,239
3,939,213	3,920,586	3,903,007	3,890,238
3,936,538	3,920,570	3,901,832	3,888,917
3,936,537	3,920,569	3,901,831	3,888,910
3,935,257	3,919,300	3,900,611	3,888,798
3,935,206	3,919,299	3,900,418	3,888,797
3,935,130	3,919,125	3,899,477	3,887,691

U.S. Patents

3,887,614
3,887,497
3,886,264
3,883,583
3,883,447
3,883,446
3,882,051
3,882,038
3,882,036
3,882,034
3,880,778
3,879,475
3,879,309
3,876,563
3,876,551
3,876,391
3,875,253
3,874,876
3,872,020
3,870,681
3,870,648
3,869,412
3,869,399
3,868,336
3,867,432
3,867,421
3,867,317
3,867,316
3,867,286
3,865,861
3,865,793
3,865,755
3,865,754
3,864,420
3,864,375
3,864,286
3,862,983
3,862,058
3,862,049
3,861,870
3,860,625
3,860,536
3,860,525
3,860,520
3,859,324
3,859,160
3,858,854
3,853,781
3,853,779
3,853,771
3,852,314
3,852,221
3,852,220
3,852,213
3,852,210
3,852,209
3,850,854
3,850,853
3,850,852
3,850,832
3,849,348
3,849,346
3,849,341
3,849,327
3,846,325
3,844,982
3,844,969
3,844,959
3,844,952
3,849,951
3,843,707
3,843,706
3,843,567
3,840,564
3,840,481
3,840,480
3,839,391
3,838,113
3,838,072
3,838,009
3,836,556
3,836,484
3,835,163
3,835,071
3,832,408
3,832,310
3,832,309
3,829,506
3,829,457
3,829,384
3,829,383
3,826,759
3,826,748
3,825,592
3,825,588
3,825,521
3,825,498
3,824,294
3,824,076
3,823,197
3,823,090
3,822,346
3,822,236
3,822,228
3,821,118
3,821,115
3,821,028
3,819,561
3,819,528
3,819,522
3,817,871
3,817,863
3,816,356
3,816,354
3,816,352
3,816,351
3,816,328
3,816,321
3,814,705
3,814,692
3,813,350
3,812,044
3,812,042
3,812,041
3,810,847
3,808,311
3,808,157
3,808,156

U.S. Patents

3,808,137	3,779,929	3,758,408	3,723,322
3,803,285	3,778,479	3,756,966	3,721,627
3,803,058	3,776,962	3,755,559	3,717,691
3,803,036	3,776,861	3,755,429	3,716,499
3,801,665	3,776,851	3,755,407	3,714,238
3,801,511	3,776,850	3,755,203	3,714,076
3,799,970	3,775,488	3,755,201	3,712,074
3,799,880	3,775,475	3,753,930	3,711,918
3,799,879	3,775,456	3,753,896	3,711,414
3,798,183	3,775,349	3,753,859	3,708,436
3,798,182	3,775,052	3,748,267	3,708,428
3,798,181	3,772,195	3,748,093	3,708,425
3,798,179	3,770,855	3,745,181	3,707,352
3,798,168	3,770,701	3,743,673	3,705,512
3,797,453	3,770,643	3,742,045	3,705,114
3,794,605	3,769,332	3,741,915	3,704,320
3,794,601	3,769,243	3,741,911	3,704,267
3,793,233	3,769,223	3,740,939	3,704,228
3,793,228	3,767,724	3,739,034	3,703,559
3,793,210	3,767,598	3,738,943	3,703,473
3,789,181	3,764,559	3,734,861	3,703,469
3,789,001	3,764,531	3,733,280	3,700,401
3,786,003	3,763,208	3,733,278	3,699,159
3,784,486	3,763,139	3,730,913	3,699,052
3,784,476	3,763,047	3,730,912	3,699,047
3,784,475	3,761,532	3,730,907	3,697,573
3,781,339	3,761,418	3,729,507	3,696,143
3,781,228	3,761,415	3,728,266	3,694,410
3,780,827	3,761,412	3,725,290	3,694,364
3,780,212	3,759,846	3,725,286	3,692,704
3,780,120	3,759,834	3,723,356	3,692,685
3,779,932	3,758,419	3,723,341	3,692,684

U.S. Patents

3,691,246
3,691,082
3,691,081
3,689,584
3,689,437
3,689,424
3,689,418
3,686,124
3,686,098
3,686,025
3,684,744
3,684,724
3,681,424
3,677,963
3,677,956
3,676,373
3,676,352
3,676,340
3,676,338
3,676,199
3,674,885
3,673,122
3,671,601
3,669,809
3,666,796
3,666,398
3,665,000
3,663,647
3,663,609
3,663,459
3,661,787
3,660,470
3,660,313
3,655,801
3,652,453
3,650,961
3,649,545
3,648,920
3,647,899
3,647,851
3,639,490
3,639,288
3,639,286
3,637,609
3,637,511
3,636,034
3,635,830
3,635,829
3,634,268
3,632,661
3,632,396
3,631,123
3,629,955
3,629,951
3,629,121
3,625,906
3,624,048
3,609,090
3,609,075
3,600,417
3,600,319
3,598,876
3,597,361
3,592,254
3,490,001
3,585,253
3,584,071
3,580,852
3,579,457
3,576,896
3,574,124
3,565,949
3,565,939
3,563,905
3,563,903
3,563,902
3,562,797
3,562,337
3,557,065
3,549,542
3,546,123
3,544,615
3,542,892
3,538,176
3,538,004
3,537,993
3,535,339
3,533,955
3,531,544
3,523,089
3,519,054
3,515,672
3,513,099
3,510,425
3,509,048
3,507,785
3,506,569
3,505,414
3,503,888
3,502,741
3,501,543
3,499,054
3,492,343
3,491,163
3,491,162
3,485,762
3,482,000
3,481,880
3,478,124
3,475,283
3,474,156
3,468,920
3,461,074
3,459,822
3,456,031
3,454,666
3,453,144
3,449,466
3,448,166
3,446,867
3,444,261
3,444,242
3,444,087
3,442,092
3,441,812
3,438,744
3,438,742
3,437,580

U.S. Patents

3,434,974
3,428,654
3,427,342
3,424,694
3,424,545
3,423,322
3,423,321
3,422,005
3,415,874
3,412,033
3,408,298
3,407,144
3,395,170
3,395,120
3,394,119
3,393,213
3,391,083
3,390,093
3,389,178
3,387,008
3,385,766
3,384,591
3,383,430
3,383,321
3,380,927
3,376,327
3,376,326
3,370,926
3,370,015
3,363,009
3,362,975
3,361,675
3,360,582
3,355,508
3,352,894
3,351,654
3,349,141
3,346,614
3,345,301
3,341,615
3,328,466
3,325,415
3,324,154
3,318,657
3,316,294
3,313,839
3,313,838
3,312,745
3,310,486
3,309,416
3,309,415
3,309,392
3,309,319
3,308,143
3,308,067
3,308,007
3,306,848
3,305,600
3,298,775
3,297,412
3,290,394
3,285,856
3,281,368
3,277,205
3,277,204
3,277,145
3,259,584
3,257,467
3,257,280
3,257,265
3,249,648
3,238,249
3,235,629
3,231,504
3,219,584
3,217,034
3,216,435
3,214,462
3,203,753
3,198,849
3,188,354
3,188,307
3,183,231
3,180,996
3,180,786
3,180,699
3,173,877
3,173,749
3,170,915
3,162,627
3,156,521
3,147,245
3,146,277
3,142,647
3,142,536
3,141,295
3,128,294
3,117,153
3,104,251
3,100,231
3,093,691
3,084,187
3,069,360
3,062,857
3,053,905
3,046,231
3,042,621
3,040,075
3,037,937
3,031,510
3,026,265
3,024,260
3,022,343
3,021,327
3,020,886
3,018,281
3,017,438
3,004,057
3,004,056
2,999,068
2,988,502
2,985,589
2,978,407
2,966,531
2,945,076
2,941,950
2,915,564
2,895,916

U.S. Patents

2,892,858	2,491,478	2,355,837	2,185,817
2,863,888	2,490,109	2,355,503	2,183,856
2,854,417	2,480,817	2,352,152	2,180,791
2,844,609	2,474,175	2,350,350	2,167,931
2,833,722	2,468,012	2,333,720	2,159,967
2,816,920	2,464,094	2,320,783	2,154,922
2,807,586	2,459,088	2,316,636	2,150,557
2,806,044	2,452,943	2,306,095	2,143,490
2,781,349	2,450,678	2,304,369	2,136,379
2,789,143	2,438,092	2,303,366	2,130,505
2,768,225	2,433,449	2,279,497	2,129,886
2,765,324	2,429,445	2,267,965	2,096,746
2,758,977	2,415,657	2,267,956	2,089,312
2,757,143	2,407,703	2,252,401	2,083,482
2,721,883	2,407,645	2,240,957	2,079,108
2,713,582	2,406,423	2,238,902	2,068,905
2,712,529	2,388,178	2,215,367	2,041,265
2,673,215	2,388,177	2,210,874	2,028,091
2,673,213	2,388,154	2,210,264	2,008,649
2,636,033	2,384,817	2,205,995	1,981,792
2,627,524	2,383,737	2,203,888	1,970,578
2,623,900	2,379,703	2,202,791	1,947,951
2,619,467	2,374,193	2,200,220	1,936,265
2,593,413	2,374,187	2,199,780	1,932,180
2,540,678	2,372,786	2,199,397	1,931,962
2,540,078	2,366,727	2,197,800	544,938
2,538,929	2,356,587		

Canadian Patents	Soviet Patents	German Patents
579,148	631,576	2,720,512
7-1959		2,422,655

German Patents

2,262,301
2,055,423
798,217
589,778
589,775
546,942

Belgian Patents

863,544
840,348
819,611
812,116

Netherlands Patents

7,403,384
7,403,381

Netherlands Patents

7,304,854
7,304,401

French Patents

1,590,710

Japanese Patents

78,114,980
7,826,782
7,401,036
7,341,707
7,110,871
6,723,124

British Patents

1,408,144
1,372,035
1,333,810
1,256,660
1,141,016
1,120,271
1,116,619
1,113,248
1,110,666
1,108,865
1,067,519
1,055,562
1,054,473
1,052,179
1,045,977
1,031,750
1,028,923
845,846

PATENTS RELATING TO DEFOAMER COMPOSITIONS

U.S. Patents

4,171,267	4,009,119	3,929,663	3,893,941
4,123,383	3,990,905	3,929,492	3,887,487
4,115,343	3,976,240	3,928,250	3,884,829
4,059,534	3,951,883	3,923,683	3,880,660
4,032,473	3,950,293	3,912,654	3,869,412
4,024,072	3,935,121	3,912,652	3,865,859
4,021,365	3,933,672	3,909,445	3,865,554
4,021,355	3,931,033	3,896,032	3,862,243

U.S. Patents

3,856,701
3,853,989
3,844,711
3,843,558
3,841,972
3,834,919
3,826,759
3,826,749
3,826,720
3,825,431
3,814,705
3,801,475
3,793,223
3,784,479
3,783,165
3,776,852
3,772,207
3,767,794
3,763,021
3,761,423
3,756,918
3,751,373
3,751,370
3,746,653
3,746,576
3,741,912
3,738,992
3,730,912
3,730,907
3,730,904
3,730,894
3,729,430
3,729,387
3,725,014
3,723,347
3,723,342
3,716,499
3,714,074
3,714,068
3,712,868
3,709,836
3,706,675
3,705,860
3,705,859
3,704,267
3,703,469
3,700,587
3,700,400
3,697,442
3,697,440
3,697,439
3,697,438
3,696,057
3,696,056
3,691,091
3,691,082
3,684,723
3,679,600
3,679,589
3,678,963
3,676,340
3,673,105
3,671,461
3,671,460
3,671,456
3,671,235
3,666,681
3,663,583
3,661,793
3,660,305
3,657,136
3,653,827
3,652,453
3,650,979
3,640,893
3,639,260
3,637,869
3,637,502
3,635,827
3,634,288
3,634,263
3,363,105
3,629,122
3,617,963
3,598,755
3,597,457
3,595,968
3,595,695
3,595,397
3,594,123
3,595,519
3,585,150
3,578,402
3,575,855
3,573,222
3,567,574
3,560,403
3,560,401
3,557,071
3,528,929
3,510,433
3,492,274
3,492,242
3,463,737
3,458,567
3,455,839
3,445,385
3,445,386
3,437,437
3,433,574
3,423,340
3,423,235
3,422,027
3,420,673
3,409,510
3,408,306
3,404,095
3,399,144
3,395,102
3,388,073
3,388,072
3,384,600
3,383,327
3,378,500
3,359,294
3,355,395
3,355,366
3,344,075

U.S. Patents

3,341,467
3,340,193
3,336,231
3,334,147
3,231,508
3,329,625
3,318,716
3,317,435
3,313,736
3,304,266
3,298,963
3,296,131
3,267,042
3,264,214
3,259,586
3,251,782
3,250,727
3,238,142
3,235,509
3,235,502
3,235,501
3,235,489
3,233,986
3,231,508
3,215,636
3,215,635
3,215,633
3,207,698
3,207,655
3,207,098
3,200,080
3,198,744
3,180,836
3,180,786
3,169,929
3,166,508
3,165,477
3,160,621
3,154,577
3,152,081
3,146,158
3,132,959
3,127,352
3,115,472
3,113,930
3,108,634
3,107,519
3,100,680
3,088,913
3,086,944
3,082,168
3,078,237
3,078,236
3,076,768
3,066,100
3,056,687
3,049,559
3,037,000
3,036,118
3,033,789
3,033,788
3,023,170
3,004,923
3,000,835
2,993,867
2,992,053
2,991,248
2,981,693
2,972,579
2,969,332
2,965,881
2,965,581
2,963,403
2,954,347
2,931,780
2,923,688
2,923,687
2,923,686
2,914,412
2,903,432
2,894,913
2,871,148
2,868,734
2,854,417
2,843,551
2,834,748
2,829,112
2,797,198
2,773,041
2,753,309
2,571,358
2,748,089
2,748,085
2,715,614
2,715,613
2,695,916
2,682,508
2,668,150
2,595,928
2,595,926
2,563,857
2,559,343
2,542,550
2,533,772
2,493,453
2,476,307
2,412,470
2,390,212
2,379,268
2,366,727
2,346,928
2,364,926
2,315,983
2,308,222
2,306,222
2,304,304
2,298,465
2,230,556
2,085,709
2,052,164
2,043,599
1,892,857
1,866,296

INDEX